Handbook of Research on Advancements in the Processing, Characterization, and Application of Lightweight Materials

Kaushik Kumar
Birla Institute of Technology, India

B. Sridhar Babu
CMR Institute of Technology, India

J. Paulo Davim
University of Aveiro, Portugal

IGI Global
PUBLISHER of TIMELY KNOWLEDGE

A volume in the Advances in Chemical and Materials Engineering (ACME) Book Series

Published in the United States of America by
IGI Global
Engineering Science Reference (an imprint of IGI Global)
701 E. Chocolate Avenue
Hershey PA, USA 17033
Tel: 717-533-8845
Fax: 717-533-8661
E-mail: cust@igi-global.com
Web site: http://www.igi-global.com

Library of Congress Cataloging-in-Publication Data

Names: Kumar, K. (Kaushik), 1968- editor. | Babu, B. Sridhar, editor. | Davim, J. Paulo, editor.
Title: Handbook of research on the advancements in the processing, characterization, and application of lightweight materials / Kaushik Kumar, B. Babu, and J. Paulo Davim, editors.
Description: Hershey, PA : Engineering Science Reference (an imprint of IGI Global), [2022] | Includes bibliographical references and index. | Summary: "This book of contributed chapters will provide the resources necessary for processing, characterization and manufacturing using lightweight materials across the globe, offering recent advances in the field of light weight material usage and its recent developments"-- Provided by publisher.
Identifiers: LCCN 2021008447 (print) | LCCN 2021008448 (ebook) | ISBN 9781799878643 (hardcover) | ISBN 9781799878667 (ebook)
Subjects: LCSH: Lightweight materials.
Classification: LCC TA418.9.L53 A38 2021 (print) | LCC TA418.9.L53 (ebook) | DDC 620.1/1--dc23
LC record available at https://lccn.loc.gov/2021008447
LC ebook record available at https://lccn.loc.gov/2021008448

This book is published in the IGI Global book series Advances in Chemical and Materials Engineering (ACME) (ISSN: 2327-5448; eISSN: 2327-5456)

British Cataloguing in Publication Data
A Cataloguing in Publication record for this book is available from the British Library.

All work contributed to this book is new, previously-unpublished material. The views expressed in this book are those of the authors, but not necessarily of the publisher.

For electronic access to this publication, please contact: eresources@igi-global.com.

Advances in Chemical and Materials Engineering (ACME) Book Series

J. Paulo Davim
University of Aveiro, Portugal

ISSN:2327-5448
EISSN:2327-5456

Mission

The cross disciplinary approach of chemical and materials engineering is rapidly growing as it applies to the study of educational, scientific and industrial research activities by solving complex chemical problems using computational techniques and statistical methods.

The **Advances in Chemical and Materials Engineering (ACME) Book Series** provides research on the recent advances throughout computational and statistical methods of analysis and modeling. This series brings together collaboration between chemists, engineers, statisticians, and computer scientists and offers a wealth of knowledge and useful tools to academics, practitioners, and professionals through high quality publications.

Coverage

- Biomaterials
- Ductility and Crack-Resistance
- Materials to Renewable Energies
- Statistical methods
- Fracture Mechanics
- Heat Treatments
- Wear of Materials
- Industrial Chemistry
- Electrochemical and Corrosion
- Multifuncional and Smart Materials

The Advances in Chemical and Materials Engineering (ACME) Book Series (ISSN 2327-5448) is published by IGI Global, 701 E. Chocolate Avenue, Hershey, PA 17033-1240, USA, www.igi-global.com. This series is composed of titles available for purchase individually; each title is edited to be contextually exclusive from any other title within the series. For pricing and ordering information please visit http://www.igi-global.com/book-series/advances-chemical-materials-engineering/73687. Postmaster: Send all address changes to above address.

Titles in this Series

For a list of additional titles in this series, please visit: www.igi-global.com/book-series

Emerging Applications and Implementations of Metal-Organic Frameworks
Shimaa Mohamed Elsaeed (Egyptian Petroleum Research Institute, Egypt) Elsayed Zaki (Egyptian Petroleum Research Institute, Egypt) and Abdel-Azim Abdel-Azim (Egyptian Petroleum Research Institute, Egypt)
Engineering Science Reference • © 2021 • 254pp • H/C (ISBN: 9781799847601) • US $225.00

Normal Partitions and Hierarchical Fillings of N-Dimensional Spaces
Gennadiy Vladimirovich Zhizhin (Russian Academy of Natural Sciences, Russia)
Engineering Science Reference • © 2021 • 280pp • H/C (ISBN: 9781799867685) • US $195.00

Applications of Nanomaterials in Agriculture, Food Science, and Medicine
Mohd Amin Bhat (Higher Education Department, Government of Jammu and Kashmir, India) Irshad Ahmad Wani (Government Degree College for Boys, Anantnag, India) and Shah Ashraf (National Institute of Technology Srinagar (NIT Srinagar), India)
Engineering Science Reference • © 2021 • 442pp • H/C (ISBN: 9781799855637) • US $215.00

Technology Development for Adsorptive Heat Energy Converters Emerging Research and Opportunities
Kostyantyn M. Sukhyy (Ukrainian State University of Chemical Engineering, Ukraine) Elena A. Belyanovskaya (Ukrainian State University of Chemical Engineering, Ukraine) and Mikhailo P. Sukhyy (Ukrainian State University of Chemical Engineering, Ukraine)
Engineering Science Reference • © 2021 • 305pp • H/C (ISBN: 9781799844327) • US $195.00

Nanotechnologies and Clusters in the Spaces of Higher Dimension Emerging Research and Opportunities
Gennadiy Vladimirovich Zhizhin (Independent Researcher, Russia)
Engineering Science Reference • © 2021 • 286pp • H/C (ISBN: 9781799837848) • US $195.00

Advanced Surface Coating Techniques for Modern Industrial Applications
Supriyo Roy (Haldia Institute of Technology, India) and Goutam Kumar Bose (Haldia Institute of Technology, India)
Engineering Science Reference • © 2021 • 342pp • H/C (ISBN: 9781799848707) • US $215.00

Handbook of Research on Advancements in Supercritical Fluids Applications for Sustainable Energy Systems
Lin Chen (Institute of Engineering Thermophysics, Chinese Academy of Sciences, China & University of Chinese Academy of Sciences, China)
Engineering Science Reference • © 2021 • 821pp • H/C (ISBN: 9781799857969) • US $415.00

IGI Global
PUBLISHER of TIMELY KNOWLEDGE

701 East Chocolate Avenue, Hershey, PA 17033, USA
Tel: 717-533-8845 x100 • Fax: 717-533-8661
E-Mail: cust@igi-global.com • www.igi-global.com

List of Contributors

Table of Contents

Section 2
Processing and Characterization

Section 3
Applications

Detailed Table of Contents

Due to the growing demand for lightweight materials in different industries, the selection and hybridization of engineering fibers and metals is becoming a promising solution as it combines the outstanding mechanical, thermal, and weathering-resistance properties from both materials. Due to their lightweight and strong mechanical properties, carbon fiber/aluminum hybrid composite-based structures have become the most dominant materials used by engineers and researchers in the recent two decades. In the present chapter, the recent development on the processing techniques and mechanical performances of these hybrid structures are reviewed in detail. In addition, the applications of these kinds of structural materials in the various industrial sectors including, automobile, aerospace, design of industrial robots, and fire protection are summarized.

The need for less weight and high-performance materials in manufacturing industries has continuously led to the development of lightweight materials through the use of advanced additive manufacturing (AM). The race of lightweight and high-performance metals continue to evolve as this continuously provides better understanding about connection existing between material processing, microstructural development, and material properties. AM technique is an interesting manufacturing process that is employed in production of engineering components with improved properties. The choice of titanium and its alloys in structural applications are attributed to their superior strength-to-weight ratio and high corrosion resistance. This chapter looked at different additive manufacturing (AM) techniques developed

The replacement of heavy metallic structures by high-performance lightweight composite materials is a prominent solution to fulfill the continuous demand in different industrial sectors. Lightweight structures based on aluminum-glass fiber reinforced plastics (GFRP) sandwich panels have been increasingly utilized in the shipbuilding, automotive, and aerospace industries for their striking mechanical and physical properties. These advantageous properties have resulted from the combination of the high tensile and flexural strengths, increased hardness, and the improved wear-resistance of aluminum laminate with the unique properties of lightweight stiffness and high strength weight ratio of glass fiber-reinforced. In this chapter, the various processing approaches, properties, and applications of these sandwich structures are summarized from a wide range of literature.

In order to take more stringent measures in fuel economy and achieve the determined performance targets, the automotive industry needs to reduce the weight of the vehicles it produces. For this reason, all automobile manufacturers have determined their own strategies. Some manufacturers use lighter aluminum, magnesium, and composite components in their cars. In this study, the joining techniques of lightweight materials such as welding and the processes of their industrial use have been examined. There is currently no single technology that can combine all metallic panels in a car body structure. However, it is known that various joining technologies are used together. With the potential to combine certain combinations of steel and aluminum, manufacturers and scientists continue to work to identify technologies with the highest potential for lightweight joining and put them into use in high-volume automobile production. Therefore, it is important to examine the weldability of light materials such as magnesium, titanium, and aluminum.

Lightweight materials were needed in many different areas, especially in order to reduce the required energy in areas such as automotive and aerospace industries. Metallic foams attract attention in lightweight material applications due to their unique properties. The pores in its structure provide advantages in many applications, both structural and functional by promising both ultra-lightweight construction, energy absorption, and damping insulation. Production techniques of metallic foams can generally be classified as liquid, solid, gas, and ionic state production according to the physical state of the metal at the beginning of the process. The production technique should be chosen according to the usage area and desired properties of the metallic foam and the suitability in terms of cost and sustainability of production. For this reason, the details of the production techniques should be known and the products

extensive usage in modern automobile parts, aircraft components, and ship structures, mainly due to their attractive properties such as low cost, high strength-to-weight ratio, excellent corrosion and wear resistance. Friction stir welding is one of the most versatile solid-state joining processes to ensure weldability between two AMC plates. In this research work, an analysis of FSW process through parameters (e.g., composition of alumina, spindle speed, feed, etc.) in joining Alumina reinforced aluminum alloy composites Al 6061 and Al 2024 together at various proportions by analyzing properties like impact strength, hardness, flatness, and ultimate tensile strength has been done. Finally, optimization is carried out to select the best possible combination using a multi-attribute decision-making technique called the complex proportional assessment of alternatives.

Chapter 15
Characterization and Spectroscopic Applications of Metal Foams From New Lightweight Materials 339
Sefa Celik, Istanbul University, Turkey
Nuray Bekoz Ullen, Istanbul University-Cerrahpasa, Turkey
Sevim Akyuz, Istanbul Kultur University, Turkey
Gizem Karabulut, Istanbul University-Cerrahpasa, Turkey
Aysen E. Ozel, Istanbul University, Turkey

Lightweight materials such as metallic foams possess good mechanical, chemical, and physical properties, which make them suitable for a wide range of functional and structural applications. Metal foams have recently gained substantial interest in both industry and academia due to their low cost, thermal conductivity, high working temperature, vibration damping, specific mechanical properties, energy absorption, and heat resistance. The use of metal foams on a large scale and successful applications depend on a detailed understanding of their characteristic properties. Metallic foams are characterized by the morphology of the porous cells (size and shape, open or closed, macro and micro), pore topology, relative density, properties of the pore wall, and the degree of anisotropy. This contribution focuses on x-ray diffraction, Fourier transform infrared (FT-IR), and Raman spectroscopic applications used for the characterization of metal foam, and also a brief of the most important applications, including a significant number of examples given.

Chapter 16
Machining of Poly Methyl Methacrylate (PMMA) and Other Olymeric Materials: A Review 363
Fredrick Mwema, University of Johannesburg, South Africa
Job Maveke Wambua, Dedan Kimathi University of Technology, Kenya

Polymers have been adopted industrially in the manufacture of lenses for optical applications due to their attractive properties such as high hardness, high strength, high ductility, high fracture toughness, and also their low thermal and electrical conductivities. However, they have limited machinability and are therefore classified as hard-to-machine materials. This study conducts a critical review on the machining of various polymers and polymeric materials, with particular focus on poly (methyl methacrylate) (PMMA). From the review it was concluded that various machining parameters affect the output qualities of polymers and which include the spindle speed, the feed rate, vibrations, the depth of cut, and the machining environment. These parameters tend to affect the surface roughness, the cutting forces, delamination, cutting temperatures, tool wear, precision, vibrations, material removal rate, and the mechanical properties such as hardness, among others. A multi-objective optimization of these machining parameters is therefore required, especially in the machining of PMMA.

Preface

We, editors, would like to present the book *Handbook of Research on Advancements in the Processing, Characterization, and Application of Lightweight Materials.* The "lightweight" term can be added to different types of materials with one common specification, and that is "lower density", much lower than normal conventional materials. Replacing cast iron and traditional steel components with lightweight materials such as aluminium and its alloys, magnesium and its alloys, carbon fiber reinforced polymer composites can directly reduce the weight of the product say i.e. automobile and aircraft body by up to 50 percent and therefore reducing fuel consumption drastically and hence increasing the efficiency. In the automotive industry, the need to reduce vehicle weight has given rise to extensive research efforts to develop aluminium and magnesium alloys for structural car body parts. In aerospace, the move toward composite airframe structures urged an increased use of formable titanium alloys. In steel research, there are also major ongoing efforts to design novel damage-controlled forming processes for a new generation of efficient and reliable lightweight steel components. All of these materials, and more, constitute today's research mission for lightweight structures. They provide a fertile materials science research field aiming to achieve a better understanding of the interplay between industrial processing, microstructure development, and the resulting material properties.

The book provides a greater knowledge and understanding of the lightweight materials in the context of processing and characterization. It covered high quality peer reviewed chapters that investigates the properties of lightweight materials, performances, trouble shooting of feedstock, processes and products. Lightweight materials science research field aiming to achieve a better understanding of the interplay between industrial processing, microstructure development, and the resulting material properties and the book chapters provides solutions for the industrial operations and also plays a significant role in inspiring break through innovations in lightweight materials from fundamentals to technological challenges and applications that are shaping era of Industry. This book also identifies the need for modern tools and techniques from multi-disciplinary approach covering all spheres of applications to address the issue above referred. Undoubtedly this book would be the panacea in all lightweight materials areas. The edited book would provide a great platform for leading Engineers, academic chairpersons of engineering and technology transfer specialists, chief research officers and vice presidents of research in government, industry, and academia around the world to exchange the latest research and ideas in their field, and be confronted and challenged by those in others. To establish individuals who can show originality in application of knowledge, together with a practical understanding of how established techniques of research and enquiry are used to create and interpret knowledge in the area of lightweight materials. The book would provide solutions for the industrial operations and also would be instrumental in inspiring break through innovations in lightweight materials from fundamentals to technological challenges and applications that are shaping era of Industry 4.0.

The chapters in the book have been provided by Researchers and Academicians working in the field and have gained considerable success in the field. The chapters of the book are segregated in three sections, namely Section 1, "The State-of-the Art"; Section 2, "Processing and Characterizations"; and Section 3, "Applications."

Section 1 contains Chapter 1 to Chapter 3, whereas Section 2 contains Chapter 4 to Chapter 10, and Section 3 carries Chapter 11 to Chapter 16.

Section 1 of the book starts with Chapter 1. Here the recent development on the processing techniques and mechanical performances of carbon fiber/aluminium hybrid composite based structures are reviewed in detail. In addition, the applications of these kinds of structural materials in the various industrial sectors including, automobile, aerospace, design of industrial robots, and fire protection are summarized. It would provide the readers an outlook on the subject.

Chapter 2 deals with a prominent technology i.e., Additive Manufacturing. Here the processing techniques for lightweight alloys and composite material especially titanium and its alloys in structural applications have been considered with different additive manufacturing techniques developed for the processing, their strengths and limitations. The chapter also highlights processing and application of titanium aluminide for high temperature applications.

Chapter 3, the concluding chapter of Section 1, deals with basic principles for thermoplastic parts finishing with impulse thermal energy method (ITEM), a modified version of Thermal Energy Method that is successfully used for finishing. The chapter focuses to highlight the basic principles of the thermoplastics treatment by acting heat fluxes inherent to ITEM providing the time-controlled production of combustion species. By means of numerical analysis, the processes of melting and healing of pores during processing are provided. A method of defining processing settings has been developed, taking into account the limitations on critical temperatures. The promising possibilities of ITEM in relation to the processing of thermoplastics parts obtained by additive technologies are also outlined.

Section 2 categorizing Processing and Characterizations of lightweight materials starts with Chapter 4, which provides Synthesis and Characterization of light weight Beryllium Chloro-silicate Phosphor. The material is developed through a solid-state reaction. X-ray Diffractometer, Photoluminescence spectra and FTIR spectra provides the authenticity of the product.

The next chapter, Chapter 5, provides information regarding the processing, properties, and uses of lightweight glass fiber/aluminium hybrid structures. Here lightweight structures based on aluminium-glass fiber reinforced plastics (GFRP) sandwich panels have been taken into consideration. In this chapter, the various processing approaches, properties, and applications of these sandwich structures are summarized from a wide range of literature.

Chapter 6 deals with joining techniques like welding in lightweight material structures. Presently there is no single technology that can combine all metallic panels in a vehicle body structure. In this chapter joining techniques of lightweight materials such as welding and the processes of their industrial use have been examined and various joining technologies are used together for the same. With the potential to combine certain combinations of steel and aluminium, manufacturers and Scientists continue to work to identify technologies with the highest potential for lightweight joining and put them into use in high-volume automobile production. Therefore, it is important to examine the weldability of light materials such as magnesium, titanium and aluminium and the chapter provides the same.

Chapter 7 provides insight on production techniques of metallic foams in lightweight materials. Today metallic foams attract attention in lightweight material applications due to their unique properties. The pores in its structure provide advantages in many applications, both structural and functional by promising both ultra-lightweight construction, energy absorption and damping insulation. Production techniques of metallic foams can generally be classified as liquid, solid, gas and ionic state production according to the physical state of the metal at the beginning of the process. This book chapter emphasizes on the production methods from past to present.

The next chapter, Chapter 8, does fatigue characterization and fractographic analysis of aluminium 6063 alloy which is widely employed in various automobile and aircraft areas because of their unique specific strength and formability. Aerospace structural components undergoes dynamic loading leading to fatigue due to mechanical stress and thermal conditions. With the above consideration, this book chapter sighted to the analysis of fatigue behaviour of Al 6063 alloy at the various total strain amplitude. The test results were quite encouraging and could identify cyclic softening during the test.

Chapter 9 maximizes of tensile strength of aluminium 6061 alloyT6 grade friction welded joints by using the desirability function. In this chapter orthogonal array L9 was used to perform the experiment utilizing rotating speed of the tool, tilting angle, and feed as the processing parameters and subsequently Taguchi approach was used to maximize the weld strength using single objective function.

The last chapter of the Section 2, Chapter 10, provides information regarding mechanical and corrosion behaviour of friction stir welded AA 6063 (Aluminium alloy). The chapter aimed to join aluminium 6063 alloy plates by FSW with an objective to provide mechanisms responsible for the formation of welds without any defects, microstructural refinement. The chapter also highlights effects of FSW parameters on resultant microstructure, mechanical and corrosion properties.

Section 3 starts with Chapter 11 dealing with fabrication of tailor-made metallic structures for lightweight applications. Tailor-made metallic structures are fabricated by welding, adhesive bonding, and mechanical joining methods. The chapter aims for not only to fabricate light-weight structures, but also to develop novel methods of joining. Developing novel methods of joining is advantageous due to the possibility of joining of dissimilar materials, improved mechanical performance and microstructures. Here, initially, tailor-welded blanks (TWB) are introduced, and after that, fabrication of TWBs by laser welding, friction stir welding, and friction stir additive manufacturing are elaborately discussed. Some critical issues in modeling the deformation during fabrication of TWBs is also discussed. A brief account of mechanical behaviour of adhesive bonded sheets and mechanical joining are presented in the later part.

Chapter 12 provides structural optimizations of different load carrying members through computational structural analysis. Load withstanding characteristics is one of the major considerations involved in Structural Engineering because the lifetime factor is directly proportional to load withstanding behaviour. Thus, this chapter computationally analyses the load withstanding behaviour of various sandwich lightweight composite materials under the given flexural load. Here four major materials are imposed under flexural loads for two different core structures namely, hexagonal cross-section and twisted cum integrated pentagonal cross-section. The major materials implemented for this comparative investigation are Aluminium Alloy, CFRP, GFRP, and KFRP. The chapter concludes with the choice of best structures with respect to their lightweight materials to withstand a high value of flexural loads.

In Chapter 13, the role of self-assembly in additive manufacturing of aerospace applications has been discussed. Additive manufacturing is a strategy to produce parts with complex geometries whose process is prohibitive in cost or impossible through subtractive or formative techniques. In order to improve the performance and reduce the cost of aerospace parts, one of the emerging design techniques is self-assembly which seeks to reduce the number of parts to produce best design practices and rules. Self-assembly represents a comprehensive strategy that improves process time, product quality, cost of materials, and printability. The purpose of this chapter is to review the technological contributions that self-assembly has had in the additive manufacturing of aerospace parts. The chapter also proposes future perspectives of the role of self-assembly in additive manufacturing.

The next chapter, Chapter 14, elaborates process evaluation and numerical optimization in friction stir welding of dissimilar AMCs. In this chapter, an analysis of FSW process through parameters viz., Composition of Alumina, Spindle speed, Feed, etc., in joining Alumina reinforced aluminium alloy composites Al 6061 and Al 2024 together at various proportions by analysing properties like Impact Strength, Hardness, Flatness and Ultimate Tensile Strength. Finally, optimization was carried out to select the best possible combination using a Multi-Attribute Decision-Making technique called the Complex Proportional Assessment of alternatives.

Chapter 15 provides insight on characterization and spectroscopic applications of metal foams from new lightweight materials. Metallic foams possess good mechanical, chemical, and physical properties, which make them suitable for a wide range of functional and structural applications. Metal foams have recently gained substantial interest in both industry and academia due to their low cost, thermal conductivity, high working temperature, vibration damping, specific mechanical properties, energy absorption and heat resistance. The use of metal foams on a large scale and successful applications depends on a detailed understanding of their characteristic properties including the morphology of the porous cells (size and shape, open or closed, macro and micro), pore topology, relative density, properties of the pore wall and the degree of anisotropy. Here, in this chapter X-ray diffraction, Fourier Transform Infrared (FT-IR), and Raman spectroscopic applications have been utilized for the characterization of metal foam. The chapter also highlights briefly on the most important applications with significant number of industrial examples.

The last chapter of the Section 3 and the book, Chapter 16, proposes machining techniques for PMMA which are adopted industrially in the manufacture of lenses for optical applications. But due to its non-conducting nature they have limited machinability and therefore classified as hard-to-machine materials. This chapter critical reviews machining of various polymers and polymeric materials, and with particular focus on poly (methyl methacrylate) (PMMA). The chapter concludes that various machining parameters affect the output qualities of polymers, and which include the spindle speed, the feed rate, vibrations, the depth of cut and the machining environment. These parameters tend to affect the surface roughness, the cutting forces, delamination, cutting temperatures, tool wear, precision, vibrations, material removal rate and the mechanical properties such as hardness, among others. Hence for an effective machining multi-objective optimization of these machining parameters is required.

First and foremost, we would like to thank God. It was your blessing that provided us the strength to believe in passion, hard work and pursue dreams. We thank our families for having the patience with us for taking yet another challenge which decreases the amount of time we could spend with them. They were our inspiration and motivation. We would like to thank our parents and grandparents for allowing us to follow our ambitions. We would like to thank all the contributing authors as they are the pillars of this structure. We would also like to thank them to have belief in us. We would like to thank all of our colleagues, friends in different part of the world for sharing ideas in shaping our thoughts. Our efforts will come to a level of satisfaction if the professionals concerned with all the fields related to lightweight materials gets benefitted.

We owe a huge thanks to all of our technical reviewers, Editorial Advisory Board Members, Book Development Editor and the team of Publisher IGI Global for their availability for work on this huge project. All of their efforts helped to make this book complete and we couldn't have done it without them.

Last, but definitely not least, we would like to thank all individuals who had taken time out and help us during the process of editing this book, without their support and encouragement we would have probably given up the project.

Kaushik Kumar
B. Sridhar Babu
J. Paulo Davim

Section 1
State of the Art

Chapter 1
Recent Advances on Smart Lightweight Carbon Fiber/Aluminum Hybrid Composite Structures

Noureddine Ramdani
Research and Development Center, Algeria

Hichem Mahres
Ecole Militaire Polytechnique, Chahid Abderrahmane Taleb, Algiers, Algeria

ABSTRACT

Due to the growing demand for lightweight materials in different industries, the selection and hybridization of engineering fibers and metals is becoming a promising solution as it combines the outstanding mechanical, thermal, and weathering-resistance properties from both materials. Due to their lightweight and strong mechanical properties, carbon fiber/aluminum hybrid composite-based structures have become the most dominant materials used by engineers and researchers in the recent two decades. In the present chapter, the recent development on the processing techniques and mechanical performances of these hybrid structures are reviewed in detail. In addition, the applications of these kinds of structural materials in the various industrial sectors including, automobile, aerospace, design of industrial robots, and fire protection are summarized.

INTRODUCTION

Fiber-Metal-Laminates (FML) exhibit superior dynamic mechanical performance combined with small densities. Due to their higher mechanical properties glass fibers, Kevlar, carbon fibers laminates reinforced polymers have been extensively utilized to strengthen metals. However, these fibers reinforced metals laminate provokes a discrepancy in their thermal expansion coefficients and the ability of gal-

DOI: 10.4018/978-1-7998-7864-3.ch001

vanic corrosion. To tackle these challenges, the fiber/metal laminate is changed by the incorporation of elastomeric interlayer, which is recommended to deal with these problems.

Aluminum Matrix Composites (AMC) represent a novel generation of light-weight, high-performance aluminum centric materials. The reinforcement of AMC could be achieved by using continuous or discontinuous fibers, in the whisker or particle forms at various volume fractions. The properties aluminum matrix can be tuned to accomplish the requirement of different industries from a suitable combination of matrix, type of reinforcing fibers and joining method. For several decades of intensive research have afforded a large scientific knowledge on the role of fiber reinforcement on the various structural, mechanical, and wear features of this metallic matrix. Recently, this matrix has been used in high-technical, structural and functional applications including aircraft building, defense, automotive, and heat management areas, sports, and recreation. It is worthy to note that research on fiber-reinforced aluminum matrix reached industrial maturity in the developed countries and is currently under the process of further ameliorating their properties (Surappa, 2003).

Owing to the increasing demand for stronger and low density structures in the industry, researchers have made strong efforts on the design of advanced prototypes for fiber-aluminum laminates. Fiber/aluminum hybrid composites produced by interlocking thin aluminum layers with fiber reinforced polymer laminates. By hybridization of the two constituents, aluminum and fiber-reinforced laminate, a combined several advantageous properties of their resulted composites can be offered like, good corrosion resistance, better thermal stability, and improved damage tolerance to fatigue crack growth and impact damage, which represents a critical requirement for aviation applications (Sinmazçelik et al., 2011). Aluminum layers and fiber-reinforced composites could be joined by several mechanically and adhesively techniques. Generally, the adhesively joined fiber/aluminum composites were demonstrated improved fatigue properties relative to those similar mechanically joined ones. The most well-known commercially in service fiber/aluminum laminates are reinforced with aramid fibers such as ARALL, those produced from high strength glass fibers, GLARE, and those strengthened with carbon fibers like CARALL.

Carbon fibers (CF) are materials comprising more than 92 wt.% of carbon and forming a fiber shape. They are very important carbonaceous material demonstrating several superior mechanical performances (tensile strength of 2–7 GPa), better physical, and chemical stability, prominent compressive strength, high Young's modulus of 200–900 GPa, small density in the range of 1750–2200 Kg/m^3, low thermal expansion, higher thermal conductivity that exceeds 800 $Wm^{-1} K^{-1}$. These fibers are much lighter than steel, aluminum, and many alloying components of manganese, zinc, and zirconium, which make them very suitable for reinforcing the different metallic plates such as magnesium or aluminum. For example, double-decker panels fabricated from fibers/epoxy resin face-sheet and aluminum foam core were manufactured as a prospective application significance in the industrial processing field. The carbon fiber was the most extensively used for reinforcing aluminum metal due to its intriguing properties, this includes lightweight and higher mechanical strength, and improved anticorrosion characteristics (Table 1). Therefore, it was widely utilized for manufacturing carbon fiber reinforced polymers (CFRP) laminates.

The production methods to join only CFRP/aluminum composite structures involve adhesive-bonding, self-piercing rivet, bolt, clinching and welding (Pramanik et al., 2017). Excluding adhesive bonding and welding techniques, the other joining processes required the infiltration of aluminum pins over joining parts and thus, surface pre-treatment is not recommended. The selection of suitable joining process primarily depends on the targeted applications. The joining technique can also predominantly influence the inherent mechanical performances of these composite structure, including the friction, impact, flexural, tensile, and fatigue characteristics. This chapter provides a comprehensive knowledge about the processing methods, bonding behavior, mechanical properties, and applications of aluminum/carbon fiber joined hybrid composites from the extensive literature.

Table 1. Some mechanical properties of carbon fiber, aluminum, and aluminum/CF composite.

Measurement	Carbon Fiber	Aluminum	Carbon/Aluminum Comparison
Modulus of elasticity (E) GPa	70	68.9	100%
Tensile strength (σ) MPa	1035	450	230%
Density (ρ) g/cm^3	1.6	2.7	59%
Specific stiffness (E/ρ)	43.8	25.6	171%
Specific tensile strength (σ /ρ)	647	166	389%

PROCESSING TECHNIQUES

Since the 1960s, carbon fiber reinforced aluminum alloy composites were prepared by a liquid-phase process where the diameter of the carbon filaments does not surpass some microns, and they could be quantified in the many thousands in one yarn, which make the utilization the solid-state techniques unusually for fabrication such hybrids. The assembly of aluminum matrix with fiber reinforced polymer composites (Al/FRPC), offers prominent hybrid structures for advanced light-weight usage. The selected process keens on the resulting physical, mechanical, and chemical interactions and thus can controlled by joining partner surface-modifications. Several problems are included in the bonding of engineering fiber-reinforced thermoplastic composites with an aluminum matrix that is recommended for the fabrication of numerous parts. This part is delighted in describing the benefits and the shortcomings of different joining techniques of aluminum alloys and carbon fiber-reinforced, including polymer adhesive bonding, mechanical fastening, welding, and some new joining approaches.

Ultrasonic Welding Technique

The ultrasonic metal welding is widely selected process for assembling lightweight hybrid composites likes aluminum or titanium alloys reinforced with fiber-reinforced thermoplastics. In this process, the polymer layer is melted over the surface to form an adhesive joint between the hybrid components. The various welding parameters of ultrasonic metal welding are collected in Figure 1 (Balle, Wagner, & Eifler, 2009). However, only when the sonotrode waves in the plane of the assembly surface the load-supported fibers are directly fused against the face of the ductile metal. The ultrasonic metal welding method was also utilized to produce aluminum sheet/CFRP joints (Balle, Wagner, & Eifler, 2007). As revealed from Figure 2, some cracks were observed between the thermoplastic and aluminum surface.

The welding during this process was undertaken in two stages: First the ultrasonic shear waves provoke a softening and dislocation of the polymeric matrix beyond the welding region, while in the second one a straight weld between the charge bearing CF of the laminated composite and the aluminum sheet was created. Balle et al. investigated the utilization of ultrasonic welding technique for bonding aluminum

Figure 1. Experimental setup for ultrasonic metal welding (1) sonotrode tip, (2) anvil, (3) clamping for Al-CFTP epoxy joint, (4) CFTP epoxy sheet, (5) AA5754 sheet. (© [2014], [Ultrasonic Plastic Welding of Carbon Fiber Reinforced Polyamide 6 to Aluminium and Steel an Experimental]. Used with permission.)

Figure 2. Thermoplastic surface after welding with uncoated aluminium adherend. Particles were eroded from the aluminium surface and impressed into the thermoplastic surface. (© [2014], [Ultrasonic Plastic Welding of Carbon Fiber Reinforced Polyamide 6 to Aluminium and Steel An Experimental]. Used with permission.)

sheet metals with CFRP at welding times bellow than four seconds. They deduced that a concentrated joint between the aluminum sheet and the carbon fibers of the CFRP laminates was generated during the ultrasonic welding technique (Balle, Wagner, & Eifler, 2009). An example of ultrasonic plastic welding of carbon fiber reinforced polyamide 6 to aluminum samples are displayed in Figure 3.

Figure 3. photos of some Carbon Fiber Reinforced Polyamide 6/Aluminum specimens produced by Ultrasonic Plastic Welding. (© [2014], [Ultrasonic Plastic Welding of Carbon Fiber Reinforced Polyamide 6 to Aluminium and Steel An Experimental]. Used with permission.)

The pressure welding process was also explored for the assembly of carbon fiber reinforced thermoplastic polymers (CFRTP) with PA66 and aluminum alloy metallic sheet. To study this processing factor, a statistical model called central composite design circumscribed (CCC) was chosen for demonstrating these hybrid joints weldability (Balle & Eifler, 2012). Balle et al. detected the appearance of mechanically deformed areas generated by ultrasonic welding, which was led to a new reorganization of the crystal structure near the mechanically pre-modified metal and they observed that that molten polyamide 66 thermoplastic can infiltrate into a hollow, nano-alumina layer, which was resulted from acidic etching of aluminum before starting the ultrasonic welding procedure (Balle et al., 2013). Wagner et al. used the same technology for elaborating high-strength, a new aluminum/CFRPs joints. They confirmed that the studied technique could be a stimulating alternate for producing dissimilar joints compared to the well-recognized joining methods, especially in engineering applications (Wagner, Balle, & Eifler, 2013). The influence of the welding energy, load parameters, and the thickness of the thermoplastic film on the adhesion bonding of their manufactured products were evaluated. A similar technique was used utilized to fabricate AA5754 aluminum/CF-reinforced epoxy including a polyamide 6 based composites where the functional surface generated in the thermosetting matrix based composite allowed a fast welding with a metallic layer (Lionetto, Balle, & Maffezzoli, 2017). Several ultrasonic spot welded composite joints were fabricated from AA5754 aluminum sheets and carbon fiber strengthened epoxy thermoset containing a co-cured thermoplastic surface sheet (Lionetto et al., 2018). Recent project aimed to developed aluminum CFRP Hybrid Composites with a PEEK based decoupling layer as shown in Figure 4.

Figure 4. A schematic sectional view of a hybrid aluminum/CFPR composite and a decoupling layer made of PEEK. (© [2020], [www.ifam.fraunhofer.de]. Used with permission.)

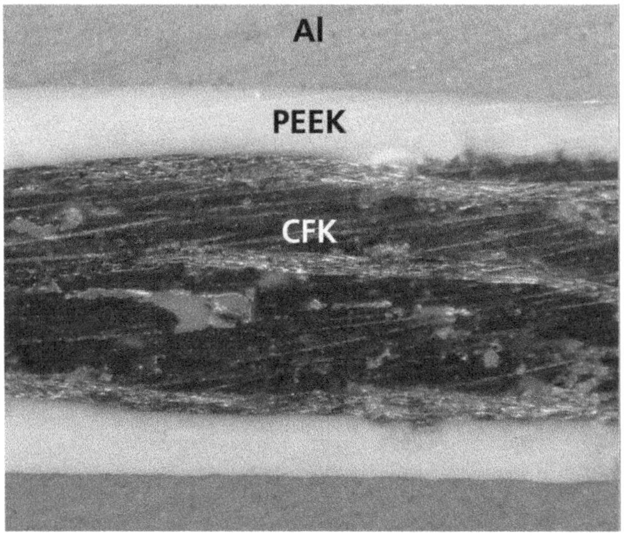

Infiltration Process

Several gas pressure infiltration methods were applied for manufacturing CF/aluminum composites, where this process allowed the fabrication of complex carbon/aluminum composites with fiber or textile reinforcement in graphite molds. However, an optimization of various parameters was required for the infiltration process in the future. (Hufenbach et al., 2009). To combine aluminum metallic sheets with carbon fabrics this process was successfully used. During this process, some aluminum pieces containing carbon fabric layers separating were firstly conductively heated up into the semi-solid form using two head-to-head electrodes of a welding gun. At a liquid portion ranged between 30% and 60%, the aluminum matrix penetrates the carbon fabric resulting in a fixed joint after solidifying this metal (Marx & Liewald, 2020).

Friction Riveting Technique

Friction Riveting (FricRiveting) is a promising joining process for producing the FRPC/metal hybrid joints. The simple pattern of this approach consists in revolving and introducing a cylindrical metal rivet into a fiber reinforced polymer composite plate. During the accelerated rotational velocities, the axial pressure generates frictional heat. Thus, a softened/melted polymeric layer is constructed underneath and around the tip of the rotating rivet for thermoplastics, during the early plunge steps, due to the presence of local enhancement in temperature. After the heating phase, the heat contribution rate surpasses the degree of heat release is attributed to the low thermal conductivity of polymers. At the tip of the rivet, this phenomenon is intensified, which plasticizes to a softer state at temperatures inferior to the alloy melting point. As the rotation is braked, the axial pressure is enhanced and the plasticized tip of the rivet is distorted forming a parabolic shape. After joining under pressure, the resulted joint is detained by the fastening forces caused by the deformed rivet tip and equally by adhesive forces at the polymer/metal interface.

The basic technique parameters are rotational speed, assembly time and assembly pressure. This processing technique offered the benefits of mechanical fastening and friction welding simultaneously, but some shortcomings should be mitigated, including directly applicable to thermoplastic polymers only, friction-riveted joints are incapacitated to disassemble and only spot-like joints can be realized. To assembly, aluminum alloys blanks and carbon fiber composites panels, a self-piercing riveting technique was employed by changing the process parameters. The various analyses and testing methods proved that this technique could be very promising for developing aluminum/carbon fiber composites (Franco et al., 2010). It is well-known that specific joining processes determined the final mechanical performance of metal/fiber composites. Franco et al. studied the influence of the space between the rivets in self-piercing riveting joints produced from aluminum/carbon fiber panels. They also investigated a mutual usage of this technique coupled with structural adhesives (Franco, Fratini, & Pasta, 2012). A new joint approach coupled fiber reinforced polymer, textile and welding processes were realized by employing titanium wire loops at the composite interface. Since the wire loops were joined with the FRP lateral by textile processes, whereas laser beam welding was used on aluminum face (Schimanski et al., 2011).

In another study, the self-piercing riveting method was modeled to highlight the residual stress state afterward piercing (Di Franco et al., 2013). A self-piercing rivets-bonded hybrid joints were developed from CFRP laminates and a 5754 aluminum alloy. The effects of the sheet thickness and the ply orientation angle of CF on the joint's performance were also evaluated, where an increase in the CF thickness was markedly enhanced the energy absorption of these composite joints (Liu & Zhuang, 2019). To further increase the processing quality of CFRP/aluminum joints, Wang et al. applied a post-curing self-piercing riveting (PC-SPR) method to assembly CFRP with aluminum pieces. (Wang et al., 2020). Friction Spot Joining (FSpJ) is a useful technique for producing metal/polymer or composite hybrid structures. It is a very talented, innovative approach to fabricate aluminum/fiber multi-material composite joints. The process parameters involved tool rotational velocity, drop depth, and the joining and force. As shown in Figure 5, this technique is included a three-part non-disposable tool, comprising a pin, a sleeve and a clamping ring (Amancio-Filho et al., 2011). The pin and sleeve can freely travel and swap of each other, whereas the clamping ring represents the fixed part. The parts are fastened together in a covering shape between the clamping ring backing and a dish to guarantee close interaction between them throughout the processing steps and for evading the parting of the joining parts under the cooling step at the end of the procedure.

The applicability of FSpJ of 2024 aluminum sheet with CF-reinforced poly(phenylene sulfide) composite laminate (CF-PPS) was effectively proven (Goushegir, dos Santos JF, & Amancio-Filho, 2014). The feasibility this technique for bonding aluminum AA2024-T3(bare and clad)/carbon fiber reinforced poly(phenylene sulfide) was also recently explored (Goushegir, dos Santos, & Amancio-Filho, 2014). A relationship between the manufacturing process parameters, bonding surface and mechanical characteristics of a newly fabricated joint using the FSpJ technique was established by Goushegir et al., where tougher joints were generated when a higher heat input was used due to a greater bonding area (Goushegir, dos Santos, & Amancio-Filho, 2015). Esteves et al. studied the parameters of the FSpJ technique for Al 6181-T4/CF-PPS hybrid composites (Esteves et al., 2015). Goushegir et al. found that the key parameters of

Figure 5. Schematic explanation of the FSpJ technique without the incorporation of film interlayer. (1) Sleeve plunging and plasticizing of the metal; (2) spot refilling and (3) joint consolidation. (© [2016], [Springer]. Used with permission.)

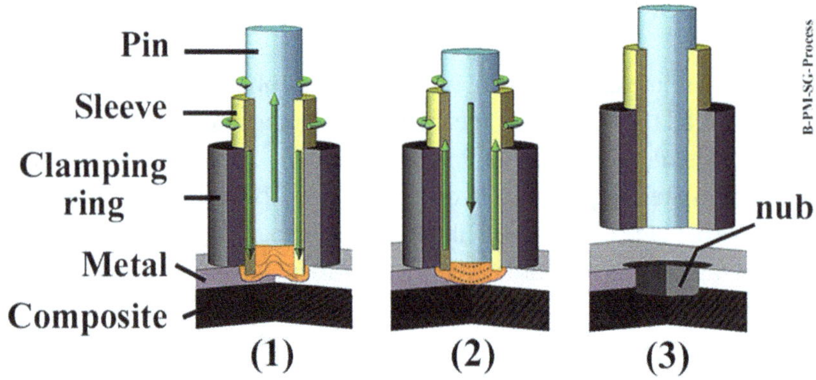

this process such as joining pressure, tool rotational speed and joining time can remarkably influence the bonding areas (Goushegir, dos Santos, & Amancio-Filho, 2015). The trend between the rotational speed and joining time was markedly determined the mechanical performance. Sand blasting was conducted on aluminum surfaces to guarantee the mechanically surface pre-treatment for improving the surface roughness. The role of several FSpJ processing factors on the performance of an aluminum AA6181-T4/ carbon fiber-reinforced poly (phenylene sulfide) (CF-PPS) double lap hybrids was scrutinized by means of both Taguchi model and analysis of variance (ANOVA) methods (Esteves et al., 2015).

A similar process was successfully employed to elaborate aluminum 2024-T3/ CF-PPS hybrid joints containing PPS film interlayer. Both micro-mechanical interlocking and adhesion forces were recognized to control bonding mechanisms in these hybrids (Andre et al., 2016). A 5052 Al alloy and plain carbon steel were previously treated using a laser-processing technique to generate rather rough, porous metallic surfaces before they welded to polyamide 6 based CFR by friction lap joining process. The remarkable improvements in the strength of the joint were related to the higher mechanical anchors and the improved chemical bonding area at the Metal/CFRP interface (Wu et al., 2020). Tool rotational speed was found to be the key parameter which majorly affected the joint shear resistance, followed by the joining time, plunge depth and joining force, while an extra joining forces provoked pressing flow of the molten layer decreasing joint strength. Interestingly, it was supposed that the numerical simulation combined with the welding experiment can effectively utilize to explore the joining process of friction stir welding between CFRTP laminated composites and aluminum metals (Jiang, Chen, & Yang, 2020).

Laser Beam Welding

The realization of the transition structures by the application of titanium wires, as transition elements, at the CFRP/aluminum interface was reported by die-casting and laser beam welding as given in Figure 6. For replacing the liquid state processes, the solid-state ones, a diffusion bonding has been a great solution. The produced hybrid showed improved load transfer for whole transition structures (Schimanski, Hehl, & Zoch, 2013). Laser direct bonding of CFRP to aluminum sheet was realized using a qualified continuous wave diode laser equipped with a line-shaped beam. The produced joints were cracked along the hybrid interface or in the melted area of carbon laminates where a fused polyamide thermoplastic was strongly joined on atomic or molecular sized level through nanometer thick aluminum oxide film (γ-Al_2O_3) located on aluminum alloy surface, reflecting a strong chemical or physical bonding between the joint components (Jung et al., 2013).

Figure 6. Scheme of laser beam welding process.

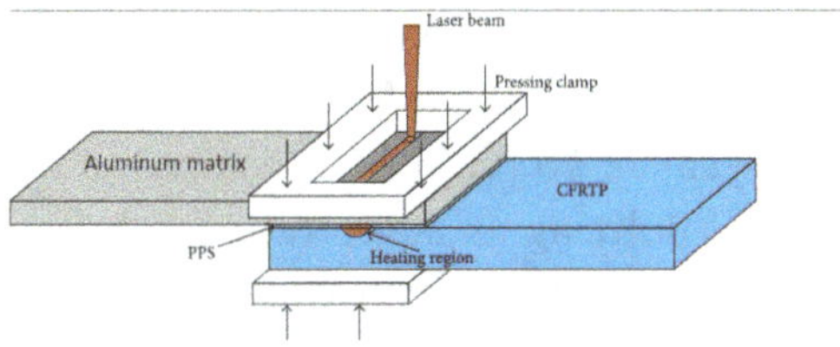

Mechanical Clinching

In order to describe the joining process of CFRP and aluminum alloy using semi-tubular self-piercing rivet process, a numerical model was established (Welf-Guntram et al., 2013). The key procedure parameters, tool geometries and material characteristics are studied by the finite element model (FEM). A comparative study between the measured and calculated data of the numerous processing and the geometrical connection parameters revealed good modeling quality for this process. The clinching technique for producing an aluminum AA6082-T6/CFRP thin piece was studied, since it was split dies with sliding segments to produce the hybrid joints (Lambiase & Ko, 2016). This investigation proved the practicability of this manufacturing method for fabricating these kinds of composite structures. In addition, the influence of the punch form, the punch diameter, and taper angle were optimized, to avoid the damage of the CFRP in the clinched joints, during this process. The two-steps clinching with split dies process was effectively utilized for assembling aluminum and CFRP joints (Lambiase & Ko, 2016). To determine the influence of the processing conditions on the mechanical strength of the joints, single lap shear experiments of simple clinched and redesigned samples were conducted. From the interesting results, the undercut was the most critical factor responsible for pullout failing the simple clinched connection.

In another study, hole clinching was a promising technique for joining CFRP and ductile aluminum metals (Lee et al., 2017). In a study conducted by Lee et al., a new clinching tool termed a spring die tool was described for improving the joinability of a CFRP/AA5083-aluminum alloy in an entire clinching process. In this technique, two pads sustained by a coil spring were utilized for increasing the formability of ductile materials and to decrease destructions to CFRP laminates through the enhancing of the compressive hydrostatic stress.

Laminate Squeeze Casting Technique

In a prominent study reported by Alhashmy, new kinds of aluminum hybrid plates were fabricated using squeeze casting process from plain weave CF preform as reinforcing phase and a 1235-H19 aluminum matrix (Alhashmy, 2012). It was concluded that the aluminum composites exhibited and improved chemical bond at the CF/aluminum interface ascribed to the construction of aluminum carbides. This technique was explored as alternate where carbon fiber layers and aluminum metals were thermally-consolidated (Alhashmy & Nganbe, 2015). Liquid infiltrates the carbon cloths from their two one-to-one adjacent aluminum layers, thus decreasing the penetration depth of the whole constituent height to a half thickness of distinct fiber layers.

Other Techniques

A heat induction joining method was selected as joining technology for the bonding of these AlMg3-CF/PA66-joint composites (Mitschang et al., 2009). Mitschang et al. studied the induction welding of AlMg3/carbon-fiber-reinforced polyamide 66 (CF-PA66) composite hybrid joints, including an extra polymeric film interlayer in place of utilizing an adhesive (Mitschang, Velthuis, & Didi, 2013). The addition of the film improved the shear strength by approximately 15% for AlMg3/CF-PA66 joints. The characterization of the joining behavior and the impact on the surface-modifications were offered by single-lap joints and microscopic analyses. In another study, a new manufacturing process was employed to produce aluminum foam by including NaCl as the space holder and argon gas was introduced into the aluminum molten metal. The salt particles were then eliminated by a dissolution technique (Hashim et al., 2014). Nestler et al. described the consolidation of semi-finished products from fiber-reinforced polyamide layers and EN AW-6082 T4 aluminum foils based hybrid composites (Nestler et al., 2016).

A new innovative joining technique was recently developed for combining carbon fabrics with aluminum layers. The woven carbon fabric was infiltrated with aluminum by using two electrodes up to the semi-solid state conductively heated two aluminum layers, where the fabric was located in-between (Marx & Liewald, 2017). An electromagnetic force-fit bonding of CFRP and aluminum tubular parts was explored, where discharging energy was kept in the pulse power, the metallic tube was compacted into a CFRP cylinder before the generation of a force-fit joint (Salamati et al., 2017). The role of each processing parameter including the discharge energy, gap between inner and outer parts, work-piece thickness and utilization of a supportive mandrel was studied. A CF/AA5052-aluminum composites were in-situ fabricated via friction stir processing technique. The inclusion of CF laminates in the aluminum layers decreased the micro-cracks nucleation and propagation (Cao et al., 2018). To surpass the disadvantages with the liquid-phase processing of CF/aluminum hybrid composites such as low wetting of the fiber with liquid aluminum and appearance of Al_4C_3 on the fibers' surface, this process was theoretically and experimentally optimized by evaluating the impact of porosity on the excreted pressure (Galyshev et al., 2019). The electro-hydraulic clinching technique was also applied to produce AA1050-O aluminum and CFRP sheets, wherein few milliseconds, the generated shock wave smashed the lower sheet and constructed the sheet to the die cavity through the pre-drilled hole in the upper one (Menghari et al., 2020).

MECHANICAL PROPERTIES

Over the last few decades, the majority of research work on CFRP/aluminum composite revealed that the assembly and configuration of CF and its bonding to aluminum matrix exhibited a significant influence on the final properties of their resulted hybrids (Shirvanimoghaddam et al., 2017). In addition, the determination of the key factors for the production of these materials was very necessary for tailoring their performances. In this section, several mechanical properties of CFRP joined with aluminum matrix are summarized.

Flexural Properties

In aluminum/CFRP sandwich structures, the bending charges are usually supported by the force couple constructed from the face sheets while the lightweight core components bear the shear loads. The critical function of the central material is to afford a greater moment of inertia. Thus, these panels exhibit improved specific mechanical performance respected to the monocoque structures under flexural test. In addition, the core fights transverse forces and maintains the laminates against total buckling and local buckling, which is responsible for enhancing its buckling resistance and its stiffness. The bending failure behavior and energy absorption capacity short square hollow section beam made of aluminum/CFRP were studied under transverse quasi-static charging. Individual three-point bending failure behaviors were depended on the lay-up arrangements (Kim, Shin, & Lee, 2013). The improvement in the specific energy absorbed was not attributed to the damage area of the developed composite beam. In a study conducted by Dhaliwal and Newaz, three-point flexural behaviors of two types of CF-strengthened aluminum composites (CARRAL) having a 3/2 configuration, which were previously produced by a vacuum press deprived of any adhesive coating, were investigated (Dhaliwal & Newaz, 2016). Crack in lower aluminum sheet, CFRP lamina fracture and delamination between upper aluminum and CFRP sheet were the main failure approaches. The bending characteristics of the hybrid structures were predicted using by modeling the progressive damage performance, where stress based composite failure and shear stress based delamination failure between neighboring layers were key features. The predicted flexural damage behavior fitted well the experimental data.

In the light of the above, the bending mechanical performance of CCF reinforced aluminum foam sandwich hybrids was evaluated by a three-point bending machine (Yan et al., 2017). The trend between a load and dislocation of the metallic matrix was recorded and deformations and failure comportments of the hybrids were discussed. The flexural load-carrying aptitude and energy absorption capability were markedly increased for the composites reinforced by CF fabric as face-sheet. Bending experiments were done on sandwich composite structures composed of CCFRP composite face sheets and the cores made from aluminum based honeycomb with various thicknesses, and the influence of core thickness on the bending properties of the sandwich panels was evaluated (Okur, 2018). The failure behavior charts under bending load of sandwich panels comprised CF reinforced facade layers in different arrangements, and four core assemblies of aluminum foam or honeycomb plates, which were manufactured by a polyurethane spraying process, were established (Rupp, Elsner, & Weidenmann, 2017). The critical forces controlling the failure mechanisms were partly changed to adequate the application of foam core composites and face layers using a porous metal. The effect of galvanic corrosion on the interface of the fiber-metal based laminates was studied, where flexural examinations were done to assess the residual flexural performance after the corrosive damage (Stoll et al., 2017). Results showed that the elastomeric interlayer constrained galvanic corrosion in the salt spray chamber, since the flexural properties of CF/aluminum containing elastomer interlayer still unchanged subsequently the corrosion tests.

The effect of surface-treatments by pure nickel for three types of are short carbon micro fibers, carbon woven fabrics, and unidirectional yarn CF-reinforced aluminum plate was evaluated. The flexural properties of the produced composites were studied room temperature using 4-point flexural tests where strong matrix/fillers interactions were observed (Valente et al., 2019). The bending failure behavior and energy captivation of the Al/CFRP hybrid tubes were explored. The results displayed that the obtained transverse energy absorption of the Al/CFRP composite tubes exceeded the sum of those pure Al tubes and pure CFRP tube. In addition, the flexural properties and energy absorption of aluminum hybrid foam core sandwich structures reinforced-CF-cold setting thermoset as facade layers were evaluated (Pandey et al., 2020). It was reported that by using a double-layer carbon fiber sheet, the flexural force bearing efficiency of the hybrids enhanced eight times respected to that of simple foam matrix. On the other part, the increase in thickness of foam core, the flexural load carrying capacity, bending stiffness and energy absorption the resulted hybrid composite can significantly increase. The role of interface bonding between aluminum and CFRP laminates on flexural comportment of the produced hybrid structures were scrutinized (Dhaliwal & Newaz, 2020). As the aluminum metal contents increased in these hybrid systems, the failure mechanism varied from rupture in the load-carrying area to gradual deformation as given in Figure 7 (Xiao et al., 2018). The stiffness and progressive damage, failure resulted from three-point bending examinations were simulated by Explicit finite element results where the predicted data well-fitted the experimental ones.

Figure 7. Damaged specimens after quasistatic bending. (© [2018], [Hindawi]. Used with permission.)

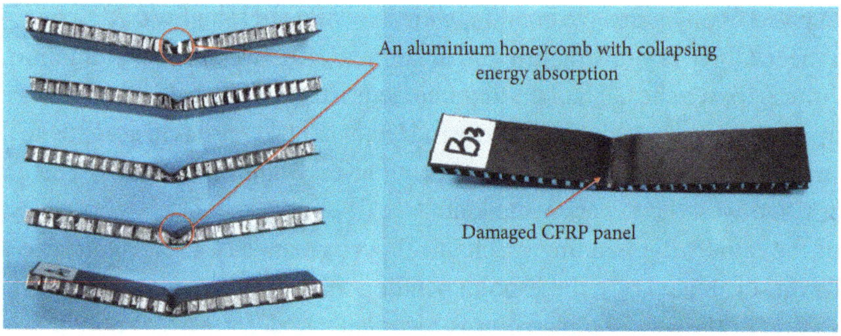

The Adhesion Strength

In one study, a nano-spike structure was deposited over the aluminum plate surface and then joined to CF reinforced thermoplastic laminates to study its drift on the adhesive bonding using single-lap joint examinations (Abe et al., 2019). It was reported that this structure hugely improved the adhesion strength. However, when this structure was coupled with silane-treatment this provoked a failure in the aluminum sheet of the single-lap specimens rather than in the interfacial region. The typical adhesion strength in these composites having an overlap length of 5 mm reached 24.9 MPa. The surface roughness influences, which was generated by different kinds of sanding rag and changing sanding times, for enhancing interfacial adhesion in aluminum/CF-reinforced epoxy hybrid materials was investigated (Kwon et al., 2019). The results showed that a proportional relationship was recorded in the surface energy of epoxy resin and epoxy-CF/Al composites. In addition, a suitable Al surface-treatment in these hybrids demonstrated real potential for promoting their properties. To produce durable adhesive joints, the effect of pre-treatments, using the ultrasonic etching process in alkaline solutions, on the aluminum substrate adhesion to CFRP was studied (Hu et al., 2019). Experimental data revealed that an improvement of 91% was observed in bond strength next an alkaline etching treatment, which ascribed to the removal of passive oxide coatings and the appearance of thin hydroxide layers through an aluminum-water feedbacks.

Impact and Crushing Properties

The crushing tests were generally conducted on the produced hybrid composite specimens at different velocities under static loading modes. The machine recorded the dislocation of the compressor head and the crushing force concurrently using a Data Acquisition System. The impact experiments were also conducted on monolith aluminum alloy, CFRP laminates, CF/aluminum laminates and composite sandwich panel made of CFRP adhered to aluminum honeycomb core panels (Zhu & Chai, 2013). Results showed that only CFRP with Nomex aluminum honeycomb core and fiber/aluminum laminate panels exhibited higher impact resistance and improved ductility. The Crashworthiness features and axial failure with damage transmission comportment were scrutinized for new aluminum/CFRP hybrids square hollow section (SHS) beam, where the specific absorbed energy and the efficiency of crush force were simultaneously enhanced in the Al/CFRP hybrid SHS beam having a $[0/90]_{2n}$ lay-up configuration, whereas a slight improvement in the thickness of the CFRP laminate was resulted (Kim et al., 2014).

By using a vacuum press, laminates composites base on a 3/2 structure carbon fiber reinforced aluminum 5052-H32 as the outer layer were developed without applying adhesive film (Dhaliwal & Newaz, 2016). Numerical simulation was done for describing the different impact stages at various energy levels. The onset impact energy at perforation failure was made in all aluminum and CF-reinforced layers was attained 31 J, and the predicted impact behavior data fitted well the experimental ones. Zhu et al. studied the axial crushing performance of an aluminium/CFRP hybrid, which was linked to plain aluminium tube and CFRP tube for exploring similar designs and crushing behaviors of the tube under static loading (Hua et al., 2017). Aiming to describe the crash behaviors and to detect crashworthiness parameters, an innovative bio-inspired CFRP panels reinforced aluminum honeycomb hybrid structures were produced (Wu et al., 2017). The various parameters including failure mode, crash behaviors, the influences of core side length, height, energy absorption, and impact velocity on the maximum load were quantified. Results showed that the crashworthiness characteristics were very dependent to the core length than to the core height; and the particular energy absorption was slightly changed when the core height was increased. It was revealed that the failure of the CFRP face-sheet comprised the matrix cracking and fiber breakage, provoking several damage approaches, likes indentation, round penetration and irregularly-shaped penetration. Additionally, the failure behavior of aluminum honeycomb matrix involved the crushing, shear and buckling in the core.

In another study, the experimental crashworthiness and the computing modeling of circular lightweight hybrid aluminum/CFRP tubes were studied (Reuter & Tröster, 2017). Relative to a pure aluminum structure, an increase of 37% in the specific energy absorption was reached. The energy absorption failures included the partial primary energy absorption modes of the CFRP and metal constituents. The simulation of axially-loaded on the hybrid structure was successfully achieved by multi shell model. Zhu et al. investigated experimentally and theoretically the crashworthiness features of aluminum/CFRP hybrid tubes subjected to various loading angles where distortion designs and several additional key factors linked to the crashworthiness were determined (Zhu et al., 2017). The energy absorption capability of all specimens was decreased in different levels with rising loading angle. This means that the hybrid tube was able to alleviate the consequences of oblique load on energy absorption capacity relative the pure CFRP tube. The energy-absorbing mechanisms of an aluminum honeycomb core covered with unidirectional and woven CFRP composite pipes were investigated (Alantali et al., 2017). Crushing tests on the multi-tube arrays have divulged that woven CFRP tubes absorbed markedly higher degrees of energy relative to their unidirectional equivalents.

To analyze the axial crushing phenomenon at low velocity axial loads, both experimental and numerical assessments were done on an aluminum column reinforced with combined carbon fiber and an aluminum honeycomb at a fixed velocity (Balaji & Annamalai, 2018). Compared to the bare aluminum hollow column, the collected data confirmed that the developed hybrids exhibited the most crashworthy with a significant enhancement of energy absorption, specific energy absorption and crush force efficiency, respectively. The low impact velocity behavior was investigated for a CF-reinforced aluminum laminates (CARAL), manufactured by hand lay-up process (Chozha Rajan et al., 2018). The recorded data demonstrated that the impact performance of the developed CARAL was improved by increasing of the CARAL laminate layers. The processing parameters by hand lay-up technique of CF-reinforced aluminum laminates were optimized (Muthu Chozha Rajan et al., 2018). The fabrication conditions such as layer thickness and matrix were critical controlled the flexural and impact responses of these composites.

To better understand crashing behavior of adhesive joints for multiple impacts from various directions, Liu et al. scrutinized the influence of impact surface and impact energy on the residual properties of the CFRP/aluminum adhesive bonded joints, which were fabricated in the hot pressing machine at precise curing temperature and pressure of the adhesive (Liu et al., 2019). The results revealed that CFRP produced improved structural integrity and reduced loss of joint strength respected to the aluminum matrix impact surface. The crushing and intrusion property of two aluminum automotive grade alloys/CFRP were studied for assessing the role of different factors such as aluminum thickness and aluminum composition, the thickness of carbon composite cover, and the capacity of composite coverage (Lebaupina et al., 2019). Results proved that intrusion properties of tubes without sacrificing the total part weight were enhanced for localized composite patches. By linking experimental and theoretical approaches, the low-velocity impact behavior and damage mechanisms of aluminum/CFRP honeycomb hybrid structures were achieved (He et al., 2019). It was reported that the face layer thickness exhibited a marked effect on the impact resistance properties, whereas a slight effect of the cell walls thickness and side length of honeycomb core on the impact load and structural stiffness of hybrids was noticed.

To evaluate the energy absorption behavior and the crushing parameters, a new type of an aluminum/CFRP/aluminum sandwich tube was produced by spinning forming (Han et al., 2020). The results revealed that due to improved contact between CFRP/Aluminum interfaces, the produced tube outperformed all hybrid structures in crashworthiness while the growth of delamination failure enhanced the energy absorption during crushing tests. Glass fiber reinforced plastics/carbon fiber reinforced plastic/aluminum (GFRP/CFRP/Al) hybrid tubes and CFRP/Al, were produced using the filament winding technique to evaluate their crushing characteristics and failure mechanisms (Cui et al., 2020). It was revealed that the failure modes of hybrid composites were prevailed by advanced brittle crack, delamination approach in CFRP layers and diamond failure mode in the metallic tube. In addition, the increase of hybrid plies augmented the explicit energy absorption, energy absorption, and the peak crushing strength.

Tensile Properties

The failure mechanisms of titanium-wire connected CFRP/aluminum composite structures under tensile loading was evaluated (Schumacher, Bomas, & Zoch, 2013). The tensile behavior of self-piercing rivets (SPRs) CFRP/aluminum 6111 T82 alloys were evaluated where a lap-shear tensile force of 3858 N was attained (Kang et al., 2016). Results showed that fall-out of the rivet head from the CFRP upper sheet was the major failure mechanism for these composites. Arranged in the order of 3/2 stacking of 0.8 mm thickness 2024 T3 aluminum sheets, unidirectional carbon fiber fabric and graphene nano-powder, modified epoxy resin, a hybrid composite was manufactured (Askin & Turen, 2019). It was concluded that incorporation of graphene nano-filler to the epoxy matrix imparted an enhancement of 9% in tensile strength and a rise of 24% in the interlaminar shear strength. The amount of permanent compressive stress in the composite increased by about 50%. The tensile properties of electromagnetic self-piercing riveting joints based on CFRP and a 5052 aluminum alloy (E-SPR) were studied in details (Liang et al., 2019). Results showed that the prepared these joints demonstrating higher tensile-shear strength properties compared to pressure self-piercing riveted (P-SPR) ones. Under stronger undercut, the shear fracture appearances reflected that the rivet of E-SPR joints remained in the metal layer while rivet of P-SPR joints stacked to the CFRP sheet.

Fatigue Properties

The fatigue properties of electromagnetically-riveted CFRP/aluminum hybrid joints were recently evaluated (Jiang et al., 2017). The failure mode of fatigue structures under every stress level was all failed at the Al layer and the fracture analysis revealed that cracks initially started near the hole in Al sheet, which was attributed by the fretting wear between Al and rivet under the dynamic cyclic loading. The fatigue properties of joints based on AA6111 aluminum reinforced CFRP were investigated (Rao et al., 2018). The fatigue life of the lap-shear joints was significantly influenced by the flushness of the rivet head against the top CFRP sheet, while in cross-tension joints, the flushness of the rivet exhibited no important role on the fatigue life of the joints. During the cyclic charging, the lap-shear joints failed because of kinked crack development in the lower aluminum layers, whereas the cross-tension joints failed when the rivet pulled was out of the upper CFRP part. In order to explore the fatigue performance of composites produced from aluminum foam core containing CF face-sheet, a high frequency fatigue machine was utilized, and the damage mechanisms of the hybrid structure were scrutinized by morphological analysis (Yan, Wang, & Song, 2020). The recorded data revealed that the fatigue life of composites reduced by enhancing the degree of loading, and the predominant life of the developed composite was its crack beginning phase.

APPLICATIONS

In the recent few years, the production of sandwich hybrid structures such as lightweight aluminum alloys/carbon fiber reinforced composites showing many outstanding physicochemical and mechanical properties which promote them as efficiently utilized by several industrial sectors as shown in Figure 8, including the automotive and aerospace. In this part, the recent applications in aerospace, automotive, robotics and ballistic protection of these materials were presented.

Automotive Industry

The need to decrease the energy consumption and to meet the severe legislation of safety for both pedestrians and passengers in automotive, has accelerated the development and usage of lightweight composites in the design of vehicles. Figure 9 reveals the different car parts produced from sandwich structures of aluminium honeycomb core and CFRP skins. For example, the bumper beams of a car which contributes to the total weight must be replaced by lightweight composites able to efficiently absorb the

Figure 8. Industrial application of sandwich composites structures.

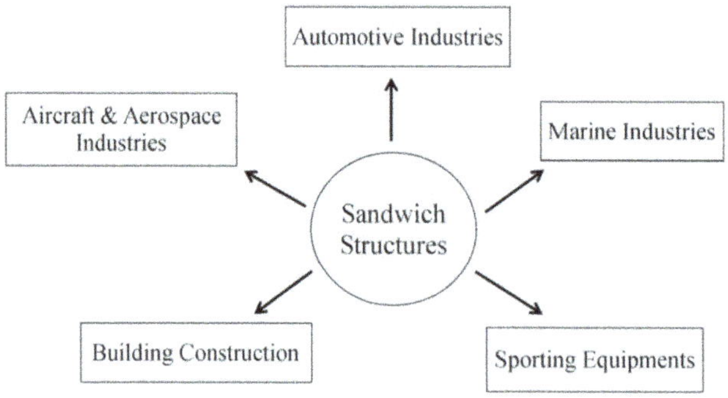

kinetic energy generated during a collision. The selection of materials for automotive components is related to a precise quantity and request process. Other critical factors are including low-cost production, security, recycling process and lightweight high-performance relationship. Aluminum metals are widely employed for fabricating panels and parts of the body structure. The various automotive components produced from advanced CFRP and aluminum composites were described and discussed with according to their advantages and risks for automotive manufacturing (Adam, 1997). Furthermore, the principal disadvantages of using advanced composites, including load applications, damage tolerance and high volume production technologies, were clarified. Technical solutions as key enablers for industrial realization were also described. The automotive aluminum compounds and their applications as castings, sheet and plate, and extrusions were described. A review was published on progress by US, European, and Japanese automobile companies in producing light-weighting vehicles based on technologies for processing aluminum alloys in various shapes (Benedyk, 2010).

In addition, carbon reinforced plastics are used for attaching on exterior of the car body. The hybrid aluminum carbon fiber reinforced polymers composites structures are mainly preferred to substitute the heavy steel and iron for reducing the overall weight. For example, multimaterial unibodies and alumi-

Figure 9. Schematic of sandwich structures with aluminium honeycomb core and CFRP skins in vehicle body.

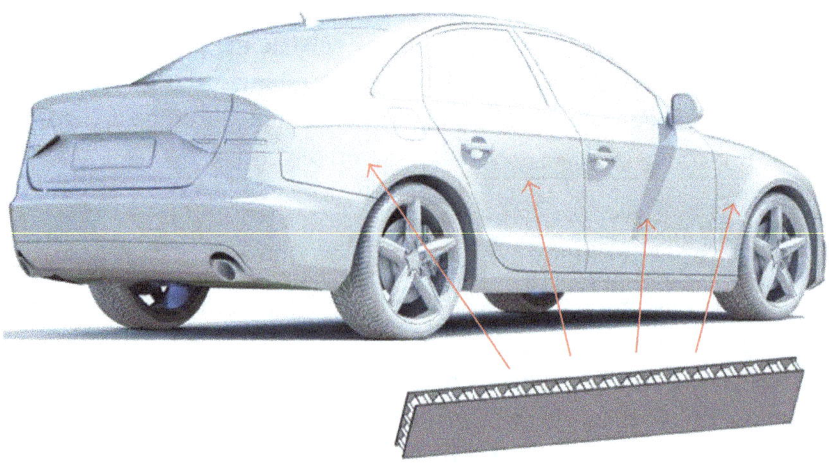

num space frames are successfully used to replace steel unibodies. Due to their durability and specific strength/stiffness, numerous conventional materials are being replaced by high-performance carbon fiber/aluminum hybrid composites in the recent few years. Instead of a beam based on steel, in modern automotive, extruded thin walled aluminum pieces were utilized in the front and rear bumpers, crash boxes, longitudinal space frames and other safety components such as side-door impact beams, engine cradles, and suspension components (Saidpour, 2004; Sun et al., 2017). Xiao et al. studied the quasistatic bending properties of aluminum honeycomb core and CFRP sandwich panels exhibiting better crashworthiness performance to use them as skins in electric vehicle body (Xiao et al., 2018).

Hybrid bonded-riveted joint was fabricated for joining CFRP and aluminum lightweight composites suitable for automotive structure applications (Chen et al., 2019). The mechanical performance of hybrid joints under temperatures from -30 to 100 °C was explored, where a maximum strength decrease of 26% and 43% was achieved under this temperature range, respectively, respected to that measured at ambient conditions. The high-velocity impact tests of a double hat bumper beam produce from aluminum (3105-H18) and carbon fiber-epoxy composites demonstrated that during collision, carbon fiber-epoxy composite absorbed equivalent energy as aluminum, which make it suitable for being utilized as a bumper beam for cars (Salifu et al., 2020). The conceptual structure shown in Figure 10 of the lightweight wheel shows the arrangement of various elements in the overall assembly; the wheel disc is taken as an integral part of a single-stage production process.

Figure 10. Concept design of the lightweight wheel (© TU Chemnitz). (© [2017], [Springer]. Used with permission.)

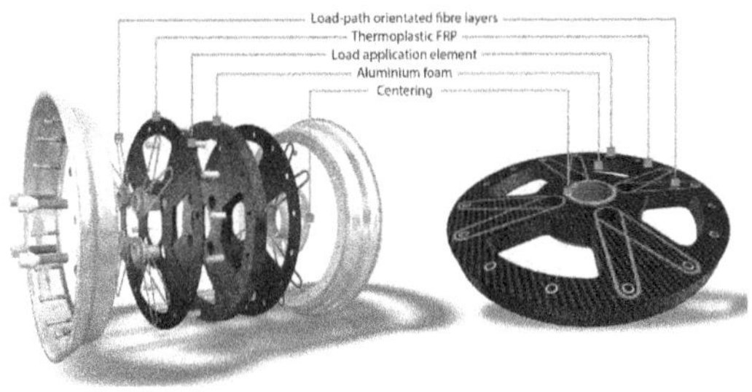

Aerospace Application

Weight decrease and enhanced damage tolerance features were the main drivers to design new kinds of materials for the aerospace industry. Thus, a new lightweight fiber/metal laminate (FML) exhibiting improved mechanical properties and low moisture absorption has been fabricated by combing the synergistic effect of each component (Botelhoet al., 2006). These hybrids are applied as fuselage skin structures of the next generation commercial aircrafts. Recently, several famous companies including EMBRAER, Aerospatiale, Boing, and Airbus were initiated many research projects relative to these materials to substitute conventional fiber reinforced polymer laminates for enhancing benefits and improving the security of their airplanes. The aluminum/CFRP hybrid structure can be manufactured as effective, lightweight structures for protection frames against collision in the aerospace area and replacing the conventional single CFRP laminates (Xu et al., 2018). The ballistic performance of unidirectional CFRP laminates and two CFRP and 2024-T3 aluminum were evaluated using a flat, hemispherical or conical nosed projectiles. Ballistic data revealed that fracture modes, ballistic limits and specific energy absorptions of the produced composites were dependent on nose forms.

Industrial Robots Design

Due to the need for increasing energy efficiency, reducing automation cost and improving automation safety solution in robotic industries, the development of lightweight composite design for industrial robot manipulators was becoming very critical. The recently designed advanced CFRP/aluminum tube materials are widely utilized for manufacturing the lightweight components of industrial robots. A sophisticated benchmark examination between CFRP and aluminum robot arms based on structural stiffen, tube thickness and material weight was described (Honarpardaz et al., 2015). The design problems of lightweight robotic arms by utilizing two types of hybrid structure composites, produced form CFRP and aluminum alloy, for minimizing the total weight of robotic arms (Yin, Liu, & Yang, 2019). Therefore, structural measurements and various layer parameters in hybrid composites were optimized, this included strength, stiffness, and dynamic constraints, where a design model and an experimental prototype were afforded with a total weight reduction of 24.32% compared to the pure metallic model. An alternative design of lightweight construction for robot arm links was proposed, which was significantly reduced the power requirements of the robot arm without compromising its stability (Karthikeyan, Sabarianand, & Suganthan, 2019).

Fire Protection

Lightweight structural materials are useful materials for fire protection application due to their high thermal insulating properties. A multi-layer polymer metal laminate constituted of aluminum and CFRP composite comprising several thin adhesively-joined aluminum foils. When exposed to fire the composite adhesive was thermally decomposed by releasing volatiles, provoking the foils to inflate and delaminate, which was responsible for reducing their thermal conductivities. The stretched composites decelerated the heat transfer from the fire into the structural support, providing lower temperatures and improved structural survivability. Christke et al. studied the fire protection behavior of two distinct thicknesses of aluminum and CFRP PML composites against a substrate. The fire exposure tests demonstrate that the substrate temperatures were decreased and the time to failure under load was markedly ameliorated (Christke et al., 2016). The protection ensured by these hybrid structures was similar or even exceeding that provided by conventional fire resistant materials, likes ceramic fiber mat or intumescent coatings.

CONCLUSION

In this chapter, the manufacturing processes, mechanical properties, and applications of aluminum alloys reinforced carbon fiber hybrid composite materials were presented by scrutinizing recent research studies. Friction lap joining, ultrasonic welding, and induction welding are main fabrication techniques used for producing aluminum/CFRP hybrids. In addition, it is found that the surface-treatment of aluminum surface can improve the adhesion strength between metallic sheets and carbon fiber reinforced laminate. Other parameters such as material deposition, cure processing, curing technique, post stretching for reserving residual stress of the hybrid composites generated during curing can equally affect the final performance of these kinds of composite structures. Respected to CFRP composites and pure aluminium metals, the aluminum/CFRP composites afforded several outstanding mechanical characteristics, including, great strength, increased elasticity modulus with enhanced toughness, better impact and crashworthiness performance, and higher fatigue-resistance features. Moreover, these hybrid composite structures exhibited better damage tolerance and lightweight characteristics which make them very suitable for engineering applications such as automotive industry, aircraft production, and lightweight robotic industries relative to their monolithic aluminum materials or CFRP laminated composites.

REFERENCES

Abe, H., Chung, J. C., Mori, T., Hosoi, A., Jespersen, K. M., & Kawada, H. (2019). The effect of nanospike structures on direct bonding strength properties between aluminum and carbon fiber reinforced thermoplastics. *Composites. Part B, Engineering*, *17*, 226–232. doi:10.1016/j.compositesb.2019.05.025

Adam, H. (1997). Carbon fibre in automotive applications. *Materials & Design*, *4*(6), 349–355. doi:10.1016/S0261-3069(97)00076-9

Alantali, A., Alia, R., Umer, R., & Cantwell, W. (2017). Energy absorption in aluminium honeycomb cores reinforced with carbon fibre reinforced plastic tubes. *The Journal of Sandwich Structures & Materials*. Advance online publication. doi:10.1177/1099636217727145

Alhashmy, H. (2012). Fabrication of Aluminium Matrix Composites (AMCs) by Squeeze Casting Technique Using Carbon Fiber as Reinforcement. *24th Materials Science Conference*.

Alhashmy, H. A., & Nganbe, M. (2015). Laminate squeeze casting of carbon fiber reinforced aluminum matrix composites. *Materials & Design*, *67*, 154–158. doi:10.1016/j.matdes.2014.11.034

Andre, N. M., Goushegir, S. M., dos Santos, J. F., Canto, L. B., & Amancio-Filho, S. T. (2016). Friction Spot Joining of aluminum alloy 2024-T3 and carbon-fiber-reinforced poly (phenylene sulfide) laminate with additional PPS film interlayer: Microstructure, mechanical strength and failure mechanisms. *Composites. Part B, Engineering*, *94*, 197–208. doi:10.1016/j.compositesb.2016.03.011

Askin, M. Y., & Turen, Y. (2019). The effect of GNP addition on mechanical and residual stress properties of 2024-T3 aluminum and carbon fiber reinforced FML. *Materials Research Express*, *6*(12), 126546. Advance online publication. doi:10.1088/2053-1591/ab575b

Balaji, G., & Annamalai, K. (2018). Crushing response of square aluminium column filled with carbon fibre tubes and aluminium honeycomb. *Thin-walled Structures*, *132*, 667–681. doi:10.1016/j.tws.2018.07.037

Balle, F., & Eifler, D. (2012). Statistical test planning for ultrasonic welding of dissimilar materials using the example of aluminum-carbon fiber reinforced polymers (CFRP) joints. *Materialwissenschaft und Werkstofftechnik*, *43*(4), 286–292. doi:10.1002/mawe.201200943

Balle, F., Emrich, S., Wagner, G., Eifler, D., Brodyanski, A., & Kopnarski, M. (2013). Improvement of Ultrasonically Welded Aluminum/Carbon Fiber Reinforced Polymer-Joints by Surface Technology and High Resolution Analysis. *Advanced Engineering Materials*, *15*(9), 814–820. doi:10.1002/adem.201200282

Balle, F., Wagner, G., & Eifler, D. (2007). Ultrasonic spot welding of aluminum sheet/carbon fiber reinforced polymer – joints. *Materialwissenschaft und Werkstofftechnik*, *38*(11), 934–938. doi:10.1002/mawe.200700212

Balle, F., Wagner, G., & Eifler, D. (2009). Ultrasonic Metal Welding of Aluminium Sheets to Carbon Fibre Reinforced Thermoplastic Composites. *Advanced Engineering Materials*, *11*(1-2), 35–39. doi:10.1002/adem.200800271

Benedyk, J. C. (2010). Aluminum alloys for lightweight automotive structures. *Materials. Design and Manufacturing for Lightweight Vehicles*, *3*, 79–113. doi:10.1533/9781845697822.1.79

Bolt, S. (2014). *Ultrasonic Plastic Welding of Carbon Fiber Reinforced Polyamide 6 to Aluminium and Steel An Experimental Study*. Master of Science Thesis.

Botelho, E. C., Silva, R. A., Pardini, L. C., & Rezende, M. C. (2006). A Review on the Development and Properties of Continuous Fiber/epoxy/aluminum Hybrid Composites for Aircraft Structures. *Materials Research*, *9*(3), 248–256. doi:10.1590/S1516-14392006000300002

Cao, X., Shi, Q., Liu, D., Feng, Z., Liu, Q., & Chen, G. (2018). Fabrication of in situ carbon fiber/aluminum composites via friction stir processing: Evaluation of microstructural, mechanical and tribological behaviors. *Composites. Part B, Engineering*, *139*, 97–105. doi:10.1016/j.compositesb.2017.12.001

Chen, Y., Yang, X., Li, M., & Mei, M. (2019). Influence of working temperatures on mechanical behavior of hybrid joints with carbon fiber reinforced plastic/aluminum lightweight materials for automotive structure. *Journal of Manufacturing Processes*, *45*, 392–407. doi:10.1016/j.jmapro.2019.07.022

Chozha Rajan, B. M., Kumar, A. S., & Sornakumar, T. (2018). Impact Response and Damage Characteristics of Carbon Fibre Reinforced Aluminium Laminates (CARAL) under Low Velocity Impact Tests. *Materials Today: Proceedings*, *5*(9), 20070–20077. doi:10.1016/j.matpr.2018.06.373

Christke, S., Gibson, A. G., Grigoriou, K., & Mouritz, A. P. (2016). Multi-layer polymer metal laminates for the fire protection of lightweight Structures. *Materials & Design*, *97*, 349–356. doi:10.1016/j.matdes.2016.02.105

Cui, Z., Liu, Q., Sun, Y., & Li, Q. (2020). On crushing responses of filament winding CFRP/aluminum and GFRP/CFRP/aluminum hybrid structures. *Composites. Part B, Engineering*, *200*, 108341. doi:10.1016/j.compositesb.2020.108341

Dhaliwal, G. S., & Newaz, G. M. (2016). Modeling Low Velocity Impact Response of Carbon Fiber Reinforced Aluminum Laminates (CARALL). *Journal of Dynamic Behavior of Materials*, *2*(2), 181–193. doi:10.100740870-016-0057-3

Dhaliwal, G. S., & Newaz, G. M. (2016). Experimental and numerical investigation of flexural behavior of carbon fiber reinforced aluminum laminates. *Journal of Reinforced Plastics and Composites*, *35*(12), 945–956. doi:10.1177/0731684416632606

Dhaliwal, G. S., & Newaz, G. M. (2020). Experimental and numerical investigation of flexural behavior of hat sectioned aluminum/carbon fiber reinforced mixed material composite beam. *Composites. Part B, Engineering*, *182*, 107642. doi:10.1016/j.compositesb.2019.107642

Di Franco, G., Fratini, L., Pasta, A., & Ruisi, V. F. (2013). On the self-piercing riveting of aluminium blanks and carbon fibre composite panels. *International Journal of Material Forming*, *6*(1), 137–144. doi:10.100712289-011-1067-2

Esteves, J. V., Goushegir, S. M., dos Santos, J. F., Canto, L. B., Hage, E. Jr, & Amancio-Filho, S. T. (2015). Friction spot joining of aluminum AA6181-T4 and carbon fiber-reinforced poly(phenylene sulfide): Effects of process parameters on the microstructure and mechanical strength. *Materials & Design*, *66*, 437–445. doi:10.1016/j.matdes.2014.06.070

Franco, G.-D., Fratini, L., & Pasta, A. (2012). Influence of the distance between rivets in self-piercing riveting bonded joints made of carbon fiber panels and AA2024 blanks. *Materials & Design*, *35*, 342–349. doi:10.1016/j.matdes.2011.09.036

Franco, G. D., Fratinib, L., Pasta, A., & Ruisi, V. F. (2010). On the self-piercing riveting of aluminium blanks and carbon fibre composite panels. *Journal Materials Form*, *3*(1), 1035–1038. doi:10.100712289-010-0947-1

Galyshev, S., Gomzin, A., Gallyamova, R., Khodos, I., & Musin, F. (2019). On the liquid-phase technology of carbon fiber/aluminum matrix composites. *International Journal of Minerals Metallurgy and Materials*, *26*(12), 1578–1584. doi:10.100712613-019-1877-7

Goushegir, S. M. (2016). Friction spot joining (FSpJ) of aluminum-CFRP hybrid structures. *Welding in the World*, *60*(6), 1073–1093. doi:10.100740194-016-0368-y

Goushegir, S. M., dos Santos, J. F., & Amancio-Filho, S. T. (2014). Friction spot joining of aluminum AA2024/carbon-fiber reinforced poly(phenylene sulfide) composite single lap joints: Microstructure and mechanical performance. *Materials & Design*, *50*, 196–206. doi:10.1016/j.matdes.2013.08.034

Goushegir, S. M., dos Santos, J. F., & Amancio-Filho, S. T. (2014). Friction Spot Joining of aluminum AA2024/carbon-fiber reinforced poly (phenylene sulfide) composite single lap joints: Microstructure and mechanical performance. *Materials & Design*, *54*, 196–206. doi:10.1016/j.matdes.2013.08.034

Goushegir, S. M., dos Santos, J. F., & Amancio-Filho, S. T. (2015). Influence of process parameters on mechanical performance and bonding area of AA2024/carbon-fiber reinforced poly(phenylene sulfide) friction spot single lap joints. *Materials & Design*, *83*, 431–442. doi:10.1016/j.matdes.2015.06.044

Hackert, A., Müller, S., & Kroll, L. (2017). Lightweight Wheel Disc with Carbon Aluminium Foam Sandwich. *Lightweight Design Worldwide*, *10*(1), 6–11. doi:10.100741777-017-0016-2

Han, Z., Li, H., Xu, X., Wang, H., Li, H., & Tao, J. (2020). Crushing characteristics of aluminum/CFRP/aluminum hybrid tubes prepared by spinning forming. *Composite Structures*, *112551*. Advance online publication. doi:10.1016/j.compstruct.2020.112551

Hashim, U. R., Jumahat, A., Ismail, M. H., & Razali, R. N. M. (2014). Fabrication and characterization of carbon fibre reinforced polymer rods with aluminium foam core. *Materials Research Innovations*, *18*(6), 204–208.

He, W., Yao, L., Meng, X., Sun, G., Xie, D., & Liu, J. (2019). Effect of structural parameters on low-velocity impact behavior of aluminum honeycomb sandwich structures with CFRP face sheets. *Thin-walled Structures*, *137*, 411–432. doi:10.1016/j.tws.2019.01.022

Honarpardaz, O., Trangärd, A., Derkx, J., & Feng, X. (2015). Benchmark of Advanced Structural Materials for Lightweight Design of Industrial Robots. IEEE, 3178-3183.

Hu, Y., Yuan, B., Cheng, F., & Hu, X. (2019). NaOH etching and resin pre-coating treatments for stronger adhesive bonding between CFRP and aluminium alloy. *Composites. Part B, Engineering*, *178*, 107478. doi:10.1016/j.compositesb.2019.107478

Hufenbach, W., Gude, M., Czulak, A., Śleziona, J., Dolata-Grosz, A., & Dyzia, M. (2009). Development of textile-reinforced carbon fibre aluminium composites manufactured with gas pressure infiltration methods. *Journal of Achievements in Materials and Manufacturing Engineering*, *35*(2), 177–183.

Jiang, B., Chen, Q., & Yang, J. (2020). Advances in joining technology of carbon fiber-reinforced thermoplastic composite materials and aluminum alloys. *International Journal of Advanced Manufacturing Technology*, *110*(9-10), 2631–2649. Advance online publication. doi:10.100700170-020-06021-2

Jiang, H., Luo, T., Li, G., Zhang, X., & Cui, J. (2017). Fatigue life assessment of electromagnetic riveted carbon fiber reinforces plastic/aluminum alloy lap joints using Weibull distribution. *International Journal of Fatigue*, *105*, 180–189. doi:10.1016/j.ijfatigue.2017.08.026

Jung, K-W., Kawahito, Y., Takahashi, M., & Katayama, S. (2013). Laser direct joining of carbon fiber reinforced plastic to aluminum alloy. *Journal of Laser Applications, 25*(3), 032003-(1) - 032003-(6).

Kang, J., Rao, H., Zhang, R., Avery, K., & Su, X. (2016). Tensile and fatigue behaviour of self-piercing rivets of CFRP to aluminium for automotive application. *Materials Science and Engineering, 137,* 1-9.

Karthikeyan, P., Sabarianand, D. V., & Suganthan, S. (2019). Investigation on Adaptability of Carbon Fiber Tube for Serial Manipulator. *FME Transactions*, *47*(3), 412–417. doi:10.5937/fmet1903412K

Kim, H. C., Shin, D. K., & Lee, J. J. (2013). Characteristics of aluminum/CFRP short square hollow section beam under transverse quasi-static loading. *Composites. Part B, Engineering*, *51*, 345–358. doi:10.1016/j.compositesb.2013.03.020

Kim, H. C., Shin, D. K., Lee, J. J., & Kwon, J. B. (2014). Crashworthiness of aluminum/CFRP square hollow section beam under axial impact loading for crash box application. *Composite Structures*, *112*, 1–10. doi:10.1016/j.compstruct.2014.01.042

Kwon, D.-J., Kim, J.-H., Kim, Y.-J., Kim, J.-J., Park, S.-M., Kwon, I.-J., Shin, P.-S., DeVries, L. K., & Park, J.-M. (2019). Comparison of interfacial adhesion of hybrid materials of aluminum/carbon fiber reinforced epoxy composites with different surface roughness. *Composites. Part B, Engineering*, *170*, 11–18. doi:10.1016/j.compositesb.2019.04.022

Lambiase, F., & Ko, D. C. (2016). Feasibility of mechanical clinching for joining aluminum AA6082-T6 and Carbon Fiber Reinforced Polymer sheets. *Materials & Design*, *107*, 341–352. doi:10.1016/j.matdes.2016.06.061

Lambiase, F., & Ko, D.-C. (2017). Two-steps clinching of aluminum and Carbon Fiber Reinforced Polymer sheets. *Composite Structures*, *164*, 180–188. doi:10.1016/j.compstruct.2016.12.072

Lebaupina, Y., Friedlib, J., Caglara, B., Piccanda, M., Pasquierb, R., & Michaud, V. (2019). Crushing and intrusion resistance improvement of aluminum beams by carbon/epoxy composite patches. *Composite Structures*, *226*, 111235. doi:10.1016/j.compstruct.2019.111235

Lee, C.-J., Kim, B.-M., Kang, B.-S., Song, W.-J., & Ko, D.-C. (2017). Improvement of joinability in a hole clinching process with aluminum alloy and carbon fiber reinforced plastic using a spring die. *Composite Structures*, *173*, 58–69. doi:10.1016/j.compstruct.2017.04.010

Liang, J., Jiang, H., Zhang, J., Wu, X., Zhang, X., Li, G., & Cui, J. (2019). Investigations on mechanical properties and microtopography of electromagnetic self-piercing riveted joints with carbon fiber reinforced plastics/aluminum alloy 5052. *Archives of Civil and Mechanical Engineering*, *19*(1), 240–250. doi:10.1016/j.acme.2018.11.001

Lionetto, F., Balle, F., & Maffezzoli, A. (2017). Hybrid ultrasonic spot welding of aluminum to carbon fiber reinforced epoxy composites. *Journal of Materials Processing Technology*, *247*, 289–295. doi:10.1016/j.jmatprotec.2017.05.002

Lionetto, F., Mele, C., Leo, P., D'Ostuni, S., Balle, F., & Maffezzoli, A. (2018). Ultrasonic spot welding of carbon fiber reinforced epoxy composites to aluminum: Mechanical and electrochemical characterization. *Composites. Part B, Engineering*, *144*, 134–142. doi:10.1016/j.compositesb.2018.02.026

Liu, X., Shao, X., Li, Q., & Sun, G. (2019). Failure mechanisms in carbon fiber reinforced plastics (CFRP)/aluminum (Al) adhesive bonds subjected to low-velocity transverse pre-impact following by axial post-tension. *Composites. Part B, Engineering*, *172*, 339–351. doi:10.1016/j.compositesb.2019.04.036

Liu, Y., & Zhuang, W. (2019). Self-piercing riveted-bonded hybrid joining of carbon fibre reinforced polymers and aluminium alloy sheets. *Thin-walled Structures*, *144*, 106340. doi:10.1016/j.tws.2019.106340

Marx, L., & Liewald, M. (2017). Semi-Solid Joining of Aluminium and Carbon Fabric. *Procedia CIRP*, *66*, 233–237. doi:10.1016/j.procir.2017.03.306

Marx, L., & Liewald, M. (2020). Joining of carbon fibres with aluminium structures – Processing and infiltration modelling. *Journal of Composite Materials*, (25), 3767–3775. Advance online publication. doi:10.1177/0021998320912317

Menghari, H. G., Babalo, V., Fazli, A., Soltanpour, M., & Ziaeipoor, H. (2020). A study on the electro-hydraulic clinching of aluminum and carbon fiber reinforced plastic sheets. *International Journal of Lightweight Materials and Manufacture*. doi:10.1016/j.ijlmm.2020.01.002

Mitschang, P., Velthuis, R., & Didi, M. (2013). Induction Spot Welding of Metal/CFRPC Hybrid Joints. *Advanced Engineering Materials*, *15*(9), 804–813. doi:10.1002/adem.201200273

Mitschang, P., Velthuis, R., Emrich, S., & Kopnarski, M. (2009). Induction Heated Joining of Aluminum and Carbon Fiber Reinforced Nylon 66. *Journal of Thermoplastic Composite Materials*, *22*(6), 687–800. doi:10.1177/0892705709105969

Muthu Chozha Rajan, B., Senthil Kumar, A., Sornakumar, T., Senthamaraikannan, P., & Sanjay, M. R. (2018). Multi Response Optimization of Fabrication Parameters of Carbon Fiber-Reinforced Aluminium Laminates (CARAL): By Taguchi Method and Gray Relational Analysis. *Polymer Composites*. Advance online publication. doi:10.1002/pc.24815

Nestler, D., Trautmann, M., Nendel, S., Wagner, G., & Kroll, L. (2016). Innovative hybrid laminates of aluminium alloy foils and fibre-reinforced thermoplastic layers. *Materialwissenschaft und Werkstofftechnik*, *47*(11), 1121–1131. doi:10.1002/mawe.201600636

Okur, M. Z. (2018). Development of Aluminum Honeycomb Cored Carbon Fiber Reinforced Polymer Composite Based Sandwich Structures. Preprints, 2018120195. doi:10.20944/preprints201812.0195.v1

Pandey, A., Muchhala, D., Kumar, R. S. S., Venkat, A. N. C., & Mondal, D. P. (2019). Flexural deformation behavior of carbon fiber reinforced aluminium hybrid foam sandwich structure. *Composites. Part B, Engineering*, *107729*. Advance online publication. doi:10.1016/j.compositesb.2019.107729

Pramanik, A., Basak, A. K., Dong, Y., Sarker, P. K., Uddin, M. S., Littlefair, G., Dixit, A. R., & Chattopadhyaya, S. (2017). Joining of carbon fibre reinforced polymer (CFRP) composites and aluminium alloys-A review. *Composites. Part A, Applied Science and Manufacturing*, *101*, 1–29. doi:10.1016/j.compositesa.2017.06.007

Rao, H. M., Kang, J., Huff, G., & Avery, K. (2018). Impact of specimen configuration on fatigue properties of self-piercing riveted aluminum to carbon fiber reinforced polymer composite. *International Journal of Fatigue*, *113*, 11–22. doi:10.1016/j.ijfatigue.2018.03.031

Reuter, C., & Tröster, T. (2017). Crashworthiness and numerical simulation of hybrid aluminium-CFRP tubes under axial impact. *Thin-walled Structures*, *117*, 1–9. doi:10.1016/j.tws.2017.03.034

Rupp, P., Elsner, P., & Weidenmann, K. A. (2017). Failure mode maps for four-point-bending of hybrid sandwich structures with carbon fiber reinforced plastic face sheets and aluminum foam cores manufactured by a polyurethane spraying process. *The Journal of Sandwich Structures & Materials*. Advance online publication. doi:10.1177/1099636217722052

Saidpour, H. (2004). Lightweight High Performance Materials for Car Body Structures. *NTI Technology Conference*, 14-19.

Salamati, M., Soltanpour, M., Fazli, A., & Zajkani, A. (2017). Parametric Study on the Electromagnetic Force-Fit Joining of Carbon Fiber Reinforced Plastic and Aluminum Tubes. *Procedia Engineering*, *207*, 986–991. doi:10.1016/j.proeng.2017.10.863

Salifu, S., Desai, D., Ogunbiyi, O., Sadiku, R., Adesina, O., & Adesina, O. (2020). Comparative study of high velocity impact response of aluminium 3105- H18 and carbon fibre-epoxy composite double hat bumper beams. *Materials Today: Proceedings*. Advance online publication. doi:10.1016/j.matpr.2020.03.828

Schimanski, K., Hehl, A., & Zoch, H. W. (2013). Failure behavior of diffusion bonded transition structures for integral FRP-Aluminum compounds. *Procedia Materials Science*, *2*, 189–196. doi:10.1016/j.mspro.2013.02.023

Schimanski, K., Schumacher, J., Lang, A., von Hehl, A., & Bomas, H. (2011). Characteristics of Titanium interface structures for advanced FRP Aluminium compounds. *Materials Science Forum*, *690*, 266–269. doi:10.4028/www.scientific.net/MSF.690.266

Schumacher, J., Bomas, H., & Zoch, H.-W. (2013). Failure behaviour of advanced seam structures for CFRP aluminium connections. *Procedia Materials Science*, *02*, 227–233. doi:10.1016/j.mspro.2013.02.028

Shirvanimoghaddam, K., Hamim, S. U., Akbari, M. K., Fakhrhoseini, S. M., Khayyam, H., Pakseresht, A. H., Ghasali, E., Zabet, M., Munir, K. S., Jia, S., Davim, P., & Naebe, M. (2017). Carbon fiber reinforced metal matrix composites: Fabrication processes and properties. *Composites. Part A, Applied Science and Manufacturing*, *92*, 70–96. doi:10.1016/j.compositesa.2016.10.032

Sinmazçelik, T., Avcu, E., Bora, M. Ö., & Çoban, O. (2011). A review: Fibre metal laminates, background, bonding types and applied test methods. *Materials & Design*, *32*(7), 3671–3685. doi:10.1016/j.matdes.2011.03.011

Stoll, M., Stemmer, F., Ilinzeer, S., & Weidenmann, K. A. (2017). Optimization of Corrosive Properties of Carbon Fiber Reinforced Aluminum Laminates due to Integration of an Elastomer Interlayer. *Key Engineering Materials*, *742*, 287–293. doi:10.4028/www.scientific.net/KEM.742.287

Sun, X., Liu, J., Lu, B., Zhang, P., & Zhao, M. (2017). Life cycle assessment-based selection of a sustainable lightweight automotive engine hood design. *The International Journal of Life Cycle Assessment*, *22*(9), 1373–1383. doi:10.100711367-016-1254-y

Surappa, M. K. (2003). Aluminium matrix composites: Challenges and Opportunities. *Sadhana*, *28*(1 & 2), 319–334. doi:10.1007/BF02717141

Valente, M., Marini, D., Genova, V., Quitadamo, A., Marra, F., & Pulci, G. (2019). Lightweight metallic matrix composites: Development of new composites material reinforced with carbon structures. *Journal of Applied Biomaterials & Functional Materials*, *17*(1S). doi:10.1177/2280800019840294 PMID:31215817

Wagner, G., Balle, F., & Eifler, D. (2013). Ultrasonic welding of aluminum alloys to fiber reinforced polymers. *Advanced Engineering Materials*, *15*(9), 792–803. doi:10.1002/adem.201300043

Wang, J., Zhang, G., Zheng, X., Li, J., Li, X., Zhu, W., & Yanagimoto, J. (2020). A self-piercing riveting method for joining of continuous carbon fiber reinforced composite and aluminum alloy sheets. *Composite Structures*, *113219*. Advance online publication. doi:10.1016/j.compstruct.2020.113219

Welf-Guntram, D., Reinhard, M., Raik, G., & Danilo, M. (2013). Numerical and Experimental Analysis of Self Piercing Riveting Process with Carbon Fiber-Reinforced Plastic and Aluminium Sheets. *Key Engineering Materials*, *554-557*, 1045–1054. doi:10.4028/www.scientific.net/KEM.554-557.1045

Wu, L. H., Xiao, B. L., Nagatsuka, K., Nakata, K., & Ma, Z. Y. (2020). Achieving strong friction lap joints of carbon-fiber reinforced plastic and metals by modifying metal surface structure via laser-processing pretreatment. *Composite Structures*, *242*, 112167. doi:10.1016/j.compstruct.2020.112167

Wu, Y., Liu, Q., Fu, J., Li, Q., & Hui, D. (2017). Dynamic crash responses of bioinspired aluminum honeycomb sandwich structures with CFRP panels. *Composites. Part B, Engineering*, *121*, 122–133. doi:10.1016/j.compositesb.2017.03.030

Xiao, Y., Hu, Y., Zhang, J., Song, C., Huang, X., Yu, J., & Liu, Z. (2018). The Bending Responses of Sandwich Panels with Aluminium Honeycomb Core and CFRP Skins Used in Electric Vehicle Body. *Advances in Materials Science and Engineering*, *2018*, 1–11. doi:10.1155/2018/5750607

Xu, M., Huang, G., Dong, Y., & Feng, S. (2018). An experimental investigation into the high velocity penetration resistance of CFRP and CFRP/aluminium laminates. *Composite Structures*, *188*, 450–460. doi:10.1016/j.compstruct.2018.01.020

Yan, C., Song, X., Zhu, H., Jing, C., & Feng, S. (2017). Flexural response of carbon fiber reinforced aluminum foam sandwich. *Journal of Composite Materials*, *52*(14), 1887–1897. doi:10.1177/0021998317735166

Yan, C., Wang, J., & Song, X. (2020). Fatigue behavior and damage mechanism of aluminum foam sandwich with carbon fiber face-sheets. *Journal of Mechanical Science and Technology*, *34*(3), 1-9.

Yin, H., Liu, J., & And Yang, F. (2019). Hybrid Structure Design of Lightweight Robotic Arms Based on Carbon Fiber Reinforced Plastic and Aluminum alloy. *IEEE Access: Practical Innovations, Open Solutions*, *7*, 64932–64945. doi:10.1109/ACCESS.2019.2915363

Zhu, G., Sun, G., Liu, Q., Li, G., & Li, Q. (2017). On crushing characteristics of different configurations of metal-composites hybrid tubes. *Composite Structures*, *175*, 58–69. doi:10.1016/j.compstruct.2017.04.072

Zhu, G., Sun, G., Yu, H., Li, S., & Li, Q. (2018). Energy absorption of metal, composite and metal/composite hybrid structures under oblique crushing loading. *International Journal of Mechanical Sciences*, *135*, 458–483. doi:10.1016/j.ijmecsci.2017.11.017

Zhu, S., & Chai, G. B. (2013). Impact of Aluminum, CFRP Laminates, Fibre-Metal Laminates and Sandwich Panels. *Composite Materials and Joining Technologies for Composites*, *07*, 199–205. doi:10.1007/978-1-4614-4553-1_21

KEY TERMS AND DEFINITIONS

Automotive Industry: All companies and activities involved in the manufacturing of motor vehicles like automobiles, buses, trucks, etc.

Composite Material: A type of material that is made by macroscopically combining two or more different materials together. The two materials work together—they do not dissolve or blend into each other—to give the composite unique properties.

Energy Efficiency: A fact that shows how efficiently energy is consumed, and it covers wide-ranging topics related to energy efficiency, energy savings, energy consumption, energy sufficiency, and energy transition in all sectors across the globe (including the automotive industry).

Chapter 2
Additive Manufacturing (AM): Processing Technique for Lightweight Alloys and Composite Material

Kamardeen Olajide Abdulrahman
University of Ilorin, Nigeria

Rasheedat Modupe Mahamood
University of Ilorin, Nigeria

Esther T. Akinlabi
Pan African University for Life and Earth Sciences Institute, Nigeria

ABSTRACT

The need for less weight and high-performance materials in manufacturing industries has continuously led to the development of lightweight materials through the use of advanced additive manufacturing (AM). The race of lightweight and high-performance metals continue to evolve as this continuously provides better understanding about connection existing between material processing, microstructural development, and material properties. AM technique is an interesting manufacturing process that is employed in production of engineering components with improved properties. The choice of titanium and its alloys in structural applications are attributed to their superior strength-to-weight ratio and high corrosion resistance. This chapter looked at different additive manufacturing (AM) techniques developed for the processing of lightweight metals, their strengths, and limitations. The chapter also looked at the role and contribution of AM to the 4th industrial revolution, processing, and application of titanium aluminide for high temperature applications.

INTRODUCTION

In this present economic time, cost reduction, time management and geometric complexity are major market demands in the development of new products. These are posing serious issues to the tooling processes that are already expensive and time consuming. Additive manufacturing (AM) and modern operational

DOI: 10.4018/978-1-7998-7864-3.ch002

structures are effective tools that can assist to reduce time and cost (Ludovico et al., 2010). The quest for lightweight high performance structural metals by major industrial sectors such as the automobile, aerospace, energy and chemical have made researchers, engineers and other major stakeholders in the manufacturing sector to continuously develop different manufacturing techniques.

Three-dimensional (3D) printing has progressed into AM process with the ability of fabricating parts directly via layer-by-layer build-up process (Abdulrahman et al., 2018). Several companies have embraced Additive manufacturing technology because of its unique capabilities such as customization, multifunctionality, improvement in product reliability, high performance rate and overall production cost reduction. These companies include Siemens and other hearing aid manufacturers that use Stereolithography and sintering machines to produce hearing aid shells, Align Technology apply stereolithography to produce molds for production of dental braces, Boeing and its suppliers employed the selective laser sintering in the production of ducts and other similar components for fighter jet F-18 (Gibson et al., 2010a). AM technique is a progress made from rapid prototyping which is now employed in the production of end-use products that now find applications in different areas such as the automobile and aerospace industries (Wholer's Report, 2016). AM processing technique produces components using 3D model data.

The main objective of the use of AM technique is to enhance the quality and performance of produced parts through cost of production, material usage and lead time (Kobryn et al., 2006). Some AM processes are undertaken with the use of lasers as the main source of energy to provide fast heating in melting the metal material thereby creating a melt pool with the capacity of enhancing diffusion rate of introduced metal powder to properly mix (Tlotleng et al., 2016).

History

Additive technology sometimes referred to as 3D printing has recently spread into new areas including medicine, energy, electronics, automobile and aerospace. The application of printing technology is becoming a highly promising manufacturing practice as it is greatly studied and its usage is fast growing (Gibson et al., 2010b). The first commercially successful 3D printing technology, the ModelMaker by Solidscape (formerly Sanders Prototype) was introduced in 1994. The ModelMaker was able to print using basic wax material as raw material. By 1996, 3D Systems joined the race of 3D printing technology with the birth of the Actua 2100 which is also based on the printing of wax material. Thermojet, an improvement on Actua 2100 was marketed in 1999 (Gornet & Wohlers, 2014). The build materials of the first set of rapid prototype machines are based on melted waxy thermoplastics which made them very useful for concept modeling and investment casting patterns (Gibson et al., 2010b).

Additive Manufacturing Techniques

Additive manufacturing (AM) is an advanced layer-upon-layer manufacturing technology that applies the use of 3D model data in the fabrication of near-net-shape components (Abdulrahman et al., 2018). AM technology helps in the fabrication of functionally graded components with specific material properties that cannot be easily manufactured using casting and other conventional manufacturing techniques (Gasper et al., 2017). The technology is fast growing because of its capability of producing complex components that seems difficult to be manufactured using the traditional production techniques such as casting and machining (Herderick, 2011). The robust capability of AM has made it popular in different areas of applications especially in aerospace, automobile, defense, energy generation, medicine and surgery. The quest for robust designs required in modern day machines and components, reduction in material wastage and lead time, environmental impact as a result of production technique and production cost have been the major advantages of why AM technique is emerging as the most preferred option among other manufacturing techniques.

There are numerous techniques employed in additive manufacturing of components. Most often, these techniques carries the name of either the method used or the company name that developed the manufacturing process (Abdulrahman et al., 2018). Basically, AM technology are used on plastic and metals. Stereolithography apparatus (SLA), Fused deposition modelling (FDM) and Selective laser sintering (SLS) are some of the AM techniques that make use of plastics while Direct metal laser sintering (DMLS), Electron beam melting (EBM), Powder bed fusion, Selective laser melting (SLM) and Laser engineered net shaping (LENS) are some of the AM techniques that uses metals for deposition process. The different classification of AM techniques employed in the production of metal components are shown in figure 1.

Figure 1. Classification of layer manufacturing techniques for metal components (Karunakaran et al., 2010)

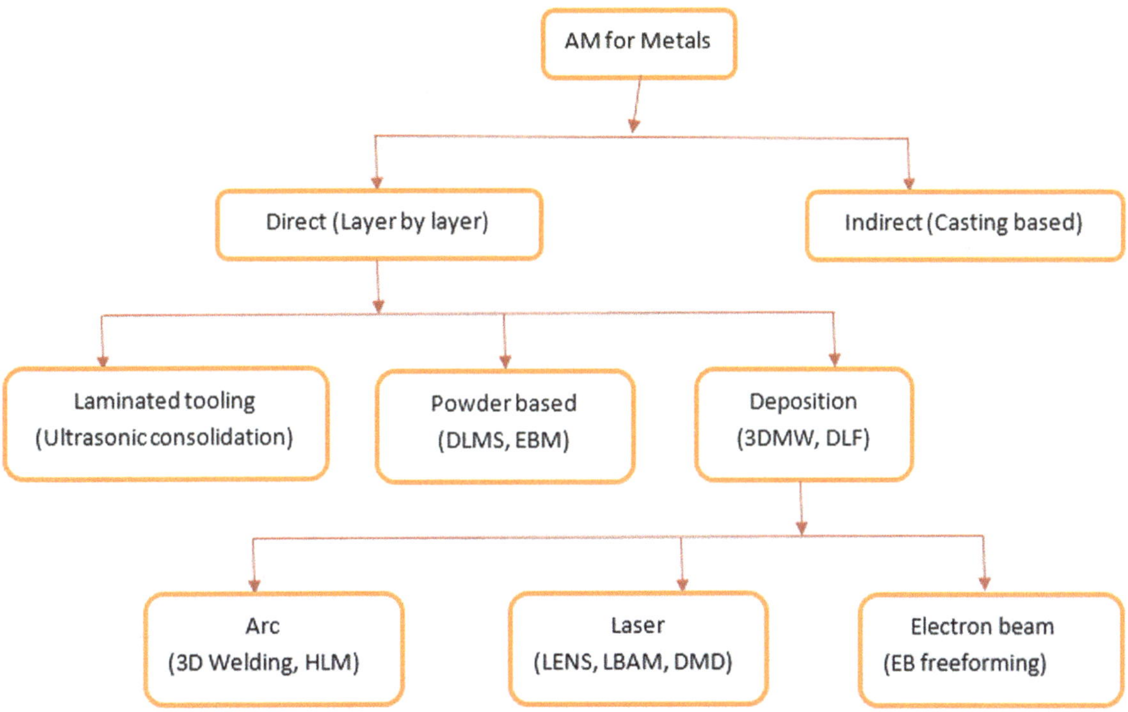

Some of the metal manufacturing processes are highlighted below:

Electron Beam Melting (EBM)

The EBM process takes place in a vacuum under high temperature condition. In this process, the energy needed for high productivity is created by high power electron beam. The powder bed for which each layer of powder is to be deposited upon at required temperature is heated by the electron beam. The microstructures of parts produced through this process do not usually contain martensitic structures and Residual stresses in the parts produced through the process are reduced due to the heating process.

Selective Laser Melting (SLM)

Selective laser melting (SLM) sometimes referred to as laser beam melting (LBM) is another type of AM process where parts of different complexities are directly fabricated from metal powder via computer aided design (CAD) files. The process as shown in figure 2, involves the division of 3D CAD model into tiny layers and the data is forwarded to the SLM machine for part production. Fine metallic powder material of about 10-45 μm sizes are then deposited layer-by-layer on the substrate. The layer-by-layer deposition process continue until the complete part is produced. The part fabricated through the process has about 100% density as standard metallic powder have been used in the process. This makes the mechanical properties of parts produced through the process to be similar or even better than parts obtained by conventional manufacturing process (Bremen et al., 2012).

Figure 2. Selective laser melting process (Bremen et al., 2012)

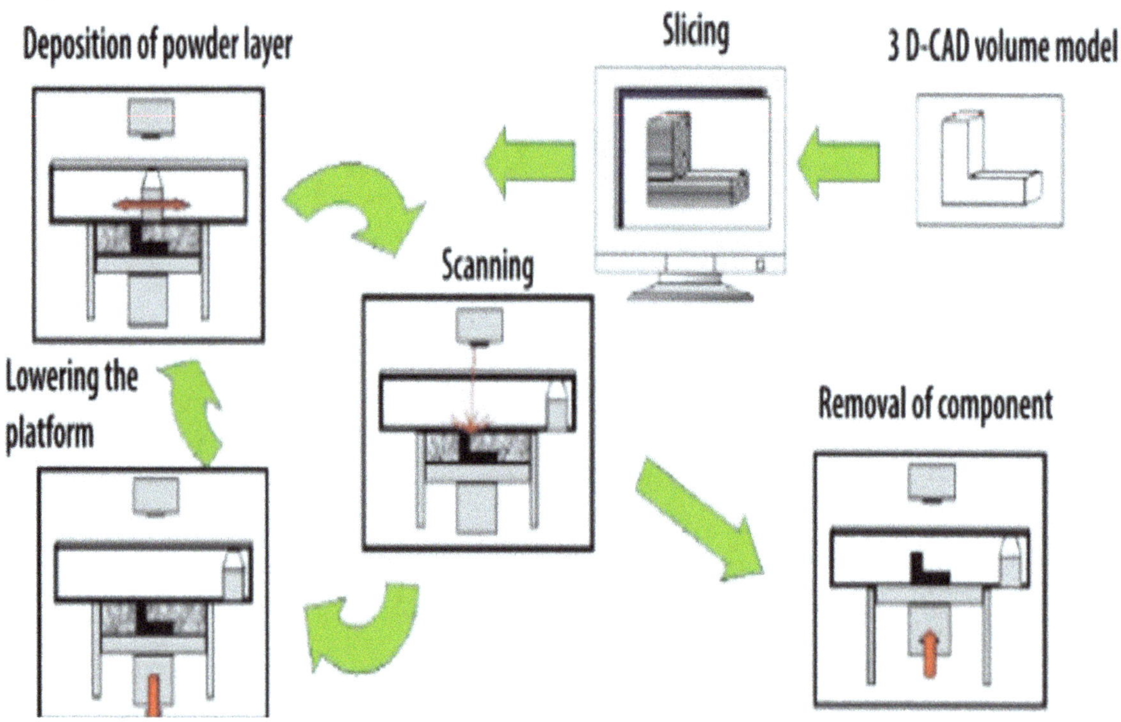

Laser Metal Deposition (LMD)

The process is sometime referred to as Direct energy deposition (DED) process. The process as shown in figure 3, is an AM metal deposition process that uses a nozzle to directly deposit melted metal material on a required surface on which it finally solidifies. Table 1 shows that the LMD process is highly productive than the SLM process. Some of the advantages of the LMD process include: the repair of complex parts that seems irreparable, ability to control material deposition, addition of new functional feature on existing part, improved metallographic quality, reduced material wastage and part dimension only limited to machine size and design.

Figure 3. Laser metal deposition process (Graf et al., 2018).

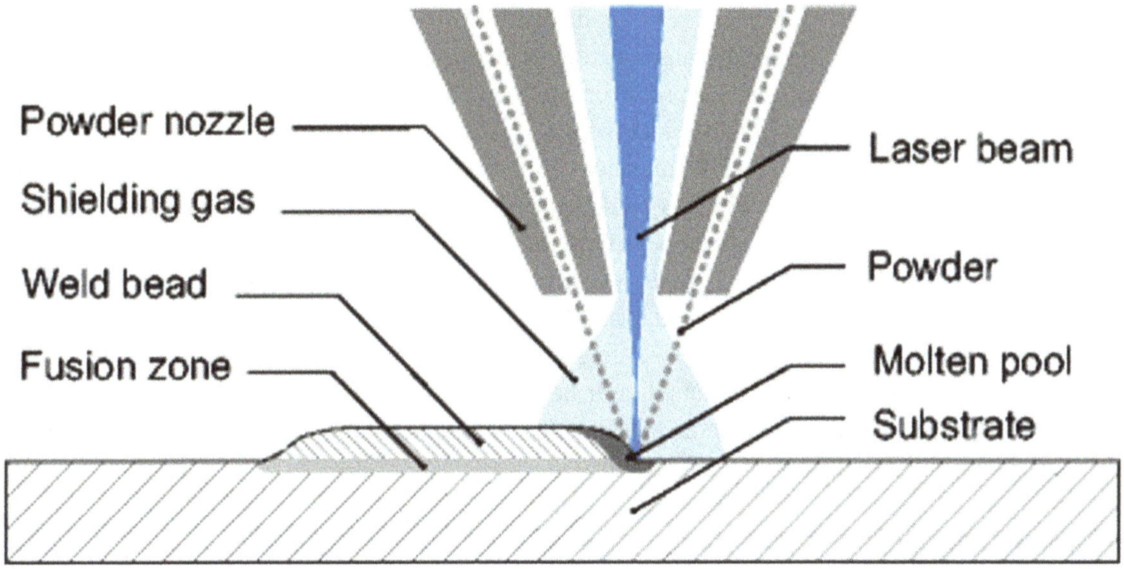

Table 1. Laser metal deposition versus Selective laser melting (Weisheit & Rolink, 2016)

Characteristics	Laser metal deposition	Selective laser melting
Materials	✓ Monolithic ✓ Gradient, hybrid	✓ Monolithic
Part complexity	Limited	Nearly unlimited
Part dimensions	Limited by handling system	Limited by the process chamber (height: 500mm, Ø:400mm)
Build-up on	✓Existing part ✓ 3D surface	✓ Flat preforms ✓ Flat surface
Build-up rate	3-140mm^3/s	1-20mm^3/s

Strengths and Limitations of AM Techniques

The production of 3D printers has been a deliberate and gradual process due to the careful processing rate of patency for commercialized AM printers, lack of open-source metal choices and high capital cost (Anzalone et al., 2013). This has limited the use of commercialized 3D printers and only made it available mainly in the production of costly finished products and rapid prototyping, and made it difficult to be employed by small research centres, small and medium scale companies especially in the developing countries that are also faced with economic challenges.

Laser Metal Deposition (LMD) Technique

Light Amplification by Stimulated Emission of Radiation simply referred to as "LASER" has become popular in recent time and make use of in different areas of applications. The far-reaching light moving characteristics of lasers made them very applicable to different areas. The application of lasers is witnessed in every day-to-day activity ranging from office equipment (as seen in laser printers, light/laser pointer, laser scanners and bar-code scanners) to medicine and surgery, construction, production, printing, and military especially in the area of locating and targeting (Abdulrahman et al., 2018).

The ability of laser to produce high intensity beam makes it highly useful to melt hard material within a short time. This is reason why it is being applied in laser AM processes as it made it visible to produce melt-pool on substrate on which deposited metal powder are built on. Laser metal deposition technique, an aspect of AM technology possesses the capability of producing near-net shape components with superior mechanical properties (Dinda et al., 2009; Kobryn et al., 2006). In laser metal deposition process, feeding metal powder moved by shielding inert gas is deposited unto a melt pool of a substrate that have been created by a well-positioned laser beam. Lightweight alloys and composites materials with improved properties have been produced using this technique (Baudana et al., 2016; Froes & Dutta, 2014; Gussone et al., 2015; Kobryn et al., 2006).

Effects of Deposition Parameters in LMD Process

Laser metal deposition process is a tough process that involves quite a number of process parameters that affect the final characteristics and quality of deposited part in terms of the physical, microstructural and mechanical properties (Shah et al., 2010). The control of deposit characteristics required adequate knowledge of the relationship between the process parameters such as the laser power, powder flow rate, scanning speed and gas flow rate. It is therefore recommended to understand the implications of deposition parameters on the physical, microstructural and the mechanical properties before careful selection of deposition parameters needed for part production. Effect of deposition parameters on the characteristics of deposited parts have remain an interesting aspect to most lightweight metal researchers and reports are being published on deposition process parameters as evident in several works (Abdulrahman et al., 2019; Akinlabi et al., 2014; Akinlabi et al., 2012a, 2012b; Gussone et al., 2015). Short descriptions of some deposition parameters are therefore provided below.

Powder Flow Rate

This is the amount of powder being released to the deposition area per unit time (Abdulrahman et al., 2018). Its unit is in revolution per minute (rpm) or gram per minute (g/min). Powder flow rate or the flowability of any metal powder is mainly dependent on the powder particle size and its morphology. The ability of spherical shaped metal powder to conveniently react to laser beam made it seem as the most preferable for laser metal deposition process (Schade et al., 2014). Powder flow rate play a serious role in material efficiency and dimensional accuracy (Kumar et al., 2014; Schade et al., 2014) of fabricated parts. Powder flow rate have also been investigated to affect the surface roughness of deposits (Shah et al., 2010) and also affect hardness, wear resistance and microstructure of deposited samples (Akinlabi & Akinlabi, 2016; Mahamood & Akinlabi, 2015; Saboori et al., 2017; Shukla et al., 2012).

Gas Flow Rate

Pityana et al., (2013) worked on the effect of gas flow rate and powder flow rate on laser metal deposited Ti6Al4V. Maintaining constant laser power of 1.8 KW and scanning speed of 0.005 m/s, gas flow rate and powder flow rate were adjusted between 2 to 4 l/min and 2.88 to 5.76 g/min respectively. The outcome of the physical, metallurgical and mechanical test carried out revealed that increase in powder flow rate led to a concurrent increase in height, width and weight of the deposited track while increase in gas flow rate led to reduction in the height, width and weight of the deposited track as a result of the high disturbance produced by the gas flow rate. It was also discovered that increase in gas flow rate led to decrease in the average microhardness of deposits, while increase in powder flow rate causes an increase in average microhardness of deposited samples.

Scanning Speed

Laser scanning speed is a deposition parameter that is used to qualify the quantity of speed at which laser moves along a specified path. Laser scanning speed has been identified to have serious impact in a deposition process that determines the quality of fabricated parts. Laser scanning speed has serious influence in laser material interaction and cooling rate in a solidification process. Laser scanning speed has been identified to affect heat affected zone in a deposition process and affect height and microhardness of fabricated parts (Sobiyi et al., 2017).

Laser Power

Laser power is a very important deposition parameter in laser deposition processes. The important of laser power is highly emphasized because of the critical role it plays in the production of melt pool and it eventual effect in the properties and qualities of produced components as well documented in literatures (Abdulrahman et al., 2019; Sharman et al., 2018; Yan et al., 2017). Some of the works that have been carried out on the effect of laser power in laser metal deposition process revealed that not only could laser power effect the mechanical and microstructural characteristics of deposited components but it also has the ability to reduce or eliminate internal cracks of deposited components if properly controlled. The effect of laser power in a deposition process are also related to laser-material interaction and the rate of cooling in the solidification process of laser deposited parts.

Laser Spot Size

The spot size of a laser is another parameter that also contribute to the quality and property of laser deposited parts. The spot size of a laser is obtained through the laser beam diameter. The laser beam diameter is used to control the concentration of laser beam on a targeted area. The smaller the spot size of laser beam, the bigger the beam intensity (Bayode et al., 2018). Hence, laser spot size is inversely proportional to the beam intensity. It is good to also note that laser spot size also determines the width of deposited tracks.

It is important to note the relationship between scanning speed, laser power and laser spot size. The relationship is necessitated because the three parameters are connected to energy input requirement in a laser deposition process. Energy density *(E)* needed in a laser deposition process is the sum of the incident energy input per unit area as shown in the equation below.

$$Energy\ density(E) = \frac{P}{VD}$$

Where energy density *(E)* is in J/mm^2, *P* is laser power usually in watt (W), *V* is scanning speed in mm/s and *D* is laser spot size in mm.

The influence of energy density has well been captured in literatures (Hu et al., 2016; Zyl et al., 2016). Increase in laser beam incident energy will usually lead to decrease in crack frequency in deposited parts. The cooling rate of any material in a deposition process is hugely factored on the laser heat input.

Applications of AM Techniques

AM techniques as earlier stated is an advanced manufacturing technique geared at mitigating the short-coming of the traditional manufacturing techniques, to meet the market demand for complex parts, reduce production cost, provide a more sustainable and environmentally friendly manufacturing process and reduced material wastage. AM technique is currently being applied to different areas of manufacturing and services such as aerospace, automotive, defence, energy, petrochemicals, mining, medicine and electronics. Some of the major areas of AM technique are highlighted below.

Aerospace

Parts manufactured by additive manufacturing technique are now employed in both commercial and military aircrafts. Some notable parts produced by AM technique for aerospace application include turbine blades, integrated piping systems, compressors and fuel nozzles. AM capabilities, advantages, benefits and results as related to the aerospace industry are highlighted in figure 4.

Automobile

Area of applications of AM technique in the production of parts used in the automobile industry are in the production of engine pistons and valves. The use of AM technique has immensely contributed to the production of lightweight functional materials now used in these industries.

Medicine

Functionally graded lightweight materials are currently being manufactured using AM technique and use in the medical and biomedical fields. Biocompatible implants that are non-toxic and meet other ba-

Figure 4. Major attribute in the integration of AM technology into the aerospace industry (Singamneni et al., 2019).

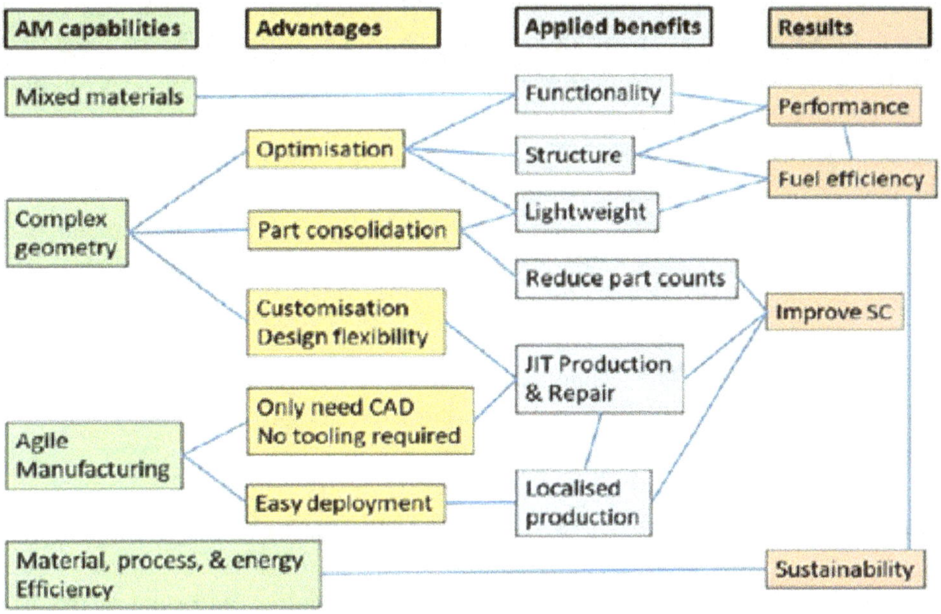

sic requirements have been produces through AM process. Dentistry and orthopedic medicine are also witnessing their share of impact of additive manufacturing.

4TH INDUSTRIAL REVOLUTION (4IR)

The 4th industrial revolution (4IR) or industry 4.0 is a new technological revolution that is fast changing the way we live, work and interact with one another. It is a revolution that comes with the possibilities of connecting billions of people through the internet and mobile phones and providing high level of processing power and data storage abilities. The influence of this industrial revolution is now being felt with new technological breakthroughs as seen in the field of robotics, the internet of things (IOT) and artificial intelligence (AI), 3D printing, biotechnology, nanotechnology, autonomous vehicles, quantum computing, energy storage, material science etc. This technological revolution is also revolutionizing industries as new business models are introduced leading to changes in the way of production, transportation, consumption and delivery systems.

The speed and scope at which these emerging technologies are transforming the industries in terms of development, adoption, complexities and interconnection across sectors have not been fully understood and that is why all global stakeholders (business, civil society, academia and governments) must come together to fully understand the incoming trend (Schwab, 2016).

History of 4th Industrial Revolution

Fourth industrial revolution is a progression from the first three industrial revolutions. The invention of steam engine and development of power loom in the 18th century were witnessed in the first industrial revolution which brought about mechanization in production. The second industrial revolution witnessed during the 19th century led to transformation in assembly line, electricity and scientific knowledge and made mass production of goods attainable. The third industrial revolution that started about 1950, comes with the introduction of computers and other digital technology. The third industrial revolution brought about huge improvement in the banking, communication, energy sectors and introduction of automation in manufacturing processes (Mcginnis, 2020). The fourth industrial revolution label is by the author of the book "The Fourth Industrial Revolution" and founder of the World Economic Forum, Klaus Schwab. Schwab noted in an article that 4IR has the ability to increase global income and better the quality of life of people around the world. The technological trend has a long-term benefit in form of efficiency and productivity. It was stressed that industry 4.0 could led to the disruption in the labour market and eventually led to serious inequality and increase social tension.

Role, Relationship and Contribution of AM Technique to 4IR

The introduction and integration of advanced information technology and state of the art intelligent production systems to better the quality of life of people around the world is highly motivating. AM is regarded as a strong tool needed to accelerate the 4th industrial revolution. The contribution of AM to 4th industrial revolution is highly critical as it has the ability to effortlessly deliver high quality goods and services at reduced price within a very short time (Dilberoglu et al., 2017). The evolvement of industry 4.0 is changing the system of manufacturing, governance and general way of life. Consumers flare for customized products as against general mass production goods is an area that the new technology is exploiting. Manufacturing companies that choose to remain highly competitive must be able to meet customers' demands. Therefore, the drive for industry 4.0 is being accelerated through breakthroughs being witnessed in areas of artificial intelligence, robotics, advanced communication system and additive manufacturing. Big data management, cloud computing, digitization of products and services are all making it possible to precisely meet customers request at very competitive prices. Additive manufacturing is a leading manufacturing technology currently changing the narrative of the way products are being manufactured as against the traditional manufacturing route. The technology gives flexibility that assist manufacturers to create better design so meet customers demand. The use of AM technology affords companies the opportunity of producing prototypes or products with the use of a single 3D printing equipment as against traditional manufacturing technique where it is not achievable as the traditional manufacturing technique require serious investment to set up. Improve flexibility and effectiveness in Productivity and customization to improve customer satisfaction are the main objective of 4th industrial revolution. AM remain an important weapon being exploited to see that the objectives of the 4th industrial revolution are well achieved.

Sustainability of AM Technique

Traditional production technique is an expensive and time-consuming process that requires special tooling materials, skilled and unskilled labour. Some of the tooling industries are saddled with serious economic problems as more awareness are being canvassed on the environmental impact of the tooling industries. The traditional manufacturing technique like machining, forging and casting are now seen as undesirable and unsustainable due to their impacts on the environment. Traditional or convectional manufacturing technique employs the like of milling machine, melting machine, heavy presses, etc. which are responsible for high energy consumption as against AM process that mainly require only a single machine for same production.

The outcome and successes recorded by additive manufacturing processes in the production of end-use products is unarguable. However, the extent of the environmental impacts of AM manufacturing processes are areas that need to be examined and understood clearly. The environmental impact by different additive manufacturing processes (selective laser sintering, electron beam melting, selective laser melting, strereolithography and fused deposition modeling), energy consumption of AM processes, material (powder) production impact and treatment of produced components are some of the aspects that have been studied (Kellens et al., 2017). The environmental impact reduction in the production of tool and die are achievable using direct laser deposition. Reduced environmental impact and cost reduction are advantages that have been recorded with Laser-based production of molds, dies and tools (Morrow et al., 2007).

The AM technology has been able to show that it is an advanced manufacturing processing technique that comes with numerous advantages such as the ability to coat, repair and produce components with improved structural and mechanical characteristics. AM has also revealed that it is possible to manufacture components from powder and wire materials. Additive manufacturing offer business leaders in production industry the opportunity to improve on existing designs and processes. AM has also been able to show that is has a comparative advantage over conventional manufacturing technique when it comes to material management and waste generation.

The raw material in AM is judiciously utilized and unused material are reusable. Whereas, the convectional subtractive manufacturing result in the production of material waste in the form of chips that are either discarded as environmental waste or reusable through the process of recycling that involves additional cost and energy consumption.

LMD PROCESSING AND CHARACTERIZATION OF TIAL ALLOY

Titanium and its alloys are increasingly becoming popular as they are now widely used in energy, automobile, aerospace, defence and other manufacturing industries because of their high strength-to-weight ratio which made them better alternative to steel. This is because titanium alloys are as strong as steels and of lesser weight when compared to steels. Titanium also possesses high melting point of about 3000^0F which made titanium and its alloys highly suitable in the production of missiles, spacecraft, armor plating and naval ships. Parts complexity, production lead-time and waste are some of the factors considered when using titanium and its alloys for part production. The high cost of titanium and its alloys begs for appropriate production method in addressing the factors geared at performance and production cost. The introduction of AM technology has helped in reducing material wastage there by saving cost of production and preserving the strength and weight advantage of the material.

Titaniumn aluminides (TiAl) are intermetallic materials that have attracted much attention because of their light weight and improved high temperature properties. Titanium and its alloys are gaining serious recognition mostly in aerospace and automotive because of their oxidation resistance, high temperature performance and lightweight which improves fuel economy.

There have been different techniques developed and used in the processing of parts especially as regards to additive manufacturing of lightweight materials such as titanium and its alloys. Norsk titanium developed an AM process called Rapid plasma deposition (RPD) process (shown in figure 5) used in the fabrication of aerospace structures. The process rapidly produces near-net-shape structures with little machining operation needed on produced parts. The process under inert argon gas atmosphere, produces structures by melting titanium wire.

Figure 5. Rapid plasma deposition technique (Norsk Titanium, 2016)

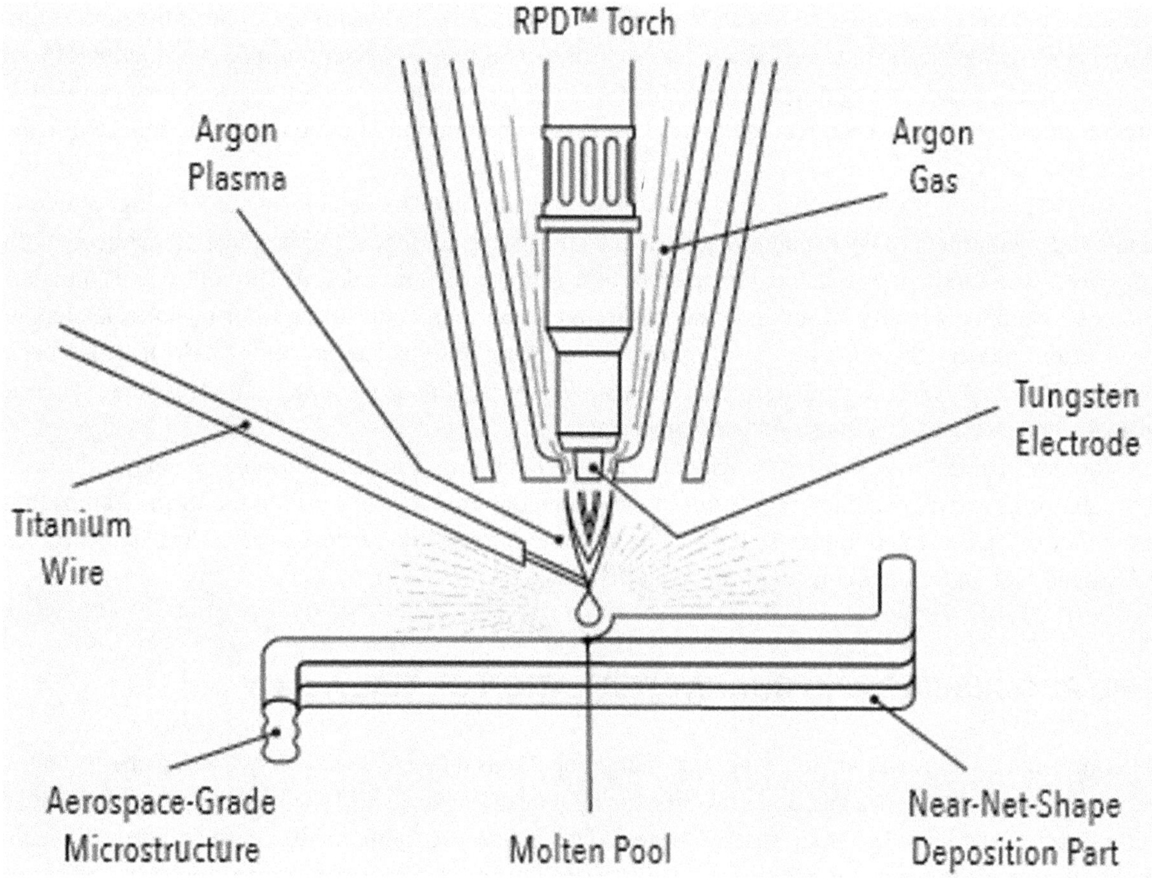

The EBM technique as earlier discussed under section 1.2 is another AM technique that is being employed in the processing of gamma titanium aluminide components. Metallic powder mesh size of -140/+325 (45-105 μm) and 3 KW electron beam power are commonly used in this particular process. Some of the advantages of this process have been investigated to include: the ability of obtaining a homogenous microstructure, fabrication of components with minimal internal defects, fine grain size and absence of residual stresses due to the high temperature at which the process takes place. Gamma titanium aluminide material have sufficient and required properties that made it suitable as structural materials for aerospace applications (Norsk Titanium, 2016).

Another interesting non-powder-based AM technique is the ion fusion formation (IFF) technique. The technique is developed at Honeywell Aerospace. This technique involves a direct metal deposition in the production of components. It uses an electronic interface to control the table position. The IFF technique in the presence of inert gas, uses a wire feedstock on arc-based welding torch to deposit metal. The process has been used in the deposition of lightweight metals and alloys such as aluminium, Ti6Al4V and 374 stainless steel. Fabrication of fully-dense deposits that do not need post treatment in the form of hot isostatic press and fast build-up speed are major advantages of IFF technique.

The experimental case study discussed in section 3.1 take a look at the characteristics of laser deposited components and went further in section 3.2 to model and simulate an exhaust valve using the properties of laser fabricated TiAl (Ti-Al-Nb-Cr) material. The simulation was performed with Solidworks 2017 software to evaluate the performance of TiAl under high temperature application.

Experimental Case Study

The experimental case study highlighted below, take a look at a study where laser metal deposition process was used in the deposition of lightweight alloy. Physical, metallographic and mechanical characterization were performed on the deposits. More details on the experimental results have been published (Abdulrahman et al., 2018; Abdulrahman et al., 2019; Abdulrahman et al., 2018; Abdulrahman et al., 2021).

Materials and Method

In this study, spherical shaped titanium aluminide alloy powder (45-150 μm particle size) manufactured by Praxair surface technologies, USA, was deposited on a commercial pure titanium (CP-Ti) substrate (size 10 x 10 x 6 mm) using 850R LENS machine available at CSIR, Pretoria, South Africa.

Table 2 highlight the elemental composition of the TiAl alloy powder that have been used in the experiment. Titanium balanced the elemental composition of the alloy with percentage composition of about 58.5%. Other materials that make up the composition are shown below.

Table 2. Other elemental composition of TiAl metal powder

Aluminum	Niobium	Chromium	Other elements
34	4.8	2.6	< 0.10

Before the deposition process, the substrates were sand blasted and cleaned with acetone to remove oil stains and impurities that may be on the surface of the substrates. Detailed procedures regarding the deposition processes employed as regard to the deposition parameters, preheating and non-preheating process and results obtained (physical, microstructural and mechanical characteristics) are now available in literatures (Abdulrahman et al., 2018; Abdulrahman et al., 2019; Abdulrahman et al., 2021).

Figure 6. SEM image of TiAl deposited (Abdulrahman et al., 2019)

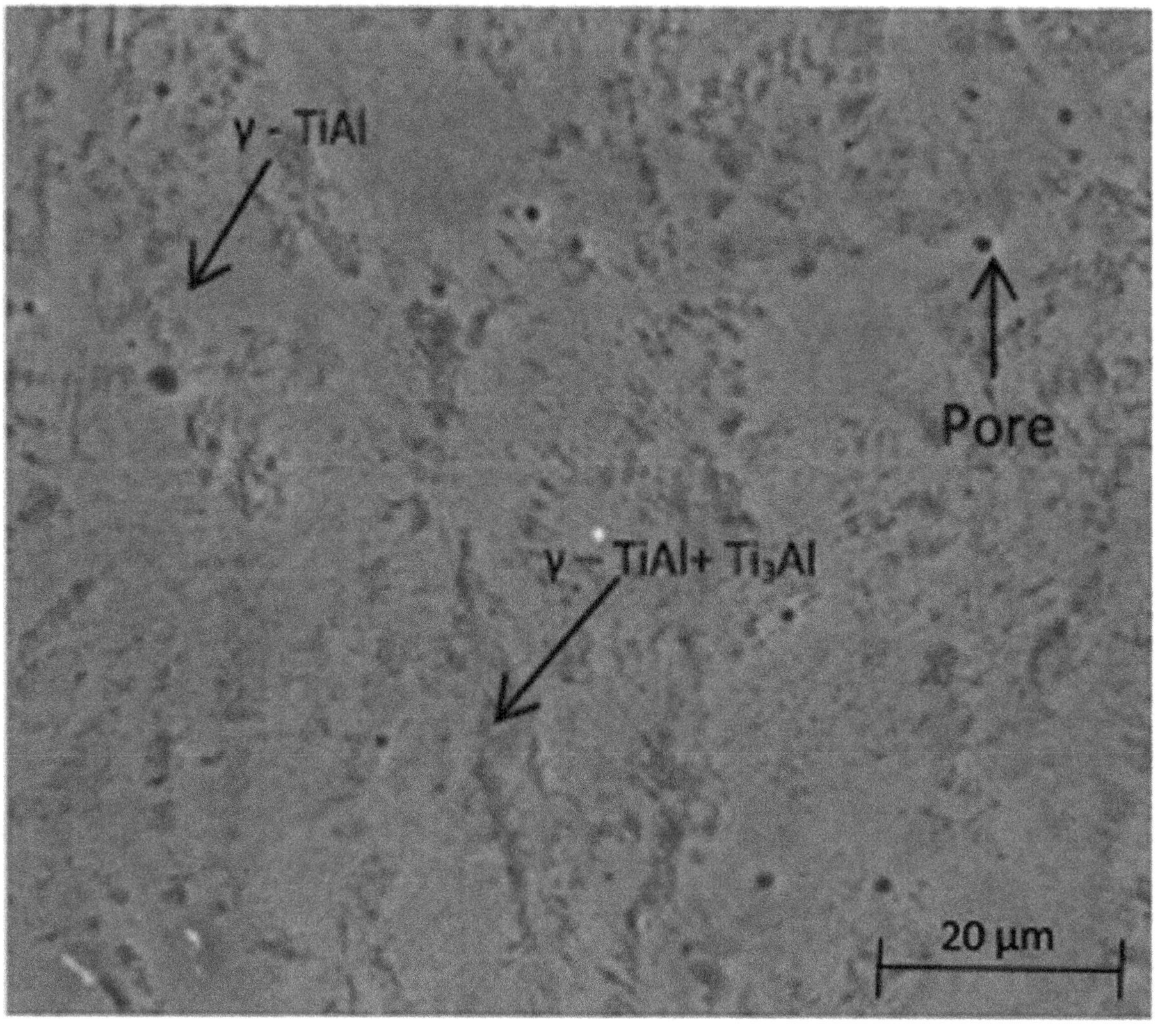

The microscopic image of TiAl sample deposited (at 400 W laser power, 4.09 g/min powder flow rate and 7.406 mm/s laser scanning speed) taken under a scanning electron microscope revealed the presence of homogenous lamellar γ-TiAl and γ-TiAl + Ti_3Al region as shown in figure 6. The deposit has fine grains and pores sparingly distributed.

Table 3. Deposition parameter effects on height and microhardness of deposits (Abdulrahman et al., 2018).

Sample	Laser power (Kw)	Scanning speed (mm/s)	Deposit height (mm)	Deposit microhardness (Hv)
A	0.4	3.17	1.6	565
B	0.45	3.17	2.1	536
C	0.4	2.65	2.3	550
D	0.45	2.65	2.7	560

Figure 7. A typical valve design (Rao & Chandu, 2014)

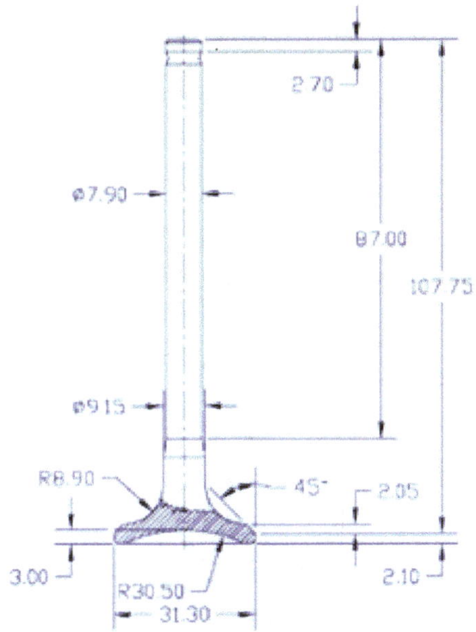

Result outcome further revealed that cracks in the laser deposited samples decreases as laser power increases and scanning speed decreases. It was also noted that change in deposition parameters has tremendous effect on heights and microhardness of deposited samples as clearly indicated in figure 3 extracted from the literature Abdulrahman et al., (2018). Other characterizations noted in the literatures are the relationship between the deposition parameters and wear rates and corrosion resistances of the deposits.

From the table above, it becomes clear that changes in the deposition parameters have a corresponding change in the quality of deposits produced via laser deposition process. This relationship was due to the laser material interaction and the cooling rate experienced in the deposited samples.

Solidworks Modeling and Simulation of Part/Component

Modeling and simulation of machine components help provide information on how well such components react when subjected to a loading conditions under certain environments. Simulations are often performed to also evaluate and optimize components and structures. Steels are one of the most common materials (martensitic and austenitic) used in the production of valves. Increase in temperature can often reduce the strength and hardness of martensitic steel and makes it applicable mainly as intake valve because exhaust valves operate at higher temperatures between 650-790 ^{0}C and sometime higher. The inlet and exhaust valves can be subjected to temperatures reaching 1930 to 2200 ^{0}C during power stroke and this is why it is very important that valves materials should be able to withstand these higher temperatures (Rao & Chandu, 2014). Exhaust valve production are basically done with austenitic alloy steels as it is stronger at higher temperatures. Materials employed in the production of valves include stainless steels, carbon-steel alloys, high strength nickel-chromium-iron alloys and titanium alloys. The close durability of titanium to stainless steel and its good strength-to-weight ratio makes its highly preferable as valve

Figure 8. Resultant temperature gradient of Ti-Al-Nb-Cr exhaust valve

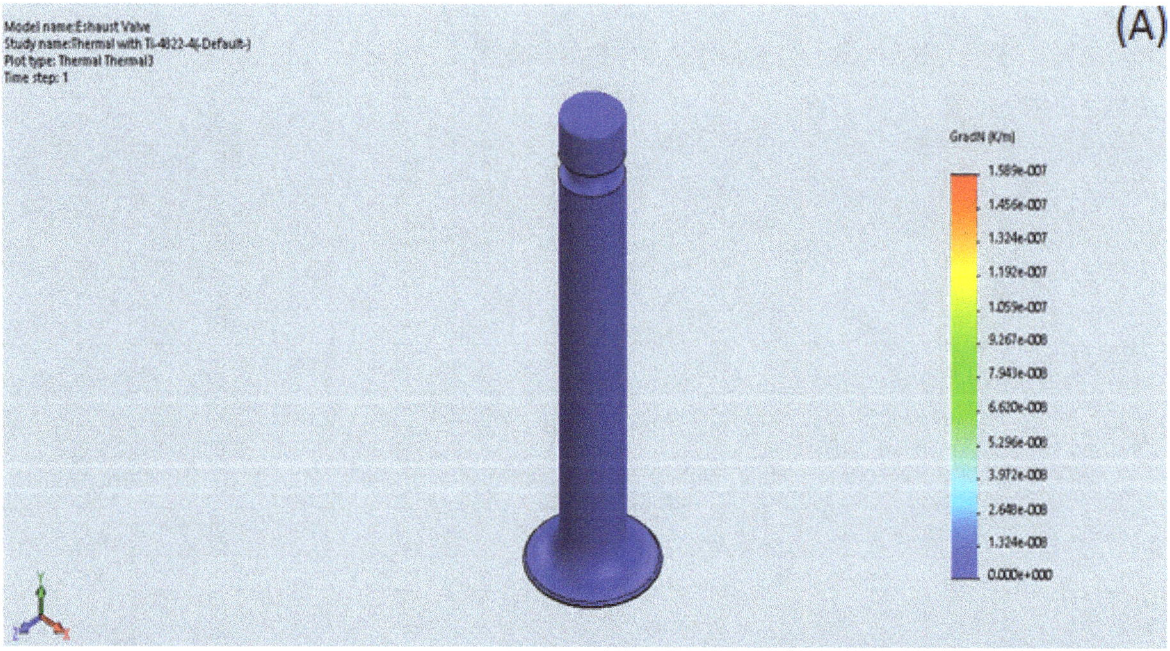

alloy material. Nonetheless, valves produced from titanium material are comparably costly (Carley, 2015). A typical valve design is shown in figure 7.

An exhaust valve was modeled and simulated (using Solidworks 2017) with Ti-Al-Nb-Cr material and the results obtained was compared with another titanium alloy (Ti-6Al-2Sn-2Zr-2Mo-2Cr-0.25Si). The basis of the comparison with a known TiAl alloy is to examine how well the laser deposited alloy with specific properties will perform under similar high temperature application. The simulations of the exhaust valve were performed to evaluate how well the titanium alloys response to specific loading conditions (thermal and buckling).

The modeled Ti-Al-Nb-Cr valve was treated as solid body having a weight of 0.914 N, yield strength of 1.4179 x 10^9 N/m^2, elastic modulus of 1.1398 x 10^{11} N/m^2 and density of 4150 kg/m^3 while that of the Ti-6Al-2Sn-2Zr-2Mo-2Cr-0.25Si valve weighs 1.024 N, yield strength of 1.07 x 10^9 N/m^2, elastic modulus of 1.23 x 10^{11} N/m^2 and density of 4650 kg/m^3. The valves of both materials were subjected to a thermal load temperature of 750 ^{0}C. Also, the two titanium alloy valves were subjected to assumed pressure of 20 N/m^2 and the results were compared.

Results and Discussion

The results (resistance temperature gradient and buckling analysis) of the simulated materials as shown in the figure 8 and 9 and figure 10 and 11 respectively.

The results revealed that both materials performed well under the loading conditions as no significant thermal trend was observed on the temperature gradients. However, there is a slight difference in their resultant heat flux results obtained. The maximum resultant heat flux that was recorded by Ti-Al-Nb-Cr alloy valve gradient was 1.112 X 10^{-6} while the maximum heat flux that was recorded on the Ti-6Al-2Sn-

Figure 9. Resultant temperature gradient of Ti-6Al-2Sn-2Zr-2Mo-2Cr-0.25Si exhaust valve

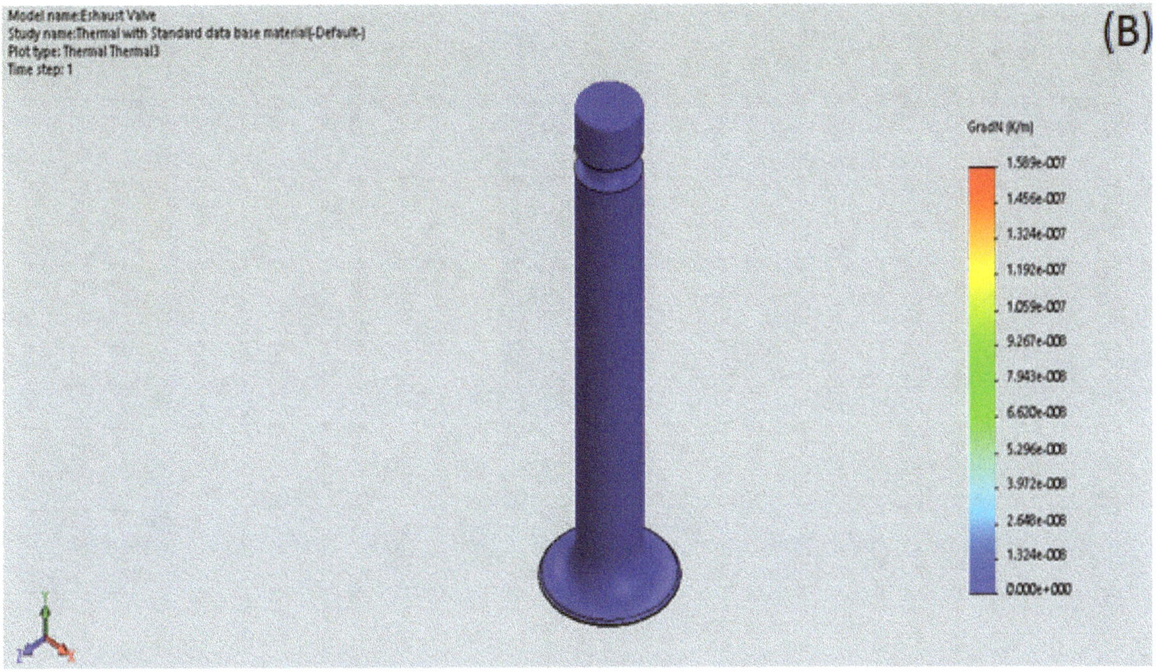

Figure 10. Resultant amplitude on Ti-Al-Nb-Cr exhaust valve

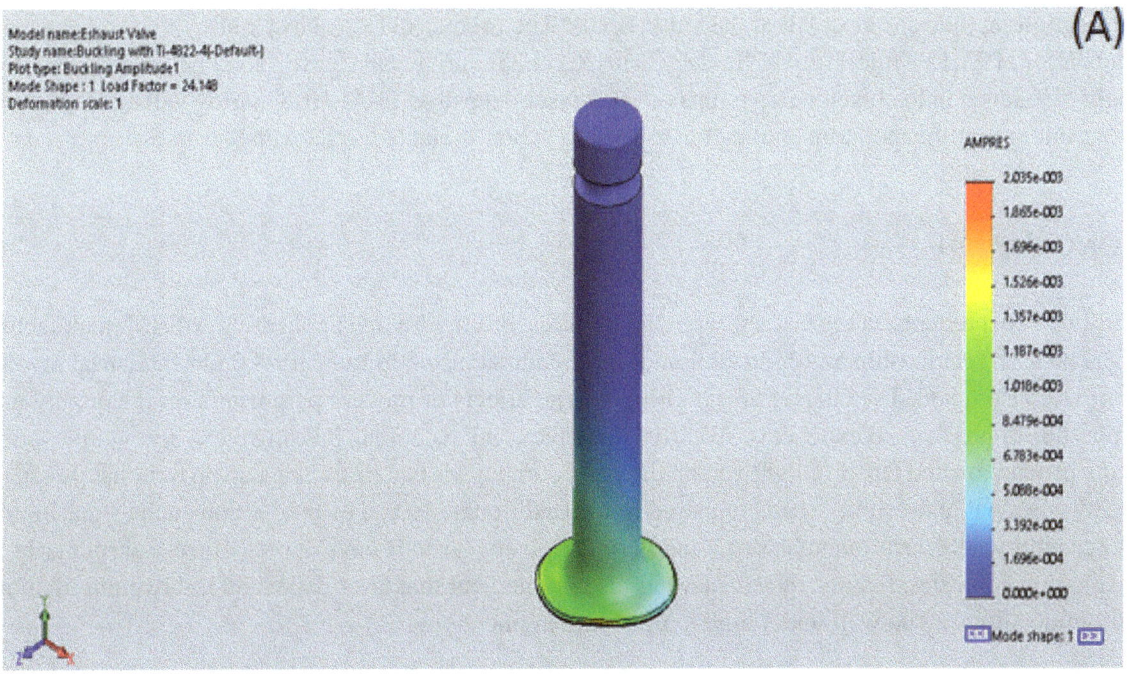

Figure 11. Resultant amplitude on Ti-6Al-2Sn-2Zr-2Mo-2Cr-0.25Si exhaust valve

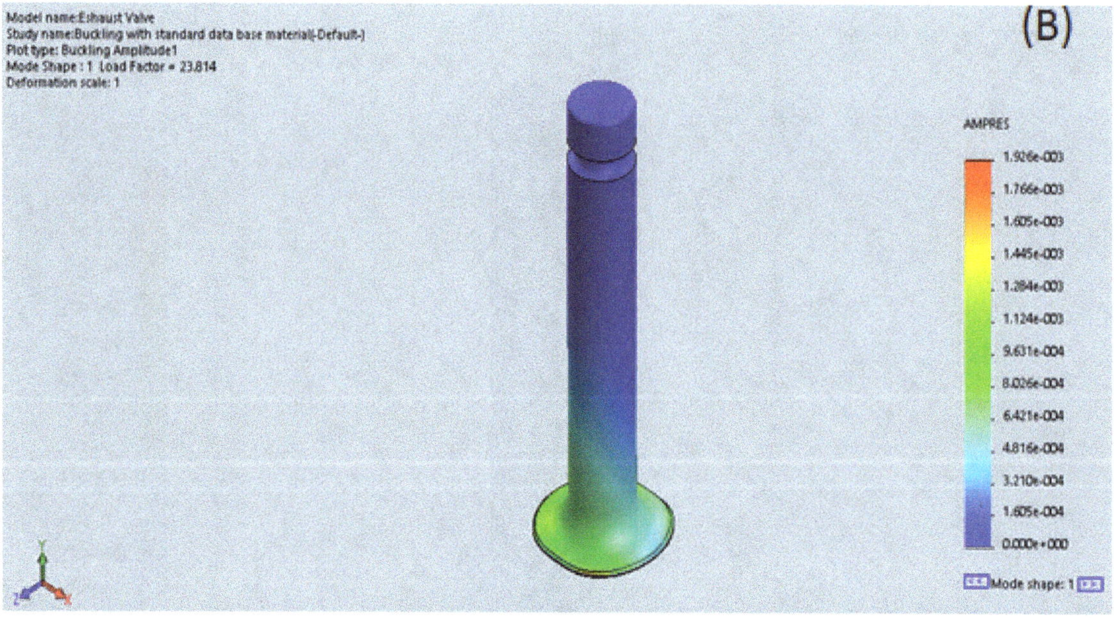

2Zr-2Mo-2Cr-0.25Si alloy valve gradient was 1.239 x 10^{-6}. Hence, the slight difference might indicate Ti-Al-Nb-Cr alloy material can also perform better under high temperature condition and can be well considered for high temperature applications. The buckling analysis results presented in figure 10 and 11 revealed maximum resultant amplitude plot of 2.035 x 10^{-3} for Ti-Al-Nb-Cr alloy and 1.926 x 10^{-3} for Ti-6Al-2Sn-2Zr-2Mo-2Cr-0.25Si alloy respectively. Even though the resultant amplitude values obtained are negligible, they are most felt at the valve heads. The produced Ti-Al-Nb-Cr alloy valve also gave a load factor of 24.15 while Ti-6Al-2Sn-2Zr-2Mo-2Cr-0.25Si alloy valve gave a load factor of 23.81. The slight difference in load factor also points out that laser deposited Ti-Al-Nb-Cr alloy valve can perform better under high-temperature application especially when subjected to buckling stresses.

CONCLUSION

This book chapter present the history, types, advantages, limitations and applications of AM manufacturing. The role, relationship and contribution of AM manufacturing to Industry 4.0 (4th Industrial revolution) was also noted. The chapter also highlighted the effects of process parameters on the quality and properties of produced components. Additive manufacturing technique has proven to be a viable option in the production and repair of lightweight alloy and composites materials and also proven that its can be highly sustainable overtime when compared to the traditional method of production such as machining and casting where such manufacturing industry are posing serious environmental treat particularly to developing countries. Finally, the chapter has also pointed out that laser deposited lightweight titanium aluminide alloy can do well under high temperature applications.

REFERENCES

Abdulrahman, K. O., Akinlabi, E. T., & Mahamood, R. M. (2018). Additive Manufacturing- Laser Metal Deposition and effect of preheating on properties of deposited Ti-4822-4 alloy. In *Additive Manufacturing Technologies from an Optimization Perspective*. Advances in Logistics, Operations, and Management Science. IGI Global.

Abdulrahman, K. O., Akinlabi, E. T., & Mahamood, R. M. (2019). Characteristics of laser metal deposited titanium aluminide. *Materials Research Express*, *6*(4), 046504. Advance online publication. doi:10.1088/2053-1591/aaf9c8

Abdulrahman, K. O., Akinlabi, E. T., Mahamood, R. M., Pityana, S., & Tlotleng, M. (2018). Laser Metal Deposition of Titanium Aluminide Composites: A Review. *Materials Today: Proceedings*, *5*(9), 19738–19746. doi:10.1016/j.matpr.2018.06.336

Abdulrahman, K. O., Mahamood, R. M., Akinlabi, E. T., & Adediran, A. A. (2021). Effect of laser power on the microstructure and mechanical properties of laser deposited titanium aluminide composite. *Advances in Materials and Processing Technologies*, *00*(00), 1–12. doi:10.1080/2374068X.2021.1945268

Akinlabi, E. T., Mahamood, R. M., Shukla, M., & Pityana, S. (2012a). Effect of Scanning Speed on Material Efficiency of Laser Metal Deposited Ti6Al4V. *International Journal of Mechanical and Mechatronics Engineering*, *6*(11), 2509–2513.

Akinlabi, E. T., Mahamood, R. M., Shukla, M., & Pityana, S. (2012b). Effect of Scanning Speed on Material Efficiency of Laser Metal Deposited Ti6Al4V. World Academy of Science, Engineering and Technology, 1531–1535.

Akinlabi, E. T., & Akinlabi, S. A. (2016). *Powder Flow Rate Influence on Laser Metal Deposited TiC on Ti-6Al-4V.* Lmd.

Akinlabi, E. T., Mahamood, R. M., Akinlabi, S. A., & Ogunmuyiwa, E. (2014). Processing parameters influence on wear resistance behaviour of friction stir processed Al-TiC composites. *Advances in Materials Science and Engineering*, *2014*, 1–12. Advance online publication. doi:10.1155/2014/724590

Anzalone, G. C., Zhang, C., Wijnen, B. A. S., Sanders, P. G., & Pearce, J. M. (2013). A Low-Cost Open-Source Metal 3-D Printer. *IEEE Access: Practical Innovations, Open Solutions*, *1*, 803–810. doi:10.1109/ACCESS.2013.2293018

Baudana, G., Biamino, S., Ugues, D., Lombardi, M., Fino, P., Pavese, M., & Badini, C. (2016). Titanium aluminides for aerospace and automotive applications processed by Electron Beam Melting: Contribution of Politecnico di Torino. *Metal Powder Report*, *71*(3), 193–199. Advance online publication. doi:10.1016/j.mprp.2016.02.058

Bayode, A., Akinlabi, E. T., & Pityana, S. (2018). Fabrication of stainless steel-based FGM by laser metal deposition. In K. Kumar & P. Davim (Eds.), *Hierarchical composite materials: Materials, Manufacturing and Engineering*. Walter de Gruyter. doi:10.1515/9783110545104-004

Bremen, S., Meiners, W., & Diatlov, A. (2012, April). Selective Laser Melting A manufacturing technology for the future? *Laser-Technik-Journal*, *9*(2), 33–38. Advance online publication. doi:10.1002/latj.201290018

Carley, L. (2015). *Valve Selection: Hot Valve Materials for Hot Engines*. https://www.enginebuildermag.com/2005/12/valve-selection-hot-valve-materials-for-hot-engines/

Dilberoglu, U. M., Gharehpapagh, B., Yaman, U., & Dolen, M. (2017). The role of additive manufacturing in the era of Industry 4. 0. *Procedia Manufacturing*, *11*(June), 545–554. doi:10.1016/j.promfg.2017.07.148

Dinda, G. P., Dasgupta, A. K., & Mazumder, J. (2009). Laser aided direct metal deposition of iconel 625 superalloy: Microstructural evolution and thermal stability. *Materials Science and Engineering A*, *509*(1), 98–104. doi:10.1016/j.msea.2009.01.009

Froes, F. H., & Dutta, B. (2014). The Additive Manufacturing (AM) of Titanium Alloys. *Advanced Materials Research*, *1019*, 19–25. . doi:10.4028/www.scientific.net/AMR.1019.19

Gasper, A. N. D., Catchpole-Smith, S., & Clare, A. T. (2017). In-situ synthesis of titanium aluminides by direct metal deposition. *Journal of Materials Processing Technology*, *239*, 230–239. doi:10.1016/j.jmatprotec.2016.08.031

Gibson, I., Rosen, D. W., & Stucker, B. (2010a). *Additive Manufacturing Technologies*. Springer., doi:10.1007/978-1-4419-1120-9

Gibson, I., Rosen, D. W., & Stucker, B. (2010b). Printing Processes. In *Additive Manufacturing Technologies*. Springer. doi:10.1007/978-1-4419-1120-9_7

Gornet, T., & Wohlers, T. (2014). *History of additive manufacturing*. http://wohlersassociates.com/history2014.pdf

Graf, B., Marko, A., Petrat, T., Gumenyuk, A., & Rethmeier, M. (2018). 3D laser metal deposition: Process steps for additive manufacturing. *Welding in the World*, *62*(4), 877–883. doi:10.100740194-018-0590-x

Gussone, J., Hagedorn, Y. C., Gherekhloo, H., Kasperovich, G., Merzouk, T., & Hausmann, J. (2015). Microstructure of γ-titanium aluminide processed by selected laser melting at elevated temperatures. *Intermetallics*, *66*, 133–140. Advance online publication. doi:10.1016/j.intermet.2015.07.005

Herderick, E. (2011). Additive Manufacturing of Metals : A Review. *Materials Science and Technology*, *176252*, 1413–1425.

Hu, Y., Wang, H., Ning, F., & Cong, W. (2016). Laser Engineered Net Shaping of Commercially Pure Titanium : Effects of Fabricating Variables. *Proceedings of the ASME 2016 International Manufacturing Science and Engineering Conference*. 10.1115/MSEC2016-8812

Karunakaran, K. P., Suryakumar, S., Pushpa, V., & Akula, S. (2010). Low cost integration of additive and subtractive processes for hybrid layered manufacturing. *Robotics and Computer-integrated Manufacturing*, *26*(5), 490–499. doi:10.1016/j.rcim.2010.03.008

Kellens, K., Mertens, R., Paraskevas, D., Dewulf, W., & Duflou, J. R. (2017). Environmental Impact of Additive Manufacturing Processes: Does AM contribute to a more sustainable way of part manufacturing ? *Procedia CIRP, 61*(3), 582–587. doi:10.1016/j.procir.2016.11.153

Kobryn, P. A., Ontko, N. R., Perkins, L. P., & Tiley, J. S. (2006). Additive Manufacturing of Aerospace Alloys for Aircraft Structures. *Meeting Proceedings RTO-AVT-139*, 3-1-3–14. http://www.rto.nato.int/abstracts.asp

Kumar, S., & Choudhary, A. K. S., & Rakesh. (2014). Effect of the Process Parameters on Geometrical Characteristics of the Parts in Direct Metal Deposition : A Review. *International Journal of Mechanical Engineering and Technology*, *5*(4), 116–122.

Ludovico, A. D., Angelastro, A., & Campanelli, S. L. (2010). Experimental Analysis of the Direct Laser Metal Deposition Process. In *New Trends in Technologies: Devices, Computer, Communication and Industrial Systems*. https://cdn.intechweb.org/pdfs/12286.pdf

Mahamood, R. M., & Akinlabi, E. T. (2015). Effect of Laser Power and Powder Flow Rate on the Wear Resistance Behaviour of Laser Metal Deposited TiC/Ti6Al4V Composites. *Materials Today: Proceedings*, *2*(4–5), 2679–2686. Advance online publication. doi:10.1016/j.matpr.2015.07.233

Mcginnis, D. (2020). *What is the Fourth Industrial Revolution*. Salesforce. https://www.salesforce.com/blog/what-is-the-fourth-industrial-revolution-4ir/

Morrow, W. R., Qi, H., Kim, I., Mazumder, J., & Skerlos, S. J. (2007). Environmental aspects of laser-based and conventional tool and die manufacturing. *Journal of Cleaner Production, 15*(10), 932–943. doi:10.1016/j.jclepro.2005.11.030

Pityana, S., Mahamood, R. M., Akinlabi, E. T., & Shukla, M. (2013). Gas Flow Rate and Powder Flow Rate Effect on Properties of Laser Metal Deposited Ti6Al4V. *Proceedings of the International Multi-Conference of Engineers and Computer Scientists, II*, 13–16.

Rao, B. S., & Chandu, D. G. (2014). Petrol engine exhaust valve design, analysis and manufacturing processes. *International Journal of Mechanical Engineering and Robotics Research, 3*(4).

Saboori, A., Gallo, D., Biamino, S., Fino, P., & Lombardi, M. (2017). An Overview of Additive Manufacturing of Titanium Components by Directed Energy Deposition : Microstructure and Mechanical Properties. *Applied Sciences (Basel, Switzerland)*, *7*(9), 883. Advance online publication. doi:10.3390/app7090883

Schade, C. T., Murphy, T. F., & Walton, C. (2014). Development of atomized powders for additive manufacturing. *World Congress on Powder Metallurgy and Particulate Materials, PM 2014*, 215–226. https://pdfs.semanticscholar.org/164c/56f1e3a3e525162cc28a65d975e2dd39e357.pdf

Schwab, K. (2016). *The Fourth Industrial Revolution*. World Economic Forum.

Shah, K., Pinkerton, A. J., Salman, A., & Li, L. (2010). Effects of Melt Pool Variables and Process Parameters in Laser Direct Metal Deposition of Aerospace Alloys. *Materials and Manufacturing Processes*, *25*(12), 1372–1380. doi:10.1080/10426914.2010.480999

Sharman, A. R. C., Hughes, J. I., & Ridgway, K. (2018). Characterisation of titanium aluminide components manufactured by laser metal deposition. *Intermetallics*, *93*, 89–92. doi:10.1016/j.intermet.2017.11.013

Shukla, M., Mahamood, R. M., Akinlabi, E. T., & Pityana, S. (2012). Effect of Laser Power and Powder Flow Rate on Properties of Laser Metal Deposited Ti6Al4V. *World Academy of Science and Technology*, *6*, 44–48.

Singamneni, S., LV, Y., Hewitt, A., Chalk, R., Thomas, W., & Jordison, D. (2019). Additive Manufacturing for the Aircraft Industry: A review. *Journal of Aeronautics & Aerospace Engineering, 8*(1). https://www.longdom.org/open-access/additive-manufacturing-for-the-aircraft-industry-a-review-18967.html

Sobiyi, K., Akinlabi, E., & Akinlabi, S. (2017). The Influence of Scanning Speed on the Laser Metal Deposition of Ti / TiC Powders. *Materials Technology*, *51*(2), 345–351.

Titanium, N. (2016). *The dawn of a new industry: Rapid plasma deposition* ™. Modem Metals.

Tlotleng, M., Masina, B., & Pityana, S. (2016). Characteristics of laser In-situ alloyed titanium aluminides coatings. *Procedia Manufacturing*, *7*, 39–45. doi:10.1016/j.promfg.2016.12.013

Weisheit, A., & Rolink, G. (2016). Influence of Process Conditions in Laser Additive Manufacturing on the Microstructure Evolution of Fe-Al Alloys- A Comparison of Laser Metal Deposition and Selective Laser Melting. pdf. *AAM Workshop, MPEI.* https://www.mpie.de/3476506/Weisheit.pdf

Wholer's Report. (2016). *3D Printing and Additive Manufacturing State of the Industry, Annual Worldwide Progress Report.* Wholer's Associates Inc.

Yan, L., Chen, X., Zhang, Y., Newkirk, J. W., & Liou, F. (2017). Fabrication of functionally graded Ti and γ-TiAl by laser metal deposition. *JOM: The Materials. Metals & Materials Society*, *69*(12), 2756–2761. doi:10.100711837-017-2582-5

Zyl, I., Van, & Yadroitsava, I., & Yadroitsev. (2016). Residual stress in ti6al4v objects produced by direct metal laser sintering. *South African Journal of Industrial Engineering*, *27*(December), 134–141.

Chapter 3
Basic Principles for Thermoplastic Parts Finishing With Impulse Thermal Energy Method

Sergiy Plankovskyy
O. M. Beketov National University of Urban Economy in Kharkiv, Ukraine

Olga Shypul
https://orcid.org/0000-0002-1356-5831
National Aerospace University "Kharkiv Aviation Institute", Ukraine

Yevgen Tsegelnyk
https://orcid.org/0000-0003-1261-9890
O. M. Beketov National University of Urban Economy in Kharkiv, Ukraine

Dmytro Brega
National Aerospace University "Kharkiv Aviation Institute", Ukraine

Oleg Tryfonov
National Aerospace University "Kharkiv Aviation Institute", Ukraine

Volodymyr Malashenko
National Aerospace University "Kharkiv Aviation Institute", Ukraine

ABSTRACT

Impulse thermal energy method (ITEM) as modification of the thermal energy method that is successfully used for finishing is considered for application to thermoplastics. The chapter focuses to highlight the basic principles of the thermoplastics treatment by acting heat fluxes inherent to ITEM providing the time-controlled production of combustion species. The properties of thermoplastics and the requirements for their treatment have the greatest impact on processing settings. Thus, the questions of the choice of the preferred fuel mixture, the type of its ignition, and combustion have been studied. By means of numerical situating, the processes of melting and healing of pores during processing are investigated. A method of defining processing settings has been developed, taking into account the limitations on critical temperatures. The promising possibilities of ITEM in relation to the processing of thermoplastics parts obtained by additive technologies are outlined.

DOI: 10.4018/978-1-7998-7864-3.ch003

INTRODUCTION

Due to the wide range of engineering and high-performance plastics (EP and HPP) application, the high demands on the accuracy are placed for thermoplastic parts quality (Drégelyi-Kiss and Horváth, 2018). Such quality indicators as reliability, lifetime, functional characteristics are mainly determined by the quality of surface cleaning and edges finishing. The formation of different contaminants at thermoplastic part production is inevitable and related with inaccurate fitting of the molds contact surfaces, deformations caused by high pressure and temperature, etc. (Muccio, 1991; Kumar and Davim, 2018; Hurina et al., 2020). At machining process, burrs and micro-particles are also inevitably formed (Jain and Jain, 2003; Altan and Altan, 2014; Aksonov et al., 2019; Kovalevsky et al., 2019; Xu et al., 2020; Adeniji et al., 2020). Finishing technologies of the thermoplastic parts, including blasting, ultrasonic vibration processing, mechanical cleaning, cryogenic deflashing, dry ice blasting and many other are widely used (Swavely, 1991; Muccio, 1999; Uhlmann et al., 2010; Woźniak, 2013; Izamshah et al., 2013; Mali et al., 2018). One of the most promising is the Thermal Energy Method (TEM) (Klocke and König, 2007; Lamikiz et al., 2011; Struckmann and Kieser, 2020) and it's variations – Impulse Thermal Energy Method (ITEM) and Impulse Thermal Energy Method with Shock Waves (ITEMSW), which provide the time-controlled production of combustion species and intensification of the heat transfer due to the action of a shock waves in the working chamber correspondingly (Plankovskyy et al., 2021a). They have unique benefits in terms of efficiency and technological capabilities. The technology for parts processing under TEM occurs in a closed chamber by burning the contaminants under the action of heat flows from the gas mixtures combustion

Advantageous features of ITEM and ITEMSW for metal processing are known and discussed in the work Plankovskyy et al. (2021a). By ITEM and ITEMSW it is possible to provide processing under the conditions of shock waves dramatically increases the heat exchange rate between the combustion products and parts, as well as treatment at deflagration combustion mode without shock waves formation. Processing of thermoplastic parts by mentioned methods has essential features. That is why the choice of the most suitable method should be based on an analysis of their capabilities and the part requirements. Apart from this, effective implementation of the processing technology involves several steps. One of the most important of them is defining processing modes, because wrong treatment regime may lead to high heat flux into the part and cause its melting and deformation. At the same time, it should be avoided soot formation and its deposition on the melted surfaces.

BACKGROUND

EP and HPP are commonly used in automotive and aerospace industries due to their low weight, high strength and corrosion resistance (Kumar and Davim, 2018; Nickels, 2019; Kondratiev et al., 2021). There are some trends in use of EP and HPP in the health care industry, numerous electronic gadgets and devices, as well as in the power electronics. For these applications, the most vital properties of thermoplastics are chemical resistance, dimensions stability and low electrical conductivity. An advantage of the thermoplastics is their ability to be recycled by melting without significant loss of their properties.

Typically, injection or compression molding are used for thermoplastics parts production process. For these manufacturing methods, flashes are often formed on casting surfaces, which should be removed during finishing operations. Flashes are formed due to inaccurate fitting of the molds contact surfaces, deformations caused by high pressure and temperature. The conditions for flash formation are well studied. Modern approaches allow predicting its formation with high reliability (Zhu and Chen, 2006). Another traditional method of manufacturing thermoplastic parts is machining. With this method, burrs are inevitably formed. The shape and quantity of burs are associated with both processing modes and used tools (Adeniji et al., 2020; Kombarov et al., 2021). The deburring process also requires additional operations.

Usually, for deburring and deflashing the same methods are used. The most common of them are blasting, ultrasonic vibration processing, mechanical treatment and thermal energy methods. The peculiarities of these methods are associated with the HPP properties and the requirements for the surface quality of the final parts. For example, high demands on the part surface smoothness or the presence of decorative texture limits the possibilities of abrasive methods. On the other hand, there are cases when it is necessary to reduce the surface roughness of the part during finishing operations, which makes abrasive methods possible to use.

Often cooling or heating are used to improve the quality of the finishing treatment of thermoplastic parts. The most relevant example of finishing treatment with the use of heating is polishing by hot air jet (Adel et al., 2018). This method aimed on 3D printed parts manufactured by means of FDM technology. The main benefit of this method is that it uses not only surface tension in melted layer but also a dynamic pressure of a fluid, which is injected from the nozzle. However, this approach is time consuming and can be not efficient for complex geometry – especially for deep holes. Treatment is possible only in case when the whole 3d path from additive process is available in other case it should be generated taking into account the layering strategy. For example, to increase efficiency of deflashing by mechanical methods, it is carried out after preliminary cooling of the parts to a brittle state using cryogenic refrigerants, such as liquid or gaseous carbon dioxide, nitrogen, freon etc. Cryogenic deflashing is non-abrasive, enough efficient and allows to remove nearly 90% of flashes (Abhilash et al., 2018; Campean and Hancu, 2019; Zindani and Kumar, 2020). Sometime to reduce production costs cryogenic deflashing can used to extend lifetime prolongation of the worned molds. In this case same additional safety measures requested while working with refrigerants and there is also a risk to damage thin elements of parts. Vibration treatment or dry ice blasting can be used to prevent damage to thin wall elements. Dry ice blasting is similar to abrasive blasting methods, but instead of using heavy abrasive material, surface-gentle dry ice is used, the particles of which are accelerated and, upon impact with the surface, remove foreign formations, creating a small explosive effect (Spur et al., 1999). This method has proved to be quite successful in removing flashes and burrs from large plastic parts, as well as cleaning molds. However, for small parts in mass production, this method is not applicable.

The high-temperature methods for removing flashes from thermoplastic parts, such as removal by a hot wire or in a high-temperature gas environment, have been used for a long time (Gillespie, 1999). Removal by a hot wire, based on wire movement in parallel to the edges and melting the root of the flash. The machining cycle time is proportional to the trim path. The disadvantages of this method are the difficulty of removing the flash for curved sections, as well as the need to adjust the wire feed speed to prevent re-soldering on the workpiece material. Another traditional high temperature method for deflashing or deburring of thermoplastics is burner flame treatment (Harper and Petrie, 2003). When the surface of the thermoplastic part briefly melts, surface tension smooths its surface. This method is most applicable to flat external surfaces and requires a lot of operator experience. Some one more method in which flashes and burrs removal occurs by combustion products of gas mixtures is the TEM (Struckmann and Kieser, 2020). At TEM deburring, the workpieces are placed in a chamber, which is filled by a fuel mixture with its further ignition. The treatment should be controlled in order to remove burrs and flashes, but not to damage the surface of the parts. More detailed description of this method and the problems associated with its application for processing thermoplastics parts will be presented below.

The idea of using heat to melt the flash is also used for infrared treatment. This technology is being developed by Heraeus Noblelight Ltd. (Baird, 2017). It was found that infrared heating can melt the flash and pull it inside the part without dripping or catching fire. The problems were found at removing large flashes. To eliminate them, it is proposed to carry out a combined treatment with an initial flash reduction using mechanical methods. It is noted that not all thermoplastics are well processed by this method. The best results were achieved for polypropylene (PP), acrinitrile butadiene styrene (ABS), polymethyl methacrylate (PMMA), and synthetic rubber. At the same time, when processing some other materials, there are known cases, when burrs were not removed, but the surface of the part was damaged. This method requires special infrared emitters, especially during treatment of the complex shapes parts.

In current state it is considered two types of finishing operations that are typical for the production of thermoplastic parts: deburring and deflashing. However, in recent years, with the spread of additive manufacturing (AM), more and more attention has been drawn to a surface polishing (Biron, 2018). This treatment usually was considered as a method of decorative finishing and only in some cases related to the production of high-precision parts and considered as one of the main stages in the production process. One of the most common methods for additive manufacturing of thermoplastic parts is fused deposition modeling (FDM). According to Wohlers (2019) over 70% of AM-related companies use the FDM technique This method is not only widely used for the manufacture of parts from monolithic plastic, but is also considered as the main method for the production of parts from fiber-reinforced thermoplastics (Valino et al., 2019; Wickramasinghe et al., 2020; Kumar et al., 2020; Gupta, 2020; Bilym et al., 2021). Some authors point out that one of limiting aspects of this technology is the sufficient surface roughness (Espach and Gupta, 2020; Hurina et al., 2020; Vambol et al., 2021). To reduce roughness and increase dimensional accuracy for parts manufactured by FDM it is required to conduct finishing operations (Taufik and Jain, 2020). Traditional CNC machining (Boschetto et al, 2016), barrel (Boschetto and Bottini, 2015) or abrasive flow (Mali et al., 2018) treatment can be used. In addition, chemical methods show good results in terms of reducing the roughness (Jin et al., 2017; Singh et al., 2017). However, as indicated in the papers Oztan et al. (2019); Zindani and Kumar, 2019; Wickramasinghe et al. (2020), on the part surfaces obtained by the FDM method, in addition to irregularities, other surface defects may appear. Figure 1*a* shows the surface roughness where the nylon filaments are separated in a line-by-line arrangement and gaps between those lines are visible. Figure 1*b* exhibits micropores on the surface of the PLA printed parts (Wickramasinghe et al., 2020). The presence of surface discontinuities has a significant effect on the mechanical characteristics of parts. However, such defects cannot be treated by the methods described early. This problem can be solved by thermal polishing methods.

The paper Chai et al. (2018) considers the possibilities of thermal polishing of thermoplastic parts printed by the FDM method using a laser. It is considered two cases of thermal polishing: surface overmelt (SOM), and surface shallow-melt (SSM). The SOM mechanism corresponds to the case when the depth of the molten bath is greater than the surface roughness as shown in Figure 2*a*. Option of the SSM is shown in Figure 2*b*, in which the melt zone depth is much smaller than that of the surface relief.

The results obtained in Chai et al. (2018) confirmed the possibility of roughness reducing with the controlled heating of the thermoplastic, while the final surface quality for various materials was significantly differ. In current research, authors did not investigate the effect of thermal polishing on the mechanical characteristics of thermoplastics, although it should be expected that it (at least in the SOM

Figure 1. Surface defect on FDM printed thermoplastic parts: (a) – surface roughness of a nylon sample; (b) – micropores on the surface of a PLA sample (Wickramasinghe et al., 2020)

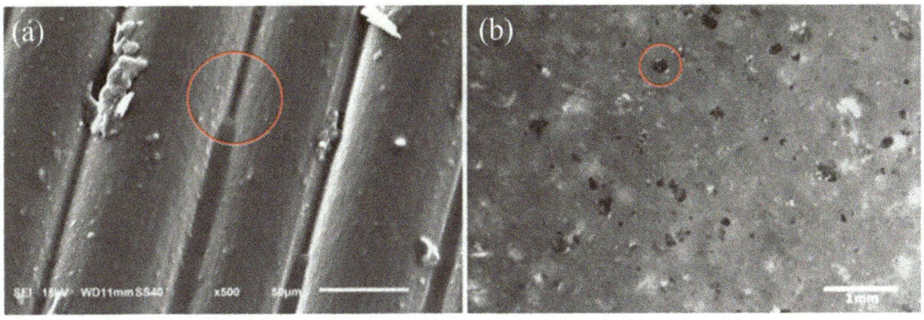

version) may remove defects such as micropores. According to the survey above, the following tasks for thermoplastic parts treatment can be highlight:

- Deburring or deflashing while maintaining the surface relief, typical for processing parts with decorative texture produced by injection or compression molding;
- Deburring and surface shallow-melt polishing, typical, for acrylic parts produced by casting and machining;
- Surface over-melt polishing for additive manufactured parts.

Figure 2. Schematic of laser surface melting regimes on FDM object surface: (a) – melted area of surface over-melt is notably deeper than surface protrusion; (b) – melted layer of surface is thinner than the surface protrusion (Chai et al., 2018)

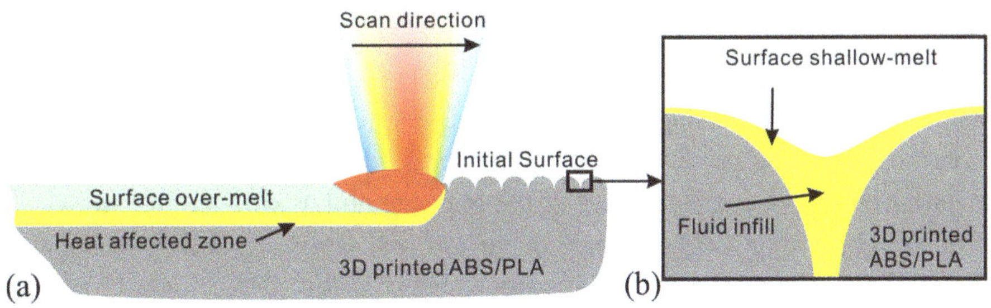

All these tasks can be solved using the thermal energy method. At TEM treatment of metal parts, the fuel gas in the oxygen excess is used (Geen and Rice, 1969; Lamikiz et al., 2011; Struckmann and Kieser, 2020). Burrs to be treated are burnt out because they have a high surface area to volume ratio, burrs are heated quicker than the main body (Figure 3*a*).

For treatment of thermoplastic parts, using mixtures with an excess of oxygen is not the optimal strategy. Thermoplastics can be carbonized in the vicinity of the combustion zone and toxic gases can be formed during the combustion of some thermoplastics (for example, polyvinyl chloride). In this case, TEM treatment with stoichiometric mixtures or mixtures with an excess of fuel is preferable. The main condition for high-quality processing in this case is precise control of the contact time of parts with

Figure 3. Various variations of TEM treatment: (a) – the scheme of the edge finishing by TEM; (b) – the scheme of the edge finishing by ITEM (Plankovskyy et al., 2021a)

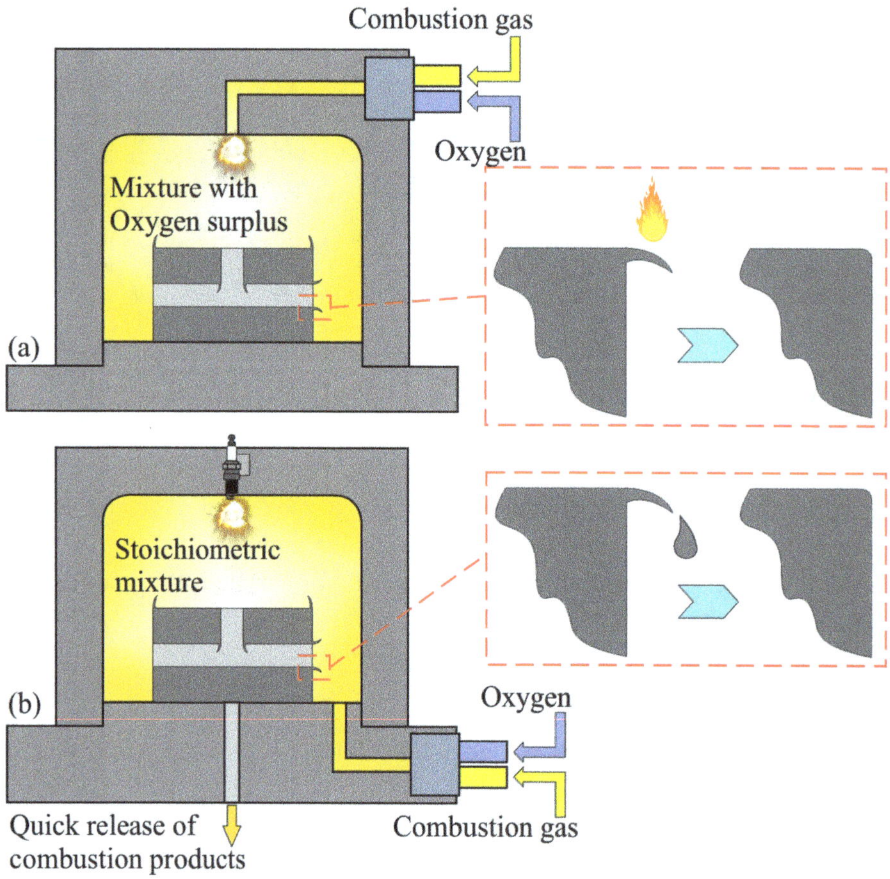

combustion products. ITEM processing meets these requirements best of all (Plankovskyy et al., 2021a). In this case, a controlled release of hot combustion products is carried out (Figure 3*b*), which avoids damage to the parts (in case of correct choice of the pressure and mixture composition).

The selection of TEM processing regimes is carried out using the design of experiment method (Fritz et al., 2012). This process is time-consuming, especially for processing a large number of parts with complex shapes. In this case, in addition to the mixture composition, initial pressure and processing time, it is necessary to determine the parts arrangement in the working chamber, which sharply increases the number of factors that should be taken into account at planning the experiment. According to the "Industry 4.0" trends, the automated assignment of ITEM processing modes based on numerical simulation looks very promising. In Plankovskyy et al. (2021a) were considered the issues related to the development of an automated method for assigning TEM processing regimes for precision metal parts.

ITEM processing of thermoplastic parts has essential features, which does not allow the direct use of the method described in Plankovskyy et al. (2021a). Therefore, in this chapter, we consider the issues of determining the parameters of ITEM processing for thermoplastic parts.

DETERMINATION OF ENERGY CHARACTERISTICS FOR ITEM EQUIPMENT

The main condition for high-quality ITEM processing is accurate prediction of the required specific heat fluxes and processing time values. The problem of sittings determination was traditionally solved by empirical relations based on the result of the factorial experiment. This approach was used for both basic TEM (Kelley and Schwarz, 1991; Fritz et al., 2012; Struckmann and Kieser, 2020) and ITEM (Bozhko et al., 1989). However, this method of parameter determining has a fundamental disadvantage, associated with the necessity to consider more than 10 factors. So, it is necessary to conduct a lot of experiments. Reduction of their number can lead to obtaining inaccurate dependencies that can't provide accurate determination of the processing regimes.

Determination of ITEM processing regimes numerically require solving the complex problem related to combustion, parts surfaces heating, taking into account variation of the material thermophysical properties, phase transitions, etc. To simplify this process, a method based on the assumption that the initial coupled problem of combustion and heat transfer into the chamber with located complex-shaped parts inside can be divided in two simple problems. The first of it in determination of the energy parameters of the ITEM treatment process (averaged specific heat fluxes, intensity and duration of shock waves action in the case their presence). The second problem is determination of the part state under the influence of the previously defined specific heat fluxes.

Due to the complexity of the processes inside the ITEM working chamber, the solution of the first problem can be solved using numerical simulation. For this reason, the described mathematical model can be used (Plankovskyy et al., 2019; 2021a).

Mathematical Model for Determining the Energy Parameters of ITEM

To describe the processes occurring in the working chamber, the model of a single-phase multicomponent flow of reacting gases can be used. The equations for calculating the required characteristics are written as follow:

$$\frac{\partial\rho}{\partial t}+div(\rho\vec{u})=0\,, \tag{1}$$

$$\rho\frac{d\vec{u}}{dt}=div\,P\,, \tag{2}$$

$$\rho\frac{de}{dt}=div(\lambda\,grad\,T)+\sum_{i=1}^{N}Q^{i}-Q_{rad}+P\cdot grad\,\vec{u}\,, \tag{3}$$

$$\rho\frac{\partial c^{i}}{\partial t}=-div\,\vec{I}^{\,i}+\dot{S}^{i},\quad i=1,\ldots,N-1\,, \tag{4}$$

where $\rho(\rho^i)$, $p(p^i)$ is the density and pressure of the mixture and species i separately; P is the stress tensor; Q^i is the heat of chemical reactions for the species i; Q_{rad} is the radiation heat; $\vec{I}^i = \rho^i \vec{w}^i$ is the diffusion flux vector; S^i is the mass rate of the species i formation; e is the internal gas energy.

The mass rate of the species i formation in the Eq. (4) is determined by summing over all reactions K in which it participates:

$$\dot{S}^i = \sum_{m=1}^{K} m^i (\nu''_{im} - \nu'_{im})\omega_m ,$$

where ω_m is the rate of reaction m, and ν''_{im}, ν'_{im} are the stoichiometric coefficients of species i before and after reaction m.

The rate of elementary reactions is determined using the finite rate chemistry model. In this case, the value of ω_m is calculated as follow:

$$\omega_m = \left(F_m \prod_{I=A,B,C,...}^{Nc} [I]^{r'_{mi}} - B_m \prod_{I=A,B,C,...}^{Nc} [I]^{r''_{mi}} \right), \tag{5}$$

where $[I]$ is species I molar fraction; F_m, B_m are constants for forward and backward reactions respectively; r_{mi} is the proportion of the component i which participates in the reaction m.

The following Arrhenius equals were used do define the constants of forward and backward reactions:

$$F_m = A_{m1} T^{\beta_{m1}} \exp\left(-\frac{E_m}{RT} \right), \tag{6}$$

$$B_m = A_{m2} T^{\beta_{m2}} \exp\left(-\frac{E_m}{RT} \right), \tag{7}$$

where A_m, β_m are the pre-exponential factor and the temperature exponent; E_m is the reaction activation energy.

Heat release/absorption rate Q^i for species i is calculated as the sum for all elementary reactions with its participation:

$$Q^i = W^i \sum_{m=1}^{K} (\nu''_{im} - \nu'_{im})\omega_m . \tag{8}$$

At determining the diffusion flux vector, this model did not take into account the effects of thermal and barodiffusion. Therefore, the diffusion rate in Eq. (4) was set by Fick's law:

$$\vec{w}^i = -D^i \, grad \ln c^i ,$$

where D^i is the diffusion coefficient of species i in a mixture of N components.

To determine the thermophysical properties of the reaction products, i.e. dependences of density, viscosity, thermal conductivity, specific heat, diffusion coefficients on pressure and temperature, the equations of state and experimental data are used. For the turbulent flow case, Eqs. (1)–(4) describe the behavior of the averaged properties, if used values of the viscosity and thermal conductivity coefficients are equal to the sum of the molecular and turbulent components. To determine these values, taking into account the features of ITEM processing, it is advisable to use the equations of the SST turbulence model (Menter, 1994; Huang et al., 2018). In this case, the convective heat flux through the contact surfaces between the combustion products and part walls is determined as follows (Kader, 1981):

$$q_w = \frac{\rho c_p \tilde{u}_\tau}{T^+}(T_w - T_f), \tag{9}$$

where T_w is the wall temperature; T_f is the temperature of combustion products in the flow core; $\tilde{u}_\tau = \left[(\tilde{u}_\tau^{vis})^4 + (\tilde{u}_\tau^{\log})^4 \right]^{0.25}$ is the velocity profile in the boundary layer; $\tilde{u}_\tau^{vis} = U_1 / \tilde{y}^+$ is the velocity in the viscous sublayer; $\tilde{u}_\tau^{\log} = U_1 / (\ln(\tilde{y}^+)/k + C)$ is the velocity in logarithmic sublayer; $\tilde{y}^+ = \max(y^+, 11.67)$, $y^+=\rho \Delta y U_1/\mu$ dimensionless distance from the first grid node to the wall; Δy is the near wall cell size; U_1 is the flow speed at the grid node closest to the wall; 11.67 is the limiting value which marks the intersection between the logarithmic and the linear profile;

$$T^+ = \Pr \cdot \tilde{y}^+ \exp(-\Gamma) + [2.12 \ln(1+\tilde{y}^+) + \beta] \exp(-1/\Gamma)$$

is the temperature profile in the near-wall layer according to the Kader solution (Kader, 1981);

$$\beta = (3.85 \Pr^{1/3} - 1.3)^2 + 2.12 \ln(\Pr); \; \Gamma = (0.01(\Pr \cdot \tilde{y}^+)^4)/1 + 5 \Pr \cdot \tilde{y}^+ .$$

Due to the short contact time of combustion products and workpieces, heating of surfaces can be neglected at determining the averaged heat flux. At the numerical calculation, this allows us to consider constant temperature of the chamber and workpieces walls. In addition, a standard slip condition for the velocity was set on the walls, as well as the initial conditions for the pressure, velocity, temperature, and composition of the mixture in the chamber.

For the numerical solution of equations (1)–(9), the commercial code ANSYS CFD was used. According to the recommendations, for accurate simulation of heat transfer between the gas and the solid part, it is necessary to provide y^+ value equal to 1 (Szczepanik et al, 2004; Travin et al., 2000). This requires generation of a fine mesh on the part surface, especially in the case of shock waves action (Palaziuk, 2018; Plankovskyy et al., 2021c) (Figure 4).

The time step at solving unsteady heat transfer problem is assigned according to Courant number requirement:

$$C = \frac{v \Delta t}{\Delta y} \approx 0{,}5 .$$

Figure 4. Section of the computational mesh with the elements refinement near the walls (Palaziuk, 2018)

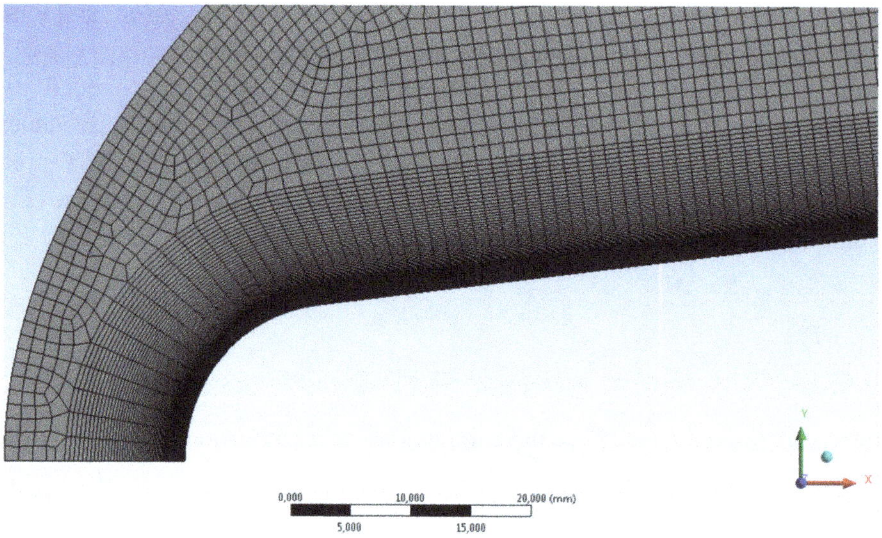

The solver settings used the AUSM scheme (Frazier et al., 2011) as well as high-precision approximations for energy, momentum, and turbulence equations.

The described model was verified on tasks that are typical for ITEM processing. The comparison of the numerical and experimental results of the combustion to detonation transition process in a detonation tube are presented in Plankovskyy et al. (2019). The installation scheme is shown in Figure 5. Figure 6 illustrates calculated and experimental pressure profiles in the sensor installation locations. Calculations and experiments were carried out for a homogeneous stoichiometric hydrogen-air mixture case. The initial pressure in the tube was 1 bar, the initial temperature was 300 K. To initiate combustion in the numerical simulation, a point ignition model was used that simulates a spark with an energy of 0.1 J.

Deep vacuum at pressure 10 Pa was provided before filling the tube to ensure homogeneity of the mixture at experimental investigations. After leak test, the pipe was filled with previously prepared fuel mixture to a pressure of 1 bar. In this case, the rate of pressure increase did not exceed 0.05 bar/s. To eliminate the effects of initial turbulence, before activating the experimental data recording system and igniting the fuel mixture, the mixture was hold for at least 1 min to completely stop the gas inside the pipe. The PCB Piezotronics 113B21 pressure sensors were used to measure the pressure. The calculated detonation velocity was compared with average velocity value based on the results of three experiments. The maximum difference was 4.02%. The difference of the maximum pressure values in the detonation front was 2.89%. Thus, based on the magnitude of the obtained deviations, it can be concluded that the described model accurately describes the combustion of gas mixtures and its transition to detonation.

Similar conclusions were made by comparing the simulation and experimental results of shock waves damping in a closed chamber. The shock waves were caused by detonation of gas mixture. The scheme of the experimental setup is shown in Figure 7. As in the previous experiment, deep vacuum at pressure 10 Pa was provided before filling the chamber with a previously prepared of stoichiometric acetylene/oxygen mixture to a pressure of $P_0 = 0.4$ and 1 bar.

The combustion was initiated in a pre-chamber. The detonation, spreading through a tube hit the end of the detonation chamber at three points, forming a regular triangle. After the transition of the detona-

Figure 5. The scheme of the installation (Plankovskyy et al., 2019)

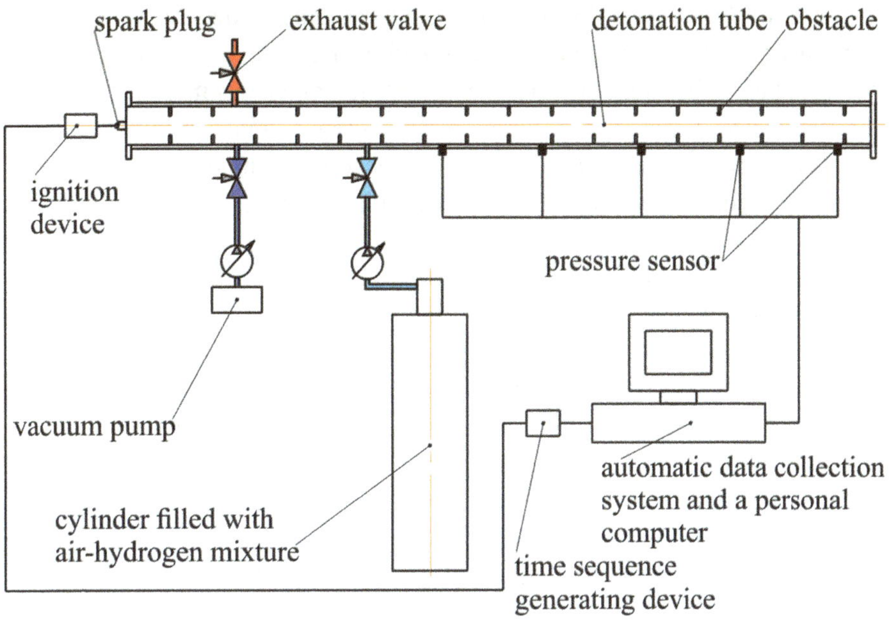

Figure 6. Calculated (top) and experimental (bottom) pressure profiles (Plankovskyy et al., 2019)

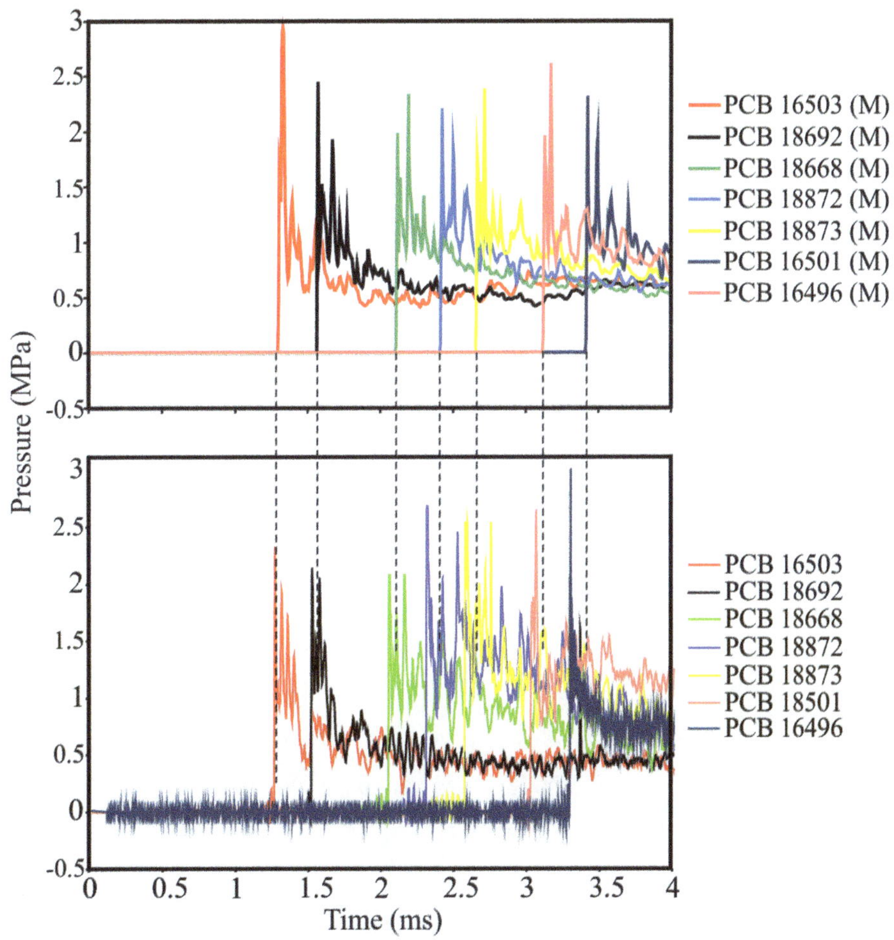

tion from the tube to the chamber, three spherical waves propagate through the volume, which form a flat detonation wave at a fairly short distance. The damping time of the shock waves was determined using a pressure sensor located at the end of the detonation chamber.

It is assumed that, before ignition, there is a stationary homogeneous mixture in the chamber, with a mass fraction of fuel in accordance to the stoichiometric composition. The initial temperature was 300 K. The presence of a detonation tube was not taken into account in the model. The presence of detonation was described by defining the separate zone with initial temperature $T = 2500$ K and a pressure $P = 1.6$ MPa.

Comparison of the simulation result with experimental data (Figure 8) shows a good agreement. In both cases, the period of the shock waves propagation corresponds to the obtained experimental value equal to 5 ms. The calculated average velocity of the shock wave propagation was 1200 m/s, which is equal to 1.3 M for the simulation conditions. This value also corresponds to the experimentally obtained value.

Thus, the mathematical model (1)–(9) accurately describes the processes of combustion and heat transfer during ITEM processing and can be used for energy parameters calculations.

Figure 7. The scheme of the experimental installation for the study of shock waves damping (Plankovskyy et al., 2021a)

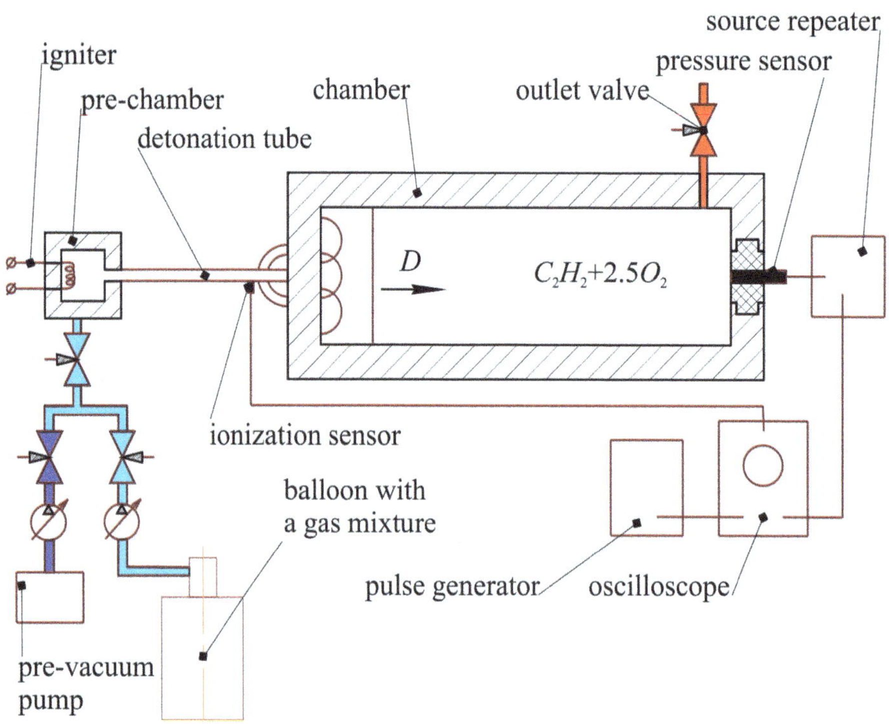

Figure 8. Comparison of the simulation result with experimental data: (a) chamber L = 0.2 m, P = 0.1 MPa; (b) chamber L = 0.4 m, P = 0.04 MPa (Plankovskyy et al., 2014)

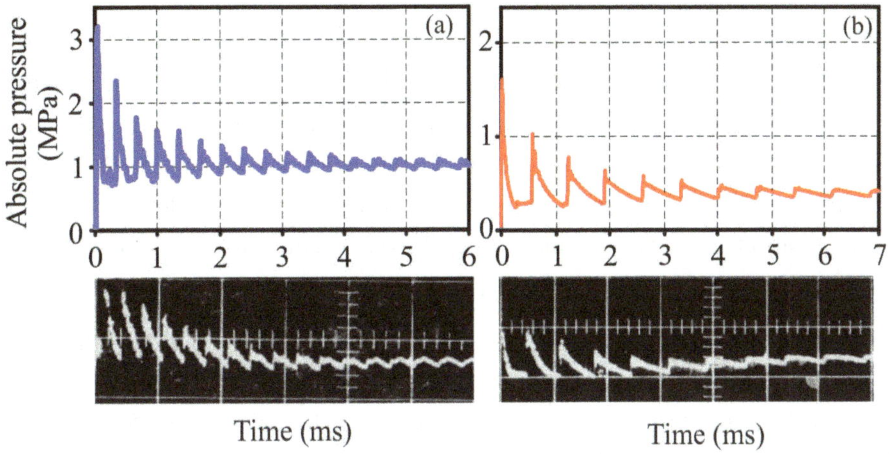

The Equivalent Chamber Method for Calculating the Energy Characteristics of ITEM Equipment

The calculation of the energy characteristics of ITEM processing requires significant computational costs, especially at conducting optimization tasks associated with the determination of processing parameters to ensure the best treatment quality. Such problems arise at determining the best (in terms of processing quality) arrangement of parts in the working chamber, at designing fixtures, etc. In order to simplify the solution of the problem and eliminate the need to recalculate the values of the heat fluxes for every specific case, the equivalent chamber method was proposed in Tryfonov (2013), and modified in Palaziuk (2018); Plankovskyy et al. (2019). According to this method, to carry out the calculation a complex shape part is replaced by a simpler one. This approach is usual in CAE simulations and used to speed up the solution. The shape of the part is simplified by removing some geometric elements that insignificantly affect the solution: grooves, chamfers, radii, etc. In the case of ITEM processing, such a shape simplification must be carried out in compliance with some restrictions.

The principle of choosing the shape of an equivalent part is described in Plankovskyy et al. (2020); Romanova et al. (2021a; 2021b) at formulating the sparest balance layout problem. Such a problem arose at searching for a method for the optimal parts placement during TEM processing. Under the action of shock waves, the magnitude of specific heat fluxes significantly depends on the interaction frequency with the parts surfaces (Plankovskyy et al., 2021b). Based on this, it was proposed a hypothesis the part in the chamber should be most evenly removed from the walls of the working chamber. It is most optimal part location. It has been shown that such arrangement is provided at coincidence of the gravity centers and main central moments directions of the unit density thin shells which are corresponds to the outer surface of the part and the inner surface of the chamber.

Based on the obtained solution, it was proposed to replace the object T_1 with a primitive object O_1. This requires the fulfillment of the following conditions:

$$J_x^{O_1} = \frac{J_x^{T_1} \cdot S_{O_1}}{S_{T_1}};\ J_y^{O_1} = \frac{J_y^{T_1} \cdot S_{O_1}}{S_{T_1}},\ J_z^{O_1} = \frac{J_z^{T_1} \cdot S_{O_1}}{S_{T_1}} \tag{10}$$

where $J_x^{O_1}$, $J_y^{O_1}$, $J_z^{O_1}$, $J_x^{T_1}$, $J_y^{T_1}$, $J_z^{T_1}$ are moments of inertia of thin unit density shells, which are coincident with the outer surfaces of the body T_1 and the equivalent body O_1 relative to their main central axes *OX, OY, OZ*; S_{T_1}, S_{O_1} are their outer surface area.

This replacement allows you to reduce the number of near wall mesh elements proportional to the ratio S_{O_1}/S_{T_1}. In addition, at calculating the averaged specific heat flux q_{eqvP_0} for bodies with the same shape, there is no need to carry out simulation at changing an initial pressure of the mixture.

It is the fact that the heat transfer coefficient $\alpha \sim \sqrt{\rho}$. For the same mixture composition and taking into account the equation of state, it gives relation $\alpha \sim \sqrt{P/T}$. Then the calculation of the averaged specific heat flux for a chamber with an equivalent part can be performed only once for arbitrary values of the initial pressure P_0 and fuel mixture temperature T_0. Based on the obtained results, the averaged specific heat flux q_p for a real part at any initial pressure P and temperature T values of the fuel mixture is determined by the formula proposed in Plankovskyy et al. (2021a):

$$q_p = q_{eqvP_0} \sqrt{PT_0/P_0T}\, \frac{S_{cham} + S_{eqv\_part}}{S_{cham} + S_{part}}, \tag{11}$$

where S_{cham}, S_{eqv_part}, S_{part} surface area of the chamber, equivalent and real part respectively.

Figure 9 shows the dependences of the heat flux averaged over the surface of the part at direct numerical simulation (dashed line) and calculated using Eq. (11) according to the simulation date, obtained for an equivalent part. In both cases, stoichiometric mixture of methane and oxygen was considered. In this case, the explosion of fuel mixture portion led to the appearance of shock waves. The maximum error at heat fluxes determination by Eq. (11) did not exceed 5%.

For the thermoplastics processing, the treatment regimens with the formation of shock waves are undesirable, since they lead to excessively high values of specific heat flux and pressure in the wave front. The consequence of this is the melting and deformation of the workpieces. Therefore, ITEM processing of thermoplastic parts is carried out under the deflagration combustion of the fuel mixture. In this case, heat flux definition has some features.

Limitation of the heat fluxes magnitude important at all stages of treatment: at the stages of fuel mixture combustion; at the stage of holding during the given time; the stage of combustion products exhausting from the working chamber. The release of combustion products begins at a supercritical pressure difference between the chamber and environment. This leads to the formation of a vacuum wave in the chamber, which can lead to undesirable increase of the averaged heat flux.

Therefore, the problem of averaged specific heat flux determination was solved within the framework of single project, which has three configurations with different boundary conditions. For the configurations, which correspond to the stages of mixture combustion and combustion products holding, the constant temperature condition was set to the chamber walls. For the configuration, which corresponds to release of combustion products, the outlet condition at the end of the exhaust channel was changed to pressure outlet with pressure 1 atm.

Figure 9. Surface averaged heat flux under the action of shock waves (Palaziuk, 2018)

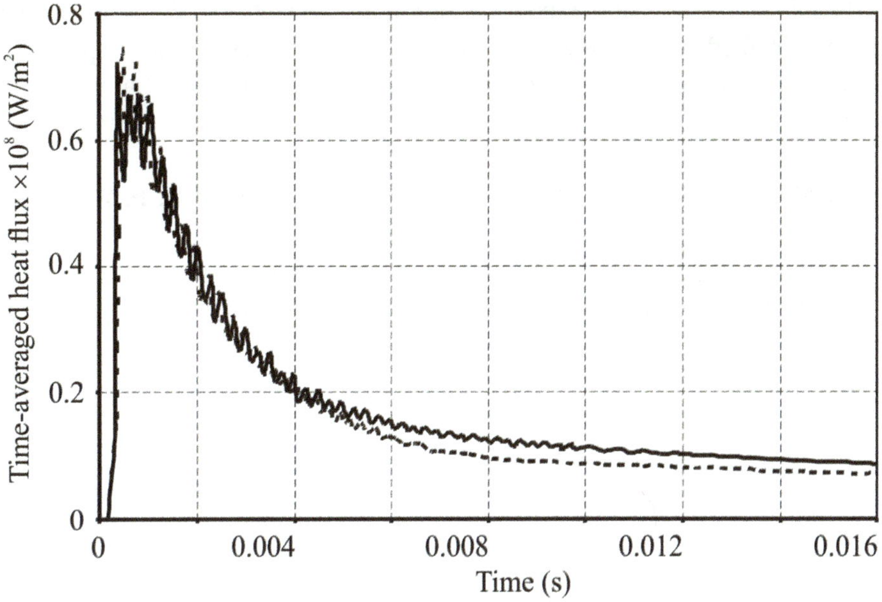

Figure 10. Calculation data of the averaged heat flux (Plankovskyy et al., 2019)

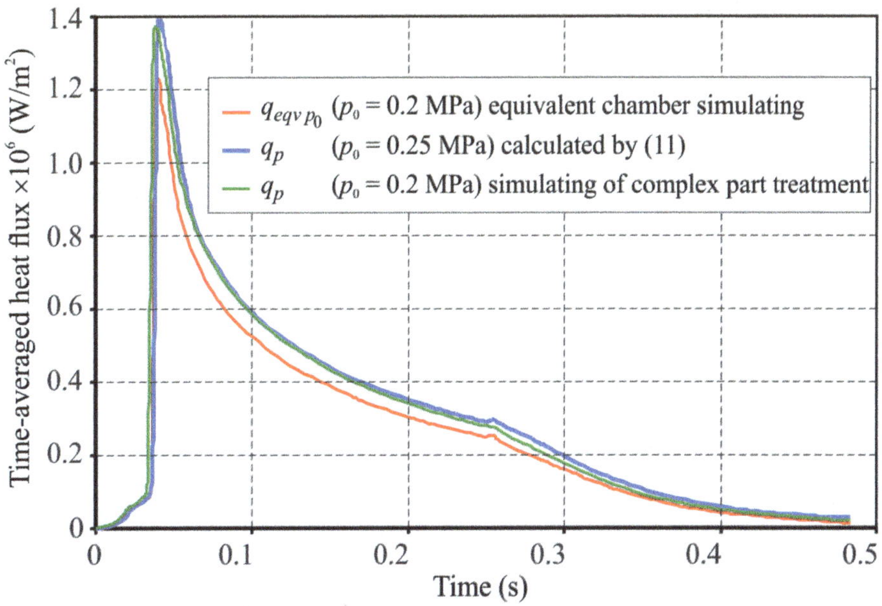

The conditions for switching between the configurations were: for the stage of combustion of the fuel mixture it is reaching of the volume averaged reaction progress to a given value (0.999); for holding the products of combustion it is achieving a predefined holding time; for the release of combustion products, it is the reaching of the volume average absolute pressure value over the of the chamber to a value 1 atm.

Figure 10 shows the dependences of the averaged specific heat flux change, obtained by numerical calculation at the initial pressures of the methane-air mixture equal to 2 and 2.5 bar. Also, this diagram shows the relation obtained for the same cases but using Eq. (11). The error of specific heat flux determination is less than 2%. At the release of combustion products, a significant (up to 30%) heat flux increase was observed.

Some Possibilities to Stabilize the Conditions of ITEM Processing

The method described in the previous section makes it possible to estimate the value of the averaged specific heat flux. In fact, the distribution of the specific heat flux over the surface of the workpieces is not uniform. In the case of processing parts of fusible materials such as thermoplastics, this can lead to different processing quality at different surface zones. Therefore, to achieve a stable quality of thermoplastic parts at ITEM processing, one of the tasks is to ensure the highest uniformity of the specific heat flux. Among the factors that should be taken into account at solving this problem, we will highlight three of them:

- Ensuring maximum homogeneity of the fuel mixture; prevent the formation of shock waves; achieve the most uniform temperature distribution of combustion products. The study of mixture formation in the ITEM working chamber can be carried out using the previously described model (1)–(9), excluding the terms and equations associated with the chemical reactions. To ensure the homogeneity of the fuel mixture, an analysis of two strategies of mixture formation was carried out;
- Sequential filling of the chamber by the mixture components, including filling the chamber by a fuel to a predefined mass; filling the chamber by an oxidizer to a predefined mass; holding during 0.1 s;
- Filling the chamber by a previously prepared fuel mixture with given composition, including filling the chamber by the mixture to a predefined mass; holding during 0.1 s.

A simulation of mixture formation in a chamber with located inside the spherical part, which consists various lengths blind holes was carried. The relationship between the fractional composition of the mixture after filling and the value of the specific heat flux after mixture combustion was investigated. For deep holes the relationship between the samples of the heat flux and the fuel (methane) mass fraction is almost linear. The diagram in Figure 11 shows the distribution of these samples for a group of holes with 2 mm diameter and 8 mm depth (Borysova, 2011).

In addition, it was carried out the simulation of the mixing process in the ITEM chamber with installed a jet engine part. The body has a complex shape with many through, intersecting and blind holes (Figure 12). Simulation was performed for two filling methods: sequential filling and filling by a previously prepared fuel mixture. At sequential filling the considerable dispersion of methane mass fraction is observed in the internal zones. This forms zones with a methane mass fraction below the ignition limit (Figure 13). At filling by previously prepared mixture, the composition in the internal zones is almost homogeneous. Everywhere the methane mass fraction is inside the range of ignition limits. It is shown that homogeneity of the mixture can be improved by increasing the mixture holding time after the end of filling (Plankovskyy et al., 2010).

Figure 11. The diagram of the heat flow and Methane mass fraction samples distribution for a group of holes with 2 mm diameter and 8 mm depth (Borysova, 2011)

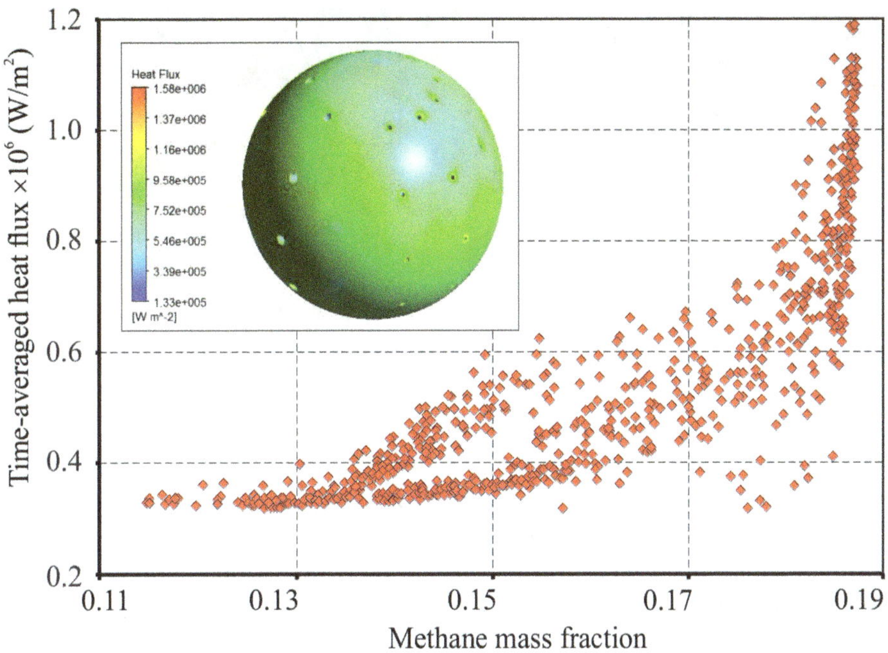

Figure 12. The body of the jet engine unit (left) and the computational grid for its internal volume (right) (Plankovskyy et al., 2010)

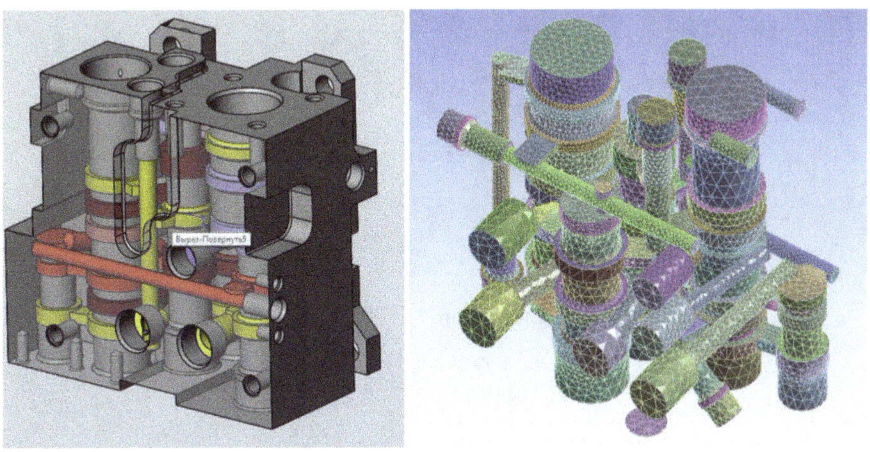

Thus, it is established that the fractional composition of the fuel mixture is a high importance factor of the TEM treatment of deep holes and complex shape cavities. To obtain a homogeneous mixture during ITEM processing, it is necessary to recommend mixture formation using filling the chamber by a previously prepared fuel mixture with increased holding time.

Figure 13. Distribution of the methane mass fraction in the internal cavities: (a) at sequential filling;(b) at filling by the prepared mixture (Plankovskyy et al., 2010)

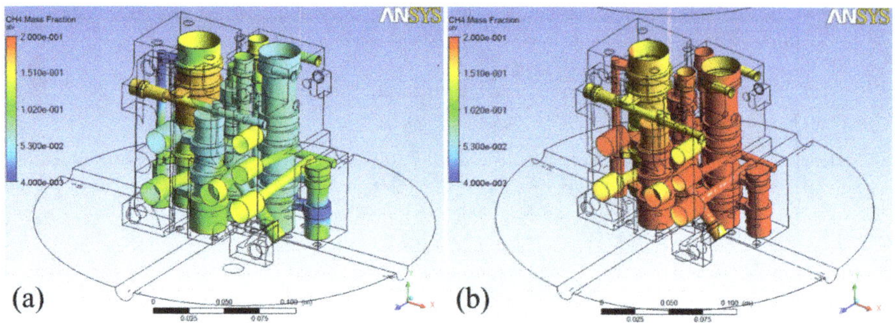

At the TEM processing of metal parts, the treatment regimes with shock waves formation are most effective. At the normal combustion of the fuel mixture, this is possible by two ways: at transition of deflagration combustion to autoignition mode with subsequent thermal explosion of the mixture, or the transition of deflagration combustion to detonation. The heat fluxes generated during such processes are significantly different. As an relevant example the results of experimental studies described in Quintens et al. (2019) can be cited. The results of the experiments showed that the time-averaged heat fluxes at autoignition mode are approximately 3 times higher than similar values at deflagration combustion (Figure 14). At the same time, the average heat flux at the detonation mode of combustion is approximately 2 times higher than at thermal explosion.

Figure 14. Heat flow on the wall for different combustion modes (Quintens et al., 2019)

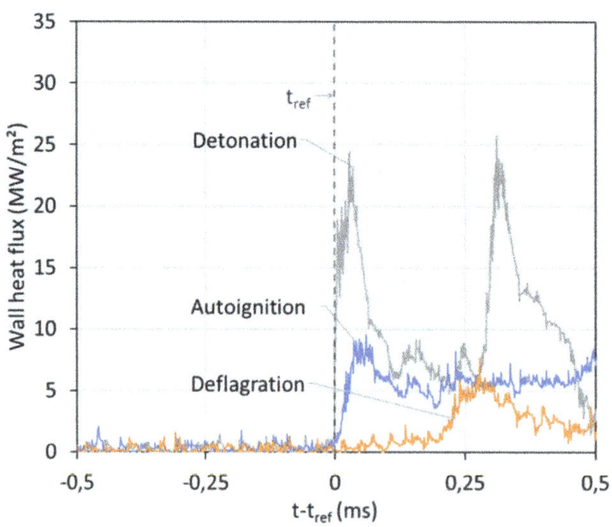

As noted earlier, shock wave modes are not suitable for processing of the thermoplastic parts due to excessive heat fluxes and pressures in the shock wave front. Therefore, for high-quality processing of thermoplastics it is necessary to use modes with deflagration combustion. To set the limits of the permissible modes of thermoplastics ITEM treatment, you can use simplified models which describe the auto ignition of the mixture during combustion in a closed volume (Senachin and Babkin, 1982; Kagan et al., 2012).

In Senachin and Babkin (1982), it was obtained the condition, which determines the possibility of autoignition of adiabatically compressed gas mixture during the flame propagation:

$$u_i \pi_e^{(\gamma-1)/\gamma-\varepsilon} \exp\left(\left(1-\pi_e^{-(\gamma-1)/\gamma}\right)/\beta_i\right) = k(\gamma-1), \tag{12}$$

where $u_i = ak_0/S_{ui}\left(p_i r_i/RT_i\right)^{\nu-1}\exp(-1/\beta_i)$ is the parameter representing the ratio of the characteristic times of frontal combustion and bulk chemical reaction; $\pi_e=1+Q_v/(c_v T_i)$; $\gamma=c_p/c_v$; $\beta_i=RT_i/E$ is the Frank-Kamenetsky parameter; Q_v is the thermal effect of a chemical reaction; k_0 and ν are preexponent rate constants and reaction order; $\varepsilon=n+m(\gamma-1)/\gamma$ is the exponent that takes into account the dependence of the normal speed of the flame on pressure and temperature; n,m are the exponents of normal speed on pressure and temperature dependence $S_r=S_u/S_{ui}=\pi^n\Theta^m$; $\pi=p/p_i$, $\Theta=T/T_i$ are the dimensionless pressure and temperature; r is the relative fraction of the missing component; a is the size of the chamber in the direction of flame propagation, and the indices u,b,i,e are referred to fresh and burnt gas, initial and current state, respectively.

Analysis of the Eq. (12) shows that there is an interval of the initial gas temperature $[T_{imin},T_{imax}]$ for which is possible the volumetric autoignition of the gas mixture ahead of the flame front. In addition, the possibility of autoignition depends on the chamber volume. Therefore, one of the ways to prevent autoignition is to reduce the gas volume by filling the chamber with technological ballast gas.

Autoignition of the gas causes its expansion, which can lead to the formation of pressure waves if the expansion rate exceeds the local speed of sound. The estimates, which were carried out in Senachin and Babkin (1982) showed that it is possible the occurrence of shock waves at a gas autoignition if the level of the fuel burnout before its occurrence does not exceed 30–50%.

A distinctive feature of combustion in a closed chamber is the uneven temperature distribution of the reaction products after combustion. This unevenness is caused by the so-called Mache effect. Due to this effect combustion products temperature is increasing according to adiabatic compression (Krainov and Moiseeva, 2018). With a spark ignition of the fuel mixture, the temperature difference of the combustion products at different points in the chamber can reach hundreds of degrees.

This is not critical for most metal processing applications, as the material allows overheating over a relatively wide range. At thermoplastics processing temperature and, as a result, heat fluxes unevenness can lead to a case, when processing is carried out with the desired quality in one zone of the chamber, and defects are formed in the other. Therefore, for the processing of thermoplastics, the uniformity of the temperature field in the ITEM chamber is a prior problem. A possible solution of this problem is the use of pre-chamber ignition. The principle of the pre-chamber ignition is to pre-ignite a small amount of the fuel in the pre-chamber with subsequent ignition of the mixture through the holes in its housing. In the working chamber, combustion is initiated in a big portion of the mixture that leads to acceleration of fuel combustion which makes temperature of the combustion products more uniform. Using described model, the combustion of methane-air mixtures with pre-chamber ignition was considered. At numerical simulation, the influence of the overflow holes on the combustion products temperature uniformity was studied. In addition, the transition of fuel mixture combustion to volumetric autoignition was tracked (Assovskii and Merzhanov, 2013).

During the simulation, the diameters of the overflow holes, their number and the angle of their location relative to the axis of the working chamber were varied. For one of the overflow holes configurations, the mixture autoignition with the formation of detonation shock waves was occurred. In other cases, it was avoided. For one of the overflow holes configurations, the uniformity of the temperature distribution was the best (the standard deviation was only 0.88%). Such pre-chamber was recommended for thermoplastic parts ITEM processing. Numerical simulation of combustion in the same working chamber using spark ignition has shown that the formation of shock waves is stable for mixtures with initial pressure value more than 0.2 MPa. In this case, a Mach effect was observed. The standard deviation of temperatures is more than 5%.

METHOD OF ASSIGNING FLASH REMOVAL MODES FOR ITEM PROCESSING OF THERMOPLASTICS

Features of Thermoplastics as an Object of ITEM Treatment

The most important properties of materials determining the possibility of being treated by the ITEM method and the corresponding processing modes are thermal conductivity, heat capacity and melting point. All material properties should be considered as temperature dependent. However, for polymers, these properties depend not only on the chemical composition and structure, but also on the molecular weight, the shape of macromolecules, the type of bond between them, branching in macromolecules, etc. (Skoglund and Fransson, 1998; Godovsky, 2012; Amigo et al., 2019; Mustafa et al., 2020). Despite the fact that ITEM treatment is accompanied by increased pressure, according to Mark (2007); Godovsky (2012), pressure weakly affects the thermophysical characteristics of thermoplastic polymers; therefore, pressure will not be considered as a main factor in the process.

The features of the thermal conductivity of amorphous polymers are due to intermolecular bonds. An increase in the number of macromolecules located parallel to the heat flow leads to an increase in the thermal conductivity of polymers. At temperatures above 100°C, many amorphous polymers turn into a liquid state (Figure 15). Experimental data show that the thermal conductivity of melted polymer is almost independent of temperature (polypropylene, polyamide, etc.) or weakly decreases with temperature rise (polymethyl methacrylate, polystyrene, etc.) (Mark, 2007; Malashenko, 2013).

According to the trend of thermal conductivity as function of temperature up to 100°C, crystalline polymers are divided into two groups. The first includes polymers, the thermal conductivity of which decreases with increasing temperature (polyethylene, polycaprolactam, polycaprolactam, etc.). Polymers of the second group are characterized by an increase in thermal conductivity with increasing temperature (polyethylene terephthalate, polychlorotrifluoroethylene, etc.). Thermal conductivity of polymers of the first group is higher than for the second group (Mark, 2007; Malashenko, 2013).

Most polymers have an anomalous temperature dependence of the heat capacity associated with structural transitions in the glassy, highly elastic, and viscous fluid state (Mark, 2007; Godovsky, 2012). With the glassy state of polymers in the temperature range above 73°C, the heat capacity is almost linear with increasing temperature. During the transition of the polymer from the glassy state to the highly elastic one, an abrupt increase in heat capacity is observed. The jump in the heat capacity in the transition region between the glassy and highly elastic states of amorphous polymers characterizes the energy

Figure 15. Thermal conductivity of some thermoplastic materials (Malashenko, 2013)

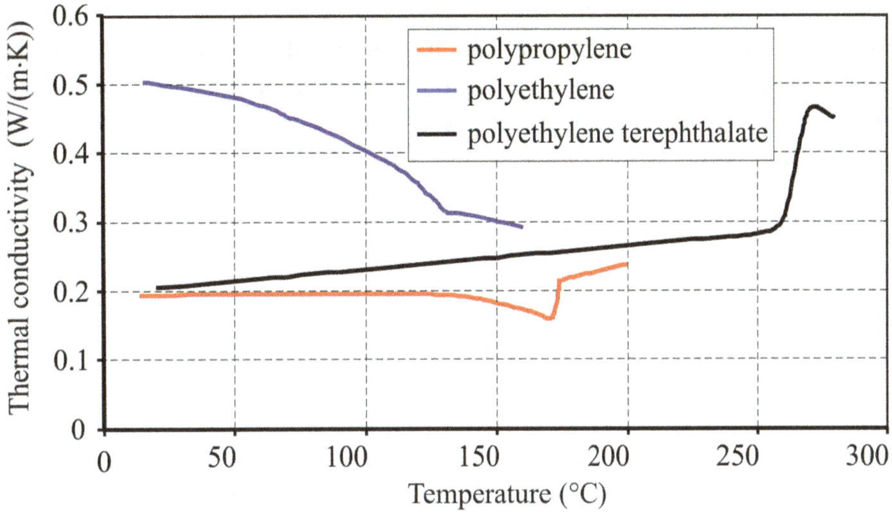

difference between these two states and significantly depends on the general inhibition of intramolecular motions (Mark, 2007; Godovsky, 2012).

A jump in heat capacity in the region of the glass transition temperature is also observed for some crystalline polymers (Figure 16). However, with an increase in the degree of crystallinity, it appears to a lesser extent. Polymers with a high degree of crystallinity, such as polypropylene and polyethylene, do not have an explicit jump in heat capacity in the region of the glass transition temperature. This increase is insignificant and lasts over a wide temperature range (Mark, 2007; Malashenko, 2013).

Figure 16. Heat capacity of some thermoplastic materials (Malashenko, 2013)

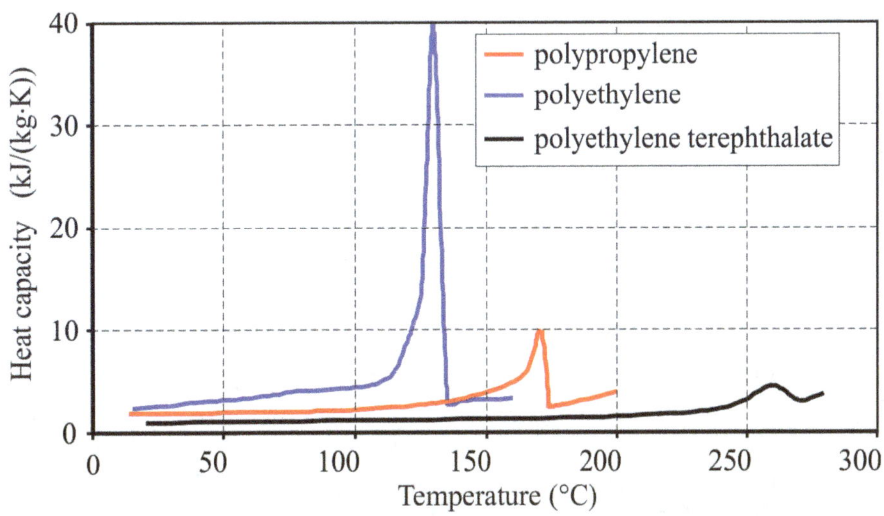

After the transition to a highly elastic state, the heat capacity of crystalline polymers increases with increasing temperature. For most amorphous polymers (polymethyl methacrylate, polystyrene, etc.), the rate of increase in heat capacity during the transition to the highly elastic state is insignificant, which can be explained by the more free state of the kinetic elements of molecules (Mark, 2007; Godovsky, 2012). Melting of polymers and their transition to a viscous fluid state occurs in a certain temperature range. The temperature dependence of the heat capacity (Figure 16) shows a slow increase firstly, and then a sharp one. The phase transition area for the vast majority of polymers observed strongly marked peak growth. During the transition of polymers to a stable molten state with increasing temperature, their heat capacity, as a rule, increases linearly (Mark, 2007; Malashenko, 2013).

In the numerical simulation of finishing processes, it is extremely important to take into account the nonlinearity of thermophysical properties of thermoplastics described above. Another important factor that needs to be taken into account when modeling flash removal is taking into account the change in its geometry during reflow. As already noted, ITEM processing of thermoplastics does not aim at flash/burr separating, but on the contrary, they should be melted and pulled into the part material by surface tension forces. That is why the numerical model of flash melting from thermoplastics should be physically and geometrically nonlinear. Most likely, due to the inaccurate setting of the properties of the processed material, the numerical model requires verification with a full-scale experiment. However, the results obtained with numerical simulating allow predicting the required processing modes with a high degree of reliability, excluding the need for a series of factorial experiments.

Numerical Simulation of Flash Melting from Thermoplastic Parts by ITEM Processing

As noted above, the peculiarities of thermoplastic polymers require nonlinear numerical models when calculating the temperature fields in the part and flash under the thermal pulse action (Kantor et al., 2001; Plankovskyy et al., 2016; Smetankina et al., 2020; Shypul and Myntiuk, 2020). Following the work Malashenko (2013) the model of the flash adjacent to the edge of the part (Figure 17) will be used. Heat capacity, thermal conductivity and density are set as user-defined functions obtained by interpolating table values.

The problem of temperature determination during flashing/deburring is considered as a non-stationary problem of heat conduction in a two-dimensional formulation:

Figure 17. The design scheme for flash/burr melting with discrete geometry change: h – flash height, l – flash length, α – flash position angle

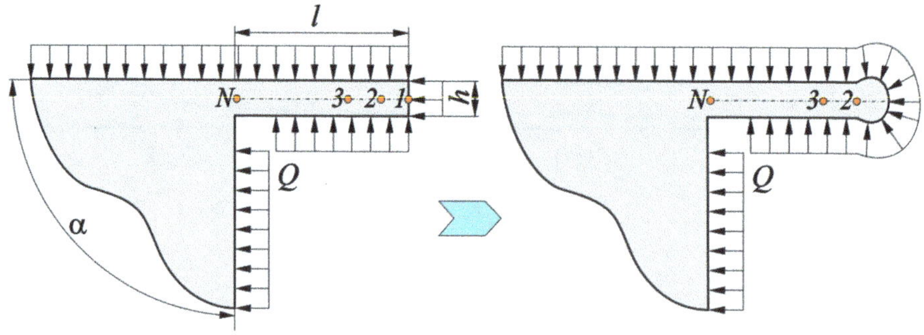

$$c_p(T)\rho\frac{\partial T}{\partial t} = div(\lambda(T)grad(T)). \quad (13)$$

The estimated area of temperature in the part is a rectangle with a side 10 times larger than the length of the flash. On the outer surface of the part and over the entire surface of the flash, the value of the heat flux is set $-\lambda\frac{\partial T}{\partial n} = Q$, on the rest of the boundary, the condition of constant temperature is set $T=T_0$, where T_0 is the initial part temperature (20°C). When the melting temperature in some part of the flash was reached, its shape was considered unchanged until the temperature at the control point, located in the middle plane of the flash, reached the melting point of the material (Figure 17). After that, it was considered that the shape of the melted part located to the right of the i-th control point acquires the shape of a circle with a radius determined by the dependence $r_i = \sqrt{l_i h/\pi}$. Herewith, the finite-element grid of the flash melted part was not rebuilt, and the change in the shape of the flash was taken into account by reducing the value of the specific heat flux on the surface of the flash to the right of the i-th control point:

$$-\lambda\frac{\partial T}{\partial n} = Q\frac{2\sqrt{\pi l_i h} - h}{2l_i + h}. \quad (14)$$

Eq. (14) describes a decrease in the heat flux through the melted part of the flash according to reduction the contact area, similar to the formation of a drop of molten material. In this way, the discrete change in the geometry of the flash is simulated during melting. The control points (Figure 17) are located in the middle of the flash height; the distance between the control points is equal to the flash height. The solution is implemented by means of a related multi-configuration formulation so as the length of the flash and, accordingly, the number of control points predetermined the number of the project configurations. In Figure 18 the results of melting modeling of a polyethylene flash with a length of 1 mm and a height of 0.2 mm, an average specific heat flux of 1.5×10^4 W/m^2 are presented. The temperature at the midpoint of the flash root N was determined for different models: 1 – non-linearity of the material properties; 2 – considering discrete flash geometry variation. Comparison of the obtained results shows their insignificant difference, about 3%. Nevertheless, taking into account the decrease in the heat flux on the reflowed part of the flash due to the reduction of the heat removal area (model 2) refines the estimated reflow time towards its increase, which can significantly affect the final processing quality.

When assigning the technological modes of ITEM treatment, it is necessary to coordinate the specific heat fluxes that occur in the ITEM chamber and the specific heat fluxes required for flash melting. In case when, after processing, the surface roughness of the part should be preserved, then there should be a limitation of reaching the melting temperature in the surface layer of the part. An additional limitation for reaching thermoplastic destruction temperature in a melted flash was used. To determine the mode of ITEM processing, using the developed mathematical model, a numerical experiment was carried out to reflow a flash with a length of 1 mm and a height of 0.2 mm for various materials. The variable parameter was the specific heat flux. In Figure 19 the simulation results are presented.

The time of the flash melting from the polypropylene part turns out to be shorter in the entire investigated range of heat flux values, despite the higher melting point than polyethylene has. This is due to the significantly higher heat capacity of polyethylene.

Figure 18. The temperature at the midpoint in the flash root over time for the models: 1 – non-linearity of the material properties; 2 – considering discrete flash geometry variation

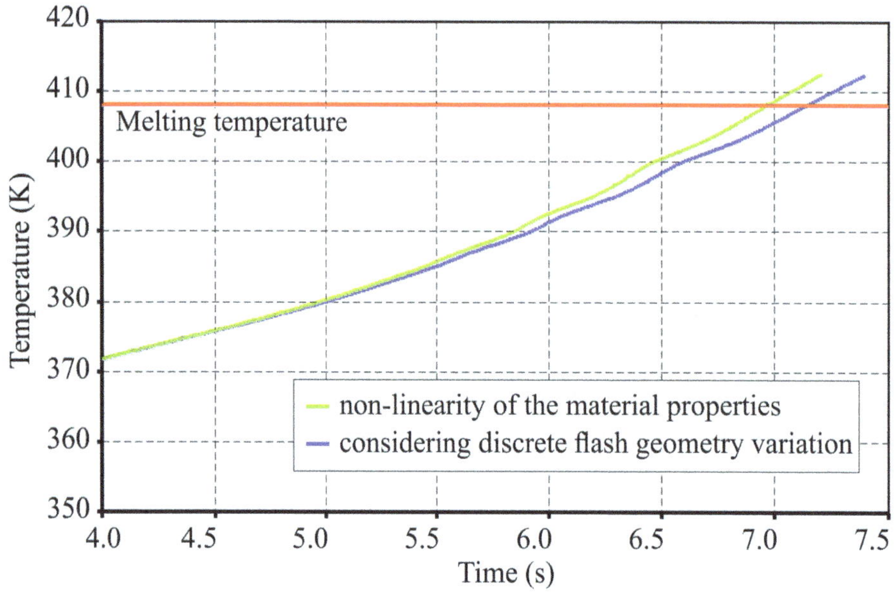

In Figure 20 shows charts for the surface temperature of the polyethylene part and the temperature of the melted part of the flash at the time when the temperature at the midpoint of the flash root *N* reached the melting point. In the studied case of processing a polyethylene part the most vital limitation condition is reaching the destruction temperature. Indeed, under the action of a specific heat flux equal to about 9×10^4 W/m^2 the droplet temperature already reaches the destruction temperature (≈573 K), while the surface temperature is even significantly lower than the melting temperature.

Figure 19. The flash melting time depending on the value of the average heat flux, h= 0.2 mm; l= 1 mm

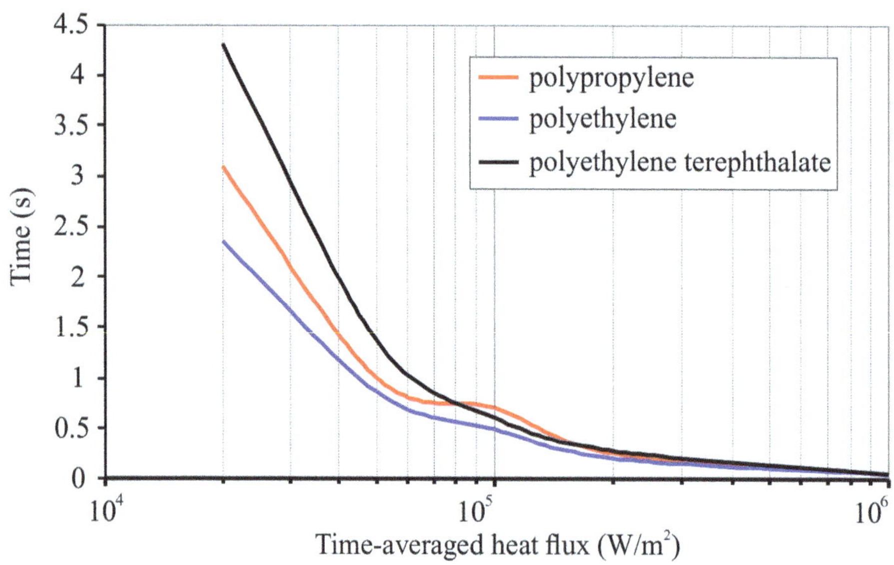

Figure 20. The temperature in the melted drop and in the polyethylene part surface layer at the flash melting moment by acting specified specific heat flux

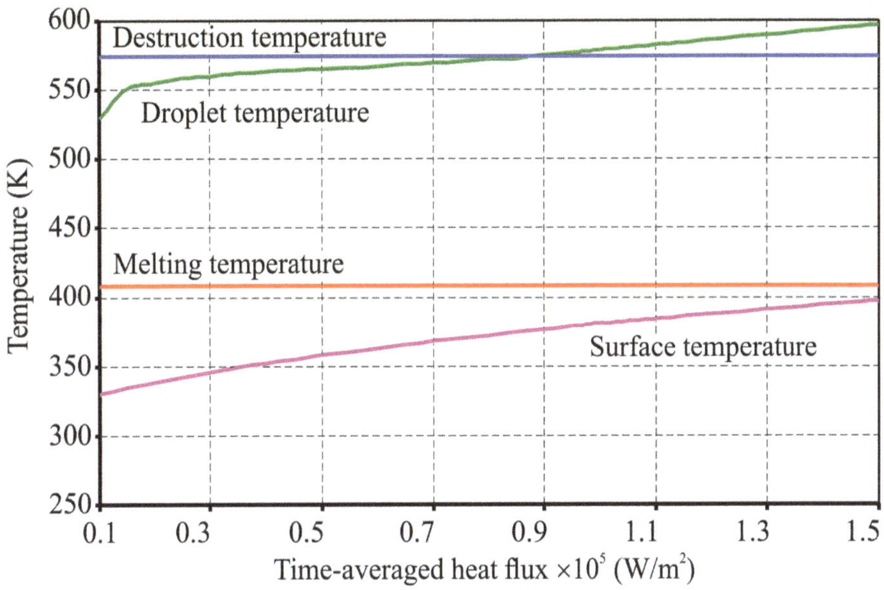

Methods for determining the ITEM treatment modes (initial mixture pressure and treatment time) for metal alloys, including those using simplified equations are well known (Plankovskyy et al., 2021a). However, their use for plastics leads to erroneous results due to significant nonlinearities of their thermal properties. Therefore, it is required to create a methodology that uses the results of numerical modeling, taking into account the features described above.

In order to determine the required ITEM processing modes for plastic parts (initial mixture pressure and processing time) several steps are required. First, according to the method described in the previous chapter, time averaged specific heat flux during the combustion of a specific fuel mixture at a given initial pressure in the ITEM chamber should be calculated. Next, for a flash of a known dimensions using the described models, the time of melting under the action of a given average specific heat flux over time should be determined $q_{melt}(t)$ (Figure 20). Due to the fact that when simulating the temperature field in the problem of melting the flash, the specific heat flux does not change over time, the calculated dependence $q_{ave}^{P_0}(t)$ must be averaged over time:

$$\bar{q}_{ave}^{P_0}(\tau) = \frac{\int_0^{\tau} q_{ave}^{P_0}(t)dt}{\tau}, 0 \le \tau \le t_{\max}. \quad (15)$$

The resulting charts must be compared (Figure 21). Then, the value of the specific heat flux is superimposed on the charts, at which the temperature limitation is reached (in current case, the temperature of the melt droplet comes over the destruction temperature).

Figure 21. The approach to determining the ITEM modes restricted by the processing temperature

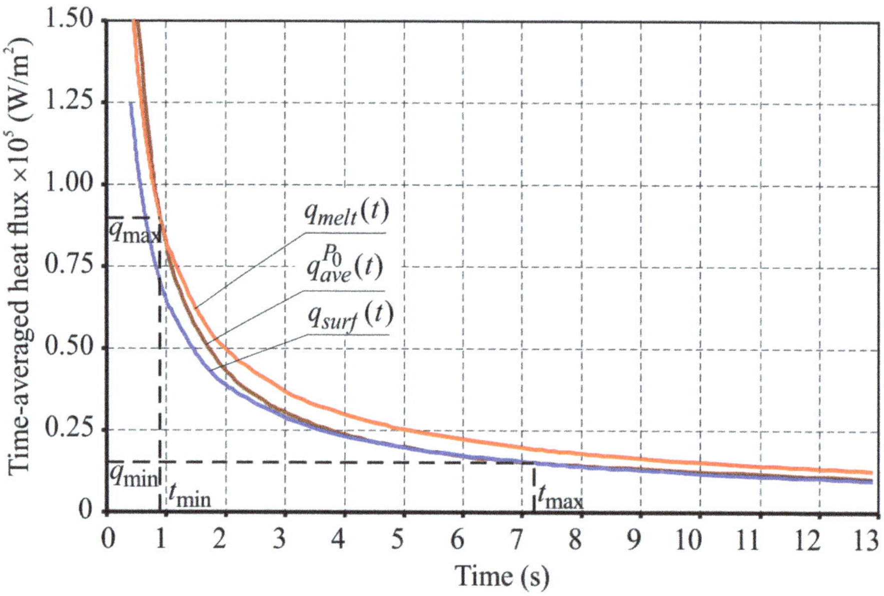

If there is no intersection between the graphs $\bar{q}_{ave}^{p_0}(t)$ and $q_{melt}(t)$ then the treatments to the treatment can not be achieved, and the initial pressure of the mixture should be adjusted. The time corresponding to the minimum distance between the curves should be determined, and appropriating initial pressure of the mixture according to the Eq. (16) can expressed as:

$$p_{\min} = p_0 \left(q_{melt}(t_{\max}) / \bar{q}_{ave}^{p_0}(t_{\max}) \right)^2 . \tag{16}$$

The maximum value of the mixture initial pressure is determined in the same way by the value of the maximum specific heat flux at which the temperature limits are reached (or by the melting temperature or the temperature of destruction):

$$p_{\max} = p_0 \left(q_{melt}(t_{\min}) / \bar{q}_{ave}^{p_0}(t_{\min}) \right)^2 . \tag{17}$$

Thus, a range of possible processing modes $((p_{min},t_{max}),(p_{max},t_{min}))$ is estimated, ensuring the flash melting without any thermal destructions and any changes in the part relief.

It should be noted that the uncertainty in the thermophysical properties of the processed materials, as well as the simplification of mathematical models, does not exclude the need of experimental verification of the calculated ITEM processing settings. However the number of experiments can be significantly reduced. Accumulated experimental results should be kept as specialized databases, which can be used to determine the technological modes of ITEM processing with high accuracy.

Numerical Simulation of ITEM Surface Treatment for 3D Printed Part by FDM Technology

In order to check the possibility for ITEM surface treatment for 3D printed parts, a two-dimensional analysis was carried. Two types of treatment were considered – surface smoothing and porosity correction. In case of FDM additive technology surface quality significantly affected by layer size. Layer height and path width depends on nozzle diameter, which can vary in a range from 0.1 to 0.3 mm. With the increase of nozzle diameter the productivity of FDM method goes up, however the general quality and accuracy decreases. In this case, finishing operations for surface correction are a must. For additive technologies, there is always a risk for pore formation. All these production issues can be treated with ITEM method.

For numerical simulation a VOF approach and melting-solidification model were used. Mathematical model was based on Vikas & Soni (2017); Kim et al. (2013). Material properties for plastic were taken from Trhlíková et al. (2016); Matbase (2021). For both cases a rectangular part with a size 10x10 mm was considered. Figure 22 shows a characteristic dimensions for layer and pore configurations.

Figure 22. Layer and pore geometrical parameters: (a) – surface smoothing; (b) – porosity correction

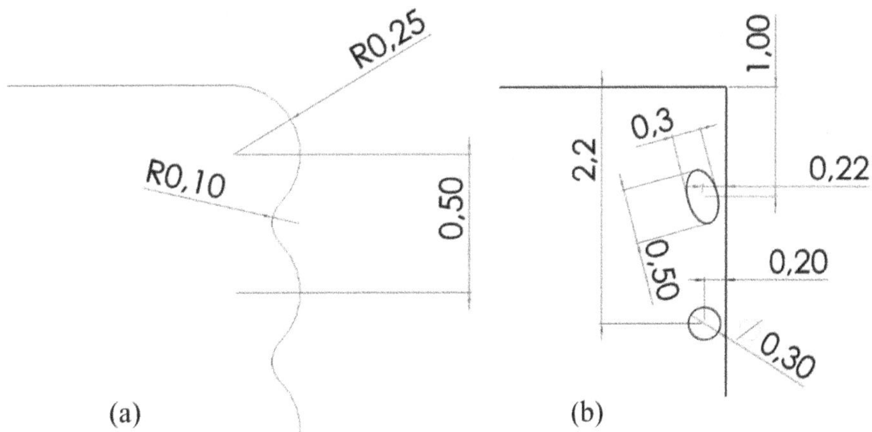

Initially a constant temperature of 25°C was set for both fluid and solid computational domain. Adiabatic wall boundary condition was applied to the walls. To simulate a heating process a constant energy source of 1×10^6 W/m^3 was used for fluid domain. Gravity was taking into account. The flow was considered as laminar. Due to the fact that VOF model is very sensitive to a time step size and mesh quality, several runs were made to estimate a required time step value and mesh size to get reasonable value for equation residuals. Figure 23 shows the final mesh resolution for both cases. Time step sized was chosen to be 1×10^{-6} s.

Several assumptions were made for mathematical model:

- Surface tension was set with a constant value for a plastic material after phase change;
- Thermal expansion was not considered.

To resolve an interface between solid and fluid material and exclude the possible inaccuracies due to a mesh resolution a dynamic mesh adaption was used.

As can be seen from results (Figure 24) ITEM can be a promising solution for surface treatment of 3D printed parts. By means of it, the shape of a part can be adjusted according to requirements for edge and surface configuration.

Figure 23. Mesh resolution: (a) – surface smoothing; (b) – porosity correction

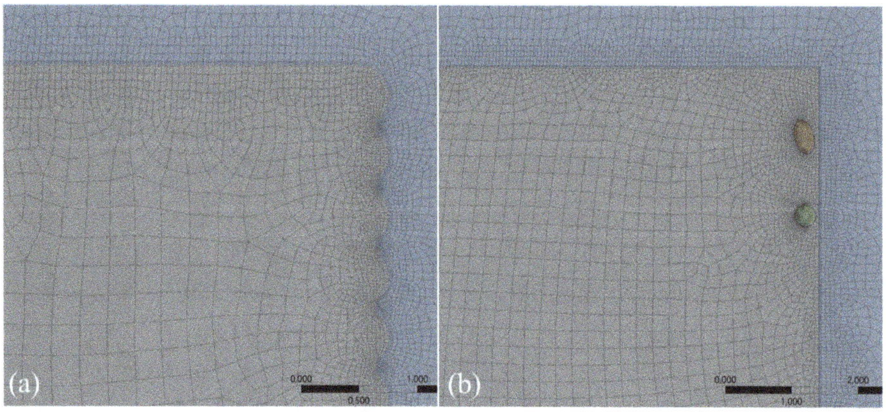

Figure 24. Volume fraction: (a) – surface smoothing; (b) – internal defects treatment

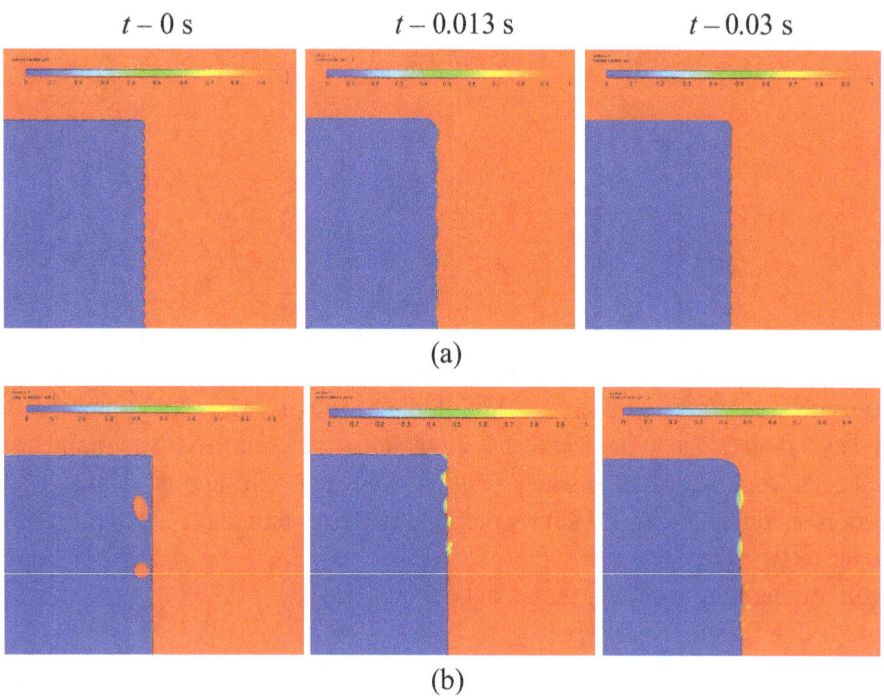

The main driving parameter for plastic parts treatment – is surface tension and viscosity for a melted material. It should be mentioned that each type of treatment requires an individual parameters for operational modes. Figure 24 shows some difference in edge configuration for surface roughness treatment and internal defects treatment with the same processing time. Current study was aimed to investigate a general effect from ITEM on 3D printed part. More detailed research on regimes and limitations will be published separately. There are several aspects which should be carefully checked:

- The size and number of pores which can be treated;
- Their position according to a processed surface;
- Combination of treatment modes.

THE PRACTICAL APPLICATIONS OF TEM FOR THERMOPLASTICS PARTS TREATMENT

There are two types of TEM processing equipment in use: machines for metal and plastic processing, which differ significantly in performance. For metal parts processing, mixtures of methane with an excess of oxygen with a pressure of up to 15 bar or more are most often used. The excess oxygen is calculated based on the amount required for the combustion reaction of the burr material.

In machines for plastics, mixtures with an excess of oxygen should not be used, this will lead to burr burning with the inevitable formation of soot on the part of the surface with the highest temperature. In addition, low pressure blends are required for processing thermoplastics as previously shown. In our opinion, the use of hydrogen as a fuel is the best option for thermoplastic parts processing. According to the Kulik (2017) patent for plastic parts, the selected ratio of oxygen and hydrogen in the fuel mixture is 1: 3 (stoichiometric ratio 1: 2). To reduce the intensity of impact on plastic parts, mixtures with a reduced initial pressure in the range from 400 to 2000 mbar are used. In the case of mixtures with an initial pressure below atmospheric pressure, it is necessary to create a vacuum in the chamber at a pressure of 200 mbar (Kulik, 2017).

ATL, Extrude Home and SGM are world leading manufacturers of TEM equipment. All modern equipment for TEM processing is equipped with CNC systems. There are equipped with various sensors to monitor the parameters of the treatment process and ensure the safe equipment operation (Gallikovà et al., 2020). The example of existing TEM deburring machine iTEM Plastics, produced by a ATL Anlagentechnik Luhden GmbH (Schulz, 2017; Struckmann and Kieser, 2020) is shown on Figure 25.

The machine has a working chamber that allows processing parts with a maximum size of 400×400×800 mm. Working pressure of a mixture of hydrogen and oxygen up to 2 bar. Applicable plastics: PMMA (polymethylmethacrylate), acryl glass, POM (polyoxymethylene), PA (polyamide), PA casting, PUR (Polyurethane), ABS (acrylnitrile-butadiene-styrene), PE (polyethylene), PP (polypropylene), silicone, laser sinter substances, NBA, Viton. Plastics applicable to a limited degree: PEEK (polyetheretherketone), PVDF, natural latex. Thermoplastics containing reinforcing fibers can only be machined to a limited extent. During deburring, they tend to melt slightly more than the reinforcing fibers. Because of this, the edge remains uneven after processing. Duroplastics or thermosetting plastics may not withstand pressure due to their fragility and break down (Struckmann and Kieser, 2020).

For thermoplastic parts processing, it is necessary to follow the general approaches typical for TEM processing. When removing it is important to ensure that the surface finish meets safety requirements.

Figure 25. iTEM Plastics thermal deburring machine by ATL Anlagentechnik Luhden GmbH (Schulz, 2017)

This means that the surface of the parts must be free of oils and greases which can cause unintentional fire and damage workpieces.

During processing, the parts should be secured at least by simple fixtures to prevent parts from flying apart in the combustion chamber during an explosion and thus damaging them. In some cases, it may be necessary to use special clamping devices for optimal position of the workpieces in the chamber. This arrangement provides conditions for the uniform impact of combustion products on all treated surfaces. With such an arrangement, it is possible to guarantee the stable quality of the finishing processing. Finally, during processing, special devices can be used to protect sensitive areas from damage by the combined action of heat flux and pressure impulses.

Additional devices for TEM processing are usually made of materials that are resistant to corrosion and retain their dimensions when heated. Most often, corrosion-resistant steels are used for these purposes. The surface of the fixture parts is involved in heat exchange with the combustion products and the processed parts, therefore, this fact should be taken into account when assigning TEM processing modes. An important element in the design of devices for TEM processing of thermoplastics is to ensure the free flow of the flame front, to exclude the formation of a jet stream, which usually leads to damage of the processed parts.

Figure 26 shows thermoplastic parts processed by TEM equipment (Schulz, 2017; Struckmann and Kieser, 2020). With the correct selected processing modes, the technology provides effective deburring operations after machining, melting the flash on a part produced by injection molding machines, and thermal polishing for outer and inner surfaces of parts. Nevertheless, the determination of technological parameters remains the main problem of TEM processing of thermoplastics when using traditional methods for selecting equipment settings using a factorial experiment. The proposed approach for determining the modes is based on the numerical simulation of the TEM processing process significantly reduces the number of experiments required. Implementation of this approach requires the use of equipment with a time-controlled release of combustion products, i.e. transition to technologies based on ITEM processing.

FUTURE RESEARCH DIRECTIONS

Considering the prospects of thermoplastics application, special attention should be paid to the study of the influence of ITEM processing on the thermoplastic mechanical properties. Correct choice of the hot gas medium, temperature-time regime of the treatment, rate of a part heating to given temperature,

Figure 26. Examples of thermoplastic parts treatment using TEM by ATL Anlagentechnik Luhden GmbH: (a) – surface and edge deburring after milling; (b) – deflashing; (c) – thermal polishing of PMMA (ATL Anlagentechnik Luhden, 2021)

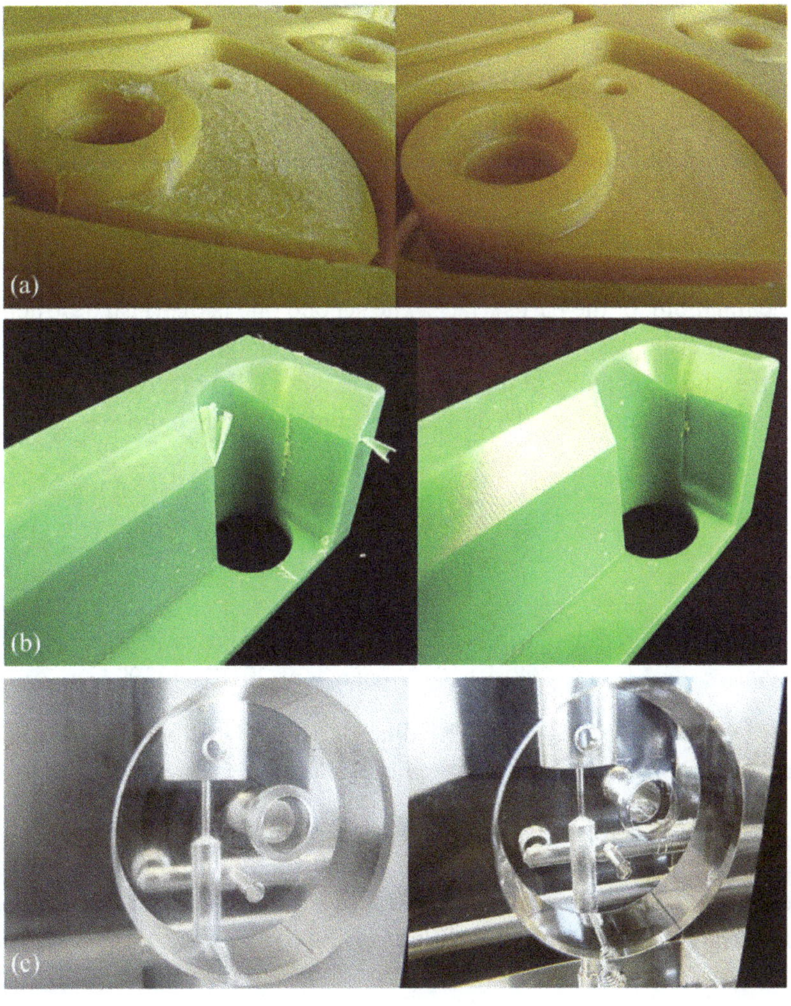

holding at this temperature, and rate of part cooling to normal temperature are most important factors, which influence on final thermoplastic mechanical properties after ITEM processing.

The molecular structure of thermoplastics is very sensitive to heat treatment. That is why, it plays key role, because make it possible to get parts with best wear resistance or provide an optimal complex of strength properties. Therefore, the choice of the processing mode should be done taking into account the thermoplastic properties, as well as design and operational purpose of the part. This will be the subject of further researches.

CONCLUSION

For finishing thermoplastics, the ITEM method is recommended. The processing of thermoplastics should be carried with the parameters ensuring deflagration mode. Volumetric self-ignition or detonation inherent to the ITEMSW method leads to excessive values of heat fluxes and pressures and, and can damage the parts.

For efficient ITEM treatment, it is necessary to provide the uniform temperature distribution of the products burned with minimization of the Mache effect. This can be done with the pre-chamber-torch ignition.

For the thermoplastics treating by ITEM, the preferred type of fuel are hydrogen and oxygen or air mixtures with a pressure up to 0.3 MPa with an excess of hydrogen. Such parameter combination excluds ignition of the thermoplastic, and the soot formation on the part surface.

With ITEM processing of thermoplastics, two finishing mods are possible. The first is the melting of the flash while retaining the surface relief. The second option is to melt the surface of the part with thermal polishing. The first case is realized with reduced pressures of the mixture and a longer processing time.

While estimating the processing modes, it is crucial to meet some limitations. The first, critical limiting condition is the melting point of the part surface. The second one is the flash destruction temperature.

ITEM polishing of thermoplastics allow not only surface smoothing but with the certain processing settings, to reduce the surface porosity of the part. This makes ITEM processing promising for finishing parts obtained by additive technologies. To develop a technique for such processing, additional research is required.

REFERENCES

Abhilash, M., Prakash, G. N., & Kasthurirengan, S. (2018). Cryogenic deflashing for rubber products. *MATEC Web of Conferences, 144, 03004.* 10.1051/matecconf/201814403004

Adel, M., Abdelaal, O., Gad, A., Nasr, A. B., & Khalil, A. (2018). Polishing of fused deposition modeling products by hot air jet: Evaluation of surface roughness. *Journal of Materials Processing Technology, 251*, 73–82. doi:10.1016/j.jmatprotec.2017.07.019

Adeniji, D., Schoop, J., Gunawardena, S., Hanson, C., & Jahan, M. (2020). Characterization and modeling of surface roughness and burr formation in slot milling of polycarbonate. *Journal of Manufacturing and Materials Processing*, *4*(2), 59. doi:10.3390/jmmp4020059

Aksonov, Y., Kombarov, V., Fojtů, O., Sorokin, V., & Kryzhyvets, Y. (2019). Investigation of processes in high-speed equipment using CNC capabilities. *MM Science Journal, 2019*(04), 3271–3276. doi:10.17973/MMSJ.2019_11_2019081

Altan, M., & Altan, E. (2014). Investigation of burr formation and surface roughness in drilling engineering plastics. *Journal of the Brazilian Society of Mechanical Sciences and Engineering, 36*(2), 347–354. doi:10.100740430-013-0089-8

Amigo, N., Palza, H., Canales, D., Sepúlveda, F., Vasco, D. A., Sepúlveda, F., & Zapata, P. A. (2019). Effect of starch nanoparticles on the crystallization kinetics and photodegradation of high density polyethylene. *Composites. Part B, Engineering, 174*, 106979. doi:10.1016/j.compositesb.2019.106979

Assovskii, I. G., & Merzhanov, A. G. (2013). Validity of experimental and theoretical modeling of combustion of high-energy materials. *Combustion, Explosion, and Shock Waves, 49*(3), 264–272. doi:10.1134/S0010508213030027

ATL Anlagentechnik Luhden GmbH. (2021). *TEM applications*. https://atl-luhden.de/en/tem-machinery-construction/tem-deburring/tem-applications/

Baird, R. (2017). *Thermal deburring of plastic components with infra-red – a new technology*. https://www.linkedin.com/pulse/thermal-deburring-plastic-components-infra-red-new-technology-baird

Bilym, P., Rogozin, A., Firsov, P., & Zolotov, S. (2021). Carbon fiber plastics for vehicles manufactured by resource-saving formation technology. *IOP Conference Series: Materials Science and Engineering, 1021*(1), 012034. doi:10.1088/1757-899X/1021/1/012034

Biron, M. (2018). *Thermoplastics and thermoplastic composites*. William Andrew.

Borysova, O. S. (2011). *Improving methods dosage energy at the finish thermopulse cleaning of precision parts of flying vehicle* [Ph.D. dissertation]. National Aerospace University "Kharkiv Aviation Institute."

Boschetto, A., & Bottini, L. (2015). Surface improvement of fused deposition modeling parts by barrel finishing. *Rapid Prototyping Journal, 21*(6), 686–696. doi:10.1108/RPJ-10-2013-0105

Boschetto, A., Bottini, L., & Veniali, F. (2016). Finishing of fused deposition modeling parts by CNC machining. *Robotics and Computer-integrated Manufacturing, 41*, 92–101. doi:10.1016/j.rcim.2016.03.004

Bozhko, V. P., Losev, A. V., Takoriants, M. P., Pleshkov, V. I., Strizhenko, V. E., Levityansky, I. A., Filippov, B. L., & Kononenko, T. I. (1989). *Thermopulse apparatus for deburring parts* (U.S. Patent No. 4 802 654). U.S. Patent and Trademark Office.

Campean, R., & Hancu, L. (2019). Proposal for cryogenic deburring equipment's design and thermal insulation. *MATEC Web of Conferences, 299, 01013*. 10.1051/matecconf/201929901013

Chai, Y., Li, R. W., Perriman, D. M., Chen, S., Qin, Q. H., & Smith, P. N. (2018). Laser polishing of thermoplastics fabricated using fused deposition modelling. *International Journal of Advanced Manufacturing Technology, 96*(9), 4295–4302. doi:10.100700170-018-1901-5

Drégelyi-Kiss, A., & Horváth, A. (2018). Investigations on the accuracy of additive and conventional manufacturing. *IOP Conference Series: Materials Science and Engineering, 393*(1), 012098. doi:10.1088/1757-899X/393/1/012098

Espach, A., & Gupta, K. (2020). Mechanical performance in fused deposition modeling manufactured parts – an additive manufacturing review. In *Proceedings of the International Conference on Industrial Engineering and Operations Management* (pp. 3201–3209). IEOM Society International. http://www.ieomsociety.org/detroit2020/papers/640.pdf

Frazier, C., Lamnaouer, M., Divo, E., Kassab, A., & Petersen, E. (2011). Effect of wall heat transfer on shock-tube test temperature at long times. *Shock Waves, 21*(1), 1–17. doi:10.100700193-010-0282-y

Fritz, A., Sekol, L., Koroskenyi, J., Walch, B., Minear, J., Fernandez, V., & Liu, L. (2012). Experimental analysis of thermal energy deburring process by design of experiment. In *Proceedings of the ASME 2012 International Mechanical Engineering Congress and Exposition* (Vol. 3, pp. 2035–2041). ASME. 10.1115/IMECE2012-88411

Galliková, J., Grenčík, J., Barta, D., & Barlok, M. (2020). FMECA analysis of thermal deburring machine EXTRUDE HONE TEM P-350. *Scientific Journals of the Maritime University of Szczecin, 63*, 9–16. doi:10.17402/434

Geen, H. C., & Rice, E. E. (1969). *Process for treating articles of manufacture to eliminate superfluous projections* (U.S. Patent No. 3 475 229). U.S. Patent and Trademark Office.

Gillespie, L. K. (1999). *Deburring and edge finishing handbook*. Society of Manufacturing Engineers.

Godovsky, Y. K. (2012). *Thermophysical Properties of Polymers*. Springer. doi:10.1007/978-3-642-51670-2

Gupta, K. (2020). A review on green machining techniques. *Procedia Manufacturing, 51*, 1730–1736. doi:10.1016/j.promfg.2020.10.241

Harper, C. A., & Petrie, E. M. (2003). *Plastics materials and processes*. John Wiley & Sons. doi:10.1002/0471459216

Huang, H., Sun, T., Zhang, G., Sun, L., & Zong, Z. (2018). Modeling and computation of turbulent slot jet impingement heat transfer using RANS method with special emphasis on the developed SST turbulence model. *International Journal of Heat and Mass Transfer, 126*, 589–602. doi:10.1016/j.ijheatmasstransfer.2018.05.121

Hurina, L., Vyshnepolskyi, Y., Pavlenko, D., & Stepanov, D. (2020). Investigation of the printing parameters influence on the bond lines length in fused filament fabrication. In *2020 IEEE 10th International Conference Nanomaterials: Applications & Properties (NAP) (pp. 02SAMA07-1)*. IEEE. 10.1109/NAP51477.2020.9309668

Izamshah, R., Azam, M. A., Hadzley, M., Ali, M. M., Kasim, M. S., & Aziz, M. A. (2013). Study of surface roughness on milling unfilled-polyetheretherketones engineering plastics. *Procedia Engineering, 68*, 654–660. doi:10.1016/j.proeng.2013.12.235

Jain, N. K., & Jain, V. K. (2003). Process selection methodology for advanced machining processes. *Journal of Advanced Manufacturing Systems, 2*(1), 5–45. doi:10.1142/S0219686703000204

Jin, Y., Wan, Y., Zhang, B., & Liu, Z. (2017). Modeling of the chemical finishing process for polylactic acid parts in fused deposition modeling and investigation of its tensile properties. *Journal of Materials Processing Technology*, *240*, 233–239. doi:10.1016/j.jmatprotec.2016.10.003

Kader, B. A. (1981). Temperature and concentration profiles in fully turbulent boundary layers. *International Journal of Heat and Mass Transfer*, *24*(9), 1541–1544. doi:10.1016/0017-9310(81)90220-9

Kagan, L. S., Gordon, P. V., & Sivashinsky, G. I. (2012). A minimal model for end-gas autoignition. *Combustion Theory and Modelling*, *16*(1), 1–12. doi:10.1080/13647830.2011.610947

Kantor, B. Y., Smetankina, N. V., & Shupikov, A. N. (2001). Analysis of non-stationary temperature fields in laminated strips and plates. *International Journal of Solids and Structures*, *38*(48-49), 8673–8684. doi:10.1016/S0020-7683(01)00099-3

Kelley, D. G., & Schwarz, K. H. (1991). Thermal Energy Deburring. Technical Paper MR91-136. Society of Manufacturing Engineers

Kim, Y., Hossain, A., & Nakamura, Y. (2013). Numerical study of melting of a phase change material (PCM) enhanced by deformation of a liquid–gas interface. *International Journal of Heat and Mass Transfer*, *63*, 101–112. doi:10.1016/j.ijheatmasstransfer.2013.03.052

Klocke, F., & König, W. (2007). Chemisches abtragen. In Fertigungsverfahren 3. VDI-Buch (pp. 127–131). Springer. doi:10.1007/978-3-540-48954-2_3

Kombarov, V., Sorokin, V., Tsegelnyk, Y., Plankovskyy, S., Aksonov, Y., & Fojtů, O. (2021). Numerical control of machining parts from aluminum alloys with sticking minimization. *International Journal of Mechatronics and Applied Mechanics*, *I*(9), 209–216. doi:10.17683/ijomam/issue9.30

Kondratiev, A., Melnikov, S., Nabokina, T., & Tsaritsynskyi, A. (2021). Effect of parameters of adhesive application by intaglio printing on honeycomb core bonding strength. In M. Nechyporuk, V. Pavlikov, & D. Kritskiy (Eds.), Integrated Computer Technologies in Mechanical Engineering – 2020 (pp. 221–233). Springer. doi:10.1007/978-3-030-66717-7_18

Kovalevsky, V. F., Skobelev, S. B., Bezzateeva, E. G., Bury, G. G., & Poteryaev, I. K. (2019). Mechanical action on edges of plastic parts using electrohydraulic effect. *Journal of Physics: Conference Series*, *1260*(11), 112016. doi:10.1088/1742-6596/1260/11/112016

Krainov, A. Y., & Moiseeva, K. M. (2018). Modeling of the combustion of a methane–air mixture in an enclosed spherical volume. *Journal of Engineering Physics and Thermophysics*, *91*(4), 918–924. doi:10.100710891-018-1817-9

Kulik, P. (2017). *Thermal deburring machine and method of thermal deburring* (WO patent 2019/054886 Al). European Patent Office.

Kumar, K., & Davim, J. P. (Eds.). (2018). *Composites and advanced materials for industrial applications*. IGI Global. doi:10.4018/978-1-5225-5216-1

Kumar, K., Zindani, D., & Davim, J. P. (2020). *Rapid prototyping, rapid tooling and reverse engineering: from biological models to 3D bioprinters*. De Gruyter. doi:10.1515/9783110664904

Lamikiz, A., Ukar, E., Tabernero, I., & Martinez, S. (2011). Thermal advanced machining processes. In J. P. Davim (Ed.), *Modern Machining Technology* (pp. 335–372). Woodhead Publishing. doi:10.1533/9780857094940.335

Malashenko, V. L. (2013). *Improving of deflashing technology of thermoplastics parts based on thermal impulse method* [Ph.D. dissertation]. National Aerospace University "Kharkiv Aviation Institute."

Mali, H. S., Prajwal, B., Gupta, D., & Kishan, J. (2018). Abrasive flow finishing of FDM printed parts using a sustainable media. *Rapid Prototyping Journal, 24*(3), 593–606. doi:10.1108/RPJ-10-2017-0199

Mark, J. E. (Ed.). (2007). *Physical Properties of Polymers Handbook.* Springer. doi:10.1007/978-0-387-69002-5

Matbase. (2021). *The free and independent online materials properties resource.* http://www.matbase.com

Menter, F. R. (1994). Two-equation eddy-viscosity turbulence models for engineering applications. *AIAA Journal, 32*(8), 1598–1605. doi:10.2514/3.12149

Muccio, E. A. (1991). *Plastic part technology.* ASM International.

Muccio, E. A. (1999). *Decoration and assembly of plastic parts.* ASM International.

Mustafa, L. M., Ismailov, M. B., & Sanin, A. F. (2020). Study on the effect of plasticizers and thermoplastics on the strength and toughness of epoxy resins. *Naukovyi Visnyk Natsionalnoho Hirnychoho Universytetu, 2020*(4), 63–68. doi:10.33271/nvngu/2020-4/063

Nickels, L. (2019). New innovations in automotive thermoplastics. *Reinforced Plastics, 63*(4), 185–188. doi:10.1016/j.repl.2019.06.041

Oztan, C., Karkkainen, R., Fittipaldi, M., Nygren, G., Roberson, L., Lane, M., & Celik, E. (2019). Microstructure and mechanical properties of three dimensional-printed continuous fiber composites. *Journal of Composite Materials, 53*(2), 271–280. doi:10.1177/0021998318781938

Palaziuk, I. S. (2018). *Method for operating conditions assignment of thermal pulse deburring of GTE parts made from heat-resistant alloys by the edge qualimetric parameters* [Ph.D. dissertation]. National Aerospace University "Kharkiv Aviation Institute."

Plankovskyy, S., Nikolaev, A., Shypul, O., Litvinchev, I., Pankratov, A., & Romanova, T. (2020). Balance layout problem with the optimized distances between objects. In P. Vasant, I. Litvinchev, J. A. Marmolejo-Saucedo, R. Rodriguez, & F. Martinez (Eds.), Data Analysis and Optimization for Engineering and Computing Problems (pp. 85–93). Springer. doi:10.1007/978-3-030-48149-0_7

Plankovskyy, S., Popov, V., Shypul, O., Tsegelnyk, Y., Tryfonov, O., & Brega, D. (2021a). Advanced thermal energy method for finishing precision parts. In K. Gupta & A. Pramanik (Eds.), *Advanced Machining and Finishing* (pp. 527–575). Elsevier. doi:10.1016/B978-0-12-817452-4.00014-2

Plankovskyy, S., Shypul, O., Tsegelnyk, Y., Pankratov, A., & Romanova, T. (2021b). Amplification of heat transfer by shock waves for Thermal Energy Method. In M. Nechyporuk, V. Pavlikov, & D. Kritskiy (Eds.), Integrated Computer Technologies in Mechanical Engineering – 2020 (pp. 577–587). Springer. doi:10.1007/978-3-030-66717-7_49

Plankovskyy, S., Shypul, O., Tsegelnyk, Y., Pankratov, A., Romanova, T., & Litvinchev, I. (2021c). Circular layout in thermal deburring. In S. Shkarlet, A. Morozov, & A. Palagin (Eds.), Mathematical Modeling and Simulation of Systems (MODS'2020) (pp. 111–120). Springer. doi:10.1007/978-3-030-58124-4_11

Plankovskyy, S., Shypul, O., Tsegelnyk, Y., Tryfonov, O., & Golovin, I. (2016). Simulation of surface heating for arbitrary shape's moving bodies/sources by using R-functions. *Acta Polytechnica*, *56*(6), 472–477. doi:10.14311/AP.2016.56.0472

Plankovskyy, S., Teodorczyk, A., Shypul, O., Tryfonov, O., & Brega, D. (2019). Determination of detonable gas mixture heat fluxes at thermal deburring. *Acta Polytechnica*, *59*(2), 162–169. doi:10.14311/AP.2019.59.0162

Plankovskyy, S. I., Shypul, O. V., Tryfonov, O. V., & Kozlov, V. G. (2010). The mixing in chamber of thermo-pulse system at cleaning of engine unit workpiece. *Aerospace Technic and Technology, 9*, 7–11. http://nbuv.gov.ua/UJRN/aktit_2010_9_3

Plankovskyy, S. I., Shypul, O. V., Tryfonov, O. V., Palazyuk, E. S., & Malashenko, V. L. (2014). The simulation of the heat transfer during shock waves damping in an enclosed chamber. *Aerospace Technic and Technology, 1*, 104–109. http://nbuv.gov.ua/UJRN/aktit_2014_1_18

Quintens, H., Michalski, Q., Moussou, J., Strozzi, C., Sotton, J., Bellenoue, M., Boust, B., Pilla, G., & Rabeau, F. (2019). Experimental wall heat transfer measurements for various combustion regimes: Deflagration, Autoignition and Detonation. In *AIAA Propulsion and Energy 2019 Forum* (p. 4381). 10.2514/6.2019-4381

Romanova, T., Pankratov, A., Litvinchev, I., Plankovskyy, S., Tsegelnyk, Y., & Shypul, O. (2021a). Sparsest packing of two-dimensional objects. *International Journal of Production Research*, *59*(13), 3900–3915. doi:10.1080/00207543.2020.1755471

Romanova, T., Stoyan, Y., Pankratov, A., Litvinchev, I., Plankovskyy, S., Tsegelnyk, Y., & Shypul, O. (2021b). Sparsest balanced packing of irregular 3D objects in a cylindrical container. *European Journal of Operational Research*, *291*(1), 84–100. doi:10.1016/j.ejor.2020.09.021

Schulz, D. (2017). Komplexe Kunststoffteile wirtschaftlich entgraten. *JOT Journal für Oberflächentechnik*, *57*(3), 48–49. doi:10.100735144-017-0263-3

Senachin, P. K., & Babkin, V. S. (1982). Self-ignition of gas in front of the flame front in a closed vessel. *Combustion, Explosion, and Shock Waves*, *18*(1), 1–5. doi:10.1007/BF00783921

Shypul, O., & Myntiuk, V. (2020). Transient thermoelastic analysis of a cylinder having a varied coefficient of thermal expansion. *Periodica Polytechnica Mechanical Engineering*, *64*(4), 273–278. doi:10.3311/PPme.14733

Singh, R., Singh, S., Singh, I. P., Fabbrocino, F., & Fraternali, F. (2017). Investigation for surface finish improvement of FDM parts by vapor smoothing process. *Composites. Part B, Engineering*, *111*, 228–234. doi:10.1016/j.compositesb.2016.11.062

Skoglund, P., & Fransson, Å. (1998). Thermophysical properties of polypentadecanolactone. *Polymer*, *39*(10), 1899–1906. doi:10.1016/S0032-3861(97)00473-4

Smetankina, N. V., Postnyi, O. V., Merkulova, A. I., & Merkulov, D. O. (2020). *Modeling of non-stationary temperature fields in multilayer shells with film heat sources. In 2020 IEEE KhPI Week on Advanced Technology (KhPIWeek) (pp. 242–246).* IEEE. doi:10.1109/KhPIWeek51551.2020.9250139

Spur, G., Uhlmann, E., & Elbing, F. (1999). Dry-ice blasting for cleaning: Process, optimization and application. *Wear, 233*, 402–411. doi:10.1016/S0043-1648(99)00204-5

Struckmann, J., & Kieser, A. (2020). *Thermal deburring*. ATL Anlagentechnik Luhden GmbH.

Swavely, D. S. (1991). Finishing and machining plastics. In M. L. Berins (Ed.), *SPI Plastics Engineering Handbook of the Society of the Plastics Industry, Inc* (pp. 657–692). Springer. doi:10.1007/978-1-4615-7604-4_23

Szczepanik, K., Ooi, A., Aye, L., & Rosengarten, G. (2004). *A numerical study of heat transfer from a cylinder in cross flow. In 15th Australasian Fluid Mechanics Conference (pp. 13–17).* https://www.aeromech.usyd.edu.au/15afmc/proceedings/papers/AFMC00212.pdf

Taufik, M., & Jain, P. K. (2020). Part surface quality improvement studies in fused deposition modelling process: A review. *Australian Journal of Mechanical Engineering*, 1–25. Advance online publication. doi:10.1080/14484846.2020.1723342

Travin, A., Shur, M., Strelets, M., & Spalart, P. (2000). Detached-eddy simulations past a circular cylinder. *Flow, Turbulence and Combustion, 63*(1), 293–313. doi:10.1023/A:1009901401183

Trhlíková, L., Zmeskal, O., Psencik, P., & Florian, P. (2016). Study of the thermal properties of filaments for 3D printing. *AIP Conference Proceedings, 1752*(1), 040027. doi:10.1063/1.4955258

Tryfonov, O. V. (2013). *The method of regime setting for the thermal pulse processing by detonable gaseous mixtures using integrated CAD/CAE-systems* [Ph.D. dissertation]. National Aerospace University "Kharkiv Aviation Institute."

Uhlmann, E., Kretzschmar, M., Elbing, F., & Mihotovic, V. (2010). Deburring with CO_2 snow blasting. In J. Aurich & D. Dornfeld (Eds.), *Burrs – Analysis, Control and Removal* (pp. 181–187). Springer. doi:10.1007/978-3-642-00568-8_19

Valino, A. D., Dizon, J. R. C., Espera, A. H. Jr, Chen, Q., Messman, J., & Advincula, R. C. (2019). Advances in 3D printing of thermoplastic polymer composites and nanocomposites. *Progress in Polymer Science, 98*, 101162. doi:10.1016/j.progpolymsci.2019.101162

Vambol, O., Kondratiev, A., Purhina, S., & Shevtsova, M. (2021). Determining the parameters for a 3D-printing process using the fused deposition modeling in order to manufacture an article with the required structural parameters. *Eastern-European Journal of Enterprise Technologies, 2*(1), 44–54. doi:10.15587/1729-4061.2021.227075

Vikas, A. Y., & Soni, S. K. (2017). Simulation of melting process of a phase change material (PCM) using ANSYS (Fluent). *International Research Journal of Engineering and Technology, 4*(5), 3289–3294.

Wickramasinghe, S., Do, T., & Tran, P. (2020). FDM-based 3D printing of polymer and associated composite: A review on mechanical properties, defects and treatments. *Polymers, 12*(7), 1529. doi:10.3390/polym12071529 PMID:32664374

Wohlers, T. (2019). *Wohlers report 2019: 3D printing and additive manufacturing state of the industry.* Wohlers Associates.

Woźniak, K. (2013). Plastic blasting and deflashing media in shot blasting treatment. *Chemik, 67*(12), 1227–1238.

Xu, J., Huang, X., Davim, J. P., Ji, M., & Chen, M. (2020). On the machining behavior of carbon fiber reinforced polyimide and PEEK thermoplastic composites. *Polymer Composites, 41*(9), 3649–3663. doi:10.1002/pc.25663

Zhu, J., & Chen, J. C. (2006). Fuzzy neural network-based in-process mixed material-caused flash prediction (FNN-IPMFP) in injection molding operations. *International Journal of Advanced Manufacturing Technology, 29*(3-4), 308–316. doi:10.100700170-005-2528-x

Zindani, D., & Kumar, K. (2019). An insight into additive manufacturing of fiber reinforced polymer composite. *International Journal of Lightweight Materials and Manufacture, 2*(4), 267–278. doi:10.1016/j.ijlmm.2019.08.004

Zindani, D., & Kumar, K. (2020). A brief review on cryogenics in machining process. *SN Applied Sciences, 2*(6), 1107. doi:10.100742452-020-2899-5

Section 2
Processing and Characterization

Chapter 4
Synthesis and Characterization of Lightweight Beryllium Chloro Silicate Phosphor

Khushbu Sharma
Shri Ramdeo Baba College of Engineering and Management, Nagpur, India

ABSTRACT

In this chapter, low weight barium-based cholorsilicate Ba5Cl6Si2O6:Eu2+ is prepared through a solid-state reaction. To confirm the structure of the synthesized phosphors, powder photographs were obtained using an x-ray diffractometer. Photoluminescence spectra and FTIR spectra were recorded. Photoluminescence spectra are studied. The emission peak is observed at 407 nm at excitation 275 nm. The intense violet-blue emission is obtained. The broad excitation band and strong emission indicate that Ba5Cl6Si2O6:Eu2+could be a good phosphor candidate for blue LED and white LEDs. Decay curve indicates the phosphor has a long afterglow feature.

INTRODUCTION

Various kinds of phosphors are already commonly used in many lighting devices, including common fluorescent tubes, electroluminescent wires, strips and surfaces, and LEDs. In a ubiquitous fluorescent tube, the inside is coated with phosphors of light weight .Electricity excites the gas-filled tube to produce shortwave light, which in turn causes the phosphors to become fluorescent and produce visible light. White LEDs, with their characteristics of compact size, high efficiency, long lifetime, low power requirement, light weight and energy savings (Zhu, 2012) can be widely used in various applications such as liquid crystal display backlighting, full-color displays, cell phones, and traffic signals. White LEDs are expected to replace conventional incandescent and fluorescent lamps in the near future (Du,2009). Eu2+-activated silicate phosphors, which have broad emission band through 5d–4f energy transition of Eu2+ activator ion, are suitable for the application of white LEDs. Eu2+ doped chlorosilicates give intense emission and which may be useful for many applications.

DOI: 10.4018/978-1-7998-7864-3.ch004

Silicate materials are useful in many applications of technological importance. Zeolites, a type of microporous aluminosilicates, are widely used as molecular sieves and catalysts. The zeolites have received less attention as potential luminescence materials. However, increasing interest in the study of guest-host composite materials has heightened research in these well-defined periodic microporous structures that are almost UV transparent and inexpensive. The lithium ceramics are promising breeder materials for fusion reactors. Rare-earth-containing silicate glasses have attracted great attention as potential materials for optical and magneto-optical devices such as an upconversion laser, a hole burning memory, and an optical switch. Though silicate-based phosphors are widely used as described in 1.6, there are some problems with synthesis and long-term use of silicates. Silicates need temperatures higher than 1000 C, even up to 1400 C for synthesis. They may get converted to glassy form. At high synthesis temperatures they can also react with low-cost crucible materials such as porcelain and china clay. During recent years, scientists have turned their attention towards chlorosilicates to get over these problems. In silicate lattices, an introduction of chloride ion can induce a red shift of excitation and emission bands of Eu2+ and Ce3+ ions because Cl ions with strong coordination effect can strengthen the crystal field splitting. Moreover, Chlorosilicates can be easily prepared by solid state reactions at temperatures below 1000 C, often as low as 700 C. They have good chemical, physical and thermal stability. Availability of cation sites with varying coordination and symmetries results in tunability of activator emission and excitation spectra. Variety of Chlorosilicates are known in chemistry and mineralogy.

Luminescence that persists after the removal of the excitation is called afterglow or persistent phosphorescence. Long-lasting phosphorescence is a phenomenon due to the thermal stimulated recombination of holes and electrons at traps which leave holes or electrons in a long-lived excited state at room temperature (Kuang, 2006). The first record of persistent phosphorescent material is in the Song dynasty of China (11th century A.D.). In the miscellaneous notes by a Song monk, of which the title is Xiang-Shan Ye-Lu, there is a story about a long-lasting phosphorescent painting. On the painting was a cow that appeared during the daytime as eating grass outside the pen, but at night as resting in it. The ink that was shown in the dark for a given duration after absorption of light is long lasting phosphorescent material. As novel functional materials, the long afterglow phosphors are drawing more and more attention in recent years because of their applications in traffic signs, emergency signage, watches and clocks, textile printing etc. As a new generation of long afterglow phosphors, lanthanide ion doped alkaline earth silicates or aluminates yield much better characteristics, such as longer duration time of the phosphorescence, brighter luminosity and improved chemical stability, than the conventional sulfide materials used earlier (Chen, 2006). Among the newly developed long afterglow materials, which have already found commercial use, akermanite structure based alkaline earth silicates R2MgSi2O7 (R = Ca, Sr, Ba) codoped with Eu2+ and Dy3+ are of special interest because of their excellent persistent luminescence combined with an easy process ability (Murayama, 1996). CaMgSi2O6: Eu, Dy, Nd and CaMgSi2O7phosphors activated by Eu2+, Dy3+ and Nd3+ with afterglow characteristics were prepared by Jiang et al. through solid-state reaction in a reducing atmosphere.

In this chapter light weight barium based chlorosilicates Ba5Cl6Si2O6:Eu2+ Phosphor is prepared. The existence of three phases $Ba_5SiO_4Cl_6$, $Ba_7Si_2O_7Cl_8$ and $Ba_5Si_2O_6Cl_6$ in the system BaO-SiO_2-$BaCl_2$ was established by Garcia et al. Winkler et al confirmed the crystal structures reported by Garcia et al., for these compounds. Later Garcia et al reported X-ray excited luminescence for $Ba_5SiO_4Cl_6:Eu^{2+}$ also. Tecotzky et al developed X-ray storage phosphors based on Eu^{2+} activated $Ba_5SiO_4Cl_6$ and $Ba_5Si_2O_6Cl_6$, Fan obtained green emission in $Ba_7Si_2O_7Cl_8:Eu^{2+}$ and suggested its use as a green emitting phosphor for solid state lighting (Fan, 2014). VUV excitation also led to similar emission which could be relevant for PDP applications (Han, 2009). Efficiency of $Ba_7Si_2O_7Cl_8$:Eu^{2+} under VUV excitation was found to be 50% of that of Zn_2SiO_4: Mn. Luminescence properties of $Ba_5SiO_4Cl_6$ activated by Eu^{2+} were first investigated by Garcia et al. Later Meijerink and Blasse investigated luminescence along with temperature dependent characteristics decay time in Eu^{2+} doped $Ba_5SiO_4Cl_6$. Abed and Buschbaum and Yang and Zhang reported the details of crystal structure and Barium coordination in $Ba_5SiO_4Cl_6$. Zeng et al observed blue emission under VUV and NUV excitation and proposed a good candidate for display and light-emitting diode devices. Recently Lai et al prepared Bi^{3+} doped $Ba_5SiO_4Cl_6$ and Zhi et al investigated-up conversion mechanism in luminescence of $Ba_5SiO_4Cl_6$ by doping different activators Yb^{3+}, Er^{3+}, Li^{+}.

SYNTHESIS PROCESS

Solid-state reaction is one of the most widely used methods for the synthesis of phosphors. Solid-solid reactions are simple to perform; starting materials are often readily available at low cost. The constituents are made to react through diffusion process. The temperature is just enough to have adequate diffusion to complete the reaction in laboratory time without melting the constituents. Reaction time and the temperature bear a sort of reciprocal relation. It may not always be possible to lower the latter sufficiently.

For the synthesis of silicates, the silicic acid was used as the silica source. Metal chlorides/carbonates/oxides are used as cationic salts. The activators are introduced in the form of the appropriate salts in the desired proportions. All constituents in the required proportions were mixed together. The mixture on thoroughly grinding was transferred to furnace for heating at desired temperature.

$Ba_5SiO_4Cl_6$ prepared through solid state reactions following recipes given in the literature. Ingredients used were $BaCO_3$, $BaCl_2$, and silicic acid (SiO_2.1.5 H_2O). For activation with Eu^{2+}, Eu_2O_3 was dissolved in dil. HCl to convert into chloride which was then added in the desired quantity to the ingredients. All ingredients $BaCO_3$, $BaCl_2$, silicic acid in required proportion 2:3:1 was mixed together. The mixture on thoroughly grinding was transferred to furnace for heating.

In the as prepared compounds europium gets incorporated in trivalent form. Useful luminescence in the studied compound, on the other hand, comes from Eu^{2+}. For reducing europium to divalent state, a simple but effective method was used. Prepared compounds were taken in covered silica crucible and kept in a stainless-steel container. The space between the crucible and the stainless-steel container was filled with activated charcoal. Tightly packed container was kept in furnace, typically at 900 C for 2 hrs. At the end of 2 hours, the container was taken out from the furnace and allowed to cool to room temperature. The silica crucible was then carefully removed and the reduced phosphor $Ba_5SiO_4Cl_6$:Eu^{2+} was used for further measurements

To confirm the structure of the synthesized phosphors, powder photographs were obtained using Rigaku Miniflex II X-ray Diffractometer. Photoluminescence spectra were recorded on Hitachi F-7200 spectro-fluorimeter with spectral slit width of 1 nm in the range 220-700 nm.

PROPERTIES OF BA5SIO4CL6

X ray diffraction pattern of $Ba_5SiO_4Cl_6$ have recorded by using Rigaku Miniflex II X-ray Diffractometer. The Rigaku Miniflex II (Fig 1) can be used for a variety of applications starting from diffraction pattern comparison of polycrystalline materials such as powder samples and metal plates to qualitative and quantitative analysis and quality management of raw materials and products. This Miniflex system also offers use of a real-time angle correction system, a single sample holder, an automated six sample changer (used in standard operation), as well as a monochromator which captures X-rays other than Cu K_α for use in analysis.

Figure 1. Rigaku Miniflex II X-Ray diffractometer

X ray diffraction pattern of $Ba_5SiO_4Cl_6$ is presented in Fig.2. The pattern is matched with ICDD data file 34-1410. A good match is seen. $Ba_5SiO_4Cl_6$ crystallizes in the monoclinic system: a = 9.45 A, b = 14.75 A, c = 11.69 A, and B = 104.39 with space group is C2/c.

Figure 2. XRD pattern of $Ba_5SiO_4Cl_6$ compared with ICDD data

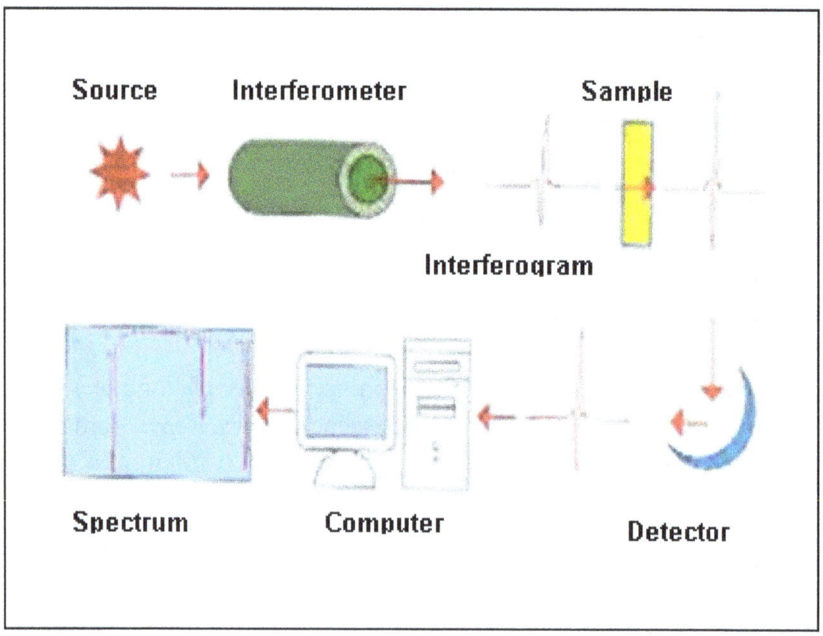

Fourier transform infrared (FTIR) spectrometry is a useful tool for identifying both organic and inorganic chemicals. It can be utilized to quantify some components of an unknown mixture and can be used to analyze liquids, solids, or gases. Data are collected and converted from an interference spectrum. Spectroscopy seeks to identify chemical properties by viewing the spectrum of emissions from a sample. The wavelength of light absorbed is characteristic of the chemical bonds in the substance. By interpreting the infrared absorption spectrum, the chemical bonds in a molecule can be determined. The spectrum of an unknown can be identified by comparing it to a known reference compound.

Figure 3. Block diagram of FTIR spectrometer

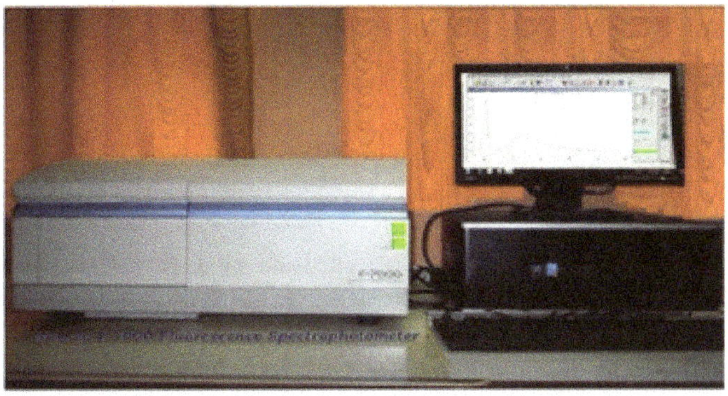

Silicates can also be characterized using FTIR. FTIR spectroscopy allows differentiation of the various types of bonds in the material to be analyzed at the molecular level. Characteristic bands for silicate systems can be observed in the FTIR spectra in the wavenumber region between 1200 and 400 cm^{-1}. FTIR spectrum for $Ba_5SiO_4Cl_6$ is shown in Fig.4. IR bands typical of SiO_4^{-4} tetrahedra can be seen in the region 760 – 1000 cm^{-1}. The strong band around 856 cm^{-1} corresponds to symmetric Si-O stretching

Figure 4. FTIR spectra for $Ba_5SiO_4Cl_6$

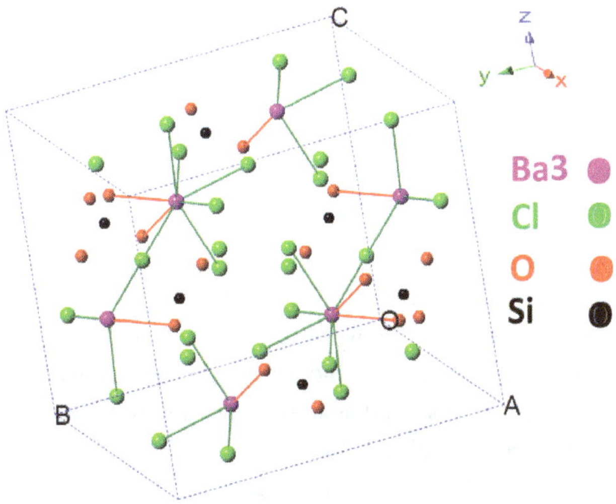

Figure 5. SEM for $Ba_5SiO_4Cl_6$

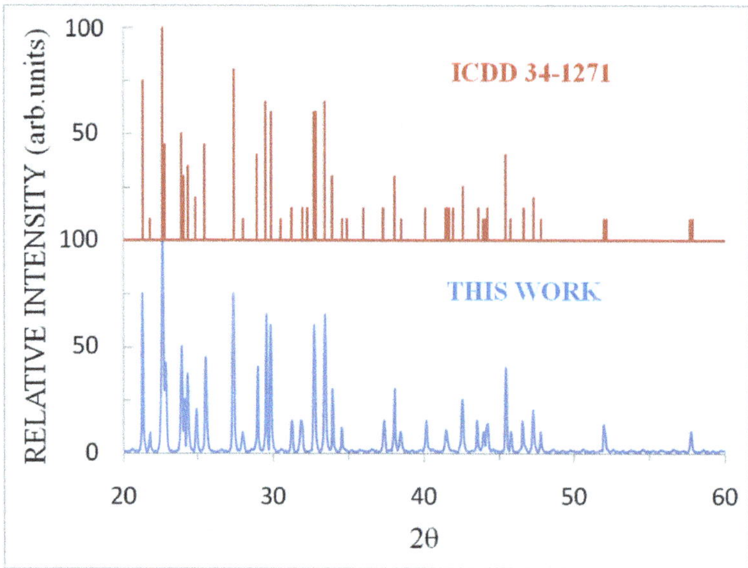

vibration mode (Sohn & Min, 2012; Zhang et al., 2009)and bands around 691 and 554 cm^{-1} are due to Si-O bending.

Figure 5 shows electron micrograph for $Ba_5SiO_4Cl_6$. Particles are rather of irregular shape. Size ranges from 5 to 50 microns.

The photoluminescence spectra (excitation and emission) were recorded on Hitachi F-7200 fluorescence spectrophotometer

The photoluminescence spectra were recorded on a Hitachi F-7200 fluorescence spectrophotometer. The spectrophotometer consists of two monochromators (one on the excitation and other on the emission side), a light source, two detectors (one for measurements and the other for monitoring), a sample holder, a data processor and graphic plotter.

The following modes are available in the Hitachi F-7200.

- Wavelength Scan
- Time Scan
- Quantitative Calculation
- Three-dimensional measurement
- Three-dimensional time scan measurement

Photoluminescence spectra were recorded with spectral slit width of 1 nm in the range 220-700 nm. The emission and excitation spectra are shown in Fig. 7. Various Eu^{2+} concentrations ranging between 0.1-2.0 mol % were tried. Maximum emission was obtained for 2 mol.%. This is consistent with earlier reports. The excitation peak is at 318 nm with shoulders around 278 and 337 nm along with various overlapping bands. Emission peak is observed at 441nm and another peak of lesser intensity at longer wavelength at 500 nm. Earlier, Garcia et al and Meijerink and Blasse observed only one emission band around 434 nm. Zeng et al also observed an emission band around 440 nm for VUV excitation, though

Figure 6. Hitachi F-7000 spectrophotometer

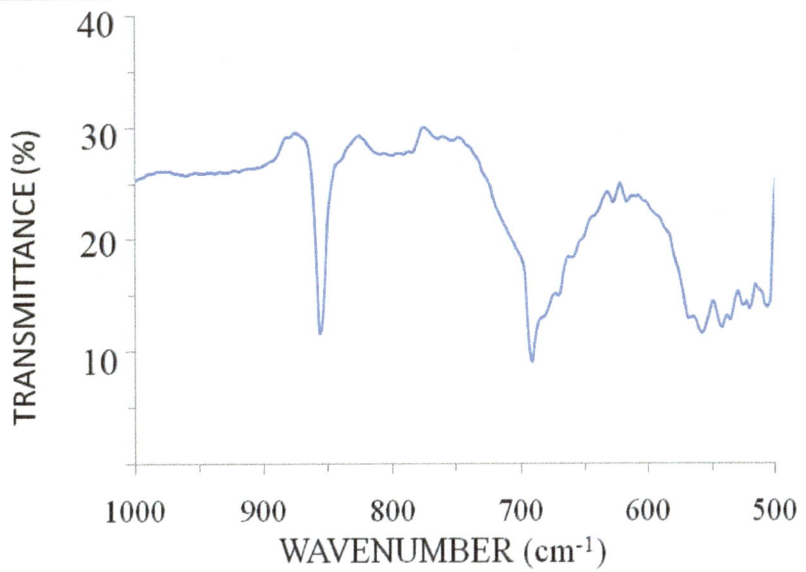

their excitation spectrum in near UV region differs from those reported in earlier works. A reason for these differences may be sought in availability of various sites for Eu^{2+}.

The ground-state electronic configuration of Eu^{2+} is $4f^7$. This results in an $^8S_{7/2}$ level for the ground state. Eu^{2+} emission arises from the lowest band of $4f^65d^1$ configuration to $^8S_{7/2}$ state of $4f^7$ configuration. The excitation arises from the transition from $^8S_{7/2}$ state of $4f^7$ configuration to the states belonging to $4f^65d^1$ configuration. The next f^7 manifold (^6P_J) lies approximately 28,000 cm^{-1} higher. The lowest lying $4f^65d$ levels begin near 34,000 cm^{-1} and are labeled ^8H_J for the free ion. The $4f^65d$ levels experience much more crystal field splitting than the $4f^7$ levels due to the increased spatial extent of the 5d orbitals and often are the metastable state, or the lowest excited state, when the Eu^{2+} ion is incorporated in a crystalline host. For cubic symmetry the effect of the crystal field on the 5d electron is to split the 5d orbitals into two components t_{2g} and e_g. For lower symmetry, the splitting can be as much as five-fold. The isotropic part of the exchange interaction between 5d and 4f electrons results in an exchange splitting into states with total spins of S=7/2 and 5/2. Thus, for the absorption spectra of Eu^{2+} in the solids, the lowest energy band arises from the state described by the notation I $4f^6(^7F_J)$ eg, S=7/2 > (Lawson & Payne, 1993) The lowest energy configuration corresponds to the situation where $^7F_J(4f^6)$ state couples to the 5d e_g orbital such that all spins are parallel. Spectral positions of these bands vary a great deal from lattice to lattice (Dorenbos, 2003). The most commonly observed emission is the dipole and spin allowed d–f-emission starting from the relaxed $4f^6(^7F_0)$ $5d^1$ level. Due to allowed nature of the transition, d–f emission is intense. The wavelength positions of the emission bands depend very much on hosts, changing from the near UV to red for example 365 nm in $BaSO_4$ to 650 nm in CaS (Blasse, 1978). This dependence is interpreted as due to the crystal field splitting of the 5d level as shown schematically in Fig. 1.7. With increasing crystal field strength, the emission band shifts to longer wavelength. The luminescence peak energy of the 5d-4f transitions of Eu^{2+} are affected most by crystal parameters indicating electron-electron repulsion. When $4f^65d$ band is higher than ^6P_j states and the f-f transitions become parity allowed, sharp lines

Figure 7. Photoluminescence spectra of $Ba_5SiO_4Cl_6$:Eu^{2+}- a) Emission for 318 nm excitation, b) Excitation for 441 nm emission

corresponding to ^6P_j ® $^8S_{7/2}$ transitions are observed. This can happen when the crystal field is weak. The sharp lines are usually observed at 360 nm (Hoffman, 1971; Hoffman, 1972) These transitions have longer decay time; of the order of several ms. In many compounds d-f emission is observed at room temperature, while at low enough temperatures sharp line f-f emission becomes dominant.

In $Ba_5SiO_4Cl_6$,there are 3 non-equivalent Barium ions 8, 9and 9. Ba(I) is nine coordinated Fig.8, having 7 Cl^- neighbours at the distances ranging from 3.171 to 3.489 Å, and 2 O^{-2} neighbours at 2.584 and 2.674 Å. Ba(II) is also nine coordinated (Fig.9), having 6 Cl^- neighbours at the distances ranging from 3.099 to 3.793 Å, and 3 O^{-2} neighbours at 2.584, 2.695 and 2.913 Å. Ba(III) is eight coordinated (Fig.10), having 6 Cl^- neighbours at the distances ranging from 3.022 to 3.350 Å, and 2 O^{-2} neighbours at 2.645 Å. Two emission bands we have observed may correspond to Eu^{2+} ions at different sites. Estimated values for Eu^{2+} emissions are 439 nm for 9-coordinated Ba1 and Ba2 sites, while 479 nm for 8-cordinated Ba3 site. In fact, co-doping with F^- resulted in populating the Ba3 site and the emission was observed at 500 nm. Thus the 500 nm band we have observed may be attributed to Eu^{2+} ions at Ba3 site.

Figure 8. Ba1 coordination in $Ba_5SiO_4Cl_6$

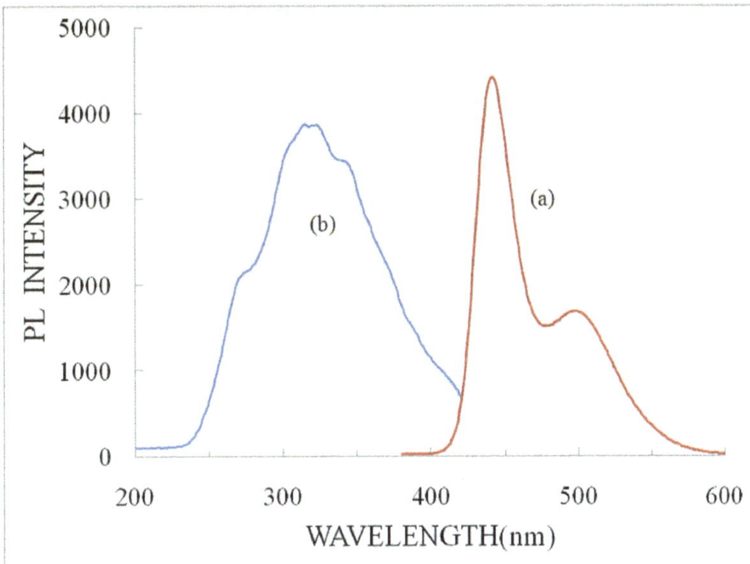

Figure 9. Ba2 coordination in $Ba_5SiO_4Cl_6$

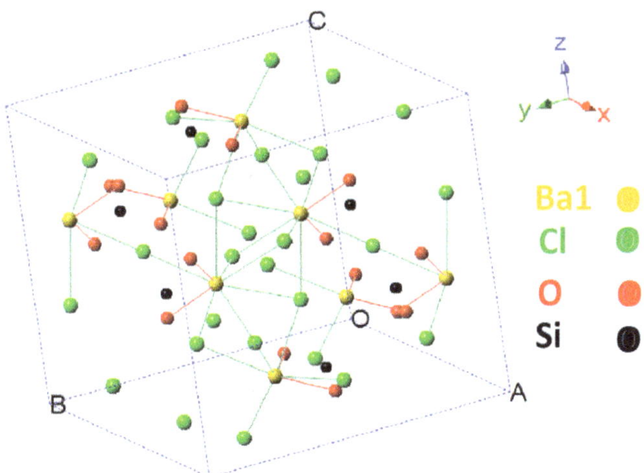

SUMMARY

-Ba5SiO4Cl6 is a new class of low weight chlorosilicates phosphor, which is easily prepared by solid state reaction and thermally stable, having higher luminescence properties and having light weight. Photoluminescence characteristic of Eu^{2+} emission was observed optimum in this host at 2mol%. Two emission band observed in pl emission may correspond to Eu^{2+} ions at different site. The emission and excitation spectra and make good agreement for use of material for long lasting phosphors The potential application areas of barium chlorosilicates phosphor for white light L are in packaging, coatings, and so on. Processing methods such as twin-screw extrusion and also for road sign applications.

Figure 10. Ba3 coordination in $Ba_5SiO_4Cl_6$

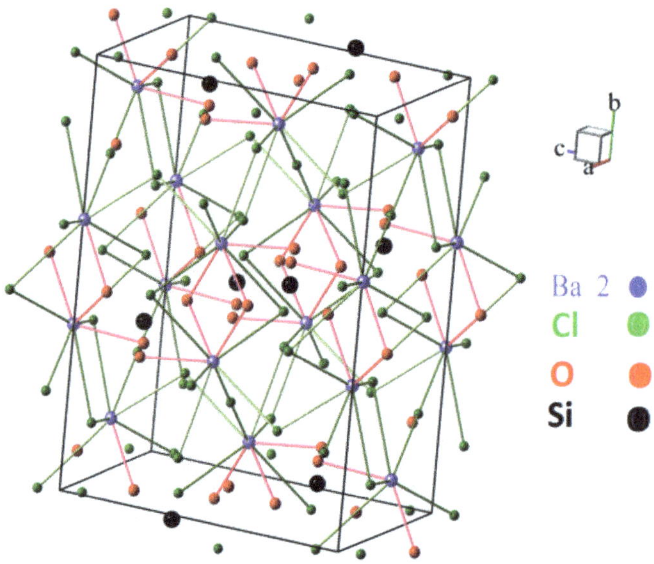

REFERENCES

Abed, M., & Mtiller-Buschbaum, H. (1992). Zur Kristallstruktur von Ba5SiO4Cl6. *Journal of Alloys and Compounds*, *190*(1), 61–66. doi:10.1016/0925-8388(92)90174-8

Blasse. (1978). Luminescence of Inorganic Solids. *Luminescence of inorganic solids: From isolated centres to concentrated systems*, 457-453.

Bo, Y. H., & Yoo, J. S. (2009). Korean. *Chemical Engineering Research & Design*, *47*, 150–156.

Chen, Y., Liu, B., Kirm, M., Qib, Z., Shi, C., True, M., Vielhauer, S., & Zimmerer, G. (2006). Luminescent properties of blue-emitting long afterglow phosphors Sr2xCax MgSi2O7:Eu2+, Dy3+ (x =0, 1). *Journal of Luminescence*, *118*(1), 70–78. doi:10.1016/j.jlumin.2005.05.011

Dorenbos, P. (2003). Energy of the first 4f7-4f65d transition of Eu2+ in inorganic compounds. *Journal of Luminescence*, *104*, 239–244.

Du, Sun, Xia, & Sun. (2009). High numerical aperture microlens arrays of close packing. *Applied Physics Letter B*, *96*, 459–463.

Garcia, Latourrette, & Fouassier. (1979). Ba5SiO4Cl6: Eu, A New Blue-Emitting Photoluminescent Material with High Quenching Temperature. *Journal of Electrochemical Society*, *126*, 1734-1741.

Garcia, C. F., & Hagenmuller, P. (1980). Charge transfer excitation of the Nd^{3+}, Sm^{3+}, Dy^{3+}, Ho^{3+}, Er^{3+} and Tm^{3+} emission in $CaGa_2S_4$ C. R. *Academy Science*, *290*, 33–36.

Hoffman, M. V. (1971). Alkaline Earth Aluminum Fluoride Compounds with Eu+ 2 Activation. *Journal of the Electrochemical Society*, *118*, 933–939.

Hoffman, M. V. (1972). Eu+ 2 emission in ternary alkaline earth aluminum fluorides. *Journal of the Electrochemical Society*, *119*, 905–911.

Jiang, L., Chang, C., & Mao, D. (2003). Luminescent properties of CaMgSi2O6 and Ca2MgSi2O7 phosphors activated by Eu2+, Dy3+ and Nd3+. *Journal of Alloys and Compounds*, *360*(1-2), 19–24. doi:10.1016/S0925-8388(03)00361-X

Kuang, J., Liu, Y., & Lei, B. (2006). Effect of RE3+ as a co-dopant in long-lasting phosphorescence CdSiO3:Mn2+ (RE =Y, La, Gd, Lu). *Journal of Luminescence*, *118*(1), 33–38. doi:10.1016/j.jlumin.2005.06.005

Lai, S., Yang, Z., Wang, R., Wu, H., Liao, J., Qiu, J., Song, Z., Yang, Y., & Zhou, D. (2013). Preparation and blue–white luminescence properties of Bi^{3+}-doped $Ba_5SiO_4Cl_6$. *Journal of Materials Science*, *48*, 8566–8571.

Lawson, J. K., & Payne, S. A. (1993). Excited-state absorption of Eu2+ doped materials. *Physical Review. B*, *47*, 14003.

Meijerink, A., & Blasse, G. (1990). Luminescence and temperature dependent decay behaviour of divalent europium in Ba5SiO4X6 (X = Cl, Br). *Journal of Luminescence*, *47*(1-2), 471–475. doi:10.1016/0022-2313(90)90052-D

Murayama, Y. (1996). Nikkei Science. *Scientific American in Japanese*, *26*, 20.

Park, J. H., Min, D. J., & Song, H. S. (2004). Amphoteric behavior of alumina in viscous flow and structure of CaO-SiO_2 (-MgO)-Al_2O_3 slags. *Metallurgical and Materials Transactions. B, Process Metallurgy and Materials Processing Science*, *35B*, 269.

Sohn, I., & Min, D. J. (2012). A review of the relationship between viscosity and the structure of calcium-silicate-based slags in ironmaking. *Steel Research International*, *83*, 611–617.

Tecotzky, M., Blasse, G., & Meijerink, A. (1992). *US Patent 5,140,163*. US Patent Office.

Von, Winkler, Ziemer, & Wieker. (1983). Article. *Journal of Inorganic and General Chemistry*, *604*, 89-92.

Yang & Zhang. (2003). The relationship between chemical bond properties and Stokes shift of Eu2+ in some silicate host lattices. *Journal of Physics and Chemistry Solids*, *64*, 213-217.

Yang, J.-Z., Qiu, J.-B., Yang, Z.-W., Song, Z.-G., Yong, Y., & Zhou, D.-C. (2015). Preparation and up conversion luminescence properties of Ba5SiO4Cl6: Yb (3+), Er (3+), Li+ phosphors. *Wuli Xuebao*, *64*, 138101.

Zeng, Q., Tanno, H., Egoshi, K., Tanamachi, N., & Zhang, S. (2006). Ba5SiO4Cl6:Eu2+ Ba5SiO4Cl6:Eu2+: An intense blue emission phosphor under vacuum ultraviolet and near-ultraviolet excitation. *Applied Physics Letters*, *88*, 051906.

Zhang, X., Wang, X., Huang, J., Shi, J., & Gong, M. (2009). Near UV-based LED fabricated with Ba5SiO4(F, Cl)6: Eu2+ as blue- and green-emitting phosphor. *Optical Materials*, *32*, 75–78.

Zhu, G., Wang, Y., Ci, Z., Liu, B., Shi, Y., & Xin, S. (2012). Ca8Mg(SiO4)4Cl2:Ce3+, Tb3+: A potential single-phased phosphor for white-light-emitting diodes. *Journal of Luminescence*, *132*(2), 531–536. doi:10.1016/j.jlumin.2011.09.029

Chapter 5
Processing, Properties, and Uses of Lightweight Glass Fiber/Aluminum Hybrid Structures

Noureddine Ramdani
Ecole Militaire Polytechnique, Chahid Abderrahmane Taleb, Algiers, Algeria

Mohammed Seddik Razali
Ecole Militaire Polytechnique, Chahid Abderrahmane Taleb, Algiers, Algeria

ABSTRACT

The replacement of heavy metallic structures by high-performance lightweight composite materials is a prominent solution to fulfill the continuous demand in different industrial sectors. Lightweight structures based on aluminum-glass fiber reinforced plastics (GFRP) sandwich panels have been increasingly utilized in the shipbuilding, automotive, and aerospace industries for their striking mechanical and physical properties. These advantageous properties have resulted from the combination of the high tensile and flexural strengths, increased hardness, and the improved wear-resistance of aluminum laminate with the unique properties of lightweight stiffness and high strength weight ratio of glass fiber-reinforced. In this chapter, the various processing approaches, properties, and applications of these sandwich structures are summarized from a wide range of literature.

INTRODUCTION

During the last few decades, a strong need in the aircraft industry for high-performance, lightweight materials has motivated an increasing trend towards the progress of improved types for fiber-metal laminates (FMLs). These materials are exhibiting several advantageous structural, physical, and mechanical characteristics as given in Table 1. These FMLs are hybrid composite materials constituted

DOI: 10.4018/978-1-7998-7864-3.ch005

from interweaving thin metallic layers and fiber-reinforced thermoplastic or thermoset-based polymer composites (FRPC) adhesives.

Table 1. Advantageous features of FML.

Aspect	Characteristics
Material behaviour	• High strength • High fracture toughness • High fatigue and impact resistance • High energy absorbing capacity
Physical properties	Low density
Durability	• Excellent moisture resistance • Good corrosion resistance • Lower material degradation
Safety	Fire resistance

Due to their outstanding mechanical and physical properties, engineering fibers such as carbon, glass, and Kevlar are widely used to reinforce metallic laminates like aluminum and titanium to produce high-performance lightweight hybrid structures. Table 2 compares the advantages and shortcoming characteristics of these fibers. The market of these sandwich composite contains several commercially-models of FMLs such as the CARALL (Carbon Reinforced Aluminium Laminate), based on carbon fibres, the ARALL (Aramid Reinforced Aluminium Laminate), containing aramid fibres, and the GLARE (Glass Reinforced Aluminium Laminate), containing stronger glass fibres. By combining the striking properties of the hybrid feature from their two main components: metals (usually aluminium) and fiber-reinforced polymer laminate, these materials demonstrated many outstanding performances including lightweight, superior mechanical tensile and bending properties, a good damage tolerance to fatigue crack development, and impact damage, especially for aeronautic part productions (Sinmazçelik et al., 2011).

Aluminium metal matrix composites were extensively produced and developed in the last few years to fit the requirement of high engineering industries such as aerospace, military, marine, and automotive. Aluminum matrix layers and fiber-reinforced polymer laminate can be assembled using various traditional processes, like mechanically and adhesively. Adhesively joined fiber aluminum laminates have been revealed to be exhibit much-improved fatigue-resistance than those mechanically joined hybrid structures (Baburaja, Venkata Subbaiah, and Kalluri, 2016). Due to the fact that machining aluminum hybrid composites to a preferred geometrical form via some optimum machine process parameters affords increased material removal rates, a small surface roughness, and the best tool wear (Garg et al., 2019.) Thus, the selection of the optimal production methods could be of great importance. Some simulation works using finite elemental analysis and various software could be critical before producing such hybrid composites to ensure the desired performances (Ananda Rao, Reddy, and Seshaiah, 2014; Kumar, Dirgantara, and Krishna, 2020). Recently, the design of automotive lightweight aluminum/ FRPC hybrid structures along with their mechanical behaviour and most relevant production techniques were presented (Kumar, Dirgantara, and Krishna, 2020). For example, the GLARE hybrid composites are now extensively applied and more progress has been achieved for providing high strength, lightweight, strong stiffness, and the recommended mechanical performances. These hybrids are developed in six main grades by using the various arrangement of glass fiber prepreg, types of aluminum matrix

Table 2. Comparison between glass, carbon and aramid fiber.

Type	Uses	Outstanding	Disadvantages
Glass fiber	Majorly utilized products of thermosets and thermoplastics. Often used in the marina area, for boats, vehicles and construction industry.	• Low price • Electrical insulating • High tensile strength • Easy to laminate	- Low stiffness - High density - Stinging grinding dust
Carbon fibre	Applications that require high stiffness and strength relative to the weight. Aviation and aerospace industry, military industry and sports products.	• High stiffness • High tensile strength • High fatigue resistance • Attractive appearance	- Expensive - Poor impact resistance - Brittle, low yield strength - Electrically conductive, form galvanic element in contact with metals
Aramid fibre	Applications that require good impact resistance and tensile strength. In ballistic laminates and reinforcements. Military industry, aviation and aerospace industry.	• Good impact resistance • High tear strength • Low density	-Expensive - Poor UV -resistance. Darken and decompose when exposed to sunlight - Low compression and bending strength due to limited wetting

laminate, and thicknesses as listed in Table 3. The advantages and shortcomings of these materials have been discussed in detail (Senthil, Raguraman, and Manalan, 2020).

Table 3. Standard grades of GLARE (Vlot, A. and Gunnink, J.W., 2001)

Glare grade	Subtitution	Aluminum sheet thickness (mm)	Aluminum alloy Type	Prepreg arrangement	Main beneficial features
Glare 1		0.3-04	7475-T761	0/0	Fatigue, strength, yield stress
Glare 2	Glare 2A	0.2-0.5	2024-T3	0/0	Fatigue, strength
Glare 2	Glare 2B	0.2-0.5	2024-T3	90/90	Fatigue, strength
Glare 3		0.2-0.5	2024-T3	0/90	Fatigue, impact
Glare 4	Glare 4A	0.2-0.5	2024-T3	0/90/0	Fatigue, strength in 0 direction
Glare 4	Glare 4B	0.2-0.5	2024-T3	90/0/90	Fatigue, strength in 90 direction
Glare 5		0.2-0.5	2024-T3	0/90/90/0	Impact
Glare 6	Glare 6A	0.2-0.5	2024-T3	+45/-45	Shear, off-axe properties
Glare 6	Glare 6B	0.2-0.5	2024-T3	-45/+45	Shear, off-axe properties

The pretreatments of aluminum metals can be even achieved by solvent wipes or complex chemical processes. Various material types such as metals, rigid ceramics, polymers, rubbers, etc., are also subjected to their own specific pretreatments. However, some surface modifications are more effective with dissimilar kinds of materials, such as silane. These molecules can importantly increase the performance of joints connecting either metals, or ceramics. With regard to the aluminum/GFRP/PP hybrid composites, the interphase can be controlled using a silane coupling agent, a bonding agent or an extra agent (Hamada, et al. 2000). Due to galvanic corrosion and unstable interface in CARALL which provokes delamination and inferior static strength, the glass fiber is preferable over carbon fiber. On the other

Table 4. Joining techniques for combining aluminum/GFRP hybrids (Vlot, A. and Gunnink, J.W., 2001)

Category	Techniques	Advantages
Mechanical fastening	- Clinching method - RIVTAC®	- Assurance of structural integrity by well-known prediction methods - Controllable volume capacity - Ease of joint inspection - Easy technology and machinery - Joint of dissimilar materials - Little surface preparation and cleaning needed - Reopenability of assembled pieces - Repair or replacement of pieces is facilitated
Adhesive bonding		- Bonding of dissimilar materials possible - Hermetic seal - Good surface finishing - Improve fatigue resistance - Low stress concentration, since the tension is spread throughout the bonded surface - Recyclability, mostly thermoplastic joints - Weight reduction, compared with other joining methods
Welding	- Friction riveting - Comeld™ - Laser joining	- Combined the advantage of mechanical and adhesive bonding techniques - No holes need to be drilled.

hand, as delamination mechanisms are major failure mode of aluminum/glass fiber laminate, this can be mitigated using pre-treatment of aluminium substrate outer surface while thanking for the recyclability, lower curing time, ease of forming, thermoset resin is substituted by thermoplastic thermosets (Patil et al., 2018). The objective of this chapter is to review the latest development on the joining techniques, surface-modification methods, mechanical properties, and applications o

JOINING TECHNIQUES

The joining techniques of glass fiber prepreg and aluminum matrix are subdivided to main categories, which are Mechanical fastening, Adhesive bonding, and Welding and hybrid. The different joining techniques for combining aluminum/GFRP hybrids are summarized in Table 4.

Mechanical Fastening: In this technique the hybrids will be joined trough a mechanical locking, such as a bolt, rivet or self-locking.

Adhesive Bonding In this method two materials will be bonded to each other by means of an adhesive. This generally required multiple stages, such as cleaning the surfaces, addition the adhesive, fastening the materials together (often under pressure), and finally curing the adhesive.

Welding and Hybrid Techniques for Glass Fiber Reinforced Plastic to Aluminum: Several kinds of welding methods for joining glass fiber reinforce plastic can also be utilized to join prepreg to aluminum metallic laminate. Friction stir welding (FSW) and induction welding are two well-known welding techniques. Friction spot joining (FSpJ) is an optional of FSW and could be involved to join aluminum to glass fiber prepreg *(Vlot, A. and Gunnink, J.W., 2001)*. In addition, Laser techniques could be similarly utilized for joining such hybrid structures.

During the recent few years, researchers have been focusing their works to manufacture new light-weight advanced materials, which could preserve the reduced weight and superior mechanical properties of aluminum alloys. Aluminum/GFRP laminated composite, represents one of the most striking combination of these two types of different materials utilized to create high-quality of the industrial parts, has been suitable to be applied in a wide range of applications in the various heavy industries (Gresham, et al. 2006). Aluminum/GFRP hybrid composites are based on alternatively layers of aluminum metal and glass fiber-reinforced polymer matrix, joined by an adhesive sheet as illustrated in Figure 1. Due to the high rigidity of glass fiber reinforced epoxy composites reinforced aluminum matrix, recent works has concentrated on thermoplastic-based glass fiber combined to aluminum matrix due to their numerous advantages including the quick manufacturing, low-cost of processing, and recycling possibility. Serval conventional joining approaches are being used for producing aluminum/GFRP hybrid structures, this includes conventional mechanical fastening and adhesive joining but hey have revealed many shortcomings especially on the bonding strength.

Figure 1. Schematic of an aluminum/GFRP sandwich laminate.

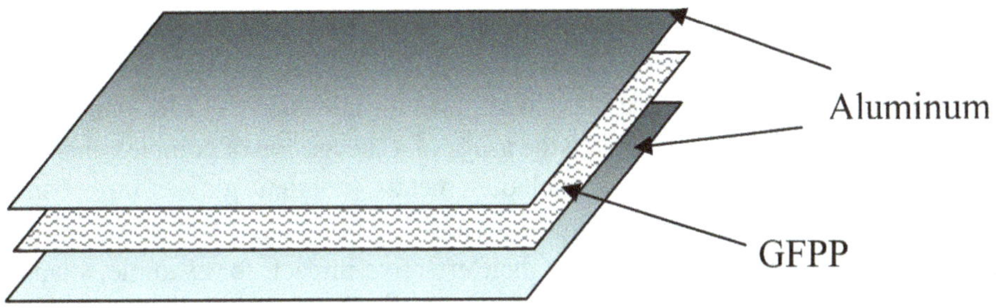

The feasibility of the Friction Spot Joining technique was discussed on aluminum 5754-H24 bonded with glass fiber and carbon fiber-reinforced poly(phenylene sulfide) based composites (Amancio, 2009). Friction spot lap joints with improved mechanical properties in the range of 16 to 29 MPa for the aluminum/composite joints were reached. The aluminum square hollow shape was joined with externally wrapped by applying an isophthalic epoxy thermoset on the glass fibers at an orientation angle of 0°/90° using a simple hand lay-up method (Paruka et al., 2015). The feasibility of mechanical clinching or press joining to bound glass fiber-reinforced plastics with aluminum sheets was investigated, where various types of tools were experienced, including split, grooved, flat dies, and rectangular ones (Lambiase, Durante, and Di Ilio, 2016). The effect of sheet thickness (thin sheets of 2 and 3 mm in thickness) and alloy composition (AA6082-T6 and AA5086) on joinability and mechanical properties of the developed joints were evaluated.

The applicability of the FricRiveting technique to join aluminum alloy 6056-T6 and glass reinforced polyamide 6 as well as and the effect of the process parameter forging force were recently demonstrated and evaluated (Proenca et al., 2012). The process was subdivided into two main steps: the first one was controlled by force and restricted by displacement, while the second stage was equally governed by force but depends on time. To manufacture new types of sandwich systems based on AL6063 T6 aluminum matrix with E-glass fiber layers the finite element analysis was performed on ANSYS by taking the various orientation of the composite material (Garre and Rohan, 2018). Ultrasonic metal welding is a promising technique for bonding light aluminum metals with glass fiber-reinforced thermoplastics

(Staab, and Balle, 2018). This process is high-reproducible, rapid times, low energy input, no extra filler materials, and the prospect of wide-ranging process data logging.

The problems related to the innovative joining technology Friction Riveting for pultruded glass fiber reinforced polyester composites riveted by Ti6Al4V aluminum rivets were investigated (Borba et al., 2014). Results revealed that a suitable degree of deformation at the tip of each rivet introduced into the composite panel resulted in improved anchoring of the rivet. Friction Press Joining (FPJ) is also a useful technique for manufacturing composites of aluminum metals and thermoplastics in lap joint configuration, which is based on modified Friction Stir Welding (FSW) (Wirth et al., 2014). During the joining process, a rotating cylindrical tool is pressed onto an aluminum surface. Friction Riveting is a well-known bonding technique for metal/polymer hybrid structures. In a work, the FricRiveting process was done for producing polyamide 6 reinforced with 30 wt.% short glass fiber (PA6-30GF) and aluminum alloy 6056-T6 (Proença, 2017).

SURFACE-TREATMENT

Serval surface-treatment techniques are applied to improve the adhesion of PFRC and aluminum matrix ranging from a straightforward solvent wipe to the usage of a succession of complex chemical modification processes. Numerous kinds of materials, including metals, inorganic glasses, composites, ceramics, elastomers, etc., be inclined to possess their unique specific pretreatments. But, some pre-treatments are generally preferred with some materials due to their effective impact, for example, silane can significantly improve the performance of joints connecting either metals or ceramics. A treated polypropylene (PP) by the inclusion of 5–30 wt. % ratio of polypropylene-grafted maliec acid (PP-g-MA) was recently developed. Reyes and Cantwell studied the adhesion between Plytron (Borealis, Norway), a unidirectional glass-fiber reinforced polypropylene, and 2024-T0 aluminum matrix by including an amorphous chromate treatment to the aluminum and filling the PP-g-MA at the interface (Reyes and Cantwell, 2000). Other researchers also used an amorphous chromate treatment process to the aluminum and incorporated a layer of PP-g-MA at the E-glass fiber/polypropylene composite-aluminum interface for providing an improved adhesion between the layers of hybrid composites (Compston et al., 2001). Chen et al. reported that pre-treatment of the aluminum matrix surfaces using an amino based silane resulted in an enhancement in the lap shear strength (Chen et al., 2007).

Pretreatment of the aluminum surface before joining it to the glass fiber reinforced thermoplastics laminate is recommended to generate a load-bearing metal-plastic joint using the thermal joining process. This surface pretreatment using laser sources is responsible for delivering high joint strengths. Till now, diverse surface-pretreatment methods were used for adhesive bonding of metallic layers and glass fiber prepreg. The increase of the adhesion features can be achieved by numerous kinds pre-treatment techniques, including microwave irradiation, chromic acid etching, surface grafting, electron beam, flame treatment, corona discharge, glow discharge, mechanical grinding modified the interlayer, and by producing in-between layers of chemically produced polythiophene (Chen, et al. 2007). Some of these surface treatment processes for metallic and non-metallic substrates in addition to a short description of the effect of the treatment on the material surface is shown in Table 5.

It was reported that the final properties of the thermoplastic and the surface topography and chemical composition of the aluminum was greatly affected the developed joint strength (Heckert, and Zaeh, 2015). The aluminum/GFPP laminates have been produced with different surface-treatment techniques such as

Table 5. Description of the surface pretreatments for metallic and nonmetallic substrates.

Substrate	Treatment Method	Effect of Treatment
Metals	Degreasing	Cleaning of the surface
Metals	Grit blast	Loose material (weak boundary) removal from the surface and increase in contact surface area
Metals	Acid etch/liquid pickling	Surface oxidation
Plastics	Corona treatment	Weak boundary layer removal and surface oxidation
Plastics	Flame treatment	Weak boundary layer removal and surface oxidation
Plastics	Chemical etching	Weak boundary layer removal and surface oxidation
fluoroplastics	Chemical etching	Surface defluorination and oxidation

silane coupling agent as described in Figure 2. Adhesion at the composite/aluminum interface was realized by surface pretreatment of aluminum using an amino-based silane coupling agent, the addition of polyolefin-based adhesive film, and treatment with PP-based film containing 20 wt. % a maleic anhydride modified polypropylene (PP-g-MA) (Guruşçu, 2009). Silane coupling molecules are usually applied as a chemically-bonded agent for aluminum substrate and polymeric chains, where they construct a tough and strong covalent bond at their interfaces. They are composed of small organic chains demonstrating numerous chemical structures on either end of each chain. Figure 2 reveals the various chemical forms of organosilane, organosilane hydrolysis, and the involved condensation reaction mechanisms.

Figure 2. Chemical structure of an organosilane, organosilane hydrolysis and condensation reaction mechanisms.

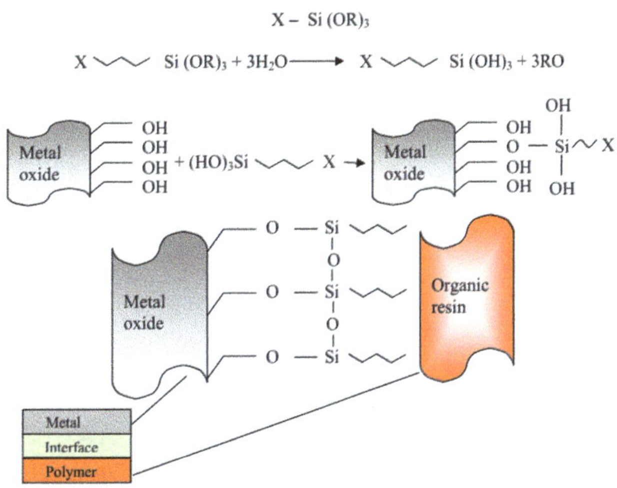

MECHANICAL PROPERTIES

The combination of aluminum and glass fiber reinforced epoxy laminate resulted in new materials known as GLARE, which exhibits much improved tensile properties such as tensile strength compared to the aluminum layer as shown in Figure 3. The mechanical properties of composite structure based on a glass fiber layer positioned at the interface of a metal panel and an aluminum foam core joined by an organic glue adhesive were evaluated (Wang et al., 2016). These composites exhibited improved comprehensive characteristic respected to the traditional aluminum foam sandwiches and the type of glass fiber was slightly influence the bending strength. The mechanical investigations of ñexible glass textiles combined into aluminum using high pressure die casting were determined, where the infiltration ratio of aluminum did slightly affect the tensile strength whereas the holding pressure and the use of the glass fiber had a marked influence (Clausen et al., 2018). A variety of failure mechanisms of these systems were detected by means of fracture surface analysis, where according to the selected type of glass textile and technique parameters. In a study, the influence of peel ply and cooling rate on the tensile properties of a short glass fiber-reinforced polypropylene/AA2024 aluminum hybrid laminate produced by hot pressing techniques laminate were scrutinized (Chen, 2018). In addition, a new cooling technique to decrease the crystallization degree of polypropylene and control the aluminum matrix deformation was offered. Aiming to stiffen the extruded aluminum alloy using lightweight glass fiber reinforced epoxy tape/mat for improving energy absorbing, the pretreatment was highly recommended (Khan, Sarang, and Hiratsuka, 2016). Thus lap shear test was used to determine bonding strength in these hybrid systems. Results showed that bending strength, energy absorption, and other mechanical properties were markedly improved by using different surface-modification processes.

Figure 3. Quantifiable strength comparison between GLARE and 2024-T3 aluminum alloy.

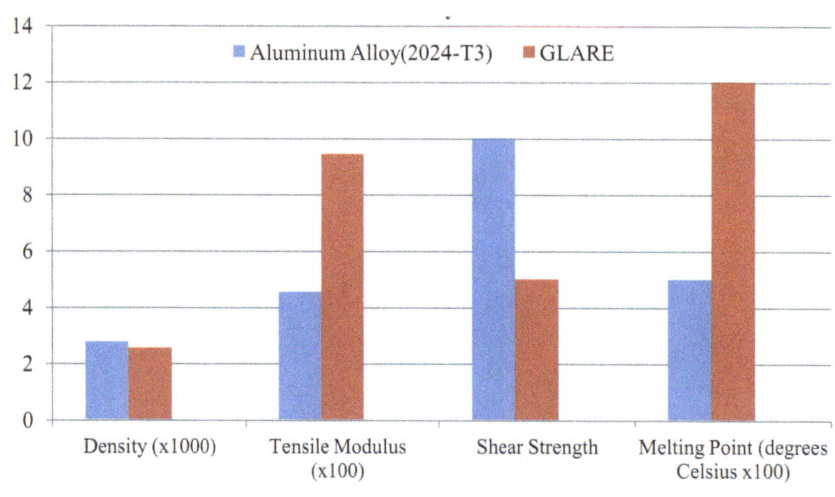

The tensile, bending, and hardness properties of edge-cracked AA 1080 aluminium alloy plates repaired with single-sided adhesively bonded E-glass fiber composite patch were examined (Pradhan et al., 2020). The recorded data confirmed that the role of patching on the mechanical properties of the cracked plate was very important. To improve the ultimate stress of aluminum metal, surface mechanical attrition treatment (SMAT) was utilized for producing a new GLARE material (Wan et al., 2015). The tensile examination and theoretical calculation revealed that the SMAT-aluminum could obviously enhance the strength of GLARE. In order to clarify the effect of heat-treatment on tensile properties of an aluminium alloy 7075 reinforced with different amounts of E-glass short fiber, a series of tensile tests was conducted whether in the as-cast or without heat-treated (T6) conditions (Sharma et al., 2003). There was an extra enhancement in both the ultimate tensile strength and hardness after aging process.

The flexural and tensile properties of the 5-layered glass fiber reinforced plastic/aluminium sheet laminate, manufactured by vacuum assisted resin transfer method (VARTM) adapted with spiral tube throughout the periphery of the prepreg, was evaluated (Vasudevan et al., 2020). In addition, the mechanical properties of the resulted hybrid composite were compared with theoretical values acquired from the FEA simulation of CAD models. These composite exhibited outstanding mechanical properties which were well-reproduced with theoretical models, while the slight discrepancy was attributed to void generated during the process. Tensile properties of aluminium/glass fibers with woven Roving laminates having alternating layers were studied based on ASTM standard (Devi and Palanikumar, 2015). The interfacial adhesion strength between the hybrid layers was also analyzed by scanning electron microscopy for the examined specimens. The mechanical characteristics of standard FML specimens manufactured from several lay ups of glass fiber/epoxy laminates with aluminum sheets were tested and compared in order to evaluate their applicability of use in the aerospace industry (Khalili., Mittal, and Gharibi Kalibar, 2005).

Tensile and flexural tests were conducted on three type of orientation of the 4/3 layer of Chopped Strand Mat (CSM), 4/3 layer of woven roving, and 4/3 layer of 45^0 stitches mat reinforced aluminum composites which were elaborated by the famous hand layup process (Mahesh and Senthil Kumar, 2010). It was demonstrated that the mechanical properties of the 4/3 layer of 45^0 stitched mat reinforced aluminum hybrid structure exhibited the best tensile and flexural strength than those produced using the other orientation. The bearing properties of recently developed hybrid glass-boron fiber/aluminum laminates, were evaluated (Yeh et al., 2012). Experimental data revealed that by fixing the joint geometry and aluminum volume ratio, the commingling of boron fibers enhanced the bearing strength of glass fiber/metal laminate. Also, the bearing properties of various glass fiber with $[0^0/90^0]$ and $[0^0/90^0/90^0/0^0]$ cross-ply commingled prepreg were also measured. In order to enhance energy absorption characteristics, partial or full wrapping of the circular aluminium tube with a 6-layers glass/epoxy composite at 90o fiber angle was demonstrated, by filament winding technique (Abbas, Ya, and Abdullah, 2016). The collected data confirmed that the SE of partially wrapped aluminium tube is 49.09% and 14.84% higher respected to wit partially wrapped steel and fully wrapped aluminium tube respectively.

To attain high strength to weight proportion, an E-glass fiber was introduced into aluminium foil matrix and they were joined by an Araldite LY556 thermoset using by hand-lay process (Veeresh Kumar and Pramod, 2017). The newly produced metal laminates were reported to exhibit increased strength, hardness, flexural strength, and improved wear resistance. The bending properties of woven shaped glass/carbon fiber/aramid fiber with a different number of layers reinforced aluminum hybrid composites were experimentally determined (Ekşi and Genel, 2015). Regardless of the used fiber type, the load bearing capacity and energy absorption capability were substantially enhanced by enhancing the number of fiber layer in these systems. In prominent research, the mechanical and tribological properties of composite material filled with black epoxy resin and aluminium tri-hydroxide using reinforcement of glass fiber were evaluated, where a significant enhancement in these hybrid properties such as stiffness, tensile strength, wear strength, and hardness was detected (Wagh and Pagar, 2018). Interestingly, the tensile strength of this material is enhanced by increasing of filler ratio and the density of the material.

Fatigue tests were also conducted on a glass fiber reinforced polyamide 6.6 (PA6.6-GF)/Aluminium joint by SPR for investigating their fatigue behavior by evaluating the effects of processing parameters and environmental aspects (Gaya et al., 2015). Results illustrated that the rivet shape has a minor effect on the joint fatigue strength compared to the other key parameters that impact the composite resistance

Figure 4. Aluminum/glass fiber layup configuration.

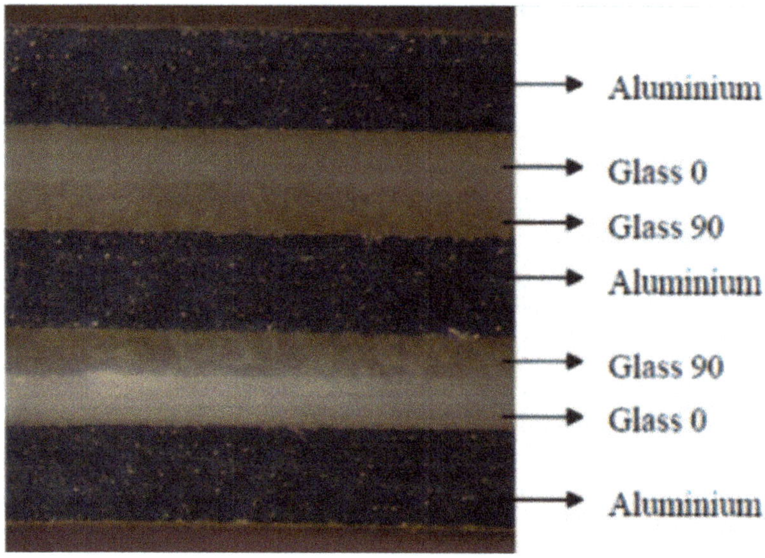

likes the type of composite and the test temperature. The influence of hydrothermal aging on the mechanical behaviour and fatigue response of a hybrid (bolted/bonded) AA7075-aluminium alloy/GRE hybrid composite single lap joints (SLJs) were studied (Mariam et al., 2019). Some kinds of hybrid joints were subjected to a humid environment at 50°C for long-duration immersion of 20, 40, 60, 80, 100, and 120 days. The authors used quasi-static and fatigue experiments to follow the evolution of the mechanical performance and the damage mechanisms of these joints during the aging tests. The dissimilar-AA7075/GRE hybrid joints exhibited the most improved joint strength and the longest failure strain while they undergone 83% and 30.2% higher fatigue strength respected to those of similar adherents of AA7075 and GRE composites, respectively.

The low-velocity impact response of aluminum alloy 7475 T761 and unidirectional S2 glass/epoxy oriented in a cross-ply configuration of 0° and 90° as shown in Figure 4, was experimentally and theoretically evaluated (Tsartsaris et al., 2011). The form and the feature of the damage inflicted by the impact were also studied using both destructive cross-sectional microphotography and non-destructive ultrasonic techniques. The prepared laminates demonstrated a higher ability to absorb energy through localized plastic deformation and through failure at the interface between the layers. The response of repetitive four low velocity impact experiments on glass fiber/epoxy-aluminum laminates (GEAML) were determined (Rajkumar et al., 2012). The influence of repeated impacts on these composites was scrutinized by evaluating the peak load, absorbed energy, slowed velocity, and the impact time exposed to a deflection at impactor static force of 52 Kgf. The highest load-bearing capacity was detected for the pure monolithic aluminium but the damage for GEAML Plate was concentrated only at the impact area which provoked inferior energy-absorbing capacity. A new type of glass fiber epoxy-based fiber metal laminate panels were produced and these composites were subjected to localized distributed air-blast loading conditions (Langdon et al., 2017). Based on the failure mode study, results reveal that both fiber configuration and bond strength participate in the blast response while the effect of bond strength was the most significant.

Figure 5. Typical configuration of a honeycomb sandwich panel.

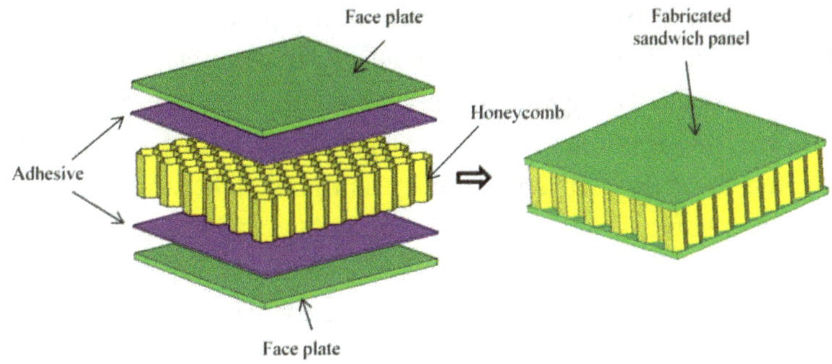

In a research study, the impact properties of aluminium foam sandwiched with glass fiber reinforced plastic (GFRP) were investigated by a drop weight impact test at velocity of 6.7 m/s by striking the specimens with and without face-sheets (Ismail, 2015). It was found that the GFRP and aluminium foam core sandwich panel demonstrated better energy absorption performances, as their specific energy absorption values were high. The failure modes and energy absorption capability of partially wrapped with 4, 6 and 8 of 90o-glass fiber/epoxy layers to form tubes were described (Abbas, Ya, at Abdullah, 2017). The experimental data showed that the partly wrapped aluminium tubes exhibited an improvement in the energy absorption of 42.54%, 47.77% and 28.91% as compared to those pure aluminium ones. The energy absorption performance and failure mechanisms of partially wrapped aluminium tubes produced by filament winding technique have been explored (Abbas, Ya, and Abdullah, 2017). The fabricated composite tubes were exposed to an axial crushing loading by means of universal testing machine, where these wrapped aluminium tubes were proved an energy absorption efficiency of 3.81%, 8.13% and 17.06% more that the metallic tubes.

The fracture and impact properties of several aluminum foam sandwich structures strongly-bonded with the glass fiber/polypropylene-based skins were determined (Kiratisaevee, and Cantwell, 2004). By analyzing the fracture surfaces, it was detected that crack propagation appeared within the composite ply in the glass fiber-aluminum laminates and along the interface of composite-aluminum foam. The energy absorption feature of these sandwich composites were evaluated where thermoplastic sandwich structures revealed that both systems can exhibit improved residual load-bearing characteristics. The impact performance of glass fibers joined aluminum laminates (GFRAL) having a thickness of 2.6 mm and 3.1 mm, was studied by using the 29 g blunt and 29 g conical nose projectiles within sub ordnance velocity range of 50 to150 m/s (Hasan and Ansari, 2013). These opposites showed improved ballistic limit for conical nose projectile respected to blunt one. Interestingly, when blunt nose projectile impacted the plate, plugs are clipped off from the metal front and from middle GFRP plate. A computational model was formed for reproducing glass fiber reinforced aluminium laminates material behavior under impact load using finite element modeling (Zaki and Mohammad, 2014). Dynamic non-linear transient analysis was realized using finite element analysis software, ABAQUS. To determine the damage area, which was slightly decreased as impact velocity rose.

The high-velocity impact response of several woven glass fibers reinforced polypropylene and an aluminum alloy were studied using a nitrogen gas gun at velocities reaching 150 m/s (Abdullah and Cantwell, 2006). The collected results revealed that the energy absorption in these hybrids occurred in the form of gross plastic deformation, membrane stretching and tearing in the aluminum plies, as well as delamination, fiber fracture, and matrix cracking in the glass fiber composite plies. In addition, the perforation resistance of these hybrid structures were well-reproduced by the Reid–Wen perforation model. Aluminum honeycomb reinforced GFRP can generally manufacture as shown in Figure 5.

The bending and the low-velocity impact response of aluminium foam sandwich bonded to the outer skins produced from glass fiber reinforced epoxy matrix and exhibiting a high capacity of energy dissipation were investigated to verify their applicability as lightweight structures in transport industries (Kara et al., 2012). The mechanical behaviour under bending and impact loading of GFRP-reinforced aluminium honeycomb panels was evaluated using a setup presented in Figure 6 (Kara et al., 2011). The experimental data showed that these panels had a good energy dissipation ability which can be further improved under bending and impact through the reinforcing with GFRP outer skins in a sandwich form.

Figure 6. Bending test of aluminum/GFRP composite specimen under load.

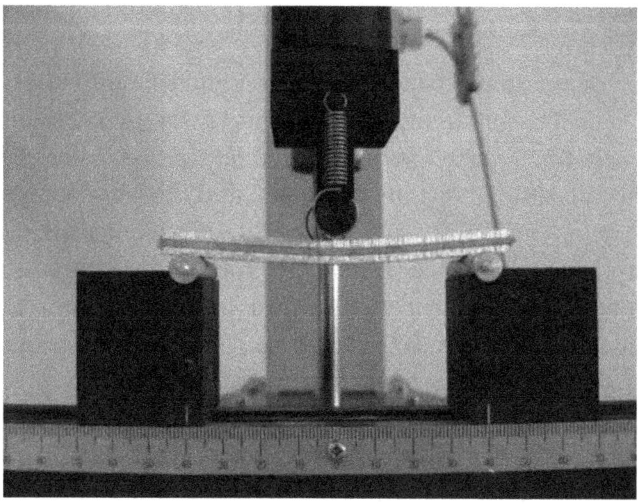

Aluminium-glass fiber reinforced plastics (GFRP) sandwich panels were produced by hand lay-up method from different aluminium thickness fractions, fiber volume fractions, and orientation and their impact behavior was conducted under low velocity impacts (Periasamy, Manickam, and Hariharasubramanian, 2012). Results proved that the bidirectional cross-ply hybrid laminate showed improved impact performance and damage resistance compared to the unidirectional hybrid laminate. An increase in aluminium thickness fraction and fiber volume ratio also generate an enhancement in the impact energy recommended for engendering cracking and perforation. The bending and the low-velocity impact response of aluminium foam sandwiches reinforced by glass fiber reinforced epoxy laminates, exhibiting higher capacity of energy dissipation, were analyzed (Crupi et al., 2014). A good agreement with the experimental data was resulted for the peak load prediction using an analytical model. A series of columnar tube composite were developed by a hand lay–up technique using epoxy–glass fiber with aluminium columnar tube as a central material (Paruka et al., 2015). The result showed the presence of many failure modes including matrix–fiber interfacial fracture, fiber breakage, and hackles, while the main failure behavior was brittle fracture including transverse shearing and splaying styles. Under the impact test of aluminium/glass fiber reinforced plastics sandwich panels, debonding of aluminumm layer and the delamination of GFRP lamina are the main the phenomena that can be detected by analyzing the composite interface using a scanning electron microscope (SEM). As shown in Figure 7, the overall morphology into laminate failure zone includes interlaminar/interlaminar crack propagation, frond bending, longitudinal splitting, flexural damage, and friction between laminates

Figure 7. SEM images of glass fiber reinforced plastics panel under impact test.

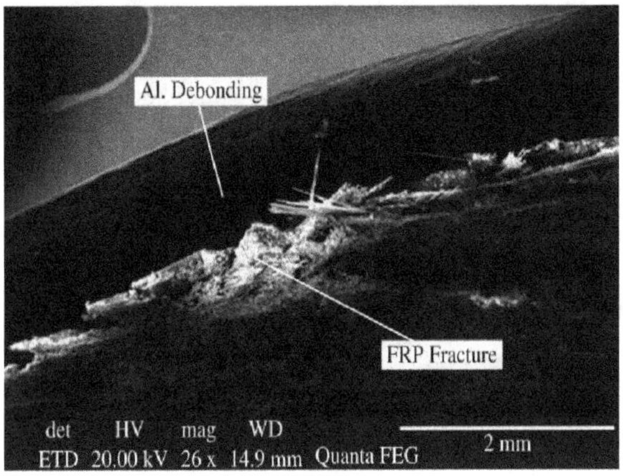

APPLICATIONS OF GFRP/ALUMINUM HYBRID COMPOSITES

Fiber/aluminum composites are constructed from interlacing layers of thin aluminum metals and fibe-reinforced adhesives. They are widely applied in building aerospace structures and automotive industries application thanking for their high specific mechanical properties and fatigue resistance (Figure 8). In the recent few years, the demand in the aircraft industry for high-performance, lightweight structures was exploded which increased the research focus on these laminated hybrid composites. The most commercially available fiber metal laminate (FMLs) is the glass reinforced aluminium laminate (GLARE), which is produced from high-strength glass fibres (Mukesh and Hynes, 2019). An example of using aluminum/GFRP in the aerospace industry is in lower and upper wings as well as in the fuselage and tail sections (Khalili, et al. 2005).

Due to the increasingly serious environmental issues of emission from the automobile and the fluctuation in fuel price, the development of lighter hybrid aluminum metal/composite sandwich materials becomes a promising alternative solution (Asnafi et al., 2000). Thus, the tendencies of producing and utilization of these composite materials in the automotive industry were considered the main efforts of the automotive industry to miniature lighter vehicles with reduced fuel consumption and reduced gas emission. Numerous car producers are adopting the use of aluminum/glass fiber laminated composites in semi-structural forms for increasing the fuel economy, preserving safety, and improving the performance of vehicles, due to their ability for reducing weight with less corrosion problematics. This can be achieved by substituting steel and cast iron conventional parts with non-heavy composite materials such as aluminum metal matrix composite, glass fiber reinforced polymer composites, which they can straightly decrease the weight of automotive pieces such as engine block and chassis (Patel et al., 2018). Glass fiber reinforced plastic (GFRP) is more practical for the automobile body applications due to

Figure 8. Applications of aluminum/glass fiber reinforced polymer sandwich composites in various industries.

lightweight compared to than steels and it is stronger than aluminium alloys. Thus, its composites were usually used to bond Al alloy sheets using various riveting techniques (Cho and Kim, 2015).

A combination of aluminum columnar parts with glass fiber composite laminate to construct a hybrid structure can be applied as collapsible energy absorbers in automotive vehicular structures to safeguard occupants and cargo. The advantage of such aluminum sandwich part in the composite is that it affords ductile and stable plastic failure mechanisms with gradual deformation in a stable way by enhancing energy absorption throughout the collision accident. Bumpers must combine strength and adequate formability; the pure aluminum alloys are insufficient for future requirements, a new aluminum/GFRP composites can offer a prominent combination of properties (Staley and Lege, 1993). Thin-walled tubular parts are generally applied as energy absorbing parts in modern automobiles which can guard occupants during an impact collision. These tubular energy absorbing elements are once plastically distorted, they are cast off and replaced. Therefore, hybrid aluminium/GFRP composite tubes were preferred for this application due to their excellent specific energy absorption characteristics (Kumar & Sundaram, 2018).

Aluminium has been the dominant material in aircraft construction for a half-century due to its outstanding grouping of light-weight strength, good ductility, improved corrosion-resistance, ease of assembly, and its low cost (Dorward, and Pritchett, 1988). Glass reinforced aluminum laminates (GLARE) can be effectively used in the aircraft upper fuselage and the leading edge surfaces of the vertical and horizontal tailplanes (Mrazova, 2013). These kinds of hybrid structures can also offer new chances for further decrease in the aircraft weight or further improvement in the engine performance (Immarigeonet al., 1995). Due to their lower thickness, good formability, high impact resistance, and high-strength aluminum sheets bonded together with glass fiber-reinforced adhesive epoxy, they reached their commercial market in the recent few years (Shankar, 1994). Nearly for a decade, aluminum/GFRP hybrid composites have started to be applied more extensively in great commercial jet airliners for the fuselage, wing as well as other structural components in place of aluminium alloys. However, high-strength aluminium alloys are still critical for airframe construction (Dursun, and Soutis, 2014).

CONCLUSION

- Glass fiber-reinforced aluminum substarte hybrid laminates were produced by bonding thin aluminum metal sheets and glass fiber/adhesive layers.
- These hybrid structures exhibited improved mechanical and fatigue properties, as well as some good impact resistance characteristics, which make them suitable for substituting metallic structures, especially in aerospace and aircraft applications.
- The surface pretreatment of aluminum matrix, as well as GFRP composites, could be strongly recommended to achieve an improved bonding strength to ensure their applicability in harsh environments.
- This chapter was sucessfuly reviewed the various joining techniques involved in the manufacturing of these composites and discussed their relevant mechanical properties and applications in the aerospace and automotive industries.

RESEARCH GAP AND FUTURE WORK

- The use of GFRP/aluminum composites in blast-resistant walls will be more increasingly studied in the future due to the high demand on such materials.
- The studies on surface-treatment using moden technique such as laser beam and plasma treatment will be also intended to intensify in the next few years.
- The inclusion of organic and inorganic materials nanomaterials in producing such hybrid composites will be also a great subject of reasrch.

REFERENCES

Abbas, T., Ya, H. H., & Abdullah, M. Z. (2016). Energy absorption capability of aluminium tube partially wrapped with glass/epoxy subjected to quasi-static loading. *Journal of Engineering and Applied Sciences (Asian Research Publishing Network), 11*(24), 14313–14318.

Abbas, T., Ya, H. H., & Abdullah, M. Z. (2017). Experimental Investigation of Energy Absorption of Partially Wrapped Thin-Walled ſ Aluminium-Glass/Epoxy ſ Tube Subjected to Quasi-Static Loading. *Materials Science Forum, 904*, 61–67. doi:10.4028/www.scientific.net/MSF.904.61

Abbas, T., Ya, H. H., & Abdullah, M. Z. (2017). Comparison of Energy Absorption of Aluminiumcomposite Tubes Subjected to Axial Loading. *Materials Science and Engineering*, •••, 205.

Abdullah, M. R., & Cantwell, W. J. (2006). The Impact Resistance of Fiber–metal Laminates Based on Glass Fiber Reinforced Polypropylene. *Polymer Composites, 27*(6), 700–708. doi:10.1002/pc.20240

Amancio, S.T. (2009). *Friction Spot Joining of Lightweight Metals and Fiber-Reinforced Polymer Hybrid Structures*. Academic Press.

Amancio-Filho, S., & dos Santos, J. (2009). Joining of Polymers and Polymer-Metal Hybrid Structures: Recent Developments and Trends. *Polymer Engineering and Science*, *49*(8), 1461–1476. doi:10.1002/pen.21424

AnandaRao, M., Reddy, K.V.K., & Seshaiah, T. (2014). Finite Element Analysis of Epoxy (Ly-556) Glass Fiber Filled Aluminium Circular Section Hollow Column. *International Organization of Scientific Research*, *4*(7), 7-12.

Asnafi, N., Langstedt, G., Andersson, C. H., Ostergren, N., & Hakansson, T. (2000). A new lightweight metal-composite-metal panel for applications in the automotive and other industries. *Thin-walled Structures*, *36*(4), 289–310. doi:10.1016/S0263-8231(00)00004-5

Baburaja, K., VenkataSubbaiah, K., & Kalluri, R. (2016). Hybrid materials of aluminium. *Materials Today: Proceedings*, *3*, 4140–4145. doi:10.1016/j.matpr.2016.11.087

Borba, N.Z., Blaga, L., Santos, J.F.d., Canto, L. B., & Filho S. T.A. (2014). *Friction Riveting of Pultruded Thermoset Glass Fiber Reinforced Polyester Composite and Ti6al4v Hybrid Joints*. Academic Press.

Chen, M.-A., Li, H.-Z., & Zhang, X.-M. (2007). *Improvement of shear strength of aluminium polypropylene lap joints by grafting maleic anhydride onto polypropylene*. Academic Press.

Chen, M.-A., Li, H.-Z., & Zhang, X.-M. (2007, April). Improvement of shear strength of aluminium-polypropylene lap joints by grafting maleic anhydride onto polypropylene. *International Journal of Adhesion and Adhesives*, *27*(3), 175–187. doi:10.1016/j.ijadhadh.2006.01.008

Chen, Y., Jin, K., Li, H., Lin, Y., Lu, Y., Hua, X., Yuan, Q., Wang, H., & Tao, J. (2018). Effect of peer ply and cooling rate on the tensile properties of Al/Gf/PP laminate prepared by hot pressing. *The Journal of Sandwich Structures & Materials*, *0*(0), 1–15. doi:10.1177/1099636218802473

Cho, H.-Y., & Kim, D. (2015). Finite element analysis for joining glass fiber reinforced plastic and aluminium alloy sheets. *Journal of Welding and Joining*, *33*(2), 78–84. doi:10.5781/JWJ.2015.33.2.78

Clausen, J., Kelch, M., Wöstmann, F. J., & Busse, M. (2018). Mechanical characterization of integral aluminum-FRP-structures produced by high pressure die-casting. *Production Engineering*, *12*(2), 269–278. doi:10.100711740-018-0811-6

Compston, P., Cantwell, W. J., Jones, C., & Jones, N. (2001). Impact Perforation Resistance and Fracture Mechanisms of A Thermoplastic Based Fiber-Metal Laminate. *Journal of Materials Science Letters*, *20*(7), 597–599. doi:10.1023/A:1010904930497

Crupi, V., Kara, E., Epasto, G., Guglielmino, E., & Aykul, H. (2014). Prediction model for the impact response of glass fiber reinforced aluminium foam sandwiches. *International Journal of Impact Engineering*.

Devi, G. R., & Palanikumar, K. (2015). Tensile Property Evaluation of Woven Glass Fiber Reinforced Plastic and Aluminium Stack. *Applied Mechanics and Materials*, *766-767*, 44–49. doi:10.4028/www.scientific.net/AMM.766-767.44

Dorward, R. C., & Pritchett, T. R. (1988). Advanced Aluminium Alloys for Aircraft and Aerospace Applications. *Materials & Design*, *9*(2), 63–69. doi:10.1016/0261-3069(88)90076-3

Dursun, T., & Soutis, C. (2014). Recent developments in advanced aircraft aluminium alloys. *Materials & Design*, *56*, 862–871. doi:10.1016/j.matdes.2013.12.002

Ekşi, S., & Genel, K. (2015). Three Point Bending Behavior of Woven Glass, Aramid and Carbon Fiber Reinforced Hybrid Composite Tube. *Acta Physica Polonica A*, *128*(2), 59–60. doi:10.12693/APhysPolA.128.B-59

Garg, P., Jamwal, A., Kumar, D., Sadasivuni, K. K., Hussain, C. M., & Gupta, P. (2019). Advance research progresses in aluminium matrix composites: Manufacturing & applications. *Journal of Materials Research and Technology*, *8*(5), 4924–4939. doi:10.1016/j.jmrt.2019.06.028

Garre, P., & Rohan, G. (2018). Design and analysis of Al6063 T6 and glass fiber reinforced composite material. *Materials Science and Engineering*, *455*, 012011–012015.

Gaya, A., Lefebvrec, F., Bergamoa, S., Valiorgueb, F., Chalandond, P., Michela, P., & Bertrand, P. (2015). Fatigue of aluminum/glass fiber reinforced polymer composite assembly joined by self-piercing riveting. *Procedia Engineering*, *133*, 501–507. doi:10.1016/j.proeng.2015.12.620

Guruşçu, A. (2009). *Joining and Interfacial Properties of Aluminum/Glass Fiber Reinforced Polypropylene Sandwich Composites*. Academic Press.

Hasan, T. U., & Ansari, R. (2013). Study of the Impact Response of Glass Fiber Reinforced Aluminium Laminates for Conical and Blunt Nose Projectiles. *International Journal of Engineering Research and Applications*, *3*(4), 580–587.

Heckert, A., & Zaeh, M. F. (2015). Laser surface pre-treatment of aluminum for hybrid joints with glass fiber reinforced thermoplastics. *Journal of Laser Applications, 27*(2), 29005-1-29005-5.

Immarigeon, J. P., Beddoes, J. C., Holt, R. T., Koul, A. K., Zhao, L., & Wallace, W. (1995). Lightweight Materials for Aircraft Applications. *Materials Characterization*, *35*(1), 41–67. doi:10.1016/1044-5803(95)00066-6

Ismail, M. F., Jumahat, A., Ahmad, N., & Ismail, M. H. (2015). Low Velocity Impact of Aluminium Foam - Glass Fibre Reinforced Plastic Sandwich Panels. *Advanced Materials Research*, *1113*, 74–79. doi:10.4028/www.scientific.net/AMR.1113.74

Kara, E., Crupi, V., Epasto, G., Guglielmino, E., & Aykul, H. (2012). Low Velocity Impact Response of Glass Fiber Reinforced Aluminium Foam Sandwich. *European Conference on Composite Materials*, 24-28.

Kara, E., Crupi, V., Epasto, G., Guglielmino, E., & Aykul, H. (n.d.). Mechanical Behaviour of Glass Fiber Reinforced aluminium Honeycomb Sandwiches. *The 19th International Conference on Composite Materials.*

Khalili, S. M. R., Mittal, R. K., & Gharibi Kalibar, S. (2005). A study of the mechanical properties of steel/aluminium/GRP laminates. *Materials Science and Engineering A*, *412*(1-2), 137–140. doi:10.1016/j.msea.2005.08.016

Khalili, S. M. R., Mittal, R. K., & Kalibar, S. G. (2005). A study of the mechanical properties of steel/aluminum/GRP laminates. *Materials Science and Engineering A*, *142*(1-2), 137–140. doi:10.1016/j.msea.2005.08.016

Khan, S., Sarang, S., & Hiratsuka, I. (2016). Study of Bending Strength for Aluminum Reinforced with Epoxy Composite. *SAE International Journal of Materials and Manufacturing*, *9*(3), 781–787. doi:10.4271/2016-01-0516

Kiratisaevee, H., & Cantwell, W. J. (2004). The Impact Response of Aluminum Foam Sandwich Structures Based on a Glass Fiber-Reinforced Polypropylene Fiber-Metal Laminate. *Polymer Composites*, *25*(5), 499–509. doi:10.1002/pc.20043

Kumar, A. P., Dirgantara, T., & Krishna, P. V. (2020). Advances in Lightweight Materials and Structures. *Springer Proceedings in Materials, 8*.

Kumar, A. P., Dirgantara, T., & Krishna, P. V. (2020). *Advances in Lightweight Materials and Structures*. Springer Nature. doi:10.1007/978-981-15-7827-4

Kumar, A. P., & Sundaram, M. S. (2018). An Axial Crushing Characteristics of Hybrid Kenaf/Glass Fabric Wrapped Aluminium Capped Tubes under Static Loading. *International Journal of Mechanical and Production*, *8*(6), 201–206.

LaGrandeur. (2011). The Effectiveness of Glass Laminate Aluminum Reinforced Epoxy. New York Institute of Technology.

Lambiase, F., Durante, M., & Di Ilio, A. (2016). Fast joining of aluminum sheets with Glass Fiber Reinforced Polymer (GFRP) by mechanical clinching. *Journal of Materials Processing Technology*, *236*, 241–251. doi:10.1016/j.jmatprotec.2016.04.030

Langdon, G.S., Klemperer, C.J.V., & Volschenk, G.F., Tonder, T.V., & Govender, R.A. (2017). The influence of interfacial bonding on the response of lightweight aluminium and glass fibre metal laminate panels subjected to air-blast loading. *Mechanical Engineering Science*, *0*(0), 1–16.

Mahesh, M., & Senthil Kumar, A. (n.d.). Comparison of mechanical properties for aluminium metal laminates (glare) of three different orientations such as CSM, woven roving and 45^0 stitched mat. *Journal of Mechanical and Civil Engineering*, 9-13.

Mariam, M., Afendi, M., Majid, M. S. A., Ridzuan, M. J. M., Sultan, M. T. H., Jawaid, M., & Gibson, A. G. (2019). Hydrothermal ageing effect on the mechanical behaviour and fatigue response of aluminium alloy/glass/epoxy hybrid composite single lap joints. *Composite Structures*, *219*, 69–82. doi:10.1016/j.compstruct.2019.03.078

Mrazova, M. (2013). Advanced composite materials of the future in aerospace industry. *Incas Bulletin*, *5*(3), 139–150. doi:10.13111/2066-8201.2013.5.3.14

Mukesh, A. M., & Hynes, N. R. J. (2019). Mechanical Properties and Applications of Fibre Metal Laminates. AIP Conference Proceedings, 2142, 100002-1–100002-6. doi:10.1063/1.5122456

Paruka, P., Mat Yasin, M., Mamat, R., Maleque, M. A., & Md Shah, M. K. (2015). Crush Zone Morphology of Epoxy–Glass Fiber–Aluminium Composite Columnar Tube due to Longitudinal Crushing Force. *Advanced Materials Research*, *1115*, 258–261. doi:10.4028/www.scientific.net/AMR.1115.258

Paruka, P., Siswanto, W. A., Maleque, M. A., & Mohd Shah, M. K. (2015). Crashworthy capacity of a hybridized epoxy-glass fiber aluminum columnar tube using repeated axial resistive force. *Journal of Mechanical Science and Technology*, *29*(5), 1941–1953. doi:10.100712206-015-0415-4

Patel, M., Pardhi, B., Chopara, S., & Pal, M. (2018). Lightweight Composite Materials for Automotive - A Review. *International Research Journal of Engineering and Technology*, *5*(11), 41–47.

Patil, N. A., Mulikb, S. S., Wangikarc, K. S., & Kulkarni, A. P. (2018). Characterization of Glass Laminate Aluminium Reinforced Epoxy-A Review. *Procedia Manufacturing*, *20*, 554–562. doi:10.1016/j.promfg.2018.02.083

Periasamy, M., Manickam, B., & Hariharasubramanian, K. (2012). Impact Properties of Aluminium - Glass Fiber Reinforced Plastics Sandwich Panels. *Materials Research*, *15*(3), 347–354. doi:10.1590/S1516-14392012005000036

Pradhan, S. S., Mishra, U., & Biswal, S. K. (2020). Experimental study on mechanical performance of cracked aluminium alloy repaired with composite patch. *Materials Today: Proceedings*, *26*, 2676–2680. doi:10.1016/j.matpr.2020.02.563

Proença, B.C. (2017). *Friction Riveting of Aluminium Alloy 6056-T6 with Short-Glassfiber-Reinforced Polyamide 6 Composite*. Academic Press.

Proenca, B. C., Blaga, L., dos Santos, J. F., Canto, L. B., Amancio-Filho, S. T. (n.d.). *Force Controlled Friction Riveting of Glass Fiber Reinforced Polyamide 6 And Aluminum Alloy 6056 Hybrid Joints*. Academic Press.

Rajkumar, G. R., Krishna, M., Murthy, H. N. N., Sharma, S. C., & Mahesh, K. R. V. (2012). Investigation of Repeated Low Velocity Impact Behaviour of GFRP /Aluminium and CFRP/Aluminium Laminates. *International Journal of Soft Computing and Engineering*, *1*(6), 50–58.

Senthil, S., Raguraman, M., & Manalan, D. T. (2020). Manufacturing processes & recent applications of aluminium metal matrix composite materials: A review. *Materials Today: Proceedings*.

Shankar, K. (1994). Fiber Metal Laminates—Matching the Best in Composites and Metals. *Materials Technology*, *9*(5-6), 114–119. doi:10.1080/10667857.1994.11785040

Sharma, S. C., Vizhian, S. P., Shashishankar, A., & Krishna, M. (2003). Influence of Heat Treatment on Microstructural and Tensile Properties of Aluminium E-Glass Short Fiber Composites. *Influence of Heat Treatment on Aluminium Composites*, *14*(4-5), 305–317. doi:10.1515/JMBM.2003.14.4-5.305

Sinmazçelik, T., Avcu, E., Bora, M. Ö., & Çoban, O. (2011). A review: Fibre metal laminates, background, bonding types and applied test methods. *Materials & Design*, *32*(7), 3671–3685. doi:10.1016/j.matdes.2011.03.011

Staab, F., & Balle, F. (2018). Ultrasonic torsion welding of ageing-resistant Al/CFRP joints: Properties, microstructure and joint formation. *Ultrasonics*. PMID:30521993

Staley, J. T., & Lege, D. J. (1993). Advances in aluminium alloy products for structural applications in transportation. *Journal de Physique. IV*, *3*(C7), 179–190. doi:10.1051/jp4:1993728

Tsartsaris, N., Meo, M., Dolce, F., Polimeno, U., Guida, M., & Marulo, F. (2011). Low-velocity impact behavior of fiber metal laminates. *Journal of Composite Materials*, *45*(7), 803–814. doi:10.1177/0021998310376108

Vasudevan, A., Kumar, B. N., Depoures, M. V., Maridurai, T., & Mohanavel, V. (2020). Tensile and flexural behaviour of glass fibre reinforced plastic – Aluminium hybrid laminate manufactured by vacuum resin transfer moulding technique (VARTM). *Materials Today: Proceedings*.

Veeresh Kumar, G. B., & Pramod, R. (2017). Investigation of mechanical properties of aluminium reinforced glass fibre polymer Composites. *AIP Conference Proceedings*, *1859*, 020084-1–020084-7.

Vlot, A., & Gunnink, J. W. (2001). *Fibre metal laminates: an introduction*. Springer. doi:10.1007/978-94-010-0995-9

Wagh, P. H., & Pagar, D. D. (2018). Investigation of mechanical and tribological behavior of composite material filled with black epoxy resin and aluminium tri-hydroxide using reinforcement of glass fiber. *AIP Conference Proceedings*, *2018*, 020025-1, 020025–15. doi:10.1063/1.5058262

Wan, Y., Wang, Z. Q., Zhou, L. M., & Chang, M. Z. (2015). Tensile Test Performance of Fibre Reinforced SMATed Aluminium Laminates. *International Conference on Material Science and Application*, 943-947. 10.2991/icmsa-15.2015.175

Wang, N., Chen, X., Li, A., Li, Y., Zhang, H., & Liu, Y. (2016). Three-point bending performance of a new aluminum foam composite structure. *Transactions of Nonferrous Metals Society of China*, *26*(2), 359–368. doi:10.1016/S1003-6326(16)64088-8

Wirth, F. X., Fuchs, A. N., Rinck, P., & Zaeh, M. F. (2014). Friction Press Joining of Laser-Texturized Aluminum with Fiber Reinforced Thermoplastics. *Advanced Materials Research*, *966-967*, 536–545. doi:10.4028/www.scientific.net/AMR.966-967.536

Yeh, P. C., Chang, P. Y., Wang, J., Yang, J. M., Wu, P. H., & Liu, M. C. (2012). Bearing strength of commingled boron/glass fiber reinforced aluminum laminates. *Composite Structures*, *94*(11), 3160–3173. doi:10.1016/j.compstruct.2012.05.001

Zaki, M. U., & Mohammad, R. (2014). Three Dimensional Simulation of Impact Response of Glass Fiber Reinforced Aluminium Laminates for Conical and Blunt Nose Projectiles. *International Journal of Emerging Technology and Advanced Engineering*, *14*(1), 480–488.

Chapter 6
Joining Techniques Like Welding in Lightweight Material Structures

Aytekin Ulutaş
https://orcid.org/0000-0002-5230-7122
Edremit School of Civil Aviation, Balıkesir University, Turkey

ABSTRACT

In order to take more stringent measures in fuel economy and achieve the determined performance targets, the automotive industry needs to reduce the weight of the vehicles it produces. For this reason, all automobile manufacturers have determined their own strategies. Some manufacturers use lighter aluminum, magnesium, and composite components in their cars. In this study, the joining techniques of lightweight materials such as welding and the processes of their industrial use have been examined. There is currently no single technology that can combine all metallic panels in a car body structure. However, it is known that various joining technologies are used together. With the potential to combine certain combinations of steel and aluminum, manufacturers and scientists continue to work to identify technologies with the highest potential for lightweight joining and put them into use in high-volume automobile production. Therefore, it is important to examine the weldability of light materials such as magnesium, titanium, and aluminum.

INTRODUCTION

The concept of the design cycle (Hevner, 2007) which is used for technological design and industrial application areas, is an artistic concept that defines our civilization development. Simon (Simon, 2019) describes the nature of this cycle as generating design alternatives and evaluating the alternatives against requirements until a satisfactory design is achieved Today's current technological designs evoke new needs in the field in which they are used. These new needs trigger new designs. At this point, the potential of realizing new designs is limited to the level of scientific knowledge offered by materials science. If the material and technical support that will bring the new design to life is provided, the new design

DOI: 10.4018/978-1-7998-7864-3.ch006

can come to life. Otherwise, it is necessary to find a new material for the new industrial design. These searches for new materials are the main driving force of development in the scientific world. However, even if a new material design has functions to meet the needs, it cannot be expected to turn into an industrial design immediately. Engineering solutions(Fergus et al., n.d.) come into play for the effective and efficient use of the material and ensure that the material developed for industrial integration becomes usable in applications.

With the rapid development of science and industry technologies, the sustainable development trend of the automobile industry such as environmental protection, energy saving and light manufacturing is increasing (Feng et al., 2016). The automotive industry, guided by European CO2 emission laws, is willing to increase energy and raw material costs. It faces the challenge of developing lighter and at the same time still rigid but affordable and crash-resistant car bodies for large-scale production. The application of weight-reduced structures does not depend solely on the availability of lightweight materials and related forming technologies. However, it also depends on cost effective and multi-material design and reliable joining technologies (Sathishkumar et al., 2021). At this stage, all of industry, such as the automotive, aerospace or maritime industry, and even every design has its own specific manufacturing conditions (Kleiner et al., 2003).

When deciding on the joining technique in manufacturing processes, many factors are evaluated together and a conclusion is r:eached. Considering these factors, such as physical and mechanical properties of the materials to be joined, outdoor conditions, joint construction and joint costs, determining the ideal joint type requires a difficult selection process. However, depending on the sector and material type, it may sometimes be necessary to stick to a single joint type, or it may be necessary to choose from a very small number of joining technologies. Figure 1 shows a graph showing the estimated percentages of all joins performed with the various options.

Figure 1. A pie chart showing the estimated percentages (based on dollar or dollar-equivalent value) of all joining accomplished by the various options; mechanical joining (including fastening and integral attachment), welding, brazing, soldering, adhesive bonding, and various hybrids of these (Messler, 2003).

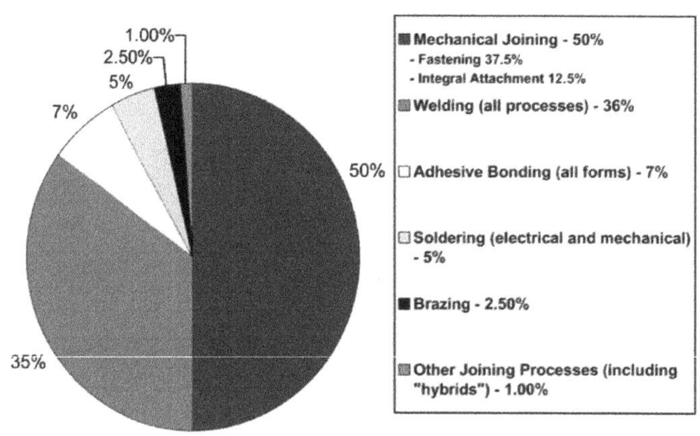

As the energy needs of the automotive and aerospace (Ward-Close & Froes, 1994) sectors increase, intensive work is being done on the mitigation of vehicles in order to improve fuel economy, reduce climate change environmental damage(Bahl et al., 2021) and negative effects on human health. These studies mostly involve the use of lightweight alloys such as magnesium, aluminum, and titanium (Wang et al., 2015)(Joost, 2012). Magnesium alloy and aluminum alloy non-ferrous metals quickly conquer the international market due to the advantages of high specific hardness, high specific strength, low density (Table 1.1), easy processing and low production cost (Schubert et al., 2001). These metals are widely used in automobile, aviation, shipbuilding, electronic equipment and chemical metallurgy. Aluminum and aluminum alloys are light metals with high corrosion resistance, ductility and strength. In addition, aluminum is relatively easy to manufacture by shaping and welding. Replacing steel plate materials with aluminum alloy, magnesium alloy and can reduce the weight of the structure by approximately 50%. It has been reported that the fuel efficiency of passenger cars can be increased by 6–8% for every 10% reduction in weight (Joost, 2012). Therefore, Mg, Al, Ti alloys and it's composites are the important way to reduce the weight of components. So, the connection of this different metals has also received an increasing interest from people (Ward-Close & Froes, 1994).

Aluminum has high electrical and thermal conductivity (Michot & Champier, 1991). It is frequently used in steel alloys and is known as the alloying element with the highest utilization rate for titanium alloys. For this reason, Aluminum alloys are a versatile metal that has applications in all production areas. Due to these superior properties, it is used to reduce weight in today's industry, especially in vehicles(Liu et al., 2013) While this situation has increased the use of aluminum in an industrial production, the necessity of welding different aluminum alloys for the same reasons has also arisen. Since aluminum and its alloys are light metal alloys, their mechanical properties can be improved by heat treatment. Magnesium and alloys are non-ferrous metals with good ductility, low density, good corrosion resistance and medium strength(Pollock, 2010). Magnesium and alloys are used in a variety of manufacturing industries, aircraft, power tool applications, and marine. Titanium and alloys are non-ferrous metals with good fatigue properties, excellent corrosion resistance, and a high strength-to-weight ratio (Szesz & Lepienski, 2017). With the advantages that titanium has, it can lead to the use of titanium and alloys in aircraft or air chassis parts, jet engine super alloy components, medical devices, corrosion-resistant chemical treatment equipment prostheses. Beryllium is known as the second metal that has the lowest density of light metal alloys. This is a useful feature in structural applications. The low density and high modulus make beryllium alloys very useful in aerospace applications. Comparison of some physical properties of Al, Mg, Ti and Be are given in Table 1.

Table 1. Comparison of some physical properties of Mg, Ti and Al

Metal	Specific gravity (gr/cm3)	Melting point (°C)	Tensile strength (MPa)	Elongation (%)	Hardness (HB)
Al	2.74	660	88	45	23
Mg	1.74	650	98	5	30
Ti	4.5	1660	241	50	152
Be	1.85	1273	370	3	163

It has unavoidably prompted difficulties in joining base materials where magnesium combinations, aluminium compounds and steels are joined. At the same time, each of these materials proposes fluctuating execution and superior properties for different segments in various areas of the new design. Accordingly, these base material mixtures will inevitably arise, which creates significant difficulties due to their contradictions during the friction welding techniques used in vehicle manufacture(Sathishkumar et al., 2021). When it comes to lightweight materials, the options are a little narrower. When the joining technologies of metals such as Al, Ti, Mg are examined, it is seen that welding technologies are used more intensively.

In the manufacturing technique, the aim is not only to obtain parts by combining materials. At the same time, the joined part must not deteriorate under operating conditions and must fulfill its task. For this reason, mergers are expected to meet some expectations.

Metals and alloys in which the welding method is used in the joining technique, with a few exceptions in practice, must be heated to a temperature close to melting in all welding methods. It should not be forgotten that the heat source and welding method used here are also effective. Due to heating to high temperatures, many chemical and metallurgical changes occur in the welding regions of the materials and in the regions adjacent to the weld (Choi et al., 2011)For example, in the welding of high carbon steel materials, a very high hardness structure emerges in the region adjacent to the weld due to the high cooling rate. For some materials, although there is no problem, special measures are needed to ensure a defect-free and satisfactory weld quality. In such cases, weld capability is mentioned.

A metal or alloy may exhibit very good weldability in one welding method, but may have very poor weldability in another method. Under all these conditions, each of the joining techniques has its own specific conditions and joining capabilities. A qualified join cannot be made before these conditions are met.

With the development in the automotive industry, new forms have been created and applications have been developed. The quality and safety of the materials used has been increased in this way. Increasing the reliability of auto parts and the durability of their operation is one of the most important problems in materials science. For this reason, both the body, suspension and some parts of the engine are made using aluminum alloys. The disadvantage of using aluminum is problems such as high conductivity and vibration in welding parts (Feng et al., 2016).

Metal-matrix composite castings such as Mg, Al have the potential to be used as an alternative to many ferrous castings in chassis components and power transmission systems. Besides, due to their different strength, tribology and ductility, light metal castings require more sophisticated design rules and advanced foundry procedures before product engineers can use them in larger quantities (G. S. Cole & Sherman, 1995.). The lightening strategy focused on aluminum-based alloys as an already available alternative to existing steel-based materials that can be used as an intermediate or long-term solution (Fadzil et al., 2021).

By using alternative technologies, the long-term reduction of CO2 emissions can be achieved by using alternative technologies (Marré et al., 2010); that is, e-mobile, in addition to the short-term fuel saving approach, through the reduction of vehicle weight. The realization of a CO2 neutral vehicle has been the vision of the future. The solutions for this are intelligent lightweight construction designs that allow to reduce vehicle weight while increasing functionality. Hybrid structures and multi-material designs have reached a promising dimension.(Contorno et al., 2006).While the development of new designs in lightweight materials continues, these designs find use in industrial applications, requiring new approaches in joining techniques.

The key to realizing innovative lightweight car bodies in multi-material design is low-cost joining technology (G. S. Cole & Sherman, 1995). The growing demand for material diversity, cost effectiveness and flexibility as well as process reliability has created a high level of innovation pressure in the field of joining technology. Since all methods have their own advantages and disadvantages, it is not possible to show a single technology to be the most appropriate application. The decision to implement more than one technology for a batch production depends on limitations such as material flexibility, costs, production volume, and required strength values. However, it is possible for serial applications to reach maturity with further development steps(Meschut et al., 2014).

JOINING TECHNIQUES IN LIGHTWEIGHT MATERIALS

Developing innovative, competitive and energy efficient products is very important for success in global markets. The use of light weight structures depends on the development of lightweight materials and associated forming technologies, as well as cost-effective and reliable joining technologies for multi-material design. At this stage, all of industry, such as the automotive, aerospace or maritime industry, and even every design has its own specific manufacturing conditions. The joining technique trends used according to the profile structure are given in Table 2

Table 2. Joining technology trends

Process	Closed Profiles/ Axisymmetric	Open Profiles/ Planar
Clinching		
Riveting		
Hemming		
Hydroforming		
Rolling		
Extrusion		
EMPT		
Legend In general not utilized Primarily utilized		

When deciding on the joining technique in manufacturing processes, many factors are evaluated together and a conclusion is reached. Considering these factors, such as physical and mechanical properties of the materials to be joined, outdoor conditions, joint construction and joint costs, determining the ideal joint type requires a difficult selection process. However, depending on the sector and material type, it may sometimes be necessary to stick to a single joint type, or it may be necessary to choose from a very small number of joining technologies. The classification of some Joining technologies is given in Table 3.

When it comes to lightweight materials, the options are a little narrower. When the joining technologies of metals such as Al, Ti, Mg are examined, it is seen that welding technologies are used more intensively (Anik, 1991). In the manufacturing technique, the aim is not only to obtain parts by combining materials. At the same time, the joined part must not deteriorate under operating conditions and must fulfill its task. For this reason, mergers are expected to meet some expectations.

Table 3. Classification joining technologies

		Mechanical Joints	
Process	**Solid State Weld**	**Force-Closed Joint**	**Form-closed Joint**
Clinching			
Electromagnetic Pulse Technology (EMPT)			
Bar Extrusion			
Hot Forging			
Cold Forging			
Friction Stir Welding			
Hemming			
Hydroforming			
Riveting			
Rolling			
Rotary Swaging			
Spining			
Legend In general not utilized Additionally utilized Primarily utilized			

Metals and alloys in which the welding method is used in the joining technique, with a few exceptions in practice, must be heated to a temperature close to melting in all welding methods. It should not be forgotten that the heat source and welding method used here are also effective. Due to heating to high temperatures, many chemical and metallurgical changes occur in the welding regions of the materials and in the regions adjacent to the weld. For example, in the welding of high carbon steel materials, a very high hardness structure emerges in the region adjacent to the weld due to the high cooling rate. For some materials, although there is no problem, special measures are needed to ensure a defect-free and satisfactory weld quality. In such cases, weld capability is mentioned. A metal or alloy may exhibit very good weldability in one welding method, but may have very poor weldability in another method. Under all these conditions, each of the joining techniques has its own specific conditions and joining capabilities. A qualified join cannot be made before these conditions are met.

Riveting Technologies

Rivet is one of the connected binding elements that cannot be dissolved. The parts to be connected to each other can be of the same type or different. Advances in welding and bonding have limited the use of the rivet. However, in recent times, with the increasing use of aluminum and magnesium alloys and composite materials in the automotive industry, the rivet bond has come to the fore again (Pickin et al., 2007).

The purpose of application of rivet constructions is to connect the parts together, provide sealing, or provide both bonding and sealing. No heat is applied to the combined elements. Therefore, there is no thermal expansion, shrinkage and deformation in the rivet bond. Although it is an easy method to apply, it weakens the part due to the hole drilled in the main parts and creates a notch effect(Çavdar & Durmuş, 2018). Electromagnetic riveting (EMR) technique can be used to rivet plates made of composite material, it is also possible to provide uniform shaft expansion in this method. Damage to the composite structure in the interference zone during riveting is seen as the main problem. As a result of the current developments in flexible electronic systems, micron-sized interconnection technologies are needed. LMR (Laser Micro Riveting) is a new technique that enables the mechanical and electrical bonding of two thin layers of different materials, so that thin film devices can be bonded on flexible polymer surfaces.

Self-piercing rivet (SPR) connections have a large share in the thin-walled structure assembly process in the metal industry, especially in the automotive industry. (Barnes & Pashby, 2000). The latter one demands for the modern solutions for both the car design and the car production technology. As an example, the self-piercing riveting is used by Audi and Jaguar for joining car body pieces. Modern joining by forming technologies such as Self-Piercing Riveting are increasingly used in sheet metal processing industries owing to their many advantages(Mucha, 2011). This is a sub-method of tightened joints, and its main benefit is that there is no drilling and that a variety of materials from different coatings can be joined, from painted and galvanized sheets to plastic materials(Pickin et al., 2007).

Besides, it demands modern solutions for car manufacturing technology and car design. Self-piercing rivets are used by manufacturers such as Jaguar and Audi to join car body parts. Joining technologies such as Self-Drilling Riveting and modern joining techniques are increasingly in demand in the sheet metal processing industries due to the advantages they provide (Mucha, 2011).

Press Joining Technology (PJT)

The usage area of PJT, which is a new sheet metal joining technology, is also expanding. It is clear that this technology will have more flexibility in terms of its mechanism. The material selection of the assembled plate components has flexibility such as different thicknesses of the plate components, the size of the junction points, the geometric shape of the plates, and the way of jointing (Nong et al., 2003). Press jointing technology (PJT) has not yet developed much as it has received little attention from researchers. As the competitive approach of automotive production increases, advanced automotive production technologies(Haghshenas & Gerlich, 2018) will be used more effectively in automobile product design and production. PJT does not use any additional joining components such as screws, nuts, pin bolts (Nong et al., 2003). The diversity of PJT paths has been determined by the characteristics of the PJT. Figure 2 shows five types of PJT techniques.

Figure 2. Five kinds of press joining technology.

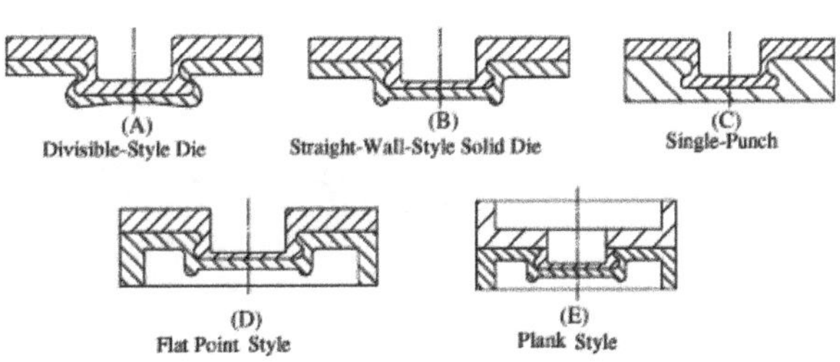

Adhesive Bonding

Adhesive bonding is a joining technique used in the manufacture and repair of a wide variety of products. Adhesive bonding is one of the basic joining processes with rivets, welding and soldering. This technique glues components together using adhesives. The benefits of adhesive bonding have been demonstrated by a number of car manufacturers in concept cars and low volume niche products, e.g. Jaguar's XJ220, Ford's AIV, Rover's ECV3, the Lotus Elise, and to a limited extent in Honda's NSX (Barnes & Pashby, 2000). Adhesive bonding can be considered a relatively low cost process in terms of equipment. However, process automation by robot, necessitates such measures as viscosity compensation, either by variable nozzle orifice size or heated supply hoses, to provide consistent application. It also requires a control system that allows the robot and adhesive dispenser system to interact. Therefore, the issue of low cost is relative.

There are also foreseeable difficulties with extensive utilisation of adhesive joints in volume production. Adhesives have a limited shelf-life and provision must be made for this by materials control. Despite increasingly sophisticated systems designed to counter problems such as increasing viscosity over time the adhesive dispensers are still likely to require regular cleaning and therefore planned routine maintenance to prevent problems. Adhesive bonding joints are susceptible to temperature and other environmental conditions. Mechanically joined composite/metal heterojunctions usually show the high susceptibility to stress concentration, while the bonding parts, such as screw bolts and rivets, are detrimental to the light weight and lead to the severe damage in the strength of the composites (Meng et al., 2019, Blaga et al., 2015).

WELDING TECHNIQUES

Welding science has a structure that includes many disciplines. With this feature, there is no doubt that industrial production is where the source, which has many scientific fields of study, sees real value. Welding techniques are used as a production technique in the emergence of many industrially acquired products that we have today. Welding can be described as the joining of two components by a coales-

Figure 3. Butt weld representation

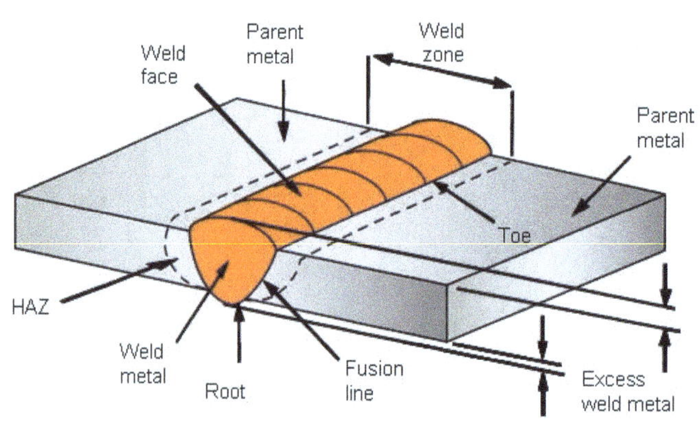

cence of the surfaces in contact with each other. This coalescence can be achieved by melting the two parts together fusion welding or by bringing the two parts together under pressure, perhaps with the application of heat, to form a metallic bond across the interface. This is known as solid phase joining and is one of the oldest of the joining techniques, blacksmith's hammer welding having been used for iron implement manufacture for some 3500 years(Mathers, 2002). Figure 3 can be examined as a traditional example of the welding process.

Welding science has a variety of different processes that have their own techniques and applications for the industry. These can be summarized as follows.

Arc Welding

Arc welding accommodates common manual, semi-automatic and automated processes. These are metal inert gas (MIG) welding, metal active gas (MAG) welding, tungsten inert gas (TIG) welding, submerged arc welding (SAW), rod welding, gas welding, flux cored arc welding (FCAW), gas metal arc welding (GMAW), shielded metal arc welding (SMAW) and plasma arc welding (Anik, 1991). Many of these techniques use a filling material. It is used to join metals such as stainless steel, aluminum and copper alloys, titanium, nickel, cobalt. Arc welding processes are widely used in all industries where metallic materials are used.

Friction Welding

Friction welding, which combines materials using mechanical friction, is a welding technique that can be performed in various ways on different materials, primarily steel, aluminum and wood. Mechanical friction generates heat that softens the mixed materials to form a bond as it cools. The way the joining occurs depends on the technique used. These techniques include linear friction welding (LFW), rotary friction welding (RFW), Friction stir welding (FSW) is known as friction stir point welding (FSSW). No filler metal, flux or shielding gas is used in friction welding. Friction welding is especially ideal for joining light aluminum alloys that cannot be welded. For this reason, it is frequently used in aviation applications. Friction Welding processes are also frequently used in industry. In addition, the processes of using it as a method for joining wood without the use of nails or glue are being explored.

Electron Beam Welding

EBW, a method of fusion bonding, uses a high-speed electron beam to join materials. The kinetic energy carried by the electrons turns into heat when they collide with the workpieces and causes the materials to melt together. Electron beam welding (EBW) is performed using a vacuum chamber to prevent the beam from scattering. It is a convenient method that can be used to join thick workpieces. It is a technique that can be applied in many industries such as aviation, automotive, railways, nuclear energy.

Laser Beam Welding

In Laser beam welding (LBW), which is mostly used to join thermoplastics and metal parts, it uses a laser to provide an ideally concentrated heat. Thus, the high welding speed at which this automated process can be performed makes it a perfect application for high volume applications. Laser beam welding can be done in air rather than vacuum.

Resistance Welding

Resistance welding is a fast and automated process that is widely used in the automotive industry. Heat is generated in the parts to be welded. Pressure is created by electrodes and jaws in the welding machine. The interface creates a resistance to the flow of current and the energy consumed turns into heat. The heat generated in this way ensures that the sheets are in full contact with the effect of the pressure applied by the electrodes. With the effect of heat and pressure, a melting occurs at the interface and a fusion is achieved. This process can be divided into two types as resistance seam welding and resistance spot welding. Seam welding involves replacing electrodes with rotating wheels to ensure a continuous leak-free welding. Spot welding, on the other hand, uses the heat transmitted between two electrodes applied to a small area while the workpieces are clamped together.

WELDING OF LIGHTWEIGHT MATERIALS

Weldability of Aluminum and Alloys

Aluminum has a privileged place among metals, with a rate of 15% in the earth. Due to its high affinity for oxygen, it is mostly found in the form of Alumina (Al2O3). Aluminum can be easily joined by welding, hot and cold soldering or mechanical bonding methods(Mathers, 2002). Some alloys are comparable to mild steel. It has the ability to deform below zero degree. Corrosion resistance is high. It is not toxic. It has good electrical and thermal conductivity. It reflects heat and light very well. It is not magnetic. Aluminum is easy to manufacture. It can be bent, rolled, pressed, drawn, bent and stretched, and rolled(Hirsch, 2010). In addition, it has a structure that can be hammered, heated and processed, or pulled from the mold to give very large shapes. With these features, the industrial usage area is expanding.

Pure aluminum melts at 660 °C. Alloys vary according to alloy content. Aluminium has high electrical conductivity, only three-quarters that of copper but six times that of steel (Mathers, 2002). Although aluminum heats up during welding or soldering, its color does not change. Therefore, it is not easy to see if the metal has reached the melting point. This is a disadvantage when resistance spot welding where the heat for welding must be produced by electrical resistance.(Mathers, 2002) High thermal conductivity compared to steel, more heat is required for the melting welding. Preheating is required for large pieces. Due to its high electrical conductivity, higher currents will be required compared to steel and the welding time will be kept short in resistance welding. More precise control of resource variables will also be required.

Aluminum alloys are the most used materials among non-ferrous materials in terms of production and consumption. Despite a long period of extensive research and practice, commercial aluminum alloys are still not fully understood in terms of phase composition and phase transformations that occur during solidification, cooling, and heating. (Belov et al., 2005).

Aluminium is non-magnetic which means that arc blow is eliminated as a welding problem.

if aluminum alloys contact with air, they immediately form an adhesive and not easily removable oxide film. In melting welding, this oxide film should be followed in order for the aluminum part and the additional metal to join properly and to have a good soldering or bonding. In cleaning agents, the oxide film is tried to be removed by the protective gas arc in the inert gas atmosphere or by mechanical or chemical methods. It is necessary to know the properties of aluminum and its alloys well in order to get positive results in welding aluminum and alloys.

Adding alloy elements in aluminum, the strength properties are increase. Major alloying elements are Magnesium, manganese, silicon, copper, zinc and sometimes lead, nickel and titanium. According to the American aluminum association, aluminum wrought alloys are classified by four letters. This classification is given in figure 4.

Figure 4. Four-digit system for aluminium and its alloys

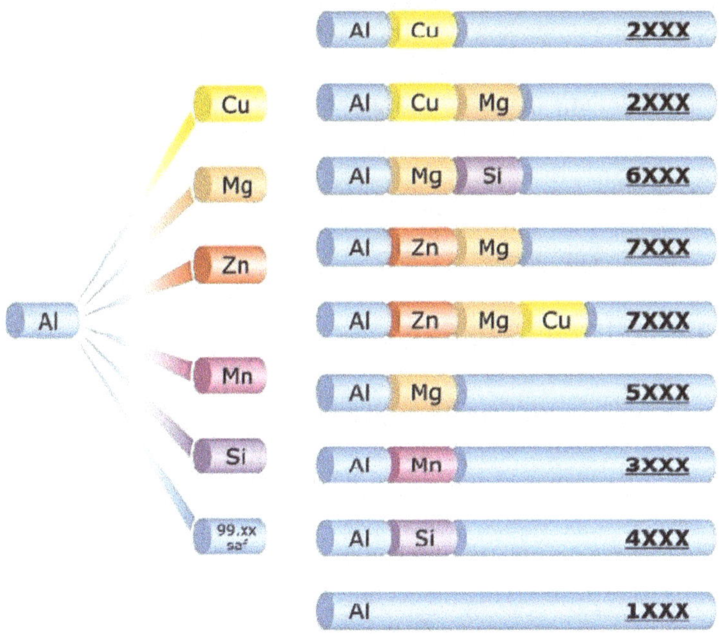

The weldability of aluminum and its alloys can be explained by the following two events:

1. The formation of alumina (Al2O3), which melts more difficultly (2030 °C) and has a higher density, on the aluminum surface causes Aluminum and its alloys to be unsuitable for welding. Its presence prevents the reached droplets from forming a bond, which will ensure that the seam is continuous. To eliminate this difficulty, a special slag-forming cover is used that melts the alumina and is easy to clean.
2. The thermal cycling during welding of some alloys causes the components in the main solid melt to precipitate in the molten zone or base material. This precipitation reduces the mechanical properties and resistance to chemical effects.

Forged alloys; It can be easily welded under shielding gas. When hand-welded to non-heat-treated alloys, their strength is lower than that of the same alloy that is not welded. As is known, three separate regions are formed in the source area.

- Melting zone
- Heat affected area
- Heat-unaffected zone (unchanged zone).

In heat treated alloys, the ability to be drawn in the form of wires decreases. If aging hardening is done after welding, the durability increases.

The coefficient of thermal expansion of aluminium is approximately twice that of steel which can mean unacceptable buckling and distortion during welding(Mathers, 2002). The thermal conductivity coefficient is six times that of steel. The result of this is that the heat source for welding aluminium needs to be far more intense and concentrated than that for steel. This is particularly so for thick sections, where the fusion welding processes can produce lack of fusion defects if heat is lost too rapidly.

Gas metal arc and gas tungsten arc welds are preferred because of their easy application and economic properties in welding high strength aluminum alloys. MIG, TIG and friction stir welding can be used as a joining method to welding of many metals such as stainless steel, as well as for welding aluminum and its alloys. The effect of welding methods used in parts, which are difficult to obtain in casting, forging and machining, on the part, and the change of heat treatments performed before welding, after welding in various aluminum and alloys are important.

Weldability of Magnesium and Alloys

Magnesium alloys are important materials for semi-finished and bulk product casting in areas such as the automobile, electronics, computer and communication industries. In addition, the ease of recycling, compared with polymers, makes magnesium alloys environmentally attractive(Pollock, 2010). For the widespread use of magnesium and its alloys in industrial fields, further development is needed for magnesium alloys regarding mechanical and tribological properties of materials, such as physical properties. At present, the commonly used methods for connecting Mg/Al dissimilar metals are ultrasonic spot welding (USW) (Jahn et al., 2007), braze welding (BW), resistance spot welding (RSW) (Han et al., 2011), laser beam welding (LBW)(Pakdil et al., 2011) and friction-stir spot welding (FSSW) (Shen et al., 2019)

In order to ensure the use of Mg and Al alloys in the industry, it is necessary to develop reliable connections between Mg and Al alloys. Fusion welding used in Mg and Aluminum alloys always produces coarse grains and large brittle intermetallic compounds (IMC) in the weld metal.(Panteli et al., 2012) In this case, it is practically not possible to combine Mg and Al alloys with melt welding (Wang et al., 2015).

Two main processes are used in welding magnesium alloys, the metal inert gas (MIG) process and the tungsten inert gas (TIG) process. Both methods only require little expense for new equipment. Moreover, both processes can be used for manual welding, which is important for repairs or small volume units. The disadvantages are low welding speeds, significantly larger heat affected zones, and higher part distortion compared to electron beam welding.

It is known that alternating current is used in TIG welding of magnesium.. Like aluminum, the negative pole reduces the heat charge on the ball electrode while the positive pole is needed to break up the oxide layer. Argon mixed with helium is used as the inert gas. Helium has a positive effect on penetration depth.

There are alloying elements that are regularly used in magnesium alloys. If magnesium alloys are obtained by casting, serious problems with the formation of porosity occur in the melt bath during welding. Scientific research on this subject shows that, as in aluminum, the dissolved process gas is released during melting. Additionally, porosities can grow in the melt. Because under high pressure, the porosities are surrounded by the solution and their exit to the outside is prevented. This growth is undesirable because it is in the form of porosities forming together in the form of foam. Therefore, the supporting cross-section of the seam is reduced and has undesirable consequences in mechanical properties. The first possible strategy to avoid this situation is to avoid joining methods involving liquid phases. Choosing a friction source as another strategy gives a better result.

The production method does not adversely affect the weldability of the wrought alloys. In addition, the applications of extruded profiles and plates will gradually increase in all areas. This will happen not only by the development of production methods, but also by the development of suitable joining methods. The heat input of the welding process affects the base material as seen in other forged materials. The width of the heat affected zone depends on the welding process. Melts, grain growth and sediment particles at the grain boundary can be identified in the heat affected zone. It is necessary to consider when the temperature of dissolutions or dislocation or recrystallization of fine grains is exceeded that may cause softening. Mechanical properties will not change with the correct combination method and additive. In fact, wrought alloys can be said to have a high degree of weldability.

Magnesium has significant potential for weight reduction, but still requires more effort in the development and research of joining technologies. The various joining processes with their specific advantages form the basis of a wide range of applications.

Weldability of Titanium and Alloys

Titanium, the ninth most abundant element in the earth's crust, is known as a metal for its high strength and weight ratio (Banerjee & Williams, 2013). It is a highly ductile, shiny and metallic white colored, low density metal with high strength. Also It is very useful as a refractory metal due to its high melting point (1.660 °C). Its paramagnetic properties and low electrical and thermal conductivity are its distinguishing features from other metals (Ezugwu & Wang, 1997).

Titanium alloys have high corrosion resistance(de Assis et al., 2006), high tensile strength / density ratio, fatigue resistance, high crack resistance, and the ability to withstand moderately high temperatures without creep(Peters et al., 2003). Because of these properties, they are used in armor coatings, aircraft, ships, spacecraft and missiles. For these applications, titanium enables the production of critical structural parts. Naturally, it is possible to add elements such as aluminum, nickel, zirconium, vanadium to the alloy in the production of these parts. As far as it is known, about two thirds of all titanium metal produced in the world is used in aircraft engines and chassis. It is known that a large number of Titanium alloys are used in the commercial field. Aluminum is the main element in titanium-based alloys.

The sensitivity of titanium and its alloys to gases limits the use of various welding methods and requires a number of additional measures. Leaving these limitations aside, it can be said that most of all Ti materials are characterized by good weldability. Some high strength alloys such as TiA17Mo4, TiA16VSn2 and TiA14MoV3 tend to form delayed (cold) cracking, stress corrosion and hardening due to their high yield point / tensile strength ratios(Rüdinger & Ismer, 1973). A stress relief annealing or complete heat treatment can help prevent them.

In summary, Titanium and all α alloys can be welded. With the α + β alloy, other lightly β stabilized alloys can also be welded (Ranatowski, 2008). However, strongly β stabilized α + β alloys become brittle in the welding process. Most β alloys can be welded successfully. However, the strengthening process by aging hardening of the welding should be applied carefully. Aged welds in some yaş alloys may become completely brittle. All grades of unalloyed titanium can usually be welded in the annealed condition rather than in the cold worked state.

Most of Titanium and its alloys can be welded mainly by TIG, plasma-arc, MIG and electron Beam Welding processes(Ranatowski, 2008). Applications and hardware are generally the same as those used in austenitic stainless steel or aluminum. However, due to the high activity of Ti and its alloys above 535 ° C, more care is required in the protection of the weld and the hot root side of the joint than with austenitic stainless steel or Al alloys. Due to their high chemical activity, titanium alloys are easy to absorb harmful gas (oxygen, hydrogen and nitrogen) and many problems arise such as low mechanical properties and unstable structures(Yunlian et al., 2000).

Argon, helium or their mixture is used for welding Ti and its alloys. The widespread use of argon is due to its cheaper and easy availability. It is especially used in thin sections where less heat is required and the arc length can be changed without a noticeable change in heat input. Helium is often used in automated work. Arc control is easy in these vehicles.

WELDING METHODS USED IN LIGHTWEIGHT MATERIALS

Metal Inert/Active Gas Welding Method (MIG-MAG)

MIG welding has a wide range of applications. It can be used in the welding of metals and alloys of all thicknesses, except for very thin plates. In MIG - MAG welding, the arc burns between a melting electrode that also acts as an additional wire and the workpiece. The shielding gas is either an inert gas such as argon or helium, or a mixture of these (MIG) or an active gas (MAG). The current flows from a wire spool to the contact tube with a leadless electrode, a wire guiding mechanism. The free wire lead is relatively short so that high current intensity can be applied to the thin electrode. One of the poles of the welding machine is connected to the electrode and the other to the work piece. Thus, the arc burns between the melting electrode and the piece.

Compared to other welding methods, MIG welding can be mechanized, work faster, use robots, can be applied easily in very complex welding constructions and can be used in any position. In addition, it provides many advantages in terms of welding all commercial metals such as carbon steel, stainless steel, aluminum and copper. In addition, the market share percentage of this method is increasing day by day. Although this technique can be applied to aluminum and alloys of any thickness, it is generally preferred for welding aluminum and alloys thicker than 3 mm. Because the welding speed and melting power in MIG welding is higher than the other gas welding method, TIG welding, very thin plates are only welded by applying the pulsed current method.

Although aluminum and its alloys melt in the temperature range between 550 - 660 °C, due to their high thermal conductivity, the heat input required for welding must be higher than that of the steel of equal thickness. The large thermal expansion coefficients of aluminum (Kim et al., 2001) and its alloys cause temperature differences and severe stresses and large distortions that occur as a result of heating and cooling in the weld zone. The structure of the MIG welding is as shown in Figure 5.

Figure 5. Aluminum welding made with MIG technique

The refractory aluminum oxide layer formed as a result of contact with air on the aluminum makes welding of aluminum and its alloys significantly difficult. In welding with direct current and reverse polarity (electrode in the positive pole), the oxide layer floating on the bath is broken up and welding is performed only with this polarity(Anik, 1991). In MIG welding of aluminum and its alloys, it is always preferred to weld with spray arc regardless of the thickness of the material. Despite the high heat input of the spray arc, the high thermal conductivity of aluminum makes it possible to weld in any position as the weld bath solidifies very quickly. Only here, left welding method should be chosen to remove the oxide layer and vertical position welding should be done from bottom to top. Thus, both the oxide layers in the areas to be welded are cleaned and the welding nozzles are melted well and welded properly.

The high thermal conductivity of aluminum alloys causes severe cooling of the welding area, especially in thick parts. In this respect, the preheat temperature applied to thick and especially cast aluminum parts should not exceed 200 ° C. In forged aluminum alloys, in general, higher current intensity and arc voltage instead of preheating provide higher heat input (Kuhlman, 2005). In this respect, a solution annealing is applied to aluminum alloys hardened by heat treatment before welding and after welding, they are hardened by applying heat treatment again. The application of the MIG welding method of Aliminium is very simple. An arc is created by connecting the ground wire to the workpiece or welding table and by touching the wire electrode at the torch tip to the welding nozzle(Anik, 1991). The machine automatically ensures the advancement of the wire and the appropriate arc length. MIG welding has become a very popular and sought-after method for welding all non-ferrous metals and alloys due to its ease of application. The arc region in the MIG method is shown schematically in Figure 6.

The most important factor determining the quality of the weld connection is the weld parameters chosen. For this reason, if the appropriate welding parameters are selected, MIG welding method can be applied successfully and comprehensively in many sectors in the welding of aluminum and its alloys, and its applicability seems to increase day by day.

Figure 6. a) MIG Welding Process b) MIG welding hardware block diagram

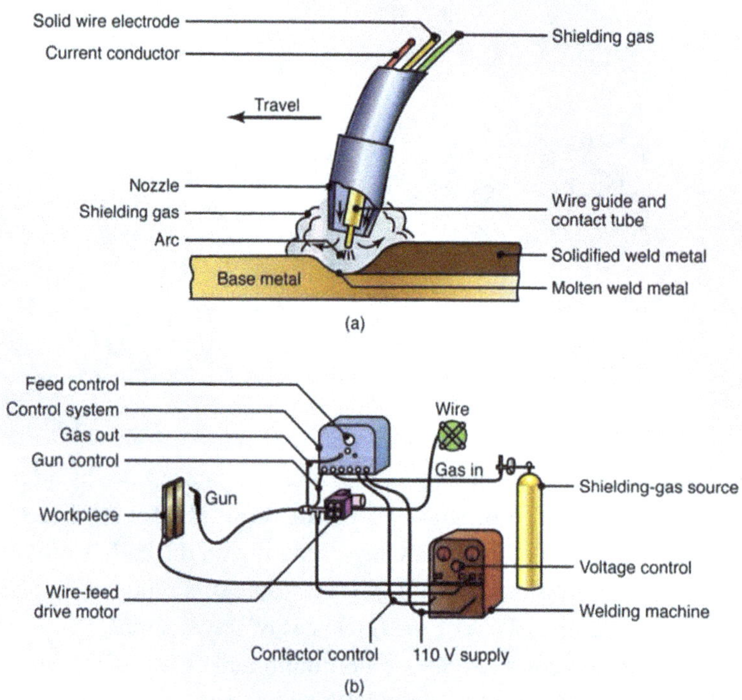

Tungsten Inert Gas Welding (TIG)

TIG welding, which is an arc welding method, is especially suitable for welding light metals such as aluminum, magnesium, titanium. Since electric current, water and gas must be continuously supplied and controlled in TIG welding method, the equipment used in this method is more complex and expensive than those used in arc welding. The word "Tungsten" in the name of this process is the non-melting electrode that enables the rear electric current to be transmitted. The word "inert" symbolizes a gas that chemically combines with other elements, and the word "gas" stands for the material covering the molten bath and arc, keeping the air surrounding the weld zone. With the TIG method, seams with superior properties are obtained compared to welding generally created by other welding methods.

In TIG welding, the arc burns freely between the tungsten electrode and the workpiece. Shielding gas consists of argon, helium or a mixture of these. One pole of the energy generator is connected to the tungsten electrode and the other to the workpiece. The arc also burns between the tungsten electrode (continuous electrode), which is only an electric conductor and arc carrier, and the part. Additional material, as a rule, is not charged. With the tungsten electrode, the molten bath and the molten tip of the additional metal are protected from the atmosphere by an inert shielding gas fed coaxially with the electrode from a shielding gas nozzle where the electrode is located (Figure 7).

A TIG torch is a specially designed electrode holder. The torch is manufactured to easily handle various sizes of tungsten electrodes and is equipped with a replaceable gas nozzle to direct the flow of shielding gas. Some torches are air-cooled, but water-cooled torches are more commonly used. As a rule, direct

Figure 7. Schematic representation of TIG welding

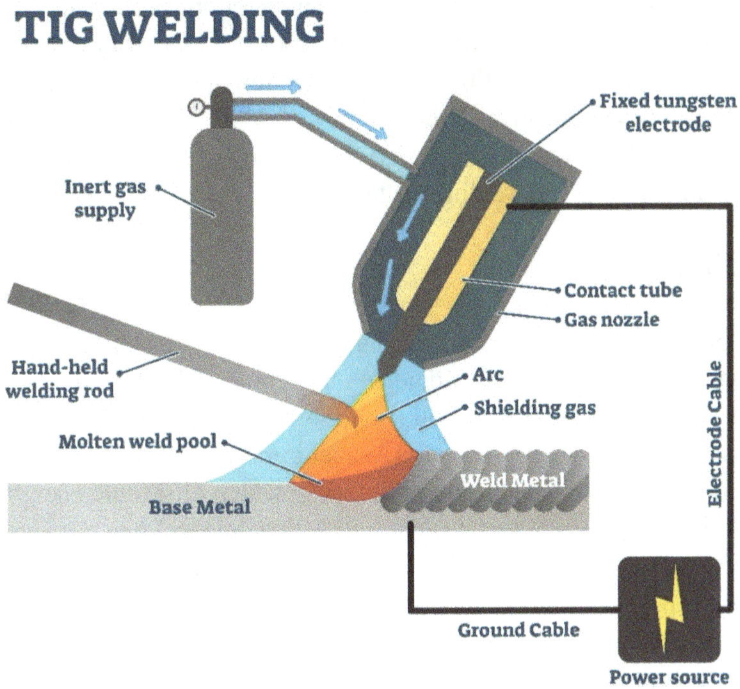

current welding is performed in TIG welding. An exception is light metals such as aluminum, magnesium and their alloys, as well as brass, beryllium, copper and copper alloys are welded with alternating current.

In materials containing an oxide layer melting at high temperatures, the solid oxide layer causes the weld pool to flow and prevents the drops from joining with the layer on which they fall. With its melting temperature of about 2050 ° C, aluminum oxide is one of these oxides that is very difficult to decompose in melting welding. Removal of this layer in TIG welding is provided by the charge carriers in the arc. Electrons move from cathode to anode where they generate heat as a result of impact. Ions move in the opposite direction. However, the kinetic energy of the ions can only be applied on the surface of the weld bath when the electrode is the anode and the cathode in the part. However, in this way the cleaning effect is significantly lower, because the strong heating of the positively polarized electrode weakens the current strength. A good average of this situation is obtained by using alternating current. Changing the pole allows the oxide layer to break up (Cathodic cleaning) and cool again when the electrode is the negative pole, respectively, as the electrode is positive pole.

Oxy-Acetylene Welding

Before the development of the capabilities offered by the arc welding technique in steel, oxy-acetylene welding was the only process capable of welding high quality in all metals. Welding capabilities included metals such as aluminum and magnesium as well as almost all alloy steels.

Oxyacetylene fusion welding is the technique that plays the most important role in the development of modern welding methods and design. (Civjan et al., 2020). Oxy-acetylene welding has limited use for simple diameter repair operations and oxy-cutting, and in a wide range of areas from smaller repair shops to large-scale heavy industry. The oxy-acetylene welding torch is used as a basic heat source for manual soldering and brazing as well as metal forming, preparation and localized heat treatment.

Oxy acetylene flame is generally used in gas melting of aluminum and its alloys. Since the oxy acetylene flame gives a temperature of 3200 ° C, it is successfully used in the welding of thick sectioned parts. Oxygen - The hydrogen flame gives a temperature of 2800 ° C and is rarely used in the welding of thin sheets. Coal gas - Oxygen (oxygen - air gas) flame also creates a temperature of approximately 2000 ° C and is especially used in the welding of thin sheets.

Ultrasonic Welding

Ultrasonic welding process is a rapid manufacturing process used to weld thin layers of metal at low temperatures and low energy consumption(Siddiq & Ghassemieh, 2008). This welding technology was invented and patented in the early 1930s. A solid-state welding technique is the formation of the bond occurs as aresult of a moderate static pressure and a süper imposed ultrasonic oscillation without reachingtheir melting points. Ultrasonic welding is a process in which ultrasonic energy is used to create a solid-state bond between twopieces of metal (Figure 8). Ultrasonic welding is a versatile and pow-erful joining technique in the microelectronic packagingindustry because of the low temperature, high yield rateand flexibility of the process(Siddiq & Ghassemieh, 2008). Compared to soldering, adhesive bonding, fusion welding or other joining techniques, ultrasonic welding is a technique that stands out with its low energy input, therefore suitable for welding in very short welding times with low temperatures in the welding area and with reasonable investment costs.

Figure 8. Schematic of the weld unit and ultrasonic welding setup.

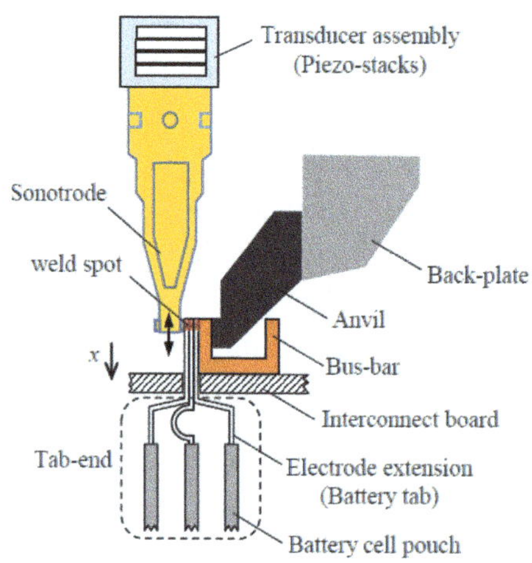

In ultrasonic metal welding processes, high-frequency ultrasonic energy is used to generate an oscillating shear force (sonotrode force) at the interface between a sonotrode and few metal sheets to produce solidstate bonds between the sheets clamped under a normal force. These forces, which influence the power needed to produce the weld and the weld quality, strongly depend on the mechanical and structural properties of the weld parts and fixtures in addition to various welding process parameters such as weld frequencies and amplitudes(Kang et al., 2013).

The ultrasonic metal welding method is characterized by short welding times (<1 sec for spot welds of metal, glass and ceramics and <4 sec for spot welds of GFRP and CFRP with metal) and consequently low energy input. This joining process is clean and environmentally friendly. It does not emit sparks, flame and smoke. The automation capacity of ultrasonic welding process is high. It is possible to monitor welding parameters electronically with data storage and statistical evaluation. These advantages of ultrasonic welding technology lead to more industrial use.

Ultrasonic metal welding is well suited to realize aluminum alloy/carbon fiber reinforced polymer (CFRP) – joints(Balle et al., 2011). For ultrasonic spot welding wedge reed and lateral derive system configuration are commonly used. In both systems the electrical energy is converted by the transducer to shear vibrations of the same frequency which are transmitted to the work pieces placed under moderate clamping force. High frequency shear vibrations (typically 20 kHz) with the combined effect of clamping force remove the oxide and contaminant layer as well as produces heat by shear deformation. As a result, nascent metals come in contact with each other and create adhesion as well as inter diffusion across the interface(Annoni & Carboni, n.d.; Shakil et al., 2014). Ultrasonic welding methods can be seen Figure 9.

Figure 9. Ultrasonic welding methods (Wagner et al., n.d.).

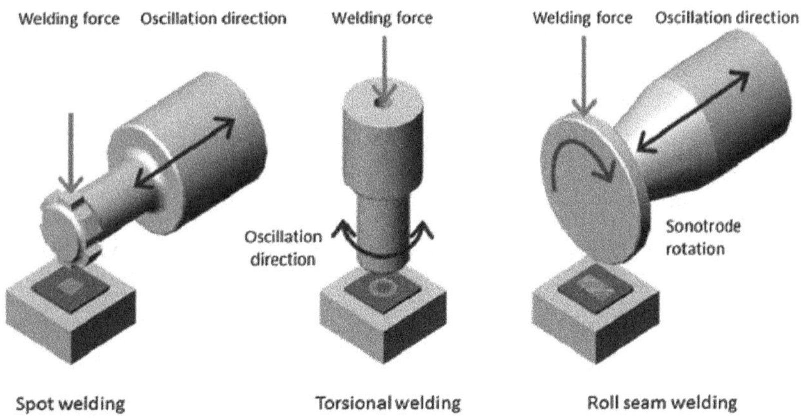

Laser Beam Welding (LBW)

Laser welding is a high-power-density fusion-welding process that produces high aspect ratio welds with a relatively low heat input compared with arc-welding processes(Kalaiselvan et al., 2021). The laser beam welding (LBW) with high energy density and welding speed is a new welding technology(Yunlian et al., 2000). Since the 1980s, Laser technology has been seen as the key technology for joining components.

Laser welding of lightweight materials for industrial applications is expanding rapidly. This trend is driven in part by the expected surge in electric vehicle market growth, as well as continued efforts to reduce weight in construction of conventional automobiles. The family of Alüminum alloy that is critical to this industry sector present considerable challenges when approached using conventional laser welding techniques. The main advantage of laser welding, due to its high energy density, is its ability to melt the area located at the edges of the joint, without affecting a large area of the part. Laser welding is a high-power-density fusion-welding process that produces high aspect ratio welds with a relatively low heat input compared with arc-welding processes(Kalaiselvan et al., 2021).

Aluminum alloys are known to be potential materials for lightweight structures in the transportation industry. Aluminum, which is preferred due to its good mechanical properties and low density, will find a wider area of use with a suitable joining technology. Laser welding offers great advantages for Aluminum at this point. Laser welding with high bond strength, low distortion and high processing speeds is unrivaled due to low local energy input. The deep penetration effect in the weld provides concentrated heat input and enables deep weld seams to be obtained (see Figure 10).

Figure 10. Deep penetration effect by laser beam welding(Schubert et al., 2001).

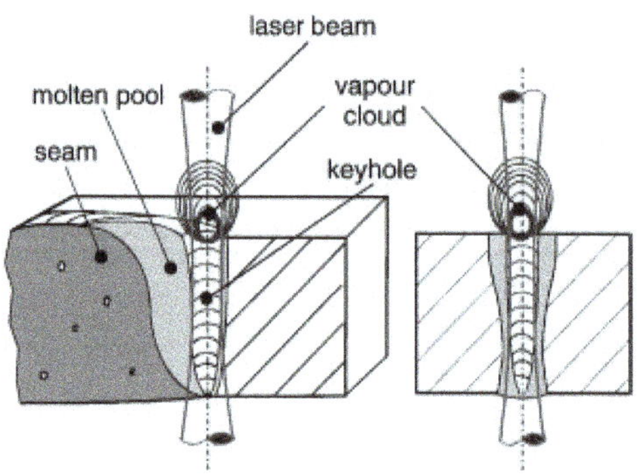

Solid-state lasers with a power output of 1 to 2.5 kW have been used for all calculations made to date. The welded plates were kept between 1.0 and 3.2 mm thickness, and the welding speed range was set between 2.5 m / min and 9.0 m / min. In this respect, magnesium's excellent absorption behavior is very important. Magnesium wrought alloys welded by laser welding are almost equivalent to the strength of the base material. Porosities and penetration cuts can be eliminated by proper adjustment of the parameters.

Laser beam welding is considered a suitable joining process for high-quality manufacturing of aircraft structures produced from aluminum alloys, which are preferred due to their favorable properties(Yunlian et al., 2000). By determining, e.g. the costs for the laser joining process it is possible to decide whether a laser based substitution structure can be economically manufactured (Schubert et al., 2001) You can see the Laser beam welding process in Figure 11.

Figure 11. Laser beam welding process

Electron Beam Welding (EBW)

EBW is a welding method performed by bombarding the joint to be welded with an accelerated dense electron beam up to 0.3-0.7 times the speed of light. This is a high energy density fusion process. As the kinetic energy of electrons hits and penetrates the workpiece, it converts into thermal energy. This energy, which causes the interface surfaces to melt, constitutes the essence of the welding process that allows the workpiece to melt and merge.(Tejedor et al., 2013).

Electron beam welding (EBW) is an important technology under normal atmospheric conditions. The process offers fast production, advanced jointing ability for the production line. In addition, it is easier to use as there is no need to create a vacuum from the electron beam source in vacuum. Additional materials may be required to prevent notches in the seam. This technique lowers the cost and makes the beam generation very high. In addition, the weld plasma is permeable to electrons and does not affect weld speed. Changes in beam efficiency due to reflection or absorption in this case prove otherwise. EBW has some advantages.

- Good bonding of the beam,
- High working efficiency,
- High process speeds,
- High potential for automation,
- High process safety and usability,
- Low reflectivity even on light metals,
- High power density,
- Lower cost than laser welding.

The electron beam with energy densities up to 10^8 Wcm^{-2} allows deep penetration and creates homogeneous, narrow V-shaped welds and small heat-affected zones (HAZs). Small weld widths and HAZs are important in small insert applications. In most situations the components are welded together without additional finishing requirements. However, by preference, the specimens should be welded from both sides—each time to half the thickness of the material—to avoid bending the specimen(van Walle, 2001).

There are some restrictions for the use of Electron Beam Welding. These are limited working distance, X-ray shield requirement, High expense costs. As a conventional Electron Beam Welding, the beam for EBW is produced in a vacuum environment. In contrast, the accelerated voltage usually starts at 150 kV or higher. The beam then passes through the special inlet containing the caps, leaving the vacuum and exiting the atmosphere. From there the inert gas or air passes through the gap and reaches the workpiece. As soon as the beam enters atmospheric gas, it reacts with molecules and expands. This effect can be reduced by using inert gas or by mixing. For more active metals, the use of helium is recommended to prevent void formation. Also, overheating of the transforming gas causes lower density, therefore the width of the beam is limited. The same behavior is observed when the beam passes through the metal vapor during welding. You can see Figure 12 for EBW.

Electron beam welding can be applied to all commercial Ti alloys that can be combined with arc welding. The ability of these alloys to weld with EBW and the way they respond to the heat cycles of the process are generally the same as against arc welding.

The vacuum environment of the electron beam welding is not affected by the atmosphere and does not require detailed and expensive protection methods of arc welding processes. Indeed, in many cases

Figure 12. Elektron beam welding

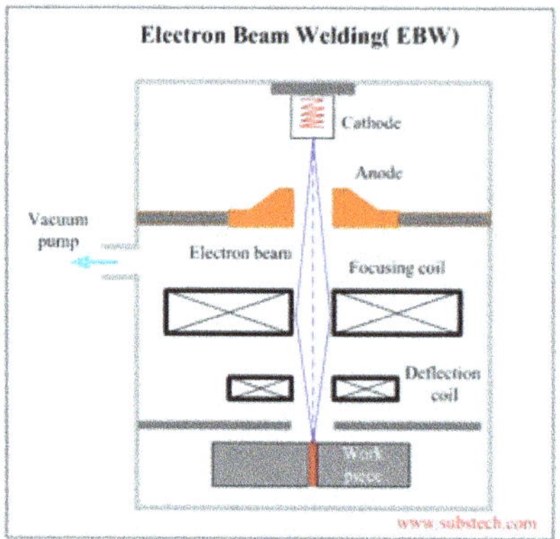

EBW is more economical than, for example, MIG welding. The most stringent, high requirements on the seam characteristics of Ti joints are met by EB welding.

Friction Welding Techniques

Friction welding (FW) is a process used to join similar and dissimilar metals and polymers (Singh et al., 2017). Friction welding is a welding method in the pressure welding group. For the welding of clamped parts, the heating process is provided without friction. By the friction obtained from the relative motion between the compressed and rotated components, the two parts are then clamped together without any additives (Figure 13). As a result, the weld shows a swelling typical for this process. When sufficient heat input is obtained, the rotation is stopped. Such heat input also results in a relatively low junction temperature (<T1). Therefore, it offers an ideal technique for materials that are difficult to join.

Although there are different types of friction welding, the most used methods are Rotary Friction Welding (RFW), Friction stir Welding. RFW has a wide range of applications, from the automotive industry to the rail and aerospace industry. The main benefits of RFW are process stability, high quality connections and good potential process automation. Friction stir welding (FSW) is a prominent solid state welding process that is widely used to obtain excellent joints in heat-treatable aluminum alloys(Rathinasuriyan et al., 2021). (FSW) is one of the green manufacturing technologies since it produces minimum emissions to the environment(Bag & Akinlabi, 2020).

The process steps of the friction welding process are as follows.

1. With the sliding friction, heat is generated and the maximum torque value is reached.
2. With the heat generated by mechanical propagation, the material is plasticized and softened. material flows radially outward.
3. A steady state is reached At this point the torque, temperature distribution and axial shortening rate are essentially constant.
4. The turn ends and upseting occurs

Heat is produced due to friction between tool shoulder and workpiece which softens the material around the pin and produces a few microstructural changes in and around weld zone. Resultant micro-

Figure 13. Friction welding process

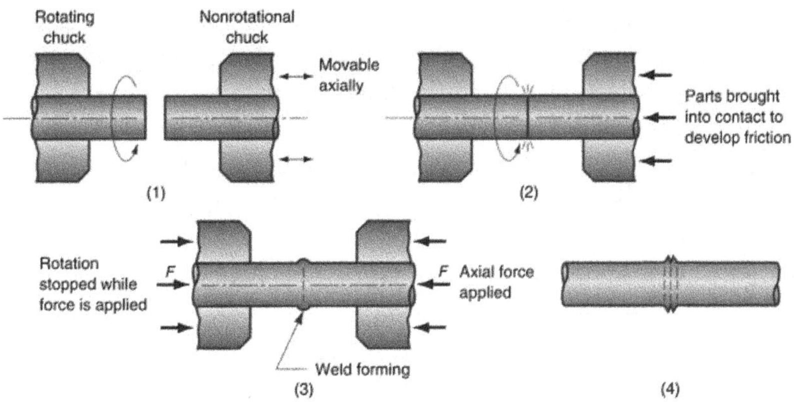

structure of the weld cross-section, the size and distribution of the precipitates in the case of precipitation strengthened alloys, mechanical properties of the weldment, and other microstructural features are dependent on the welding parameters, tool dimensions, initial microstructure of workpiece material, and temperature distribution during welding(Mishra & Sidhar, 2017).

In addition, the process allows the welding of different types of materials as it enables the formation of intermetallic phases without the formation of a liquid phase. For steel and aluminum combinations, this method is currently used on a large scale and has wide application for composite design. The advantages of friction welding are listed below.

- No additives,
- No inert gas,
- Provides high capability for automation,
- There is no fluid phase,
- Composite production is possible,
- Provides minimum effect on the base material.

Diffusion Welding

Due to the increase in the variety of materials used in today's industry, the necessity of different metal connections in places requiring different properties, and the increasing importance of economic factors especially in recent years, it has become necessary to combine materials with different properties. The joining of advanced materials, which are mostly used in the aerospace and aircraft industries, is possible with methods known as solid state welding techniques, including diffusion welding.

Diffusion welding is preferred in metals and alloys with dissimilar metals and alloys, when the strength of materials is reduced due to the formation of brittle intermetallic phase in the joint area after welding. The diffusion welding method, which can be applied at low temperatures, can combine different materials and has other advantages, which cannot be obtained with conventional welding methods, is a joining method that is developed day by day and will find a much wider application area in the future.

Diffusion bonding is a solid state bonding process in which two prepared surfaces are combined at high temperature ranging from 0.5 to 0.8 Tm under pressure. The process has a significant impact on the design and production of workpieces. Bonding similar or different materials provides significant advantages for a joining technique(Wu et al., 2007). It is possible to bind all materials with different chemical and metallurgical properties with diffusion bonding. The process depends on various parameters, especially time, applied pressure and bonding temperature to promote microscopic atomic motion to ensure full metallurgical bonding(Aydin et al., 2012).

As is explicitly indicated by the designation of this joining method, diffusional processes are essential in establishing a metallic bond between faying surfaces. For these transport processes to be operative, surfaces which are to be joined must be brought in intimate contact. Also, for them to be effective, atomic mobilities must be sufficiently high to cause mass transport across and along the interface. In practical terms these requirements can be satisfied by the application of pressure to bring about better surface-surface contacts and by increasing the temperature to enhance atomic mobilities

Diffusion plays a key role in the superplastic deformation process that progresses at very slow temperatures and high deformation temperatures. Therefore, it is possible to combine diffusion bonding and super plastic forming in one process. This can be extremely interesting for magnesium alloys. Otherwise, they are only castabFigle materials due to their limited deformation capabilities. Today, the diffusion bonding method is used in the aerospace industry, especially for titanium alloys that are difficult to handle. In this case, even complex components can be produced from titanium sheets and profiles, which is not possible with conventional methods. The diffusion welding process sample can be seen in figure 14.

Figure 14. Diffusion bonding process

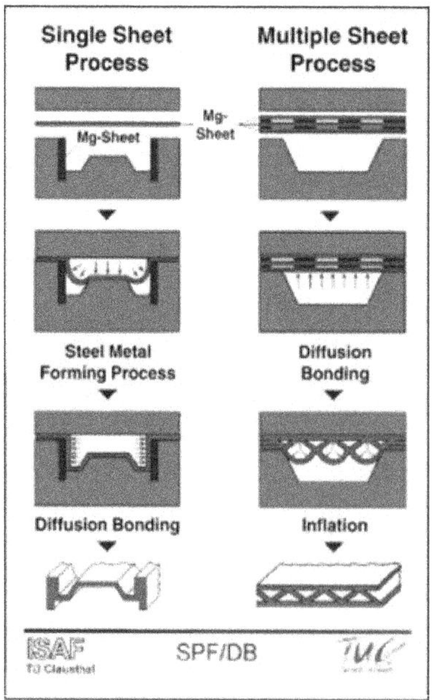

To predict the bonding process precisely, the actual information of roughness of the intended bonding surfaces and the kinetic conditions of operating mechanisms in diffusion bonding should be considered (Wu et al., 2007). There are four important control factors in affecting the bonding characteristics. These are Bonding surface roughness, bonding time, bonding temperature and bonding pressure.

An intermediate layer is generally used in the application of diffusion welding, especially when joining different metals and alloys. It is possible to reduce the temperature, pressure and time values required for welding by using metal foil or coating-like intermediate layers. Interlayer minimizes the heterogeneity in the weld area and facilitates the formation of the joint. The intermediate layers can be in the form of foils, electrolytic coating and spraying.

The functions of the intermediate layer can be listed as follows;

- Balances parts with different physical properties, such as thermal expansion rates,
- It prevents the formation of fragile, hard metal phase formation,
- Improves surface convergence with plastic deformation,
- It reduces the preparation stages of the parts to be joined,
- They reduce the working temperature.

As is known, cohesion forces occur in crystalline materials due to the attraction between atoms. Normally, the equilibrium condition for atoms is that the force acting on them is zero. However, when the solid material is stretched by the effect of external forces, the atoms leave their equilibrium state and create a tension within the crystal that is balanced by the external force. The force of attraction between atoms increases in proportion to the degree of divergence from each other, decreases again after passing through a maximum. The attraction between the surfaces of two different solid materials is defined as adhesion. While cohesion occurs between solid surfaces of the same type due to cohesion forces, adhesion work is done between different solid surfaces due to adhesion forces. Both cohesion and adhesion are a function of the free energy of the surface. Also for a joint consisting of two brittle materials or a brittle ductile material (metal-ceramic) adhesion is a measure of the bond strength. It is necessary to ensure that the material surfaces are in close contact with which an atomic bond can be formed in order to ensure unification between the material surfaces, that is, the formation of adhesion and cohesion forces. Another force that significantly affects the joint behavior of two materials by diffusion is the Van Der Wallis forces acting between molecules depending on the chemical structure of the materials.

Elektroslag Welding (ESW)

In recent years, the use of electroslag welding (ESW) has been increasing. Electroslag welding does not require highly skilled labor and produces relatively error-free connections at fast deposition rates. Another advantage of the process is that it does not require close connection equipment. As the plate thickness used in welding increases, it becomes more efficient. Studies have shown that the weak fracture toughness caused by the severe thermal cycle in the heat-affected zone in the welding of electro-slag originated materials is also due to grain coarsening.

In the electrocurf welding method, a heat occurs as a result of the resistance shown against the electric current passing through the molten slag. With this heat, the slag is kept in a molten state, the filler metal (electrode) is melted and the joining edges of the main part to be joined are melted and the welding process is performed. The weld pool is protected by this molten slag. The molten slag moves with the weld pool along the joint section. Liquid slag is kept in a liquid state with the heat generated as a result of the resistance of the slag against the current while the electric current passes between the work bar and the electrode.

In the method, a rectangular space is created for the welding process by closing between the plates to be welded with water-cooled copper shoes. The surfaces coming out of the cutter are placed between the lugs and welded directly. The depth of the slag bath is between 30-50 mm. The excessive slag thickness makes it difficult to pass the current, reducing the depth of the working seam and lowering the temperature of the weld puddle. Approximately 5 kg of powder is needed for 100 kg of weld metal in the method. When using flux cored electrodes, there is no need for dust for slag. The required powder is obtained from the core of the wire. Figure 15 shows a schematic picture of the method.

Figure 15. Elektroslag welding process

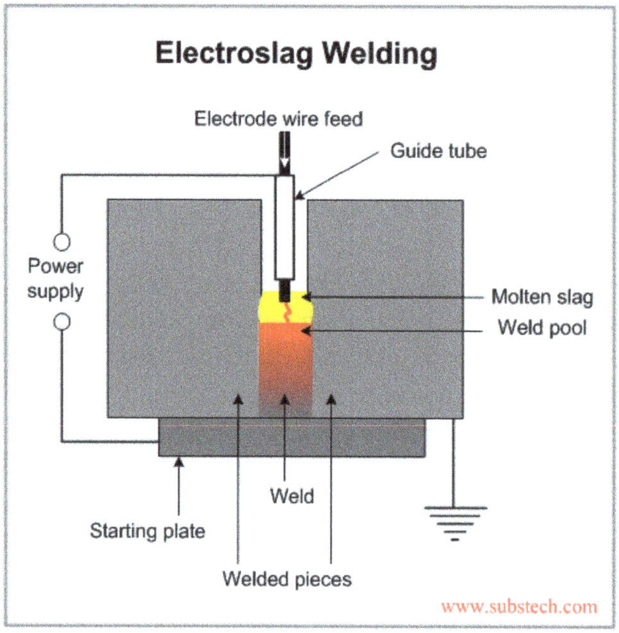

Thick (> 40 mm) Ti or Ti alloy parts can be best joined by electroslag welding. Short seams in Ti up to 400 mm thick are commonly welded with plate electrodes. It has been customary to protect the slag bath under a blanket of pure argon to keep gaseous impurities out of the weld.

RESULTS

The weight of components is an important issue in all areas of daily life (Brumm & Bürkner, 2015). This inevitably involves the application of lightweight materials, e.g., magnesium, aluminum, and titanium based alloys and composites(Pollock, 2010). However, in practice, within the framework of economic criteria. The use of steel to provide high strength is still the most is the preferred choice of material (Han et al., 2011). In this case lower density of aluminum and titanium alloys, must be used with steel. However, joining light materials to be used with steel (Nong et al., 2003) is a challenging process. For example, the use of welding for multi-material joints is not directly possible due to different physical properties of dissimilar materials, such as melting temperatures and micro-structural incompatibilities

Although there are a wide variety of techniques used in joining light materials(Kalaiselvan et al., 2021) different methods are used for each material, and it is seen that hybrid technologies are used together with traditional techniques in general (Troschitz et al., 2020). As all of these methods have individual advantages and disadvantages, a general recommendation for one technology is not possible. The automotive and aviation industry's need for these joining technologies leads to more scientific studies in this area.

Rivet, which is an unsolvable/shape-bound bonding element between the same or different types of elements, has come to the forefront again in recent years with the increasing use of aluminum and magnesium alloys and composite materials in the automotive and aerospace industries(Çavdar & Durmuş, 2018). However, the rapid development and new designs in composite materials will enable more effective use of bonding techniques (Barnes & Pashby, 2000) in joining processes with their numerous advantages.

Welding technologies find a widespread use almost as the primary method in production. However, with the developing technologies, new welding techniques are also being formed. Wide range of welding technologies have become a solution point for almost all types of metallic materials.

In addition to the weight advantage of lightweight materials, it is also important to improve their mechanical properties by heat treatment. The innovations in coupling technologies that accompany them will soon enable us to get to a better point in reducing carbon emissions. It is clear that we will continue to follow new designs and applications in the automotive and aviation sector.

REFERENCES

Anik, S. (1991). *Gedik Eğitim Vakfı Kaynak Teknolojisi Eğitim Araştırma ve Muayene Enstitüsü*. Gedik Education Foundation.

Annoni, M., & Carboni, M. (n.d.). *Science and Technology of Welding and Joining Ultrasonic metal welding of AA 6022-T4 lap joints: Part I-Technological characterisation and static mechanical behaviour*. doi:10.1179/1362171810Y.0000000014

Aydin, K., Kaya, Y., & Kahraman, N. (2012). Experimental study of diffusion welding/bonding of titanium to copper. *Materials & Design*, *37*, 356–368. doi:10.1016/j.matdes.2012.01.026

Bag, S., & Akinlabi, E. T. (2020). Eco Friendly Aspects in Hybridization of Friction Stir Welding Technology for Dissimilar Metallic Materials. In *Encyclopedia of Renewable and Sustainable Materials* (pp. 225–236). Elsevier. doi:10.1016/B978-0-12-803581-8.11153-1

Bahl, S., Singh, T., Kumar, V., Sehgal, S., & Bagha, A. K. (2021). A systematic review on recent progress in advanced joining techniques of the lightweight materials. *AIMS Materials Science*, *8*(1), 62–81. doi:10.3934/matersci.2021005

Balle, F., Huxhold, S., Wagner, G., & Eifler, D. (2011). Damage monitoring of ultrasonically welded aluminum/CFRP-joints by electrical resistance measurements. *Procedia Engineering*, *10*, 433–438. doi:10.1016/j.proeng.2011.04.074

Banerjee, D., & Williams, J. C. (2013). Perspectives on Titanium Science and Technology. *Acta Materialia*, *61*(3), 844–879. doi:10.1016/j.actamat.2012.10.043

Barnes, T. A., & Pashby, I. R. (2000). Joining techniques for aluminum spaceframes used in automobiles. Part II - adhesive bonding and mechanical fasteners. *Journal of Materials Processing Technology*, *99*(1), 72–79. doi:10.1016/S0924-0136(99)00361-1

Belov, N. A., Eskin, D. G., & Aksenov, A. A. (2005). Introduction. In *Multicomponent Phase Diagrams* (pp. v–vi). Elsevier. doi:10.1016/B978-008044537-3/50000-7

Blaga, L., dos Santos, J. F., Bancila, R., & Amancio-Filho, S. T. (2015). Friction Riveting (FricRiveting) as a new joining technique in GFRP lightweight bridge construction. *Construction & Building Materials*, *80*, 167–179. doi:10.1016/j.conbuildmat.2015.01.001

Brumm, S., & Bürkner, G. (2015). Gas Metal arc Pulse Welding with Alternating Current for Lightweight Materials. *Materials Today: Proceedings*, *2*, S179–S187. doi:10.1016/j.matpr.2015.05.008

Çavdar, K., & Durmuş, A. (2018). Effects of pressure and time on radial riveting process. *Journal of the Faculty of Engineering and Architecture of Gazi University*, *33*(1), 312–322. doi:10.17341/gazimmfd.406803

Choi, D. H., Lee, C. Y., Ahn, B. W., Choi, J. H., Yeon, Y. M., Song, K., Hong, S. G., Lee, W. B., Kang, K. B., & Jung, S. B. (2011). Hybrid Friction Stir Welding of High-carbon Steel. *Journal of Materials Science and Technology*, *27*(2), 127–130. doi:10.1016/S1005-0302(11)60037-6

Civjan, S. A., Guihan, T., & Peterman, K. (2020). Testing of oxyacetylene weld strength. *Journal of Constructional Steel Research*, *168*, 105921. Advance online publication. doi:10.1016/j.jcsr.2019.105921

Cole, G. S., & Sherman, A. M. (1995). Light weight materials for automotive applications. *Materials Characterization*, *35*(1), 3–9. doi:10.1016/1044-5803(95)00063-1

Contorno, D., Filice, L., Fratini, L., & Micari, F. (2006). Forming of aluminum foam sandwich panels: Numerical simulations and experimental tests. *Journal of Materials Processing Technology*, *177*(1–3), 364–367. doi:10.1016/j.jmatprotec.2006.04.028

de Assis, S. L., Wolynec, S., & Costa, I. (2006). Corrosion characterization of titanium alloys by electrochemical techniques. *Electrochimica Acta*, *51*(8–9), 1815–1819. doi:10.1016/j.electacta.2005.02.121

Ezugwu, E. O., & Wang, Z. M. (1997). Titanium alloys and their machinability—A review. *Journal of Materials Processing Technology*, *68*(3), 262–274. doi:10.1016/S0924-0136(96)00030-1

Fadzil, M., Abdullah, A. B., Samad, Z., Yusof, F., & Manurung, Y. H. P. (2021). Application of lightweight materials toward design for sustainability in automotive component development. In *Design for Sustainability* (pp. 435–463). Elsevier. doi:10.1016/B978-0-12-819482-9.00006-X

Feng, M., Li, Y., Zhao, C., & Luo, Z. (2016). Mechanical properties and interface morphology of Mg/Al ultrasonic spot weld bonding welds. *Science and Technology of Welding and Joining*, *21*(8), 688–699. doi:10.1080/13621718.2016.1144262

Fergus, J. W., Mishra, B., Anderson, D., Sarver, E. A., & Neelameggham, N. R. (Eds.). (2015). *Engineering solutions for sustainability : Materials and resources II : John Wiley & Sons*. Retrieved August 25, 2021, from https://books.google.com/books/about/Engineering_Solutions_for_Sustainability.html?hl=tr&id=XV0avgAACAAJ

Haghshenas, M., & Gerlich, A. P. (2018). Joining of automotive sheet materials by friction-based welding methods: A review. *Engineering Science and Technology, an International Journal, 21*(1), 130–148. doi:10.1016/j.jestch.2018.02.008

Han, L., Thornton, M., Li, D., & Shergold, M. (2011). Effect of governing metal thickness and stack orientation on weld quality and mechanical behaviour of resistance spot welding of AA5754 aluminium. *Materials & Design, 32*(4), 2107–2114. doi:10.1016/j.matdes.2010.11.047

Hevner, A. R. (2007). A Three Cycle View of Design Science Research. *Scandinavian Journal of Information Systems, 19*(2).

Hirsch, J. (2010). Aluminium sheet fabrication and processing. In Fundamentals of Aluminium Metallurgy: Production, Processing and Applications (pp. 719–746). Elsevier Ltd. doi:10.1533/9780857090256.3.719

Jahn, R., Cooper, R., & Wilkosz, D. (2007). The effect of anvil geometry and welding energy on microstructures in ultrasonic spot welds of AA6111-T4. *Metallurgical and Materials Transactions. A, Physical Metallurgy and Materials Science, 38*(3), 570–583. doi:10.100711661-006-9087-0

Joost, W. J. (2012). Reducing Vehicle Weight and Improving U.S. Energy Efficiency Using Integrated Computational Materials Engineering. *JOM, 64*(9), 1032–1038. Advance online publication. doi:10.100711837-012-0424-z

Kalaiselvan, K., Dinaharan, I., & Murugan, N. (2021). Routes for the Joining of Metal Matrix Composite Materials. In *Reference Module in Materials Science and Materials Engineering*. Elsevier. doi:10.1016/B978-0-12-803581-8.11899-5

Kang, B., Cai, W., & Tan, C. A. (2013). Dynamic response of battery tabs under ultrasonic welding. *Journal of Manufacturing Science and Engineering, 135*(5), 051013. Advance online publication. doi:10.1115/1.4024535

Kim, B. G., Dong, S. L., & Park, S. D. (2001). Effects of thermal processing on thermal expansion coefficient of a 50 vol.% SiCp/Al composite. *Materials Chemistry and Physics, 72*(1), 42–47. doi:10.1016/S0254-0584(01)00306-6

Kleiner, M., Geiger, M., & Klaus, A. (2003). Manufacturing of Lightweight Components by Metal Forming. *CIRP Annals, 52*(2), 521–542. doi:10.1016/S0007-8506(07)60202-9

Kuhlman, G. W. (2005). Forging of Aluminum Alloys. In ASM Handbook (Vol. 14A, pp. 299–312). doi:10.1361/asmhba0003996

Liu, A., Tang, X., & Lu, F. (2013). Study on welding process and prosperities of AA5754 Al-alloy welded by double pulsed gas metal arc welding. *Materials & Design, 50*, 149–155. doi:10.1016/j.matdes.2013.02.087

Marré, M., Gies, S., Maevus, F., & Tekkaya, A. E. (2010). Joining of lightweight frame structures by die-less hydroforming. *International Journal of Material Forming, 3*(S1), 1031–1034. doi:10.100712289-010-0946-2

Mathers, G. (2002). *The welding of aluminium and its alloys*. CRC Press.

Meng, X., Huang, Y., Xie, Y., Li, J., Guan, M., Wan, L., Dong, Z., & Cao, J. (2019). Friction self-riveting welding between polymer matrix composites and metals. *Composites. Part A, Applied Science and Manufacturing*, *127*, 105624. doi:10.1016/j.compositesa.2019.105624

Meschut, G., Janzen, V., & Olfermann, T. (2014). Innovative and highly productive joining technologies for multi-material lightweight car body structures. *Journal of Materials Engineering and Performance*, *23*(5), 1515–1523. doi:10.100711665-014-0962-3

Messler, R. W. (2003). Joining comes of age: From pragmatic process to enabling technology. *Assembly Automation*, *23*(2), 130–143. doi:10.1108/01445150310471365

Michot, G., & Champier, G. (1991). Physical metallurgy of P/M aluminium alloys. *Memoires et Etudes Scientifiques de La Revue de Metallurgie*, *88*(7–8), 425–439. doi:10.1016/B978-0-08-099431-4.00002-6

Mishra, R. S., & Sidhar, H. (2017). Friction Stir Welding. In *Friction Stir Welding of 2XXX Aluminum Alloys Including Al-Li Alloys* (pp. 1–13). Elsevier. doi:10.1016/B978-0-12-805368-3.00001-7

Mucha, J. (2011). A study of quality parameters and behaviour of self-piercing riveted aluminium sheets with different joining conditions. *Strojniski Vestnik. Jixie Gongcheng Xuebao*, *57*(4), 323–333. doi:10.5545v-jme.2009.043

Nong, N., Keju, O., Yu, Z., Zhiyuan, Q., Changcheng, T., & Feipeng, L. (2003). Research on press joining technology for automotive metallic sheets. *Journal of Materials Processing Technology, 137*(1-3), 159–163. doi:10.1016/S0924-0136(02)01083-X

Pakdil, M., Çam, G., Koçak, M., & Erim, S. (2011). Microstructural and mechanical characterization of laser beam welded AA6056 Al-alloy. *Materials Science and Engineering A*, *528*(24), 7350–7356. doi:10.1016/j.msea.2011.06.010

Panteli, A., Robson, J. D., Brough, I., & Prangnell, P. B. (2012). The effect of high strain rate deformation on intermetallic reaction during ultrasonic welding aluminium to magnesium. *Materials Science and Engineering A*, *556*, 31–42. doi:10.1016/j.msea.2012.06.055

Peters, M., Kumpfert, J., Ward, C. H., & Leyens, C. (2003). Titanium Alloys for Aerospace Applications. *Advanced Engineering Materials*, *5*(6), 419–427. Advance online publication. doi:10.1002/adem.200310095

Pickin, C. G., Young, K., & Tuersley, I. (2007). Joining of lightweight sandwich sheets to aluminium using self-pierce riveting. *Materials & Design*, *28*(8), 2361–2365. doi:10.1016/j.matdes.2006.08.003

Pollock, T. M. (2010). Weight loss with magnesium alloys. *Science*, *328*(5981), 986–987. doi:10.1126cience.1182848 PMID:20489013

Ranatowski, E. (2008). Weldability of Titanium and its Alloys - Progress in Joining. *Advances in Materials Sciences*, *8*(2). Advance online publication. doi:10.2478/v10077-008-0033-2

Rathinasuriyan, C., Muniamuthu, S., Mystica, A., & Senthil Kumar, V. S. (2021). Investigation of heat generation during submerged friction stir welding on 6061-T6 aluminum alloy. *Materials Today: Proceedings*, *46*, 8320–8324. Advance online publication. doi:10.1016/j.matpr.2021.03.310

Rüdinger, K., & Ismer, A. (1973). Recent Development in the Application of Titanium for Motorcars. *Titanium Science and Technology*, 185–199. doi:10.1007/978-1-4757-1346-6_16

Sathishkumar, G. B., Sethuraman, P., Chanakyan, C., Sundaraselvan, S., Joseph Arockiam, A., Alagarsamy, S. V., Elmariung, A., Meignanamoorthy, M., Ravichandran, M., & Jayasathyakawin, S. (2021). Friction welding of similar and dissimilar materials: A review. *Materials Today: Proceedings*. Advance online publication. doi:10.1016/j.matpr.2021.03.089

Schubert, E., Klassen, M., Zerner, I., Walz, C., & Sepold, G. (2001). Light-weight structures produced by laser beam joining for future applications in automobile and aerospace industry. *Journal of Materials Processing Technology*, *115*(1), 2–8. doi:10.1016/S0924-0136(01)00756-7

Shakil, M., Tariq, N. H., Ahmad, M., Choudhary, M. A., Akhter, J. I., & Babu, S. S. (2014). Effect of ultrasonic welding parameters on microstructure and mechanical properties of dissimilar joints. *Materials & Design*, *55*, 263–273. doi:10.1016/j.matdes.2013.09.074

Shen, Z., Ding, Y., & Gerlich, A. P. (2019). *Advances in friction stir spot welding*. doi:10.1080/10408436.2019.1671799

Siddiq, A., & Ghassemieh, E. (2008). Thermomechanical analyses of ultrasonic welding process using thermal and acoustic softening effects. *Mechanics of Materials*, *40*(12), 982–1000. doi:10.1016/j.mechmat.2008.06.004

Simon, H. A. (2019). The Sciences of the Artificial (3rd ed.). MIT Press. doi:10.7551/mitpress/12107.001.0001

Singh, R., Kumar, R., & Hashmi, M. S. J. (2017). Friction Welding of Dissimilar Plastic-Based Material by Metal Powder Reinforcement. In *Reference Module in Materials Science and Materials Engineering*. Elsevier. doi:10.1016/B978-0-12-803581-8.04160-6

Szesz, E. M., & Lepienski, C. M. (2017). Anodic bonding of titanium alloy with bioactive glass. *Journal of Non-Crystalline Solids*, *471*, 19–27. doi:10.1016/j.jnoncrysol.2017.04.038

Tejedor, T. A., Singh, R., & Pilidis, P. (2013). Maintenance and repair of gas turbine components. In Modern Gas Turbine Systems: High Efficiency, Low Emission, Fuel Flexible Power Generation (pp. 565–634). Elsevier Ltd. doi:10.1533/9780857096067.3.565

Troschitz, J., Vorderbrüggen, J., Kupfer, R., Gude, M., & Meschut, G. (2020). Joining of thermoplastic composites with metals using resistance element welding. *Applied Sciences (Switzerland)*, *10*(20), 1–12. doi:10.3390/app10207251

van Walle, E. (2001). Mechanical Test Specimens, Reconstitution of. In Encyclopedia of Materials: Science and Technology (pp. 5265–5268). Elsevier. doi:10.1016/B0-08-043152-6/00918-9

Wagner, G., Balle, F., & Eifler, D. (n.d.). *Ultrasonic Welding of Hybrid Joints*. doi:10.1007/s11837-012-0269-5

Wang, S. Q., Patel, V. K., Bhole, S. D., Wen, G. D., & Chen, D. L. (2015). Microstructure and mechanical properties of ultrasonic spot welded Al/Ti alloy joints. *Materials & Design*, *78*, 33–41. doi:10.1016/j.matdes.2015.04.023

Ward-Close, C. M., & Froes, F. H. (1994). *Developments in the Synthesis of Lightweight Metals*. Academic Press.

Wu, G. Q., Li, Z. F., Luo, G. X., Li, H. Y., & Huang, Z. (2007). Dynamic simulation of solid-state diffusion bonding. *Materials Science and Engineering A*, *452–453*, 529–535. doi:10.1016/j.msea.2006.10.115

Yunlian, Q., Ju, D., Quan, H., & Liying, Z. (2000). Electron beam welding, laser beam welding and gas tungsten arc welding of titanium sheet. *Materials Science and Engineering A*, *280*(1), 177–181. doi:10.1016/S0921-5093(99)00662-0

Chapter 7
Production Techniques of Metallic Foams in Lightweight Materials

Nuray Beköz Üllen
https://orcid.org/0000-0003-2705-2559
Istanbul University-Cerrahpasa, Turkey

Gizem Karabulut
https://orcid.org/0000-0003-0930-5380
Istanbul University-Cerrahpasa, Turkey

ABSTRACT

Lightweight materials were needed in many different areas, especially in order to reduce the required energy in areas such as automotive and aerospace industries. Metallic foams attract attention in light-weight material applications due to their unique properties. The pores in its structure provide advantages in many applications, both structural and functional by promising both ultra-lightweight construction, energy absorption, and damping insulation. Production techniques of metallic foams can generally be classified as liquid, solid, gas, and ionic state production according to the physical state of the metal at the beginning of the process. The production technique should be chosen according to the usage area and desired properties of the metallic foam and the suitability in terms of cost and sustainability of production. For this reason, the details of the production techniques should be known and the products that can be obtained and their properties should be understood. In this respect, this chapter emphasizes the production methods from past to present.

INTRODUCTION

Metal foams are a class of engineering materials developed for light-weight material applications (Dukhan, 2013). Metallic foam consists of a rigid frame and air-containing internal and external pores, which gives the material different characteristics. Due to the porous structure, metallic foams provide significant

DOI: 10.4018/978-1-7998-7864-3.ch007

advantages in terms of vibration resistance, energy, and thermal absorption as well as lightness. Metal foam structures have attracted attention due to their high stiffness-to-weight and strength-to-weight ratios. Metallic foams are becoming an interesting and important field of research in recent times. It can be used in many structural and functional applications in many fields such as aviation, railway, building, and biomedical industries and especially in the automotive industry (Yilong et al., 2016; Claar et al., 2000; Andure et al., 2012; Qin et al., 2016; Bisht et al., 2019). People encounter porous structures in many places in nature (wood, bone, corals, pumice, lava, etc.). The porous form observed in the structure of lightweight but strength materials in nature has attracted the attention of scientists and they have studied material production in these forms (Banhart, 2001; Bauer et al., 2013; Liu and Chen, 2014). Manmade porous materials that people encounter at many points in daily life can basically be polymeric, ceramic, and metallic. Porous plastics are found in many different applications such as foam cups, food packaging, and airbags. Polymer foams can't show rigidity under loading and are not resistant to high temperatures. Ceramic foam structures are often preferred in filtering applications. This material is limited in use because it is brittle under suddenly loading and is difficult to machining (Qin et al., 2016; Sivertsen, 2007; Yilong et al., 2016). For these reasons and owing to the unique characteristics of metallic ones have the potential for many applications. Since the pore structure can be controlled, the usage areas of metallic foams are varied (Garcia-Moreno, 2016). Basically, the pore structure of metallic foams is divided into two categories and usage areas are accordingly diversified. Pores are named according to whether they are connected or not: open (through) or closed (Qin et al., 2016). Foams with open porous structures are used in heat exchangers and absorbers, especially implants and filters, due to their thermal and permeability properties (Gülsoy and German, 2008). Closed porous metallic foams are preferred in structural applications due to their specific mechanical properties (Vendra et al., 2011). It is used in applications such as crash absorbers in vehicles and sound absorbers in machines with the increase in the thickness of the pore walls and the development of energy absorption properties compared to open pores structures (Vendra et al., 2011; Andure et al., 2012; Bauer et al., 2013; Yilong et al., 2016).

Metallic foam production technique should be selected by considering factors such as usage area, desired properties, cost and continuity. Therefore, all production techniques should be known in detail and the relationship between product-feature-structure should be understood. Studies on the production of metallic foams were first made by Sosnick in 1943. Benjamin Sosnick melted the mixture of aluminium and mercury in a high-pressure vessel. Then, due to the melting temperature of aluminium with the releasing of the pressure, mercury was evaporated and formed a foamy structure (Banhart and Weaire, 2002). Since then, many new methods have been discovered in metallic foam production. New techniques are explored for reasons such as cost of production, commercialization, being able to apply to different metals, and better quality of products. This chapter of the book, it is aimed to create an up-to-date source by dealing with production methods from past to present and focusing on developments.

Production Techniques

As a new type of material, metallic foams have attracted the attention of scientists. "Foam" is defined in the literature as the structures obtained by the distribution of the gas phase in a liquid or solid phase (Banhart, 2001). Metallic foam means in the solid phase, liquid metallic foam can only be during the production stage. In the literature, porous structures have been given different names according to different pore morphologies such as cellular, porous, sponge and foam (Banhart, 2016). In this chapter, porous structure production techniques of metals are generally mentioned in order to provide a broad perspective and a detailed resource.

Direct Foaming of Metals

These methods are based on the foaming of liquid metal. Foam structure can be obtained in two different ways: by direct gas blowing and by using a foaming agent. The production method of metallic foam by gas injection is based on the foaming with gas bubbles injection from an external source into the molten metal. However, gas bubbles tend to rise to the surface due to the high buoyancy of the liquid metal (Banhart, 2001). In order for the gas bubbles to dissipate and remain in the liquid metal, they must be stabilized by increasing viscosity. This can be achieved by adding alloying elements or fine ceramic particles (Rajak and Gupta, 2020). For this purpose, powders such as aluminium oxide, silicon carbide or magnesium oxide are added to the molten metal (Banhart, 2001; Banhart, 2004; Bhatt et al., 2015). Schematic representation and steps of the process are shown in Figure 1. The first step of the process is to prepare the molten metal and the stabilizers. It is important that stabilizers are wetted by the molten metal and distributed evenly in it. Then, the melt is foamed by blowing air, argon or nitrogen gas into the melt from an external source using a rotating impeller or vibrating nozzles (Banhart, 2000). Thanks to the use of this impeller or nozzle, fine gas bubbles distributed uniformly in the melt are obtained (Degischer et al., 2002). This is important for the mechanical properties of the product. A mixture of gas bubbles and molten metal is obtained. The resulting mixture is taken from the upper surface of the liquid with the help of a conveyor belt, and then it is cooled and solidified so that a porous structure is obtained (Benouali et al., 2002; Bhatt et al., 2015). Then, the desired product is obtained by applying secondary processes such as flattening and cutting to the semi-solid foam (Banhart, 2001).

Figure 1. Porous metallic part production with gas injection technique (Kranzlin and Niederberger, 2015)

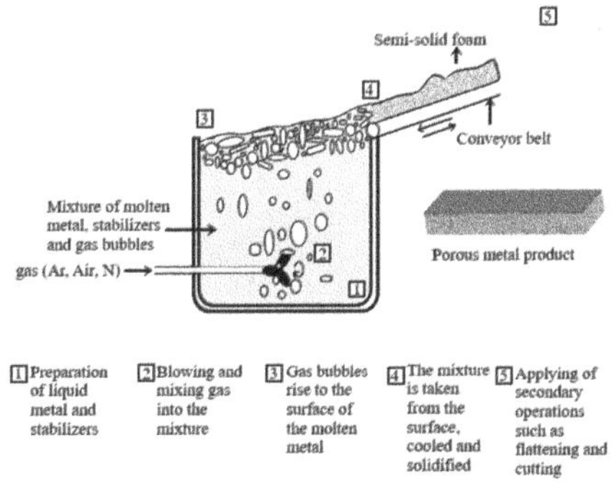

Production of aluminium foams by gas injection method is licensed by CYMAT Aluminium Corporation. Aluminium is melted with stabilizer additives such as SiC, aluminium oxide, titanium diboride, zirconium oxide, magnesium oxide and other production steps are applied (Wang et al., 2007; Kulshreshtha and Dhakad, 2020). A Closed foam structure is obtained and an open pore structure may occur on the surface due to the cutting process (Rajak and Gupta, 2020; Banhart, 2001). Zinc foam and other metals with low melting temperatures can be produced by this method (Banhart, 2001). The method is cost-effective compared to other metallic foam production methods, easy to apply, and suitable for continuous production. However, the size and distribution of the created gas bubbles cannot be controlled and thus the mechanical properties of the products are not superior since gas bubbles are not distributed uniformly. It can only be produced in slab form and is suitable for the production of low-density metals.

A new technique has been developed by ALUINVENT to improve the distribution of pores and to reduce the pore size of metallic foams produced by the gas injection method. In the technique, while gas bubbles are given into the liquid metal, gas bubbles are separated with the help of ultrasonic sound waves (Garcia-Moreno, 2016).

In the other method, the liquid metal is directly foamed by adding blowing agents to the molten metal. Metal hydrides (TiH2, CaH2, ZrH2, etc.) are generally preferred as blowing agents (Matijasevic-Lux, 2006). As a result of the using of hydride blowing agents, hydrogen gas is released from these particles and bubbles are formed in the molten metal (Banhart, 2001; Kulshreshtha and Dhakad; 2020). Metal oxide or metal carbonate compounds can also be used as blowing agents (Patel et al., 2018). Figure 2 shows schematically the foaming of aluminium with this method patented by ALPORAS. In the process, calcium is added and mixed to the molten aluminium metal in order to increase its viscosity, and the molten metal becomes stabilized due to the formation of some oxide compounds (CaO, CaAl2O4) or intermetallics (Banhart, 2003; Degischer et al., 2002). The hydride blowing agent is then added, and with the release of gas, the melt expands slowly. Foaming occurs at constant pressure and the vessel is filled with a liquid foam structure. Then, while the vessel is cooled, the structure is solidified. The foam removed from the vessel in block form then can be given into the desired shape with other processes (Srivastava and Sahoo, 2007; García-Moreno, 2016; Gábora and Mankovits, 2019). Zinc, magnesium, copper alloys, and steel foams are manufactured in a similar way to aluminium foam production with this method (Banhart, 2001; Srivastava and Sahoo, 2007; Banhart, 2013).

Figure 2. Schematic depiction of the method using blowing agents for aluminium foam production (Banhart, 2000)

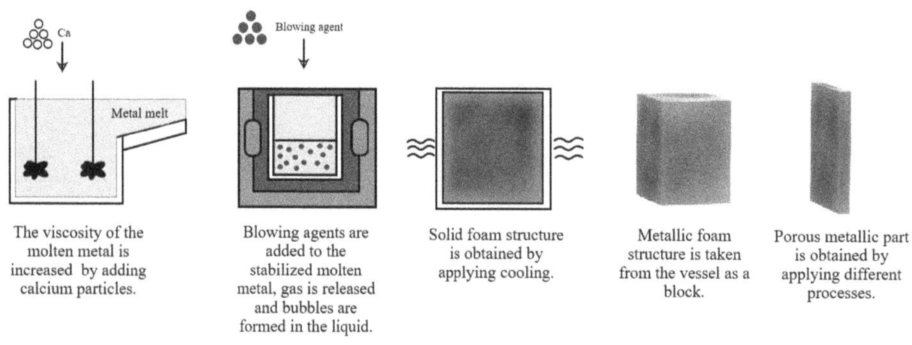

It is an easily applicable method for aluminium foam production. A relatively uniform structure can be obtained and the process can be controlled. However, it is a high-cost method due to the use of metal hydrides. Complex metal foam forms cannot be produced. It is difficult to control porosity and pore sizes, cracks may occur in the pore walls due to the rapid cooling process and the presence of hydrogen.

Solid-Gas Eutectic Solidification

This method is mainly based on the difference in gas solubility of metals of the liquid and solid phase. Firstly, the molten metal is charged with gas (Banhart, 2001). This is achieved by melting the metal under high pressure in a gas atmosphere. Then the temperature is lowered below the eutectic transition temperature. Thus, the charged gas precipitates (Babscan et al., 2003). When suitable conditions are provided, gas bubbles become trapped in the molten metal. Later, with the solidification of the structure, long pores are formed depending on the solidification direction (Banhart, 2003; Babscan and Banhart, 2006). This technique is also known as "Gasar Process" and a regularly oriented (lotus-type) porous structure can be obtained (Zhang et al., 2013a; Zhang et al., 2013b). Schematic representation of lotus-type porous structure production is given in Figure 3.

Figure 3. Regularly oriented porous structure production using solid-gas eutectic solidification technique (Shapovalov and Boyko, 2004)

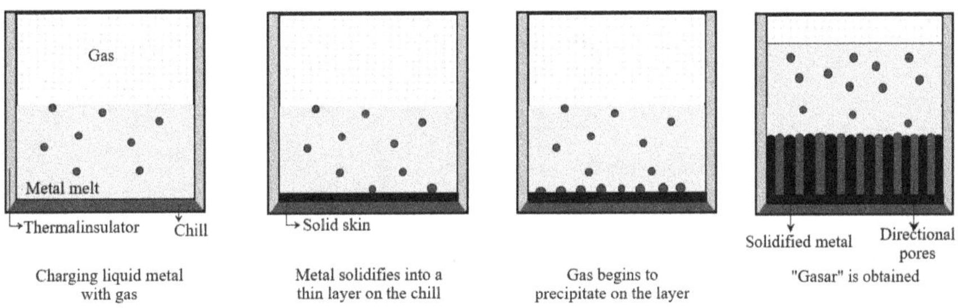

The method is cost-effective as no blowing agent is used, but the required equipment is complex and can be costly. It is advantageous in that many metals can be produced and porosity can be controlled. Single-type metallic foam production is possible and relatively slow. It is not a suitable method for high production capacities. Base metals and alloys such as aluminium, copper, magnesium, nickel, chromium, and iron are produced by this method (Kulshreshtha and Dhakad, 2020).

Foaming Powder Compacts

One of the starting materials is metal powders; it can be in the form of elemental metal powders, mixtures of different metal powders or alloy powders (Babcsán et al., 2003). Then metal powders mixed with the appropriate blowing agent. After that, the mixture is compressed to make a precursor material for the foaming process using a compression method suitably selected according to the starting materials (Banhart, 2001; García-Moreno, 2016). Any residual porosity is not desired in the structure after compression as it will cause defects in the foam structure (Babcsán et al., 2003). The precursor material is expanded by applying heat treatment and finally the foam structure is obtained. It is raised to a temperature near the melting point of the metal in order to decompose the blowing agent in the semi-solid precursor material (Banhart, 2003). As the metal begins to melt, the metal expands with the gas released from the blowing agent. After the foam formation is completed, the final product is obtained with cooling (Ashby et al., 2000; Benouali et al., 2002; Claar et al., 2018; Rajak and Gupta, 2020). Since the expansion in the product obtained by this method cannot be controlled exactly, it is ensured that the foaming process is carried out in a mold to obtain the desired shape of foam product. The foam structure is either injected into the mold, or the parts coming from the extrusion are foamed in the mold (Banhart, 2001). Schematic representation of the foaming powder compacts technique is given in Figure 4.

Figure 4. Diagram of powder compact melting technique for producing metal foam (Banhart, 2001; Ashby et al., 2000)

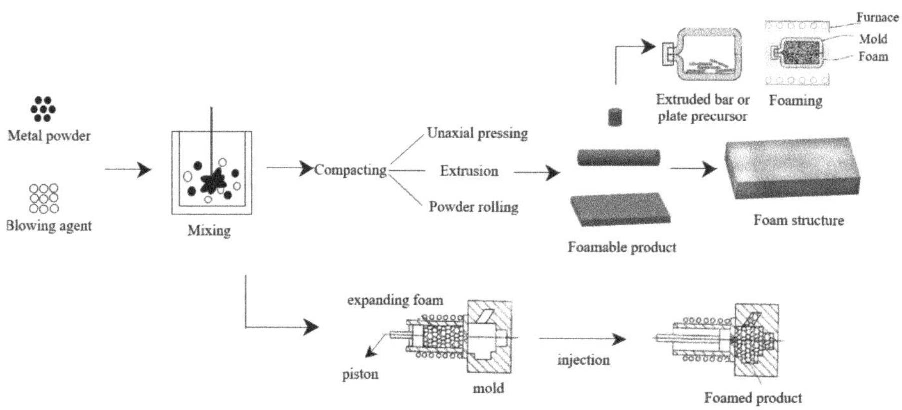

Many metals such as aluminium, tin, brass, zinc, lead, gold, bronze, and steel can be produced in foam form by this method. However, it is necessary to choose the appropriate blowing agent. Metal foams which are produced with this method have closed porous structure. Even complex shapes can be produced in high quality and in many metal foam forms with this method. However, it is high-cost manufacturing (Banhart, 2001; Srivastava and Sahoo, 2007; Claar et al., 2018; Kulshreshtha and Dhakad, 2020).

Casting Techniques

Metallic foam production with casting methods is based on the same point among themselves. A volatile material, which is the negative image of the porous structure to be produced with liquid metal, is the common point of the methods. The methods differ from each other in terms of the temporary material used. This material can be wax slurry or a space holder.

In the production of porous materials by casting using space holders, space holding fillers are placed in a mold and then molten metal is poured. Then, the porous structure is obtained by removing the space

Figure 5. Production steps of casting around space holder technique (Banhart, 1999)

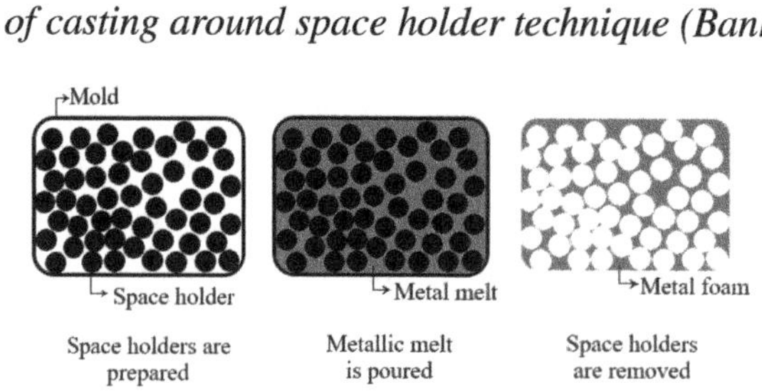

holder. Organic or inorganic granules can be used as space-holding filler. Leaching or heat treatment is required to remove them. Aluminium, magnesium, tin, zinc and lead metal foams are produced by this technique (Banhart, 1999; Banhart, 2001; Garcia- Moreno, 2016; Degischer et al., 2002; Ashby et al., 2000; Srivastava and Sahoo, 2007; Patel et al., 2018; Sutygina et al., 2020; Rajak and Gupta, 2020). A schematic description of the casting around the space holder method is given in Figure 5.

This method is cost-effective compared to other methods. It is an environmentally friendly method. It provides easy control for pore size and geometry, and thus net-shape and near-net-shape production is possible. Thus, it may not require a secondary production process. But metals with high surface tension are difficult to wet the particles, thus making it difficult to fill the mold uniformly.

When using the polymer foam method, a wax slurry made of heat-resistant materials is infiltrated and dried. Then, the polymer foam is volatilized by applying heat. Thus, liquid metal is poured into the remaining space. Metallic foam is obtained by removing the mold (Dukhan, 2013). The method is known as foam replication or investment casting. According to the shape and properties of the polymer foam, the properties of the metallic foam can be changed. Porous structure up to 97% porosity can be produced. Metallic foams are open-porous structures. Magnesium, copper, stainless steel and aluminium and alloys foams are produced by this method (Bauer et al., 2013; Salvo et al., 2014; Garcia-Moreno, 2016; Patel et al., 2018; Duan et al., 2019; Kulsheshtha and Dhakad, 2020). In Figure 6, a schematic representation of the method is given. High porosity can be achieved and the pore quality is superior with this method. It is an easy-to-apply method. However, filling the mold homogeneously can be difficult and the fine pore walls can be damaged during the removal of the wax slurry.

Figure 6. Investment casting technique for porous structure production (Atwater et al., 2018)

Deposition Techniques

Metal foams can be produced by different deposition methods. This can be spray deposition, electro-deposition or chemical vapor deposition. Schematic presentations of these techniques are given in Figure 7.

In the spray deposition technique, the metal is in liquid form as the starting material. Firstly, the molten metal is continuously atomized. The metal droplets are micron in size (Srivastava and Sahoo, 2007). Then, the blowing agent is added to this atomized metal spray. These droplets are deposited on a substrate material formed in the desired shape of the foam structure. During the deposition of the droplets, pores are formed in the deposited part due to the gas release of the blowing agent. Metals and alloys or metal matrix composites can be produced by this method (Banhart, 2001; Banhart, 2016). It is important to choose the appropriate production parameters and blowing agent for each metal. Metal foams produced by this method have fine equiaxed grain and low oxide content however pores are not distributed homogeneously. Less porosity is obtained compared to other methods (~%60). Steel and

Figure 7. Deposition methods in production of porous metal (Banhart, 2001; Ashby et al., 2000; Rajak and Gupta, 2020)

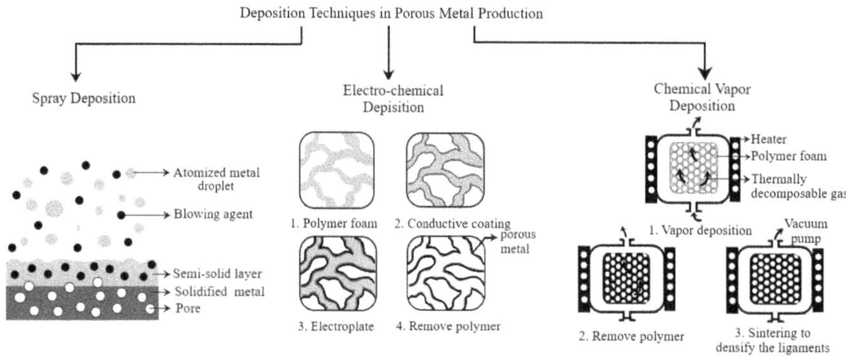

copper alloy foams can be produced with this technique (Banhart, 2001; Srivastava and Sahoo, 2007; Patel et al., 2018).

In the electro-deposition method, the starting materials are metal in an ionic state and polymer substrate. The used polymer foam must have electrical conductivity. If the foam is a nonconductor, directly electrodeposition cannot be applied. It is one of the most important steps of the method. The polymer substrate is made electrically conductive by various methods such as vacuum evaporation, cathode sputtering, arc ion plating, electroless plating, and electrically conductive slurry coating (Banhart, 2001). The slurry coating method can be preferred because it is simple, cheap and applicable. However, it is difficult to obtain a batch coating with this method. In this respect, electroless plating may come to the fore (Tian and Guo, 2010). The ionic solution can be electrically conductive based on carbon black or graphite. After the electrically conductive coating is made, the metal ions are deposited with the electrodeposition technique. Later, the polymer foam is removed by thermal treatment, so that a porous structure is obtained. Copper, nickel, zinc, gold, titanium, and nickel-chromium foams can be produced with this technique (Banhart, 2001; Patel et al., 2018; Salvo et al., 2014). The foams are open-porous structures. The high cost of the process is its disadvantage. However, it is a method used in certain industries such as the production of porous electrodes (Ashby et al., 2000; Atwater et al., 2018; Duan et al., 2019; Kulshreshtha and Dhakad, 2020).

In the chemical vapor deposition method, the starting materials are metal or metallic compounds in the gaseous phase and a porous substrate material with the pore properties of the porous metal to be produced. The material used as substrate material can also be polymer foam or lattice block materials (Banhart, 2001; Patel et al., 2018). In this method developed by INCO, the metal vapor is created in a vacuum chamber. It is then allowed to deposit on the cold substrate. After the deposition process is completed, cooling carried out. Then the polymer substrate is removed and open-porous metal foam is obtained. After removing the polymer, sintering is applied to densify the ligament (Ashby et al., 2000; Salvo et al., 2014; Kranzlin and Niederberger, 2014). The metal can be chosen for this method is limited as it must be obtained in the vapor phase. Nickel, copper, iron, titanium, and tantalum metals can be produced as metal foams with this technique (Degischer et al., 2002; Garcia- Moreno, 2016; Atwater et al., 2018; Ryan et al., 2006; Rajak and Gupta, 2020). This technique is not economical. However, this method industrially preferred due to obtaining a porous product with high chemical purity and smooth pore structure (Kulshreshtha and Dhakad, 2020; Pasarin et al., 2003).

Figure 8. Gas entrapment process for porous metal production (Kennedy, 2016; Banhart, 2001)

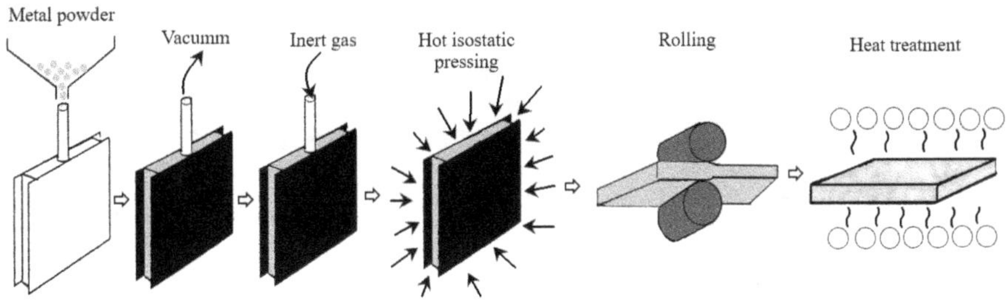

Sintering Metal Powders and Fibers

In this method, which is frequently used in industry for the production of porous metallic materials, metallic powders or fibers are partially sintered at different temperatures and porosity can be obtained at varying rates. The amount and shape of pores vary according to the morphology and amount of the particle used (Banhart, 2001). The method has attracted attention by the industry due to its various advantages such as its simplicity, that it does not require additional processes and materials, and that it can be applied to all metals that can be obtained in powder form or as a fiber (Rechtin et al., 2018). In addition, it is one of the advantages of the method to work at lower temperatures, since the metal will not reach the molten state. The interconnectivity of the metal powders depends on the sintering success and size of the particles (Ryan et al., 2006; Kranzlin and Niederberger, 2014). Optimization of the sintering temperature is important. Interconnecting is less at low temperatures and the mechanical properties of the product may not reach the desired values. Porosity gradually decreases at high temperatures. The porosity amount obtained by the method is around 20-50% (Srivastava and Sahoo, 2007; Kennedy, 2016; Patel et al., 2018). Nickel and alloys, bronze, steel, titanium, and alloys can be manufactured in this technique (Banhart, 2001; Degischer et al., 2002; Oppenheimer et al., 2004).

Gas Entrapment Technique

It is one of the methods based on powder metallurgy for porous metal production. In the method, an inert gas is used instead of a blowing agent. It is benefited from the very low solubility of the inert gas in the metal (Ashby et al., 2000). The first step of the production process is to fill the gases into a metal can and vacuum. The can is vacuumed until to remove all oxygen, this is important in order not to damage the metal (Kennedy, 2016). Then, can is backfilled with inert gas at between 3-5 atmosphere. Hot isostatic pressing (HIP) process is applied to the metal can be filled with inert gas and metal powders (Banhart, 2001). During this step, the metal densifies and the gas disperses homogeneously in the structure. The number of pores in the densified sample by the HIP process is relatively low compared to the initial step of the process. For this reason, a rolling process is applied to create small pores and to create a more homogeneous distribution, thereby improving the structure (Ashby et al., 2000). During heat treatment, the inert gas expands and creates pores in the structure, thus porous metal is formed. Sandwich structure is produced due to the use of cans in the process. These structures are important for light-weight applications. It is a method applicable to all metal powders. 20-40% porosity as closed type is achieved. The

Figure 9. Porous metal production via blowing agents and metal powders sintering (Jain et al., 2020)

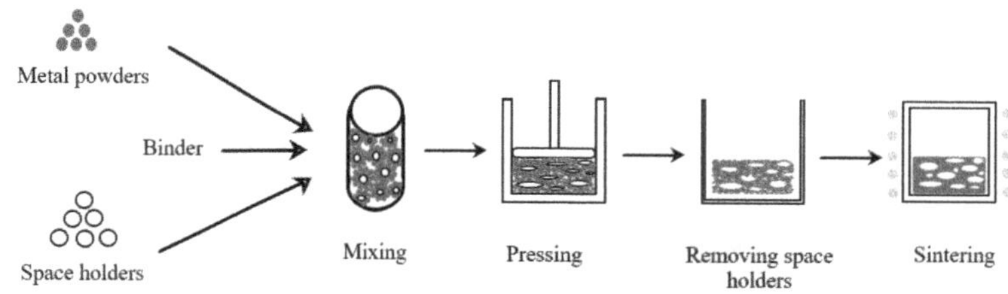

technique is often used for porous structures produced in Ti6Al4V alloy (Kennedy, 2016; Degischer et al., 2002; Kulshreshtha and Dhakad, 2020). Figure 8 shows a schematic representation of the process.

Sintering Hollow Spheres

The method is based on the principle of obtaining hollow metallic spheres and forming a porous metallic structure by sintering these spheres together. Hollow metallic spheres can be obtained as many different metals and alloys by many different methods (Degischer et al., 2002; Rajak and Gupta, 2020). After the spheres are obtained, compression and sintering processes are applied to obtain a structure with high porosity. It is important that the compression process is chosen properly so that the metallic spheres are not damaged (Ashby et al., 2000). After sintering, open and closed pores are formed. The porosity of up to 97% can be achieved (Srivastava and Sahoo, 2007; Kennedy, 2016). The most important advantage of the method is that the pore size can be controlled. One of the advantages of this technique is various materials such as superalloys, metals, and alloys and intermetallic compounds can be used in this method. Steel (Smith et al., 2012), aluminium (Duan et al., 2019), copper, nickel, cobalt, nickel-chromium, or titanium hollow spheres and thus porous structures can be obtained with this technique (Banhart, 2001; Bauer et al., 2013).

Sintering Blowing Agents and Metal Powders

It is an important method of metallic foam production by traditional powder metallurgy (Garcia- Moreno, 2016). Porous metal can be produced with metallic powders and many types of blowing agents (space holders, ceramic particles, polymer grains, salts, and even metals, etc.) (Kennedy, 2016). Using a mixer, metallic powders are coated around the blowing agent by means of a binder. Pore properties can be controlled by using different sizes, shapes, and amounts of blowing agents (Aşık and Bor, 2014). This is the most important advantage of the method. The green structure formed by the blowing agent coated with metal powders is compressed in order to provide better strength (Atwater et al., 2018). Then it is necessary to remove the used blowing agent. The removal process varies according to the amount of metal powder used. Completely removal may not occur due to closed type porosity. It can be removed by heat treatment, leaching, or with a solvent solution (Banhart, 2001). Finally, sinter neck growth is created by applying the sintering process between metallic powders (Ryan et al., 2006). Thus, a porous metallic structure is obtained. Both open and closed pores can be obtained (Smith et al, 2012; Jain et al., 2020). A wide range porosity rate (%20-80) can be obtained depending on the amount of space holder. This method is advantageous because of the ability to obtain desired pore properties and near-net-shaped or net-shaped products (Sazegaran and Nezhad, 2021). Titanium (Esen and Bor, 2007) and its alloys (Esen and Bor, 2011), steel (Bekoz and Oktay, 2013; Jain et al., 2021), magnesium (Wen et al., 2001; Hao et al., 2009), aluminium (Hakamada et al., 2007) metals can be produced by this technique as a porous form. The schematic representation of the method is given in Figure 9.

Additive Manufacturing

Additive manufacturing (AM) is a new method compared to traditional manufacturing methods. It is an increasingly widespread technology due to its advantages and potential. It attracts attention because it allows the design and application of different pore types especially for porous structure production, and the control of various pore properties such as the amount, shape, and size (Atwater et al., 2018; Chen et al., 2020). Porous material production with AM includes a 3D printing process (Patel et al., 2018). AM processes are used to fabricate three-dimensional complex geometries. This feature removes the design limit and AM technologies represent an interesting process to realize metal foams (Almonti and Ucciardello, 2019). The processes can be separated into two groups according to the way of processing the powder layer: powder deposition and powder bed techniques (Wang et al., 2017b). Schematic representation of the techniques is given in Figure 10.

In AM with powder deposition methods, metallic powders are directly added to the system with a nozzle and processed layer by layer with the laser (Frazier, 2014). The biggest advantage of these methods is also allowing the production of large parts. They can also be used to repair damaged parts. Directed Energy Deposition, Laser Metal Deposition, and Laser Engineered Net Shaping (LENS) techniques are used for porous metal production (Wang et al., 2017b; Shim et al., 2018).

Likewise, layer by layer production is carried out by using a laser beam in the production with the powder-bed system. However, the powder is transferred layer by layer through a moving system to form a powder bed in the production area (Wang et al., 2017b). The powders are contained in a chamber that can move up and down. There is a moving mechanism in the chamber to transfer the powders to the production chamber. The unit containing the powders moves up and the moving mechanism transfers the powders to the other unit. Powders are as a thin layer transferred to the production chamber is processed with a laser or electron beam (Frazier, 2014; Changdar and Chakraborty, 2021). The production chamber also moves down and this process is repeated. After the latest layer is carried out, the unfused powders in the bed are removed. Thus, a porous part is obtained (Kennedy, 2016; Sola and Nouri, 2019; Dani et al., 2020). Selective laser melting, electron beam melting, direct metal laser sintering, and selective laser sintering techniques are powder bed systems (Takezawa et al., 2017; Wang et al., 2017b; Hernández-Nava, 2015; Yuan et al., 2019).

The high cost of production is its main disadvantage (Patel et al., 2018). However, it is important because it provides precision in the production of special parts according to its use. It is used in the production of important materials such as aircraft parts and biomedical implants. Therefore, studies are focused on titanium and its alloys (Koike et al., 2018; Li et al., 2006; Hernández-Nava et al., 2015; Taniguchi et al., 2016; Shim et al., 2018). In addition, porous parts produced by additive manufacturing are used in applications such as micro battery architectures, microbial fuel cells, heat exchanger, and filters (Guddatti et al., 2019). Pure magnesium and Mg alloys, ferrous and Fe alloys, zinc and Zn alloys, Co-Cr alloys and many metals and alloys can be produced as porous metal form with AM techniques (Patel et al., 2018; Yuan et al., 2019; Li et al., 2020a; Xu et al., 2020; Ma et al., 2020).

Figure 10. Porous product production with powder deposition and powder bed techniques (Guddati et al., 2019)

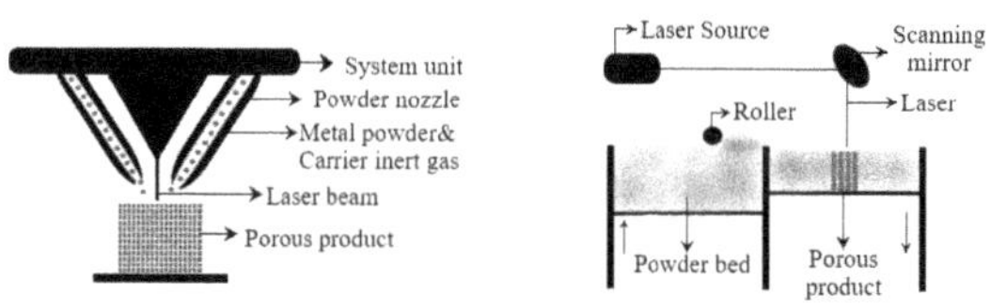

Table 1. Studies of different production methods of common metallic foams

Metals	Methods	References
Aluminium	Direct foaming with gas injection	(Wang et al., 2017a)
	Direct foaming with blowing agent	(Huang et al., 2019)
	Solid-gas eutectic solidification	(Komissarchuk et al., 2018)
	Foaming powder compacts	(Papantoniou et al., 2019)
	Sintering blowing agents and metal powders	(Sun et al., 2021)
	Casting techniques	(Dahil et al., 2020)
	Additive manufacturing	(Huo et al., 2020)
Copper	Sintering blowing agents and metal powders	(Guo et al., 2018; Parvanian et al., 2014)
	Sintering hollow spheres	(Sundarram, 2013)
	Casting techniques	(Nakajima, 2010; Nassef and El-Hadek, 2016)
	Foaming powder compacts	(Shum, 2020; Pop et al., 2017)
	Deposition techniques	(Gao et al., 2016; Balzarotti et al., 2019)
Gold	Deposition techniques	(Cherevko and Chung, 2011; Dong et al., 2013)
	Casting techniques	(Hodge et al., 2006; Li et al., 2017)
	Sintering blowing agents and metal powders	(Cox and Dunand, 2011)
Iron	Sintering blowing agents and metal powders	(Čapek et al., 2015; Sharma and Pandey, 2018)
	Casting techniques	(Rabiei et al., 2004)
	Additive manufacturing	(Sharma and Pandey, 2018; Li et al., 2019)
Steel	Casting techniques	(Mapelli et al., 2013)
	Deposition techniques	(Aal and Aly, 2009)
	Sintering metal powders and fibers	(Yuan et al., 2012)
	Gas entrapment technique	(Mad Rosip et al., 2015; Farhadi, 2018)
	Sintering hollow spheres	(Szyniszewski et al., 2014)
	Sintering blowing agents and metal powders	(Hsu et al., 2021; Tian et al., 2016; Bekoz and Oktay, 2012; Bekoz and Oktay, 2014)
	Additive manufacturing	(Meenashisundaram et al., 2020; Koike et al., 2020; Caiazzo et al., 2017)
Magnesium	Direct Foaming with blowing agent	(Lu et al., 2013)
	Casting techniques	(Gupta and Meenashisundaram, 2015)
	Sintering blowing agents and metal powders	(Hao et al., 2009; Aida et al., 2017)
	Additive manufacturing	(Allavikutty et al., 2021; Kopp et al., 2019)
Nickel	Sintering blowing agents and metal powders	(Unver et al., 2013)
	Additive manufacturing	(Cheng and Li, 2020; Kaur et al., 2020)
Titanium	Sintering blowing agents and metal powders	(Jakubowicz et al., 2013; Jha et al., 2013; Xie et al., 2017)
	Casting techniques	(Müller et al., 2013; Lee et al., 2010; Hasirci and Hasirci, 2018)
	Gas entrapment technique	(Tang et al., 2015; Prabhu et al., 2015; Murray and Dunand, 2004)
	Sintering hollow spheres	(Wally et al., 2015)
	Deposition techniques	(Carcel et al., 2009; Raihan et al., 2021)
	Additive manufacturing	(Harun et al., 2018; Matsushita et al., 2017)
Zinc	Casting techniques	(Mondal et al., 2014)
	Sintering blowing agents and metal powders	(Sadighikia et al., 2015; Mohbe and Mondal, 2021)
	Additive manufacturing	(Qin et al., 2020; Li et al., 2020b)

CONCLUSION

Recent advancements in technology have led to the increased availability of manufacturing processes for various elemental or alloy metallic foams. As seen in the literature, relative densities ranging from 50% to 98% have been achieved in both open and closed cell foams, with anisotropic or uniform cell morphologies. Today, it is possible to obtain light structures that have 50-90% porosity with open and closed pores using the current production methods. According to production methods, different metallic foams are summarized in Table 1. Cheaper methods for metallic foams production increase the preference in the sector. Metallic foams have structural and functional applications in many industrial sectors. Research aimed at improving manufacturing processes focus on producing better quality metallic foams in a more economical way. It is possible to ensure the morphological and structural homogeneity of the porous structure with the quality production process. As understood from the research in the literature; there is no applicable theoretical or numerical model that allows estimation of the effect of changing parameters for most production processes. Improving production methods can be accelerated with the well-designed process models. Metallic foam production is currently on a relatively small scale, but manufacturers are trying to improve their processes and reduce costs. Some methods have demonstrated industrial-scale manufacturing potential through continuous production processes. Using cheaper starting materials, shortening processing steps, and reducing scrap during production are common strategies for reducing production costs. Developments in production techniques of metallic foams; it is very promising both on an industrial and academic scale.

REFERENCES

Aal, A. A., & Aly, M. S. (2009). Electroless Ni–Cu–P plating onto open cell stainless steel foam. *Applied Surface Science*, *255*(13-14), 6652–6655. doi:10.1016/j.apsusc.2009.02.073

Aida, S. F., Zuhailawati, H., & Anasyida, A. S. (2017). The effect of space holder content and sintering temperature of magnesium foam on microstructural and properties prepared by sintering dissolution process (SDP) using carbamide space holder. *Procedia Engineering*, *184*, 290–297. doi:10.1016/j.proeng.2017.04.097

Allavikutty, R., Gupta, P., Santra, T. S., & Rengaswamy, J. (2021). Additive Manufacturing of Mg alloys for Biomedical Applications: Current Status and Challenges. *Current Opinion in Biomedical Engineering*, *18*, 100276. doi:10.1016/j.cobme.2021.100276

Almonti, D., & Ucciardello, N. (2019). Design and thermal comparison of random structures realized by indirect additive manufacturing. *Materials (Basel)*, *12*(14), 2261. doi:10.3390/ma12142261 PMID:31337088

Andure, M. W., Jirapure, S. C., & Dhamande, L. P. (2012). Advance automobile material for light weight future—a review. In *IJCA Proceedings on International Conference on Benchmarks in Engineering Science and Technology* (pp. 15-22). Academic Press.

Ashby, M. F., Evans, T., Fleck, N. A., Hutchinson, J. W., Wadley, H. N. G., & Gibson, L. J. (2000). *Metal foams: a design guide*. Elsevier.

Aşık, E. E., & Bor, Ş. (2015). Fatigue behavior of Ti–6Al–4V foams processed by magnesium space holder technique. *Materials Science and Engineering A*, *621*, 157–165. doi:10.1016/j.msea.2014.10.068

Atwater, M. A., Guevara, L. N., Darling, K. A., & Tschopp, M. A. (2018). Solid state porous metal production: A review of the capabilities, characteristics, and challenges. *Advanced Engineering Materials*, *20*(7), 1700766. doi:10.1002/adem.201700766

Babcsán, N., & Banhart, J. (2006). Metal foams: towards high-temperature colloid chemistry. *Colloidal Particles at Liquid Interfaces*, 445-499.

Babcsán, N., Banhart, J., & Leitlmeier, D. (2003). Metal foams–manufacture and physics of foaming. In *Proceedings of the International Conference Advanced Metallic Materials* (pp. 5-15). Academic Press.

Balzarotti, R., Beretta, A., Groppi, G., & Tronconi, E. (2019). A comparison between washcoated and packed copper foams for the intensification of methane steam reforming. *Reaction Chemistry & Engineering*, *4*(8), 1387–1392. doi:10.1039/C9RE00125E

Banhart, J. (2000). Manufacturing routes for metallic foams. *JOM*, *52*(12), 22–27. doi:10.100711837-000-0062-8

Banhart, J. (2003). Aluminum foams: On the road to real applications. *MRS Bulletin*, *28*(4), 290–295. doi:10.1557/mrs2003.83

Banhart, J. (2004). Industrialisation of aluminium foam technology. In *Proceedings of the ninth International Conference on aluminium alloys* (pp. 764-770). Academic Press.

Banhart, J. (2013). Light-metal foams—History of innovation and technological challenges. *Advanced Engineering Materials*, *15*(3), 82–111. doi:10.1002/adem.201200217

Banhart, J. (2018). Production of Metal Foams. *Comprehensive Composite Materials*, *II*, 347–363. doi:10.1016/B978-0-12-803581-8.09976-8

Banhart, J., & Weaire, D. (2002). On the road again: Metal foams find favor. *Physics Today*, *55*(7), 37–42. doi:10.1063/1.1506749

Bauer, B., Kralj, S., & Bušić, M. (2013). Production and application of metal foams in casting technology. *Tehnicki Vjesnik/Technical Gazette*, *20*(6).

Bekoz, N., & Oktay, E. (2012). Effects of carbamide shape and content on processing and properties of steel foams. *Journal of Materials Processing Technology*, *212*(10), 2109–2116. doi:10.1016/j.jmatprotec.2012.05.015

Bekoz, N., & Oktay, E. (2013). Effect of heat treatment on mechanical properties of low alloy steel foams. *Materials & Design*, *51*, 212–218. doi:10.1016/j.matdes.2013.03.098

Bekoz, N., & Oktay, E. (2014). The role of pore wall microstructure and micropores on the mechanical properties of Cu–Ni–Mo based steel foams. *Materials Science and Engineering A*, *612*, 387–397. doi:10.1016/j.msea.2014.06.064

Benouali, A. H., Froyen, L., Delerue, J. F., & Wevers, M. (2002). Mechanical analysis and microstructural characterisation of metal foams. *Materials Science and Technology*, *18*(5), 489–494. doi:10.1179/026708302225002056

Bhatt, A., Khanna, M., & Pimoli, B. S. (2015). Metal foaming of aluminium alloys. *IOSR Journal of Mechanical and Civil Engineering*, *12*(1), 40–44.

Bisht, A., Patel, V. K., & Gangil, B. (2019). Future of metal foam materials in automotive industry. In *Automotive Tribology* (pp. 51–63). Springer. doi:10.1007/978-981-15-0434-1_4

Caiazzo, F., Campanelli, S. L., Cardaropoli, F., Contuzzi, N., Sergi, V., & Ludovico, A. D. (2017). Manufacturing and characterization of similar to foam steel components processed through selective laser melting. *International Journal of Advanced Manufacturing Technology*, *92*(5), 2121–2130. doi:10.100700170-017-0311-4

Čapek, J., Vojtěch, D., & Oborná, A. (2015). Microstructural and mechanical properties of biodegradable iron foam prepared by powder metallurgy. *Materials & Design*, *83*, 468–482. doi:10.1016/j.matdes.2015.06.022

Carcel, B., Carcel, A. C., Perez, I., Fernandez, E., Barreda, A., Sampedro, J., & Ramos, J. A. (2009, April). Manufacture of metal foam layers by laser metal deposition. In *XVII International Symposium on Gas Flow, Chemical Lasers, and High-Power Lasers* (Vol. 7131, p. 713123). International Society for Optics and Photonics. 10.1117/12.816702

Changdar, A., & Chakraborty, S. S. (2021). Laser processing of metal foam-A review. *Journal of Manufacturing Processes*, *61*, 208–225. doi:10.1016/j.jmapro.2020.10.012

Chen, H., Han, Q., Wang, C., Liu, Y., Chen, B., & Wang, J. (2020). Porous Scaffold Design for Additive Manufacturing in Orthopedics: A Review. *Frontiers in Bioengineering and Biotechnology*, *8*, 609. doi:10.3389/fbioe.2020.00609 PMID:32626698

Cheng, D., & Li, L. (2020). An overview of laser-based multiple metallic material additive manufacturing: From macro-to micro-scales. *International Journal of Extreme Manufacturing*.

Cherevko, S., & Chung, C. H. (2011). Direct electrodeposition of nanoporous gold with controlled multimodal pore size distribution. *Electrochemistry Communications*, *13*(1), 16–19. doi:10.1016/j.elecom.2010.11.001

Claar, T. D., Yu, C. J., Hall, I., Banhart, J., Baumeister, J., & Seeliger, W. (2000). Ultra-lightweight aluminum foam materials for automotive applications. *SAE Transactions*, 98–106. doi:10.4271/2000-01-0335

Cox, M. E., & Dunand, D. C. (2011). Bulk gold with hierarchical macro-, micro-and nano-porosity. *Materials Science and Engineering A*, *528*(6), 2401–2406. doi:10.1016/j.msea.2010.11.072

Dahil, L., Katirci, R., & Sümbül, H. İ. (2020). Effect of Artificial Aging Process on Aluminum Foam Made of Etial 160. *Transactions of the Indian Institute of Metals*, *73*(11), 2739–2745. doi:10.100712666-020-02081-w

Dani, I., Drossel, W. G., Milaev, N., Korn, H., Hannemann, C., Hohlfeld, J., & Wertheim, R. (2020). Sustainability of industrial components using additive manufacturing and foam materials. *Procedia Manufacturing*, *43*, 10–17. doi:10.1016/j.promfg.2020.02.102

Degischer, H. P., Körner, C., Singer, R. F., Banhart, J., Baumgärtner, F., Rausch, G., Arnold, M., Thies, M., San Marchi, C., Mortensen, A., Andersen, O., & Stephani, G. (2002). Material definitions, processing, and recycling. Handbook of Cellular Metals: Production, Processing, Applications, 5-70.

Dong, Y., Shang, W., Yang, J., Zhang, L., Zhang, W., Li, Z., Guo, L., Zhan, X., Du, H., Deng, B., & Pu, Y. (2013). The impact of low-Z impurities on x-ray conversion efficiency from laser-produced plasmas of low-density gold foam targets. *Physics of Plasmas*, *20*(12), 123305. doi:10.1063/1.4859215

Dukhan, N. (Ed.). (2013). Metal foams: fundamentals and applications. DEStech Publications, Inc.

Esen, Z., & Bor, Ş. (2007). Processing of titanium foams using magnesium spacer particles. *Scripta Materialia*, *56*(5), 341–344. doi:10.1016/j.scriptamat.2006.11.010

Esen, Z., & Bor, Ş. (2011). Characterization of Ti–6Al–4V alloy foams synthesized by space holder technique. *Materials Science and Engineering A*, *528*(7-8), 3200–3209. doi:10.1016/j.msea.2011.01.008

Farhadi, S. (2018). *Dynamic Characterization and Modelling of Metallic Foam Material* (Doctoral dissertation). Politecnico di Torino.

Frazier, W. E. (2014). Metal additive manufacturing: A review. *Journal of Materials Engineering and Performance*, *23*(6), 1917–1928. doi:10.100711665-014-0958-z

Furumoto, T., Koizumi, A., Alkahari, M. R., Anayama, R., Hosokawa, A., Tanaka, R., & Ueda, T. (2015). Permeability and strength of a porous metal structure fabricated by additive manufacturing. *Journal of Materials Processing Technology*, *219*, 10–16. doi:10.1016/j.jmatprotec.2014.11.043

Gábora, A., & Mankovits, T. (2019). Quality control of closed-cell metal foam produced by direct foaming. *IOP Conference Series. Materials Science and Engineering*, *659*(1), 012037. doi:10.1088/1757-899X/659/1/012037

Gao, X., Zhou, J., Du, R., Xie, Z., Deng, S., Liu, R., Liu, Z., & Zhang, J. (2016). Robust superhydrophobic foam: A graphdiyne-based hierarchical architecture for oil/water separation. *Advanced Materials*, *28*(1), 168–173. doi:10.1002/adma.201504407 PMID:26551876

García-Moreno, F. (2016). Commercial applications of metal foams: Their properties and production. *Materials (Basel)*, *9*(2), 85. doi:10.3390/ma9020085 PMID:28787887

Guddati, S., Kiran, A. S. K., Leavy, M., & Ramakrishna, S. (2019). Recent advancements in additive manufacturing technologies for porous material applications. *International Journal of Advanced Manufacturing Technology*, *105*(1), 193–215. doi:10.100700170-019-04116-z

Gülsoy, H. Ö., & German, R. M. (2008). Sintered foams from precipitation hardened stainless steel powder. *Powder Metallurgy*, *51*(4), 350–353. doi:10.1179/174329008X286703

Guo, C. Q., Sun, Y. D., Zhou, Y., Xie, B., Wang, T. Y., & Zuo, X. Q. (2018). Fabrication, structure and property of copper foam. *Materials Science Forum, 933*, 41–48. doi:10.4028/www.scientific.net/MSF.933.41

Gupta, M., & Meenashisundaram, G. K. (2015). Synthesis of Magnesium-Based Biomaterials. In *Insight into Designing Biocompatible Magnesium Alloys and Composites* (pp. 17–34). Springer.

Hao, G. L., Han, F. S., & Li, W. D. (2009). Processing and mechanical properties of magnesium foams. *Journal of Porous Materials, 16*(3), 251–256. doi:10.100710934-008-9194-y

Harun, W. S. W., Kamariah, M. S. I. N., Muhamad, N., Ghani, S. A. C., Ahmad, F., & Mohamed, Z. (2018). A review of powder additive manufacturing processes for metallic biomaterials. *Powder Technology, 327*, 128–151. doi:10.1016/j.powtec.2017.12.058

Hasirci, V., & Hasirci, N. (2018). Metals as Biomaterials. In *Fundamentals of Biomaterials* (pp. 35–49). Springer. doi:10.1007/978-1-4939-8856-3_3

Hernández-Nava, E., Smith, C. J., Derguti, F., Tammas-Williams, S., Léonard, F., Withers, P. J., Todd, I., & Goodall, R. (2015). The effect of density and feature size on mechanical properties of isostructural metallic foams produced by additive manufacturing. *Acta Materialia, 85*, 387–395. doi:10.1016/j.actamat.2014.10.058

Hodge, A. M., Hayes, J. R., Caro, J. A., Biener, J., & Hamza, A. V. (2006). Characterization and mechanical behavior of nanoporous gold. *Advanced Engineering Materials, 8*(9), 853–857. doi:10.1002/adem.200600079

Hsu, C. M., Tzeng, Y. C., Chen, S. F., Chen, Y. L., & Lee, H. L. (2021). Fabrication of 17-4PH Stainless Steel Foam by a Pressureless Powder Space Holder Technique. *Advanced Engineering Materials, 23*(6), 2001202. doi:10.1002/adem.202001202

Huang, R., Ma, S., Zhang, M., Xu, J., & Wang, Z. (2019). Dynamic deformation and failure process of quasi-closed-cell aluminum foam manufactured by direct foaming technique. *Materials Science and Engineering A, 756*, 302–311. doi:10.1016/j.msea.2019.04.050

Huo, P. C., Hao, Z. Y., Bai, P. K., Zhang, L. Z., Wu, L. Y., Du, W. B., & Han, B. (2020). The Compressive Behavior of Porous AlSi10Mg Prepared by Selective Laser Melting (SLM). Lasers in Engineering, 47.

Jain, H., Mondal, D. P., Gupta, G., & Kumar, R. (2021). Effect of compressive strain rate on the deformation behaviour of austenitic stainless steel foam produced by space holder technique. *Materials Chemistry and Physics, 259*, 124010. doi:10.1016/j.matchemphys.2020.124010

Jain, H., Mondal, D. P., Gupta, G., Kumar, R., & Singh, S. (2020). Synthesis and characterization of 316L stainless steel foam made through two different removal process of space holder method. *Manufacturing Letters, 26*, 33–36. doi:10.1016/j.mfglet.2020.09.005

Jakubowicz, J., Adamek, G., & Dewidar, M. (2013). Titanium foam made with saccharose as a space holder. *Journal of Porous Materials, 20*(5), 1137–1141. doi:10.100710934-013-9696-0

Jha, N., Mondal, D. P., Majumdar, J. D., Badkul, A., Jha, A. K., & Khare, A. K. (2013). Highly porous open cell Ti-foam using NaCl as temporary space holder through powder metallurgy route. *Materials & Design*, *47*, 810–819. doi:10.1016/j.matdes.2013.01.005

Kaur, G., Pulagara, N. V., & Lahiri, I. (2020). Three-Dimensional Graphene Materials for Supercapacitors. *Graphene as Energy Storage Material for Supercapacitors*, *64*, 77–128. doi:10.21741/9781644900550-4

Kennedy, A. (2012). Porous metals and metal foams made from powders. *Powder Metallurgy*, 124.

Koike, R., Matsumoto, T., Aoyama, T., & Kondo, M. (2020). Fabrication method for stainless steel foam block in directed energy deposition. *CIRP Annals*, *69*(1), 173–176. doi:10.1016/j.cirp.2020.04.060

Koike, R., Matsumoto, T., Kakinuma, Y., Aoyama, T., Oda, Y., Kuriya, T., & Kondo, M. (2018). A basic study on metal foam fabrication with titanium hydride in direct energy deposition. *Procedia Manufacturing*, *18*, 68–73. doi:10.1016/j.promfg.2018.11.009

Komissarchuk, O., Hao, H., Zhang, X. L., & Karpov, V. (2018). Fabrication of Al–Si Gasar by mold casting technique. *International Journal of Materials Research*, *109*(4), 332–340. doi:10.3139/146.111612

Kopp, A., Derra, T., Müther, M., Jauer, L., Schleifenbaum, J. H., Voshage, M., & Kröger, N. (2019). Influence of design and postprocessing parameters on the degradation behavior and mechanical properties of additively manufactured magnesium scaffolds. *Acta Biomaterialia*, *98*, 23–35. doi:10.1016/j.actbio.2019.04.012 PMID:30959185

Kränzlin, N., & Niederberger, M. (2015). Controlled fabrication of porous metals from the nanometer to the macroscopic scale. *Materials Horizons*, *2*(4), 359–377. doi:10.1039/C4MH00244J

Kulshreshtha, A., & Dhakad, S. K. (2020). Preparation of metal foam by different methods: A review. *Materials Today: Proceedings*, *26*, 1784–1790. doi:10.1016/j.matpr.2020.02.375

Lee, J. H., Kim, H. E., Shin, K. H., & Koh, Y. H. (2010). Improving the strength and biocompatibility of porous titanium scaffolds by creating elongated pores coated with a bioactive, nanoporous TiO2 layer. *Materials Letters*, *64*(22), 2526–2529. doi:10.1016/j.matlet.2010.08.038

Li, J. P., de Wijn, J. R., Van Blitterswijk, C. A., & de Groot, K. (2006). Porous Ti6Al4V scaffold directly fabricating by rapid prototyping: Preparation and in vitro experiment. *Biomaterials*, *27*(8), 1223–1235. doi:10.1016/j.biomaterials.2005.08.033 PMID:16169073

Li, Y., Jahr, H., Pavanram, P., Bobbert, F. S. L., Puggi, U., Zhang, X. Y., Pouran, B., Leeflang, M. A., Weinans, H., & Zadpoor, A. A. (2019). Additively manufactured functionally graded biodegradable porous iron. *Acta Biomaterialia*, *96*, 646–661. doi:10.1016/j.actbio.2019.07.013 PMID:31302295

Li, Y., Jahr, H., Zhou, J., & Zadpoor, A. A. (2020a). Additively manufactured biodegradable porous metals. *Acta Biomaterialia*. PMID:32853809

Li, Y., Pavanram, P., Zhou, J., Lietaert, K., Taheri, P., Li, W., San, H., Leeflang, M. A., Mol, J. M. C., Jahr, H., & Zadpoor, A. A. (2020b). Additively manufactured biodegradable porous zinc. *Acta Biomaterialia*, *101*, 609–623. doi:10.1016/j.actbio.2019.10.034 PMID:31672587

Li, Z., Luo, J., Tan, X., Fang, Q., Zeng, Y., Zhou, M., Wu, W., & Zhang, J. (2017). X-ray nanotomography characterizations of gold foams. *Materials Letters, 205*, 215–218. doi:10.1016/j.matlet.2017.06.056

Liu, P., & Chen, G. F. (2014). *Porous materials: processing and applications*. Elsevier.

Lu, G. Q., Hai, H. A. O., Wang, F. Y., & Zhang, X. G. (2013). Preparation of closed-cell Mg foams using SiO2-coated CaCO3 as blowing agent in atmosphere. *Transactions of Nonferrous Metals Society of China, 23*(6), 1832–1837. doi:10.1016/S1003-6326(13)62667-9

Ma, S., Tang, Q., Han, X., Feng, Q., Song, J., Setchi, R., Liu, Y., Liu, Y., Goulas, A., Engstrom, D. S., Tse, Y. Y., & Zhen, N. (2020). Manufacturability, mechanical properties, mass-transport properties and biocompatibility of triply periodic minimal surface (TPMS) porous scaffolds fabricated by selective laser melting. *Materials & Design, 195*, 109034. doi:10.1016/j.matdes.2020.109034

Mad Rosip, N. I., Ahmad, S., Jamaludin, K. R., & Mat Noor, F. (2015). Morphological analysis of SS316L foam produced by using slurry method. *Advanced Materials Research, 1087*, 68–72. doi:10.4028/www.scientific.net/AMR.1087.68

Mapelli, C., Mombelli, D., Gruttadauria, A., Barella, S., & Castrodeza, E. M. (2013). Performance of stainless steel foams produced by infiltration casting techniques. *Journal of Materials Processing Technology, 213*(11), 1846–1854. doi:10.1016/j.jmatprotec.2013.05.010

Matijasevic-Lux, B. (2006). *Characterisation and optimisation of blowing agent for making improved metal foams* (Doctoral Thesis). Technische Universität Berlin, Fakultät III – Prozesswissenschaften.

Matsushita, T., Fujibayashi, S., & Kokubo, T. (2017). Titanium foam for bone tissue engineering. In *Metallic Foam Bone* (pp. 111–130). Woodhead Publishing. doi:10.1016/B978-0-08-101289-5.00004-4

Meenashisundaram, G. K., Xu, Z., Nai, M. L. S., Lu, S., Ten, J. S., & Wei, J. (2020). Binder Jetting Additive Manufacturing of High Porosity 316L Stainless Steel Metal Foams. *Materials (Basel), 13*(17), 3744. doi:10.3390/ma13173744 PMID:32847089

Mohbe, M., & Mondal, D. P. (2021). Properties of Zn foam filled with cenosphere microballoons. *Materials Today: Proceedings, 46*, 7448–7451. doi:10.1016/j.matpr.2021.01.073

Mondal, D. P., Goel, M. D., Bagde, N., Jha, N., Sahu, S., & Barnwal, A. K. (2014). Closed cell ZA27–SiC foam made through stir-casting technique. *Materials & Design, 57*, 315–324. doi:10.1016/j.matdes.2013.12.026

Müller, D. W., Matz, A. M., & Jost, N. (2013). Casting open porous Ti foam suitable for medical applications. *Bioinspired. Biomimetic and Nanobiomaterials, 2*(2), 76–83. doi:10.1680/bbn.12.00023

Murray, N. G. D., & Dunand, D. C. (2004). Effect of thermal history on the superplastic expansion of argon-filled pores in titanium: Part I kinetics and microstructure. *Acta Materialia, 52*(8), 2269–2278. doi:10.1016/j.actamat.2004.01.039

Nakajima, H. (2010). Fabrication, properties, and applications of porous metals with directional pores. *Proceedings of the Japan Academy. Series B, Physical and Biological Sciences, 86*(9), 884–899. doi:10.2183/pjab.86.884 PMID:21084772

Nassef, A., & El-Hadek, M. (2016). Microstructure and mechanical behavior of hot pressed Cu-Sn powder alloys. *Advances in Materials Science and Engineering*, *2016*, 2016. doi:10.1155/2016/9796169

Oppenheimer, S. M., O'Dwyer, J. G., & Dunand, D. C. (2004). Porous, superelastic NiTi produced by powder-metallurgy. *TMS Lett*, *1*, 93–94.

Papantoniou, I., Kyriakopoulou, H., Pantelis, D., & Manolakos, D. (2019). Metal foaming by powder metallurgy process: investigation of different parameters on the foaming efficiency. *Frattura ed Integrità Strutturale, 13*(50), 497-504.

Parvanian, A. M., Saadatfar, M., Panjepour, M., Kingston, A., & Sheppard, A. P. (2014). The effects of manufacturing parameters on geometrical and mechanical properties of copper foams produced by space holder technique. *Materials & Design*, *53*, 681–690. doi:10.1016/j.matdes.2013.07.047

Paserin, V., Marcuson, S., Shu, J., & Wilkinson, D. S. (2003). The chemical vapor deposition technique for Inco nickel foam production–manufacturing benefits and potential applications. *Cellular Metals and Metal Foaming Technology*, 31.

Patel, P., Bhingole, P. P., & Makwana, D. (2018). Manufacturing, characterization and applications of lightweight metallic foams for structural applications. *Materials Today: Proceedings*, *5*(9), 20391–20402. doi:10.1016/j.matpr.2018.06.414

Pop, M. A., Geaman, V., Radomir, I., & Bedo, T. (2017). Capacity of energy absorption by flick through shock in cooper foams. *Journal of Porous Media*, *20*(5), 405–415. doi:10.1615/JPorMedia.v20.i5.30

Prabhu, S., Raja, V. K., & Nikhil, R. (2015). Applications of Cellular Materials–An Overview. *Applied Mechanics and Materials*, *766*, 511–517. doi:10.4028/www.scientific.net/AMM.766-767.511

Qin, J., Chen, Q., Yang, C., & Huang, Y. (2016). Research process on property and application of metal porous materials. *Journal of Alloys and Compounds*, *654*, 39–44. doi:10.1016/j.jallcom.2015.09.148

Qin, Y., Wen, P., Xia, D., Guo, H., Voshage, M., Jauer, L., Zheng, Y., Schleifenbaum, J. H., & Tian, Y. (2020). Effect of grain structure on the mechanical properties and in vitro corrosion behavior of additively manufactured pure Zn. *Additive Manufacturing*, *33*, 101134. doi:10.1016/j.addma.2020.101134

Rabiei, A., O'Neill, A. T., & Neville, B. P. (2004). Processing and development of a new high strength metal foam. *MRS Online Proceedings Library, 851*.

Raihan, M. M., Otsuka, Y., Tsuchida, K., Manonukul, A., Ohnuma, K., & Miyashita, Y. (2021). Damage evaluation of HAp-coated porous titanium foam in simulated body fluid based on compression fatigue behavior. *Journal of the Mechanical Behavior of Biomedical Materials*, *117*, 104383. doi:10.1016/j.jmbbm.2021.104383 PMID:33596530

Rajak, D. K., & Gupta, M. (2020). An Insight Into Metal Based Foams. Springer.

Rechtin, J., Torresani, E., Ivanov, E., & Olevsky, E. (2018). Fabrication of titanium-niobium-zirconium-tantalium alloy (TNZT) bioimplant components with controllable porosity by spark plasma sintering. *Materials (Basel)*, *11*(2), 181. doi:10.3390/ma11020181 PMID:29364165

Ryan, G., Pandit, A., & Apatsidis, D. P. (2006). Fabrication methods of porous metals for use in orthopaedic applications. *Biomaterials*, *27*(13), 2651–2670. doi:10.1016/j.biomaterials.2005.12.002 PMID:16423390

Sadighikia, S., Abdolhosseinzadeh, S., & Asgharzadeh, H. (2015). Production of high porosity Zn foams by powder metallurgy method. *Powder Metallurgy*, *58*(1), 61–66. doi:10.1179/1743290114Y.0000000109

Salvo, L., Martin, G., Suard, M., Marmottant, A., Dendievel, R., & Blandin, J. J. (2014). Processing and structures of solids foams. *Comptes Rendus Physique*, *15*(8-9), 662–673. doi:10.1016/j.crhy.2014.10.006

Sazegaran, H., & Nezhad, S. M. M. (2021). Cell morphology, porosity, microstructure and mechanical properties of porous Fe-C-P alloys. *International Journal of Minerals Metallurgy and Materials*, *28*(2), 257–265. doi:10.100712613-020-1995-2

Shapovalov, V., & Boyko, L. (2004). Gasar—A new class of porous materials. *Advanced Engineering Materials*, *6*(6), 407–410. doi:10.1002/adem.200405148

Sharma, P., & Pandey, P. M. (2018). Morphological and mechanical characterization of topologically ordered open cell porous iron foam fabricated using 3D printing and pressureless microwave sintering. *Materials & Design*, *160*, 442–454. doi:10.1016/j.matdes.2018.09.029

Shim, D. S., Seo, J. Y., Yoon, H. S., Lee, K. Y., & Oh, W. J. (2018). Additive manufacturing of porous metals using laser melting of Ti6Al4V powder with a foaming agent. *Materials Research Express*, *5*(8), 086518. doi:10.1088/2053-1591/aad117

Shum, T. T. C. (2020). *Investigating mass flow in copper foam* (Doctoral dissertation). RMIT University.

Sivertsen, K. (2007). *Polymer foams. In Polymer Physics, Spring*. Massachusetts Institute of Technology.

Sola, A., & Nouri, A. (2019). Microstructural porosity in additive manufacturing: The formation and detection of pores in metal parts fabricated by powder bed fusion. *Journal of Advanced Manufacturing and Processing*, *1*(3), e10021. doi:10.1002/amp2.10021

Srivastava, V. C., & Sahoo, K. L. (2007). Processing, stabilization and applications of metallic foams. Art of science. *Materials Science Poland*, *25*(3), 733–753.

Sun, L., Wang, Y., Wang, L., Wang, F., Xu, H., Huang, W., & You, X. (2021). Preparation and properties of controllable aluminum foam. *Materials Research Express*, *8*(2), 026526. doi:10.1088/2053-1591/abe5ef

Sundarram, S. (2013). *Fabrication and characterization of open celled micro and nano foams*. Doctoral Dissertion, Faculty of the Graduate School of The University of Texas at Austin.

Sutygina, A., Betke, U., Hasemann, G., & Scheffler, M. (2020). Manufacturing of Open-Cell Metal Foams by the Sponge Replication Technique. *IOP Conference Series. Materials Science and Engineering*, *882*(1), 012022. doi:10.1088/1757-899X/882/1/012022

Szyniszewski, S. T., Smith, B. H., Hajjar, J. F., Schafer, B. W., & Arwade, S. R. (2014). The mechanical properties and modeling of a sintered hollow sphere steel foam. *Materials & Design, 54*, 1083-1094.

Takezawa, A., Kobashi, M., Koizumi, Y., & Kitamura, M. (2017). Porous metal produced by selective laser melting with effective isotropic thermal conductivity close to the Hashin–Shtrikman bound. *International Journal of Heat and Mass Transfer*, *105*, 564–572. doi:10.1016/j.ijheatmasstransfer.2016.10.006

Tang, H. P., Wang, J., & Qian, M. (2015). Porous titanium structures and applications. In *Titanium powder metallurgy* (pp. 533–554). Butterworth-Heinemann. doi:10.1016/B978-0-12-800054-0.00028-9

Tian, D. R., Pang, Y. H., Yu, L., & Sun, L. (2016). Production and characterization of high porosity porous Fe-Cr-C alloys by the space holder leaching technique. *International Journal of Minerals Metallurgy and Materials*, *23*(7), 793–798. doi:10.100712613-016-1293-1

Tian, Q. H., & Guo, X. Y. (2010). Electroless copper plating on microcellular polyurethane foam. *Transactions of Nonferrous Metals Society of China*, *20*, s283–s287. doi:10.1016/S1003-6326(10)60057-X

Unver, I., Gulsoy, H. O., & Aydemir, B. (2013). Ni-625 Superalloy Foam Processed by Powder Space-Holder Technique. *Journal of Materials Engineering and Performance*, *22*(12), 3735–3741. doi:10.100711665-013-0702-0

Vendra, L. J., Brown, J. A., & Rabiei, A. (2011). Effect of processing parameters on the microstructure and mechanical properties of Al–steel composite foam. *Journal of Materials Science*, *46*(13), 4574–4581. doi:10.100710853-011-5356-4

Wally, Z. J., Van Grunsven, W., Claeyssens, F., Goodall, R., & Reilly, G. C. (2015). Porous titanium for dental implant applications. *Metals*, *5*(4), 1902–1920. doi:10.3390/met5041902

Wang, N., Chen, X., Li, Y., Liu, Z., Zhao, Z., Cheng, Y., Liu, Y., & Zhang, H. (2017a). The cell size reduction of aluminum foam with dynamic gas injection based on the improved foamable melt. *Colloids and Surfaces. A, Physicochemical and Engineering Aspects*, *527*, 123–131. doi:10.1016/j.colsurfa.2017.05.023

Wang, Q., Xu, F. M., Xu, Q. Y., & Xiong, S. M. (2007). Study on an aluminum foam by gas injection foaming. *Foundry*, *56*, 814–818.

Wang, Z., Wang, C., Li, C., Qin, Y., Zhong, L., Chen, B., Li, Z., Liu, H., Chang, F., & Wang, J. (2017b). Analysis of factors influencing bone ingrowth into three-dimensional printed porous metal scaffolds: A review. *Journal of Alloys and Compounds*, *717*, 271–285. doi:10.1016/j.jallcom.2017.05.079

Wen, C. E., Mabuchi, M., Yamada, Y., Shimojima, K., Chino, Y., & Asahina, T. (2001). Processing of biocompatible porous Ti and Mg. *Scripta Materialia*, *45*(10), 1147–1153. doi:10.1016/S1359-6462(01)01132-0

Xie, B., Fan, Y. Z., Mu, T. Z., & Deng, B. (2017). Fabrication and energy absorption properties of titanium foam with CaCl2 as a space holder. *Materials Science and Engineering A*, *708*, 419–423. doi:10.1016/j.msea.2017.09.123

Xu, Z., Zhu, Z., Wang, P., Meenashisundaram, G. K., Nai, S. M. L., & Wei, J. (2020). Fabrication of porous CoCrFeMnNi high entropy alloy using binder jetting additive manufacturing. *Additive Manufacturing*, *35*, 101441. doi:10.1016/j.addma.2020.101441

Yilong, L., Guibao, Q., Yang, Y., Xuewei, L., & Chenguang, B. (2016). Preparation and compressive properties of magnesium foam. *Rare Metal Materials and Engineering*, *45*(10), 2498–2502. doi:10.1016/S1875-5372(17)30022-X

Yuan, L., Ding, S., & Wen, C. (2019). Additive manufacturing technology for porous metal implant applications and triple minimal surface structures: A review. *Bioactive Materials*, *4*, 56–70. doi:10.1016/j.bioactmat.2018.12.003 PMID:30596158

Yuan, W., Tang, Y., Yang, X., & Wan, Z. (2012). Porous metal materials for polymer electrolyte membrane fuel cells–A review. *Applied Energy*, *94*, 309–329. doi:10.1016/j.apenergy.2012.01.073

Zhang, X., Li, Y., Liu, Y., & Zhang, H. (2013). Fabrication of a bimodal micro/nanoporous metal by the Gasar and dealloying processes. *Materials Letters*, *92*, 448–451. doi:10.1016/j.matlet.2012.10.090

Zhang, X., Li, Y., Zhang, H., & Liu, Y. (2013). Fabrication of a three-dimensional bimodal porous metal. *Materials Letters*, *106*, 417–420. doi:10.1016/j.matlet.2013.05.077

Chapter 8
Fatigue Characterization and Fractographic Analysis of Aluminium 6063 Alloy

Sreearravind M.
School of Mechanical Engineering, SASTRA University (Deemed), India

Ramesh Kumar S.
School of Mechanical Engineering, SASTRA University (Deemed), India

Ahilan C.
Department of Mechanical Engineering, Sri Venkateswara College of Engineering and Technology, India

ABSTRACT

Aluminium and its alloy are widely employed in various automobile and aircraft areas because of their unique specific strength and formability. Al alloys that have been employed in aerospace structural components will undergo dynamic loading, which leads to fatigue due to mechanical stress and thermal conditions. Considering studies toward the low cycle fatigue behaviour of Al alloys are significantly narrowed, this chapter sighted to the analysis of fatigue behaviour of Al 6063 alloy at the various total strain amplitude (TSA) of 0.4% and 0.8%, which performed through the low cycle fatigue testing machine at the frequency rate of 0.2 Hz. The test results show that for 0.4% TSA, the number of cycles to failure (N) is 1786, whereas as the TSA increases, N got reduced. For 0.8% TSA, the cycle to failure is 291 and samples undergone cyclic softening during the test. The rate of cyclic plastic strain raised up with the increase in the TSA. Crack propagation was observed along with the quasi-cleavage fracture for 0.4% TSA and cleavage fracture for 0.8% TSA.

DOI: 10.4018/978-1-7998-7864-3.ch008

INTRODUCTION

Aluminium is one of the most commonly used lightweight materials in the automotive industry and the aerospace industry. The aluminium density is 2.71g/cc, which implies Al is lighter than Iron and Copper. The aluminium's corrosion resistance is excellent, and thin, adherent oxide layers contribute to these characteristics. Additionally, aluminium is also preferred in the aerospace industry because of its high specific stiffness and specific strength, fracture toughness, fatigue resistance, and ductility. The other important aspect of the aluminium alloy is the ability to fabricate since aluminium foil can be into micron thickness. In the automotive industry, aluminium alloys have been rooted since the nineteenth century. In modern days, aluminium alloys have replaced many steel parts to reduce vehicle weight and improve efficiency. By the early 1970s, the share of aluminium in a car only accounted for 35kg. The share of aluminium drastically improves to 153kg, and it is estimated that by 2025 the share will be 250kg, shown in figure 1. In aerospace, aluminium plays a vital role in producing weight-efficient aircraft components (P. C. Angelo & B. Ravisankar, 2018). This leads the aerospace industry to move ahead in fuel-efficient transportation and high speed and altitude flying. Among other lightweight materials such as magnesium, titanium, and fibre reinforcement materials, aluminium and its alloy account for more than 60% in the recent aircraft industry, as shown in figure 1. The below chart represents the approximate accountancy of various materials in aerospace manufacturing (Payne, 1976).

Figure 1. The contribution of aluminium alloys in car manufacturing.

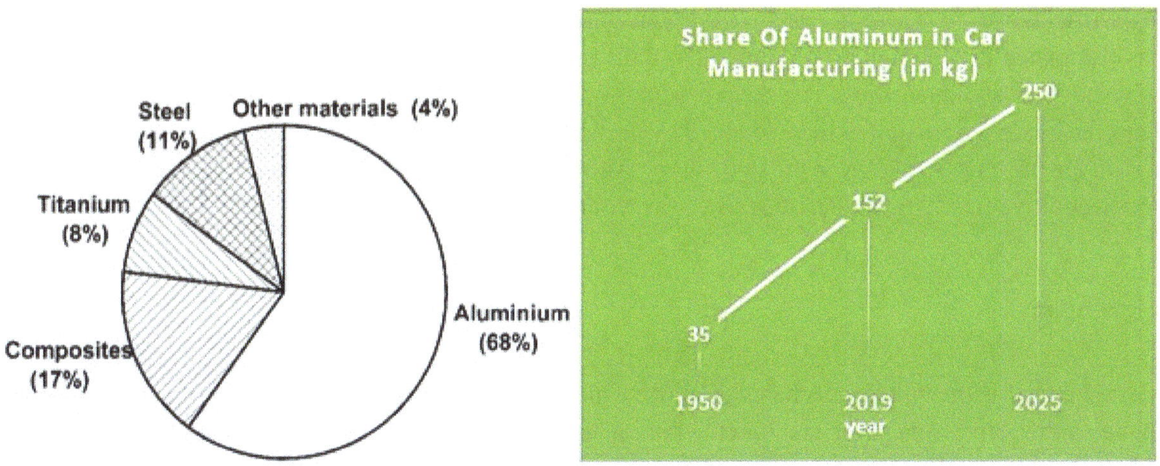

Aluminium and its alloys are classified into wrought and casting alloys. Among the wrought alloys, some alloys alter their mechanical properties through a proper heat-treatment process. This series of wrought alloys are having a yield strength range of 450 to 600 MPa. This alloy series is used in semi-structural and structural parts on aircraft. Some wrought alloys did not react to heat treatment, so that those series may be subjected to cold working. This non-heat-treatable has a yield strength below 300MPa, so the application is restricted in the automotive and aerospace industry. Some particular cases of alloy respond to heat treatment, and also, it can be cold worked. The application of cast aluminium products in the aerospace and automotive industry is very low compared to wrought aluminium alloys due to load-bearing ability (Srivatsan, 2002).

Aluminium – magnesium- silicon alloys are a moderate strength alloy in the aluminium series. This series is mostly utilised in the wrought form, and the 6xxx series is heat treatable. The existence of magnesium and silicon as major alloying elements in this series. These alloying elements form the intermetallic compounds, which enhance the strength of the alloy. Mg and Si enable to forms of an essential intermetallic compound, Mg_2Si, which hardens the alloys. The magnesium and silicon content balanced between the composition of 0.75% and 1.3%. In 6xxx alloys, according to the equilibrium diagram, the atomic ratios of Magnesium and Silicon are always maintained close to 2:1. However, in parallel experimental evidence, the ratio of 1:1 for Magnesium and Silicon is sufficient to enhance the strength of the aluminium alloy through the precipitation hardening. They can usually be aged at elevated temperatures. Precipitate Mg_2Si formation is the most complex intermetallic complex in this aluminium alloy. The formation of GP zones influences precipitation stages by the cluster of Si atoms (Hardouin Duparc, 2010).

During the commercial operation, mechanical properties can be modified within the delay at room temperature between the artificial ageing and quenching. For example, if the alloy contains one per cent of Mg_2Si, 10% of its tensile property can be reduced in the alloy within 24 hours of delay compared to other alloys, which properly ages hardened. On the other hand, if the presence of Mg_2Si is less than 0.9%, tensile properties are enhanced (Chang et al., 2003). The phenomenon has been contributed by the atomic clustering and vacancies which occur at room temperature. In this series, copper is also added in a small amount, i.e., lesser than 0.25% (Bahl et al., 2020), which helps retrieve the alloy's tensile property during artificial ageing. These alloys are readily available in the form of an extruded rod so that this alloy has eliminated the solution heat treatment as a distinct process. The finished products will be generally subjected to water quenching. The sheet products obtained from the wrought process will be air-cooled. Thus the moderate strength will be retained by the heat treatment process at elevated temperature. In this series, the most utilised alloy is the 6063 alloy. Aluminium 6063 alloy has been utilised in architectural and decorative finishing applications (SreeArravind et al., 2019).

FATIGUE

Metal Fatigue is the type of mechanical failure which appears in various forms. The frequent variation in the applied stress and strain result in the fatigue failure of a material. When the cyclic load acts in the material at a high temperature may consider as thermomechanical fatigue. The fatigue has various stages until the final fracture in the components. The stages of fatigue start with permanent damage caused by nucleation and led to substructural change. Followed by the microstructural change, the micro-cracks emerge in the medium. The formed cracks propagate along the surface of the material (McCullough et al., 2019). The dominant crack initiates the stable propagation intensively, leading to the component's catastrophic failure, as illustrated in figure 2.

Mechanical fatigue is classified into high cycle and low cycle fatigue, based on the number of cycles to failure. The type of deformation during the loading has given in figure 3 (Akano & Fakinlede, 2013; Chandran, 2005). In high cycle fatigue, the material deforms within the elastic region controlled by constant stress. In the low cycle, the materials deform plastically under constant strain.

Figure 2. A representation of fatigue crack initiation comprising the occurrence of a slip: (a): sample before fatigue testing (b): mechanism in the sample during cyclic loading - fatigue crack initiation and (stage I and II) propagation of the crack.

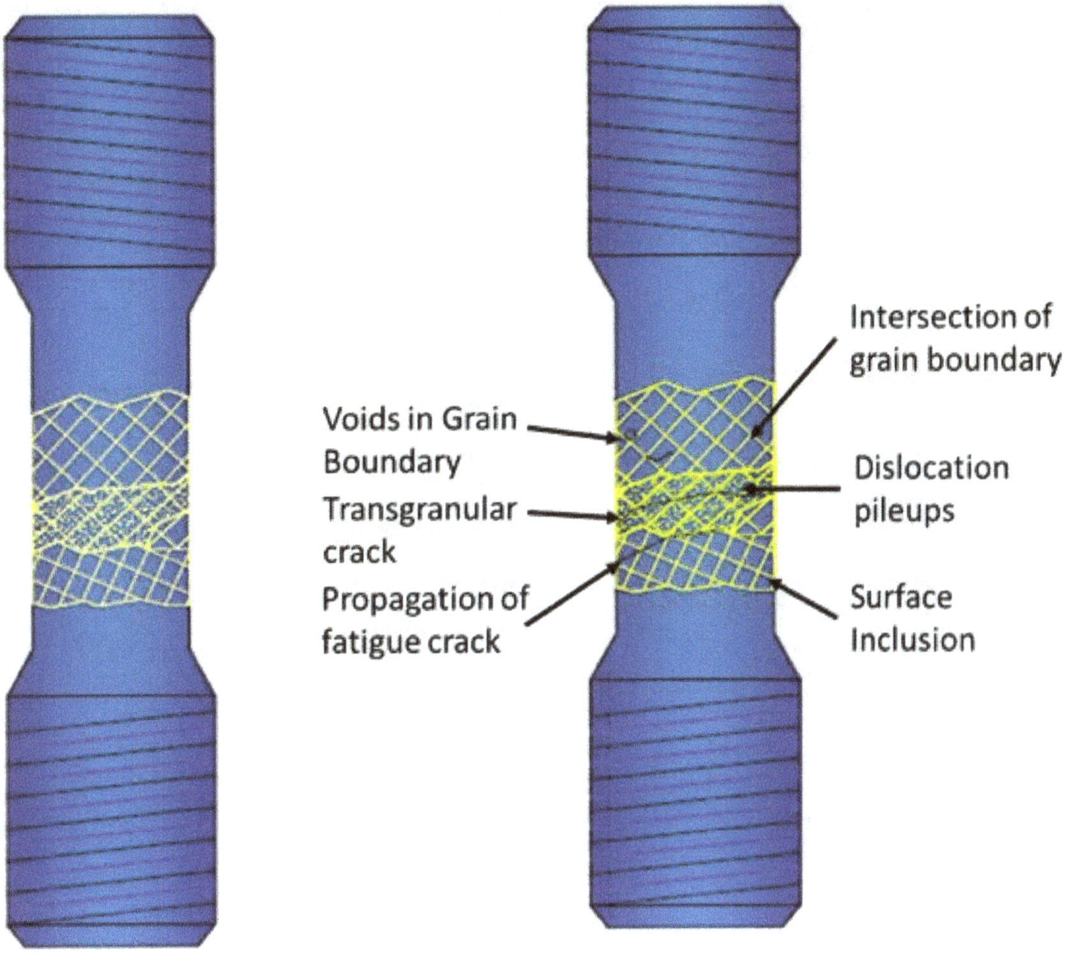

LOW CYCLE FATIGUE

Low cycle fatigue could also be classified into Low cycle fatigue (LCF) – for less than 10^4 cycles to failure and Very low life cycles - for less than 100 cycles to failure. The very low cycle regime is well-known now. It is titled "extremely low cycle fatigue" (ELCF) to distinguish from the narrow portion of the life cycle curve that fits the Manson–Coffin law, which is occasionally introduced as the "low cycle fatigue" in more specific reasoning. The failure paths are also dissimilar in these two regimes. For instance, in push-pull fatigue tests of cylindrical bars, the rupture in the ELCF range often occurs at the centre of the round bar, while in the strict LCF regime, the crack often starts from the surface. Attempts have been made to control Manson– Coffin law's difficulty in the ELCF regime. More recently, a group of researchers inferred a power exponential connection by raising the logarithm of the plastic strain amplitude to a non-unity power and showed good accordance with trial data (Dieter, 1986).

Figure 3. Different failure modes for a structure

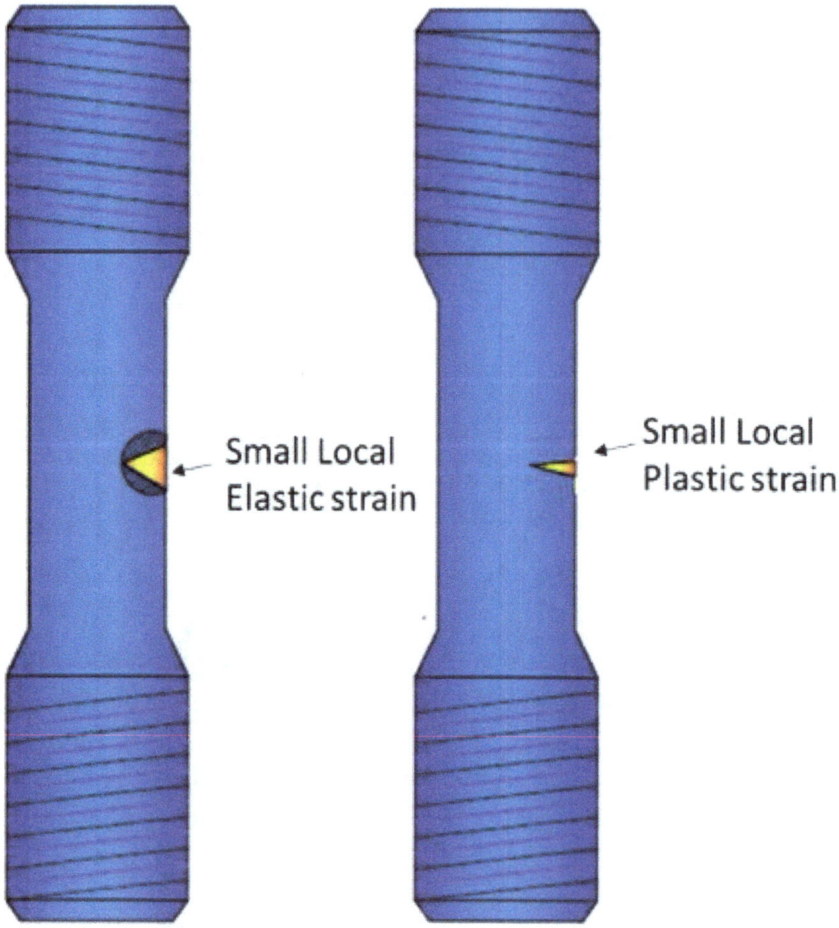

Low cycle fatigue is the repeated cyclic loadings that generate exceptional plastic distortion in a material. Low cycle fatigue usually happens because of repeated localised yielding near stress raisers, such as holes, fillets, and notches, despite the elastic deformation occurring over the component's bulk. Uniaxial testing is performed on several smooth (unnotched) specimens under various cyclic deformation levels in a typical low cycle fatigue test (Dursun & Soutis, 2014). The mode of testing is strain control instead of stress control. A stress response is observed while cyclic loading and the number of cycles to failure is reported for these tests. Several tests are essential to ascertain the cyclic stress-strain curve, including the strain life curve figure 4 (Dieter, 1986).

Under the cyclic loading within the elastic region, stress is directly propositional to strain through elasticity's modulus. However, plastic strains produced by cyclic loading and responses are more complicated and create a hysteresis loop in figure 5 from point O to point A, and the section is in tension. On removing the load of the strain, the specimen's reaction comes after the curve from A to the D. At D, the section is under no stress. As the specimen is subjected to compressive stress, the strain reaction come after in order the curve from D to point B. delivering the compressive stress from B and reapplying

Figure 4. Low cycle fatigue curve ($\Delta\varepsilon_p$ vs.N).

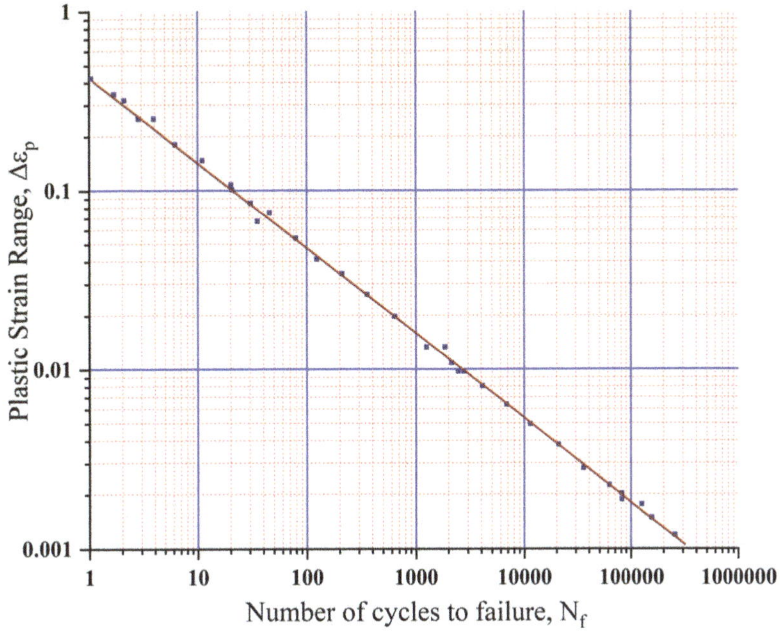

the tensile stress, the specimen stress-strain condition comes back to point A along the curve explain by B, C, and A. Points A and B represent the cyclic stress and strain limits. The total strain $\Delta\varepsilon$, consists of both elastic and plastic components:

$$\Delta\varepsilon = \Delta\varepsilon_e + \Delta\varepsilon_p \tag{1}$$

Where $\Delta\varepsilon_e$ is the elastic strain and $\Delta\varepsilon_e=\sigma_s/E$, and $\Delta\varepsilon_p$ is the plastic strain, and $\Delta\varepsilon_p$ is the width of the loop at its centre, i.e., the distance CD in figure 5. The area of the hysteresis loop is equal to the energy loss or the work done per cycle.

Low-cycle fatigue test data are commonly shown as a plot of the plastic strain range, $\Delta\varepsilon p$, versus cycles to failure, N. When plotted on log-log coordinates, a straight line is obtained the Coffin-Manson relation describes that:

$$\frac{\Delta\varepsilon_p}{2} = \varepsilon_f^{'}(2N)^c \tag{2}$$

Where $\Delta\varepsilon_p$=2 is the plastic strain amplitude, and $\acute{\varepsilon}_f$ is the fatigue ductility coefficient defined by the strain intercept at 2N=1. For many metals, $\acute{\varepsilon}_f$ is roughly equal to the true fracture strain, ε_f. 2N is the number of strain turnaround to failure, and c is the fatigue ductility exponent, which usually varies between -0.5 and - 0.7. A smaller value of c results in longer fatigue lives.

Figure 5. Stable cyclic stress versus strain hysteresis loop

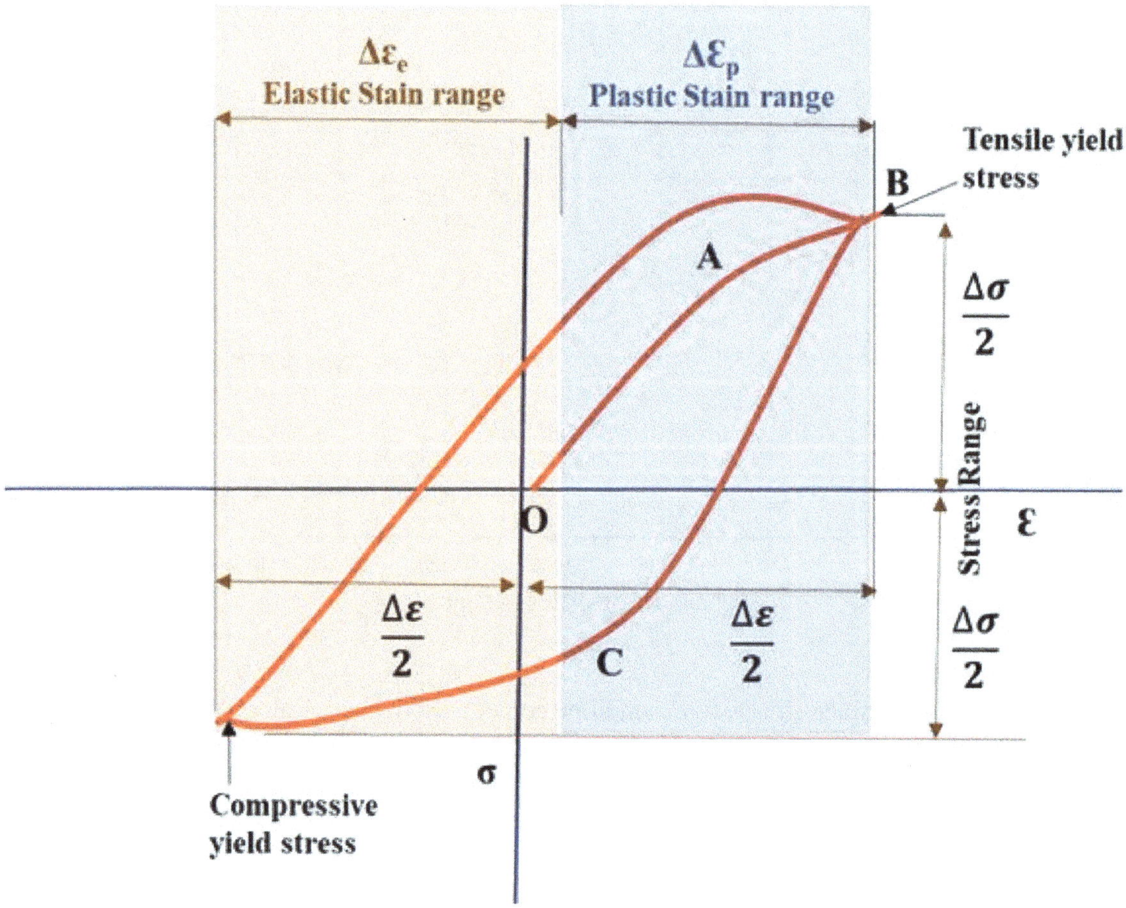

FRACTOGRAPHY

Fatigue failure is a progressive failure process that occurs under cyclic loading, and it is comprised of three distinct stages: Stage I is related to the crack nucleation at 45o to the load direction (following slip planes); Stage II is the continuous crack growth, perpendicular to the stress up to the point when the remaining cross-section can no longer withstand the applied load; and, finally, Stage III, which is the instant ultimate fracture due to overload (De Freitas et al., 2015; Totten et al., 2003). The rate of crack propagation during Stage III fracture is almost equal to half of the speed of sound in the material. The majority of the mechanical components (shaft, gears, turbine blades, bearings, rolls, etc.) are subjected to cyclic/periodic loading conditions and, hence, the fatigue failure mode is the predominant fracture mechanism (Pantazopoulos, 2019; Papadopoulou et al., 2020)

Figure 6 shows the major surface features seen on almost every fatigue face. The origin is where the crack actually started. The crack then grew slowly across the fatigue zone, with a typical growth rate. During this slow crack growth, there were variations in the load that resulted in corresponding variations in the crack growth rate that appear as progression marks (Mecholsky et al., 2020; Piñeiro-Jiménez et al., 2007). Eventually, the crack reached the point where the remaining material was overstressed, and

Figure 6. Macroscopic features of the fracture surface.

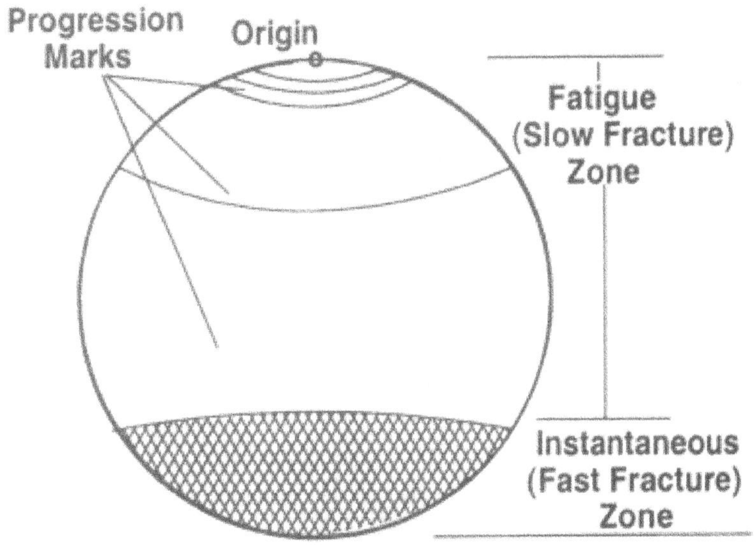

the overload zone resulted. In the overload zone, most cracks grow as macroscopically brittle fractures, and the crack growth rate is an approximately half time of the speed of sound in the piece. However, the overload zone may develop by either ductile or brittle fracture mechanisms. A single origin usually indicates a failure with low overstress, while the presence of multiple origins may be the result of either high stress or high-stress concentrations (Papakyriacou et al., 2001).

The fatigue zone is the area of slow crack growth. There are low-cycle fatigue failures where the crack growth occurs over relatively few cycles, frequently less than a hundred. The progression marks show how the crack has grown and are only present in fractures where there have been substantial variations in the component stress as the crack grew across the piece. There are actually two mechanisms that generate progression marks. Most commonly, they are seen in the older portion of a failure, where they show gross changes in load, such as start-up and shutdown forces. However, in the latter stages of a fracture life, they show the individual stress cycles. Fatigue striations show each stress cycle experienced by the part and are generally visible only under extremely high magnification, while progression marks are visible to the naked eye (Sangid, 2013).

Table 1. Elemental composition of Al6063 extruded rod

Elements Wt.%	Si	Fe	Cu	Mg	Mn	Zn	Ti	Cr	Al	Other elements
Standard	0.2-0.46	0.35 maz	0.1 max	0.45-0.9	0.1 max	0.1 max	0.1 max	0.1 max	Balance	-
Tested	0.426	0.192	0.051	0.516	0.048	0.058	0.013	0.020	Remainder	0.050

Figure 7. BISS nano plug n play fatigue testing machine with the capacity of +25kN

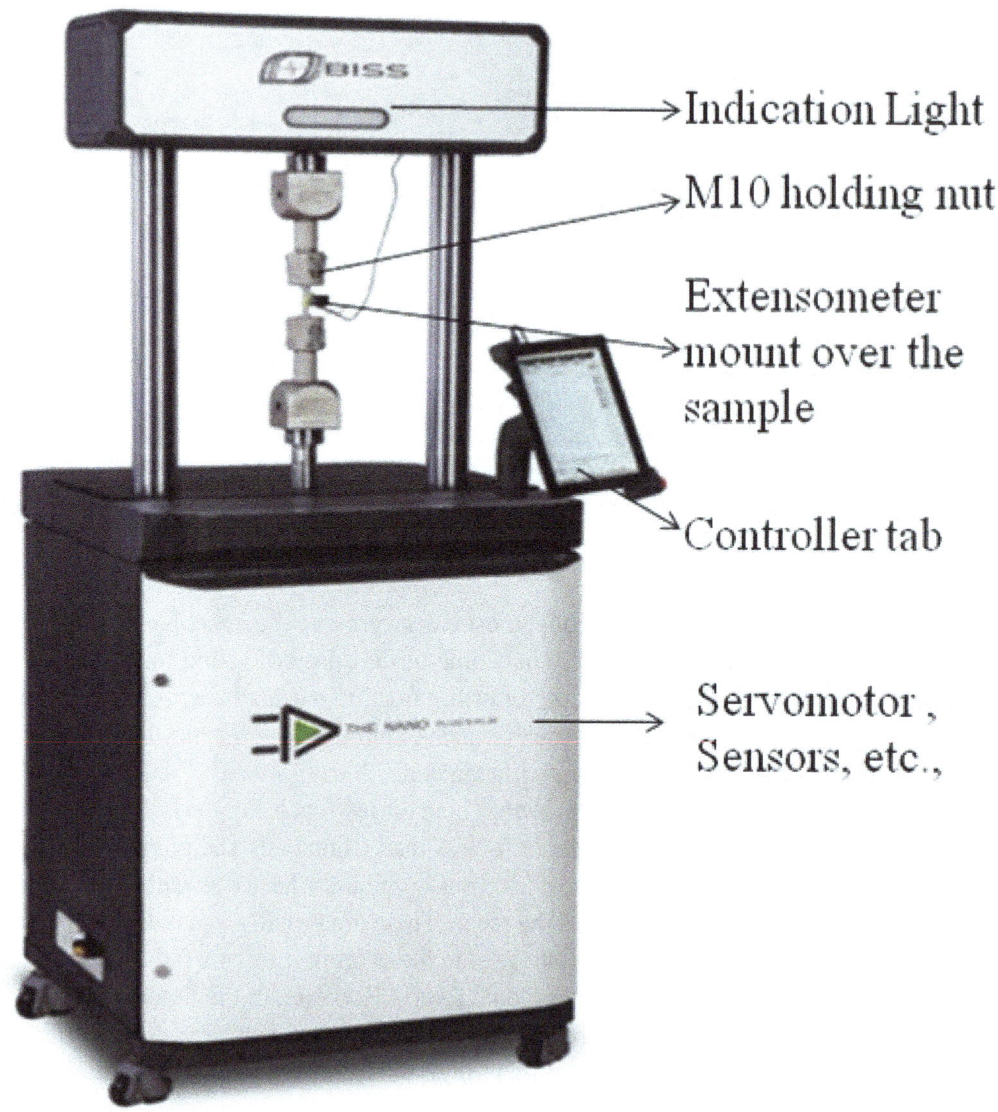

LOW CYCLE BEHAVIOUR OF ALUMINIUM ALLOY 6063

Aluminium 6063 alloy subjected to the low cycle fatigue test. The elemental material composition with percentage is given in table 1, which is observed through the XPS method.

The low cycle fatigue test was performed in the BISS Nano Plug n Play fatigue testing machine with the capacity of +25kN, the setup shown in figure 7.

The test was performed at a total strain amplitude of 0.4% and 0.8% with a cyclic frequency of 0.2Hz. Al6063 roads were machined according to the ASTM E 606 standard. A testing sample with dimensions is illustrated in figure 8. In figure 9, the prepared Al6063 sample has shown with scale reference

Figure 8. The dimension of the Al6063 sample is according to ASTM E606 standards.

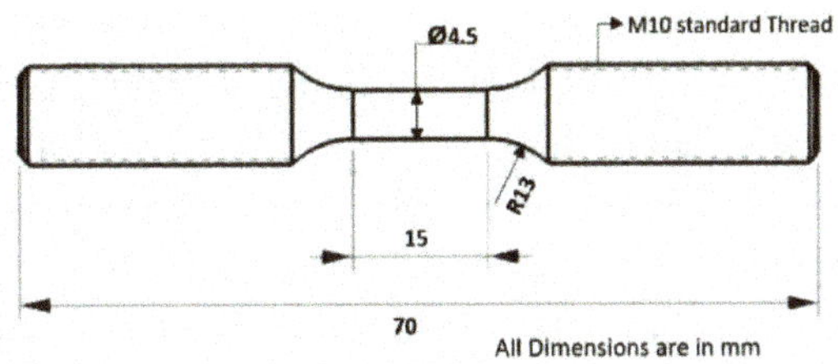

Figure 9. a) Al6063 samples prepared for low cycle fatigue test b) Al6063 sample after fatigue failure

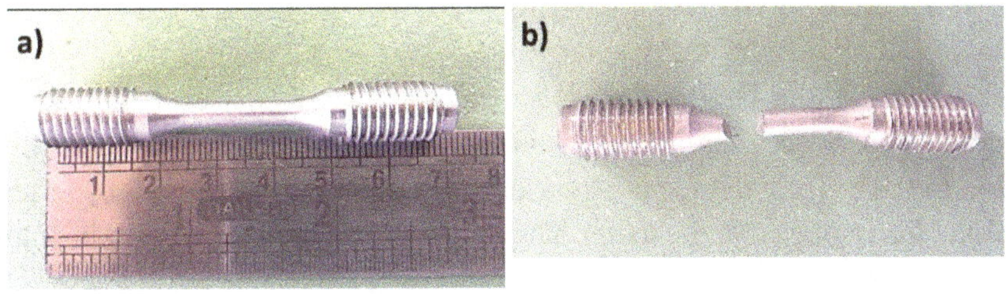

Cyclical Stress Response of Aluminium 6063

The low cycle fatigue response of Al6063 alloy failed under the Total strain amplitude of 0.4%, and 0.8% with the cyclic frequency of 0.2Hz were discussed below. From figure 10, The SN curves illustrate the behaviour of the aluminium alloy under strain-controlled fatigue. For the high total strain amplitude of 0.8%, the number of cycles to failure is 291 for Al6063, shown in figure 10. From the observation, the brittle fracture was obtained on the surface. The high total strain amplitude leads the sample to immediate failure due to a high degree of deformation. The stress amplitude response of the sample for 0.8% TSA

Figure 10. S-N Curve for the Al6063 with a total strain amplitude of 0.4% and 0.8%

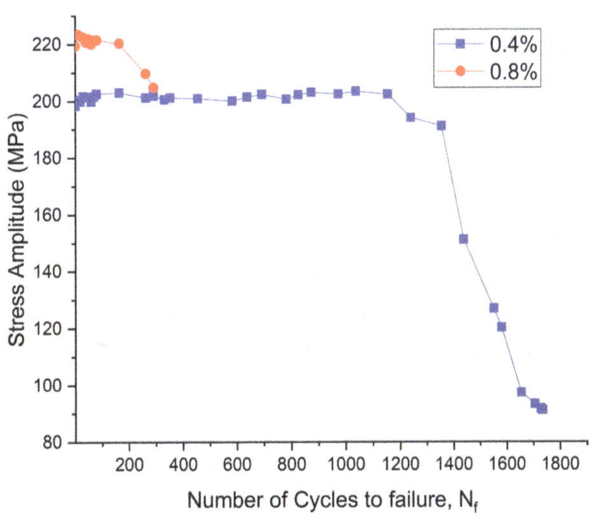

is illustrated in figure 10. The stress amplitude response also indicates that the sample has undergone cyclic softening (Yahya et al., 2015).

In the case of 0.4% TSA, Al6063 has 1786 cycles to the ultimate failure. The stress response of the Al6063 is given in figure 10. The sample exhibits cyclic stability for 1100 cycles. Followed by the stability region, the crack starts to nucleate microscopically. Then the crack propagation starts to propagate in the planar surface, and a macro crack was formed, which leads to the ultimate failure of the sample at 0.4% TSA. The fractured surface shows the ripple formation, i.e., fatigue striation caused due to the cyclic loading, as shown in figure 10. When comparing both 0.4% and 0.8%, the increased TSA reduces the cycle to failure.

Influence of Mean Stress on the Fatigue Cycle

The mean stress influences the fatigue life of a material. The response of the mean stress implies whether the material experiences hardening or softening. In figure 11, the mean stress response of Al6063 has given. In 0.4% TSA, the mean stress of the sample initiates in a tensile manner, which is about +05.734Mpa This implies that the material responded well to the given strain amplitude (Sharma et al., 2014; Srivatsan, 1991).

Figure 11. Mean stress response of Al6063 alloys concerning strain amplitude of 0.4% and 0.8%

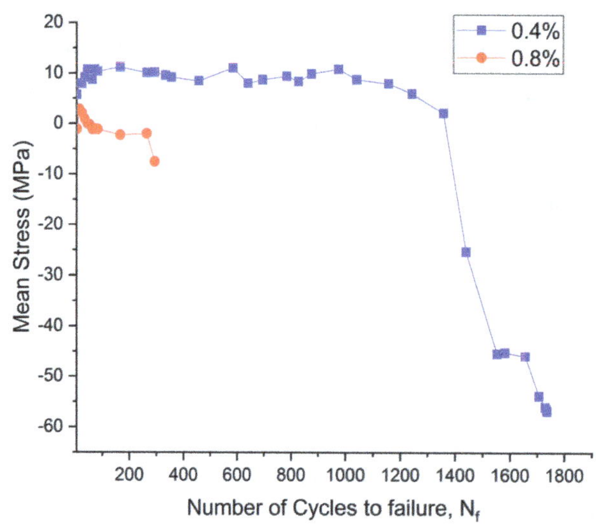

In the strain amplitude of 0.8%, the mean stress was maintained positively up to 30 cycles. This positive mean stress longs up to 1335 cycles. By following that, the mean stress gets into negative mean stress, which implies the crack started to propagate in the medium. This negative mean stress also means that the fatigue life of the material is nearing the fail. At the crack propagation point, negative mean stress dominates the region (Mirza et al., 2013).

Stress-Strain Hysteresis Loop and Plastic Strain Amplitude Response

In determining the low cycle fatigue life of Al6063, the stress-strain curve plays a vital role in analysing the deformation behaviour until failure, which is shown in figure 12. The specimen undergoes expansion and contraction during the dynamic loading. In 0.4% of TSA, the material undergoes cyclic hardening for the initial ten cycles, then cyclic stability formed until the crack initiation, which is up to 1100 cycles, after that material experienced the cyclic softening. This phenomenon is induced during the propagation of a crack in the sample due to the dynamic loading (Mirza et al., 2017).

Figure 12. Cyclic stress-strain loop for (a) total strain amplitude of 0.4% and (b) total strain amplitude of 0.8%

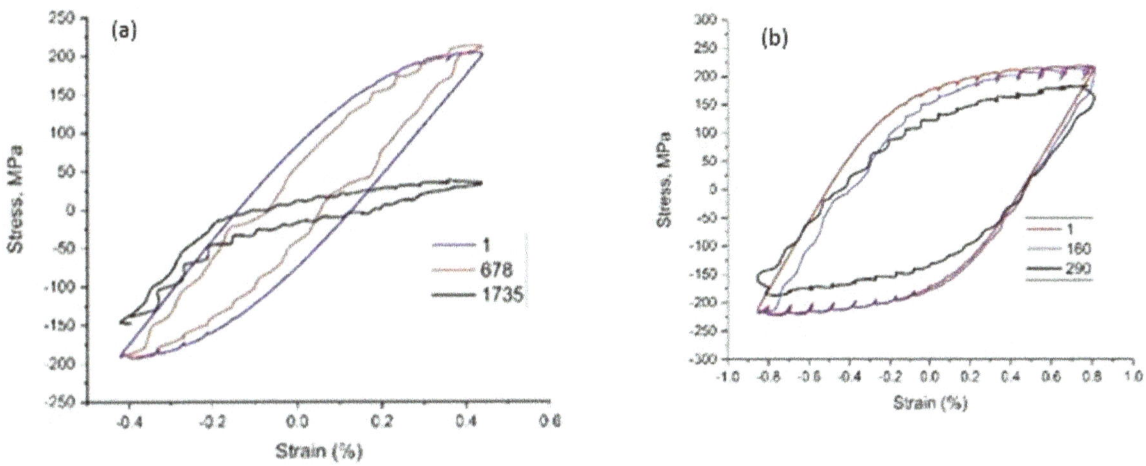

In 0.8% TSA, the stress-strain response of the material implies that the material undergoes cyclic softening until failure. Due to the high strain amplitude rate, there is no cyclic stability recorded. Materially fails at a very early stage compared to 0.4% TSA. The hysteresis loop of stress-strain is defined by the following equation,

$$\Delta\sigma = K'(\Delta\varepsilon_p)^{n'} \tag{3}$$

In the equation hardening exponent for the cyclic strain is given by n'. Coefficient of cyclic strength represented by K'. The Bauschinger effect contributes to the cyclic hardening and softening phenomenon occurring in the material. The result has defined an increase in yield strength in one direction by decreasing the yield strength in the opposite direction (Golański & Mroziński, 2013). The life of the material decreases when the total strain rate increases, influencing the plastic strain rate shown in figure 13.

The plastic strain amplitude response of material during the fatigue test reveals the deformation rate of the material. For 0.4% TSA, the plastic strain amplitude follows a stable trend until the failure. The slope developed narrowly after an increase in strain rate—the stability of the material achieved in the range of 0.05% of plastic strain amplitude. For 0.8% TSA, plastic strain response has changed dramatically due to the higher strain rate in which the sample failed below 300 cycles (Kosturek et al., 2020).

Figure 13. Plastic strain amplitude of Al6063 alloys for the strain amplitude of 0.4% and 0.8%

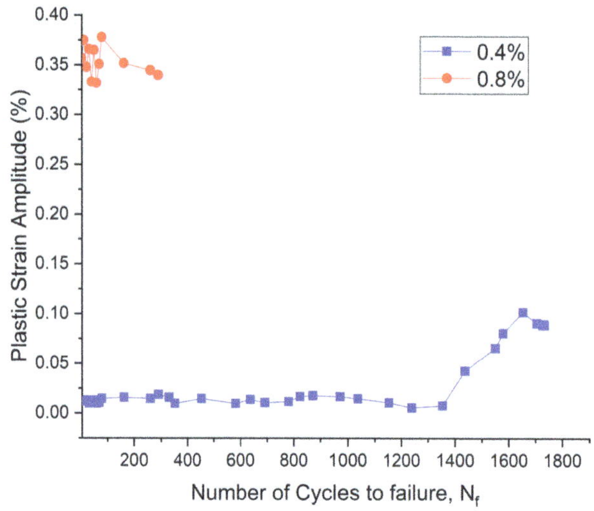

Modulus of Elasticity Response during Dynamic Loading

The elastic modulus changes during the dynamic loading are caused due to frequent changes in yield stress in both directions (Grigore & Popp, 1991). The response is calculated through the unloading modulus' feedback since the increment of strain rate is directly linked with the phase of unloading modulus.

Figure 14. The response of the unloading modulus for different total strain amplitude

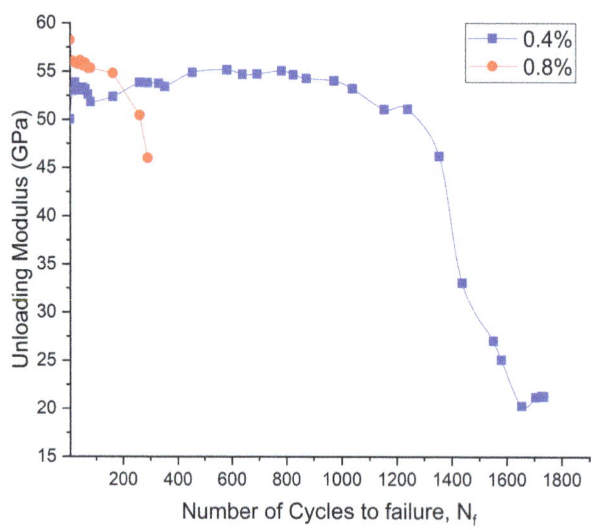

Figure 15. Fractography of 0.4% total strain amplitude, which shows quasi cleavage fracture surface.

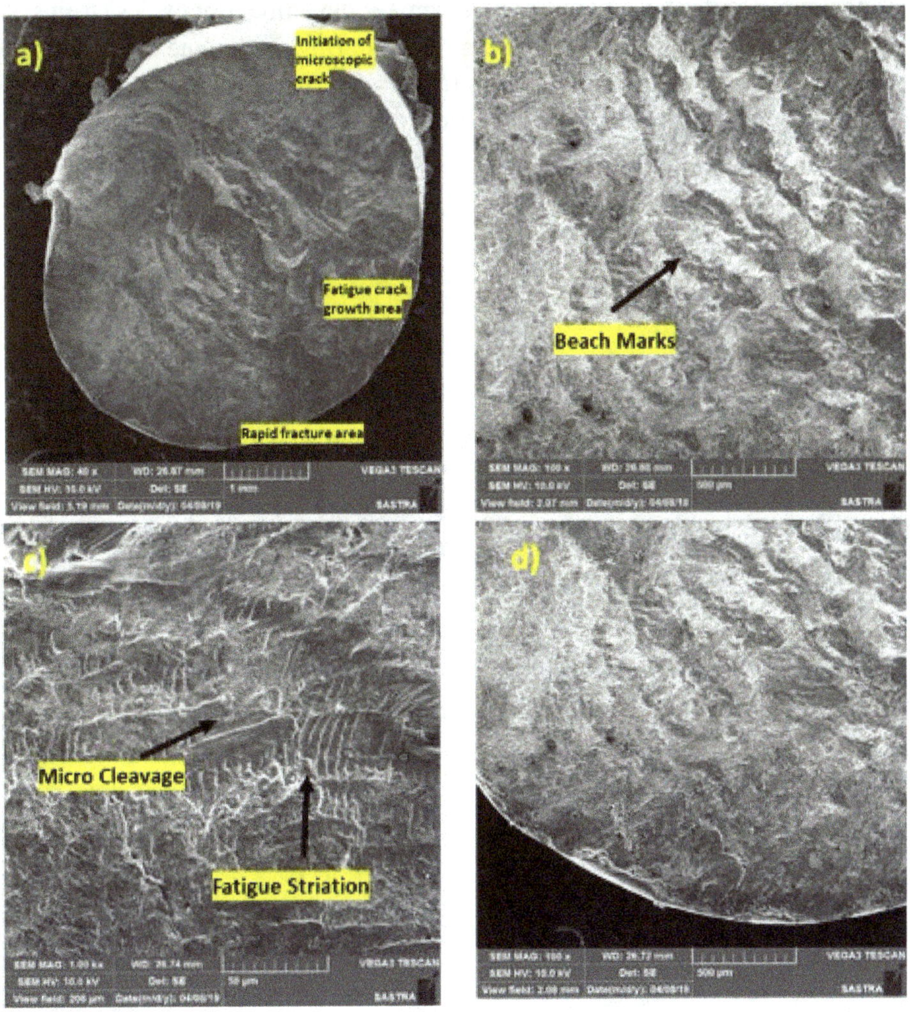

In TSA of 0.4% and 0.8%, there is an insignificant rise in the modulus at the initial stage, shown in figure 14. For 0.4% TSA, there is modulus stability between the range of 10 to 1000 cycles, which shows the materials' stable deformation. For a higher total strain amplitude of 0.8%, modulus followed the decreasing trend until it fails (Salerno et al., 2007).

Fractography of Al6063 Sample

The fractured surface of Al6063 is analysed through the scanning electron microscope. The fractography can reveal the crack initiation region and propagation direction of the specimen—the fractography of Al6063, shown in figure 15. In the fractography for the 0.4% TSA, the crack was initiated from the surface, which can be witnessed through the cleavage region nearby, as shown in figure 15 (a), marked as microscopic crack initiation. The fatigue crack propagates along the direction of crack nucleation, which can be witnessed through the beach marks, as shown in figure 15 (b). The cyclic loading caused beach marks on the specimen's surface, in that each mark denotes the number of cycles (Agius et al., 2017).

The ripple formation is due to extrusion and intrusion during the fatigue testing, which can be observed as secondary cracks in figure 15 (c). Figure 15 (d) rapid fracture region is obtained, representing the final planar fracture after macro crack propagation at stage II (Begum et al., 2008).

Figure 16. Fractography of 0.8% total strain amplitude which exhibits cleavage fracture

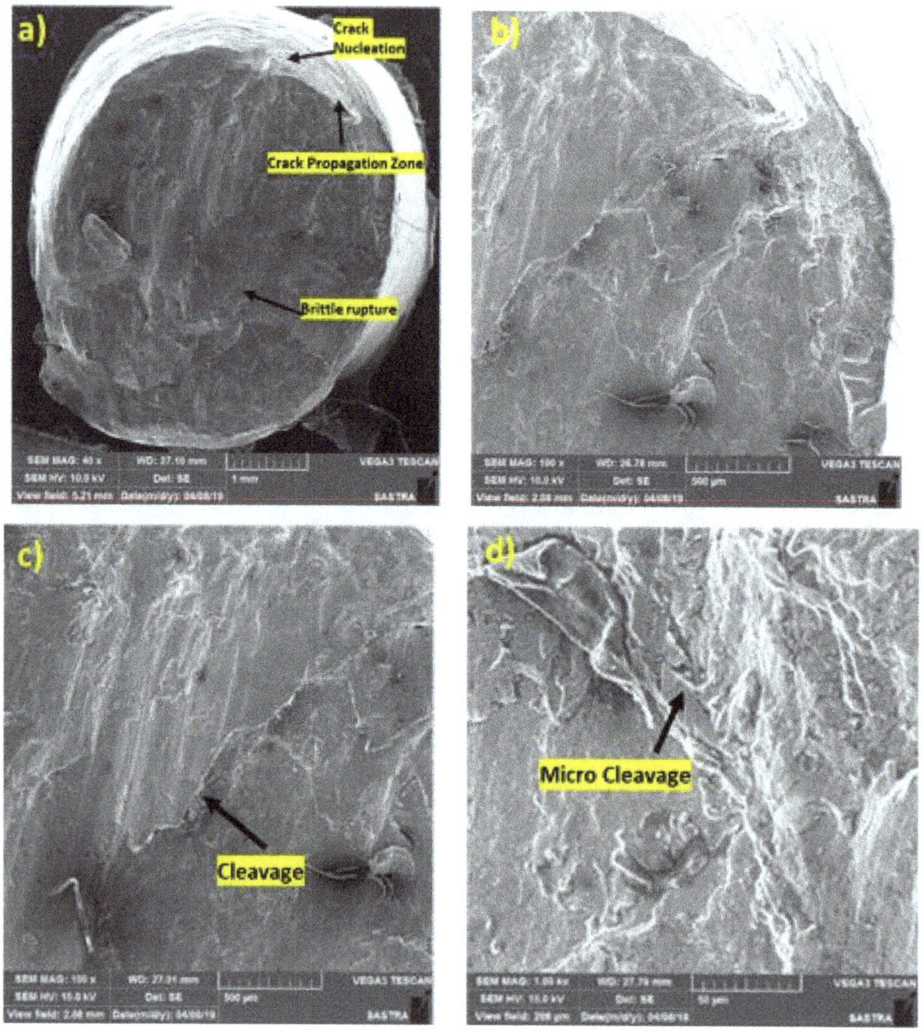

Figure 16 (a) shows no formation of dimple fracture for 0.8%TSA due to the high strain rate. The progressive crack due to fatigue is witnessed along with the crack nucleation site. I crack in a planar mode during the stage due to a higher strain rate, shown in figure 16 (b). The material turned brittle and exhibited a cleavage fracture, which is shown in figures 16 (c) and (d) because there is no stabilised crack growth occurred in the sample (Srivatsan, 1991).

SUMMARY

Strain controlled low cycle fatigue tests were carried out on an extruded 6063 Aluminium alloy followed by the test fatigue parameters, cyclic deformation characteristics such as unloading modulus along with fracture mode of the Al alloy were investigated. The following conclusions can be drawn: Cyclic hardening was witnessed in the higher total strain amplitude of 0.8% only, which could be due to the formation and interaction of twins. At low strain amplitudes, the cyclic stress amplitude basically remained constant and cyclic softening happened after crack initiation. Basuchinger effect was obtained in all strain amplitudes, leading to a big difference between the tensile and compressive yield stresses and thus, asymmetric hysteresis loops were formed. This is basically attributed to the activity of twinning in compression during unloading and subsequent de-twinning in tension during loading. The mean stress depends on the total strain amplitude. Unloading modulus during cyclic deformation, being constant at the low strain amplitude, was observed to decrease with increasing total strain amplitudes and increase with an increasing number of cycles at the high strain amplitudes due to the presence of pseudoelastic behaviour. Fatigue lifetime increased with decreasing strain amplitude. The fatigue parameters used to determine the fatigue life following the Coffin-Manson law and Basquin's equation were evaluated. SEM examination revealed that fatigue cracks initiated from the specimen surface due to the larger grain sizes near the specimen surface. The initiation site contained cleavage and quasi cleavage in respect to the total strain amplitude. The crack Propagation of the fatigue was basically categorised by fatigue striation-like features, coupled with the occurrence of the secondary cracks.

The fatigue behaviour of aluminium alloy is one of the main concerns to employing it in engineering applications. Aluminium alloys are heat treatable and respond to severe plastic deformation and other work-hardening processes, improving mechanical properties. The potential of aluminium alloys has to be explored in a full-fledged manner. Many kinds of research have focused on improving aluminium stuff and making them fit the applications like aerospace and automobile to reduce the weight to improve efficiency.

ACKNOWLEDGMENT

This work was supported by the "Research and Modernization fund, SASTRA University" grant number R&M/0035/SoME-008/2015-16. The authors thank SASTRA Deemed University for their financial assistance.

REFERENCES

Agius, D., Kourousis, K. I., Wallbrink, C., Hu, W., Wang, C. H., & Dafalias, Y. F. (2017). Aluminum Alloy 7075 Ratcheting and Plastic Shakedown Evaluation with the Multiplicative Armstrong–Frederick Model. *AIAA Journal*, *55*(7), 2461–2470. doi:10.2514/1.J055833

Akano, T. T., & Fakinlede, O. A. (2013). Fatigue Failure Model for Polymeric Compliant Systems. *ISRN Polymer Science*, *2013*, 1–11. doi:10.1155/2013/321489

Angelo & Ravisankar. (2018). *Non Ferrous Alloys: Structures, Properties, and Engineering Applications*. Cengage.

Bahl, S., Hu, X., Sisco, K., Haynes, J. A., & Shyam, A. (2020). Influence of copper content on the high temperature tensile and low cycle fatigue behavior of cast Al-Cu-Mn-Zr alloys. *International Journal of Fatigue, 140*, 105836. doi:10.1016/j.ijfatigue.2020.105836

Begum, S., Chen, D. L., Xu, S., & Luo, A. A. (2008). Strain-Controlled Low-Cycle Fatigue Properties of a Newly Developed Extruded Magnesium Alloy. *Metallurgical and Materials Transactions. A, Physical Metallurgy and Materials Science, 39*(12), 3014–3026. doi:10.100711661-008-9677-0

Chandran, K. S. R. (2005). Duality of fatigue failures of materials caused by Poisson defect statistics of competing failure modes. *Nature Materials, 4*(4), 303–308. doi:10.1038/nmat1351

Chang, S. Y., Lee, K. S., Lee, S. H., Hong, S. K., Park, K. T., & Shin, D. H. (2003). Effect of Al Content and Pressing Temperature on ECAP of Cast Mg Alloys. *Materials Science Forum, 419–422*, 491–496. . doi:10.4028/www.scientific.net/MSF.419-422.491

De Freitas, W. J., Takeya, G. S., Casteletti, L. C., Neto, A. L., & Totten, G. E. (2015). *Influence of sintering atmosphere in the hardness and corrosion resistance of 17-4 PH stainless steel shaped from powder injection molding process*. Academic Press.

Dieter, G. E. (1986). *Mechanical metallurgy*. McGraw-Hill.

Dursun, T., & Soutis, C. (2014). Recent developments in advanced aircraft aluminium alloys. *Materials & Design, 56*, 862–871. doi:10.1016/j.matdes.2013.12.002

Golański, G., & Mroziński, S. (2013). Low cycle fatigue and cyclic softening behaviour of martensitic cast steel. *Engineering Failure Analysis, 35*, 692–702. doi:10.1016/j.engfailanal.2013.06.019

Grigore, D. R., & Popp, O. T. (1991). Erratum: The complete classification of generalised homogeneous symplectic manifolds. *Journal of Mathematical Physics, 32*(2), 564–564. doi:10.1063/1.529395

Hardouin Duparc, O. B. M. (2010). The Preston of the Guinier-Preston Zones. Guinier. *Metallurgical and Materials Transactions. B, Process Metallurgy and Materials Processing Science, 41*(5), 925–934. doi:10.100711663-010-9387-z

Kosturek, R., Śnieżek, L., Torzewski, J., & Wachowski, M. (2020). Low Cycle Fatigue Properties of Sc-Modified AA2519-T62 Extrusion. *Materials 2020, 13*(1), 220. doi:10.3390/ma13010220

McCullough, R. R., Jordon, J. B., Allison, P. G., Rushing, T., & Garcia, L. (2019). Fatigue crack nucleation and small crack growth in an extruded 6061 aluminum alloy. *International Journal of Fatigue, 119*, 52–61. doi:10.1016/j.ijfatigue.2018.09.023

Mecholsky, J. J. Jr, DeLellis, D. P., & Mecholsky, N. A. (2020). Relationship between fractography, fractal analysis and crack branching. *Journal of the European Ceramic Society, 40*(14), 4722–4726. doi:10.1016/j.jeurceramsoc.2019.12.061

Mirza, F. A., Chen, D. L., Li, D. J., & Zeng, X. Q. (2013). Low cycle fatigue of a rare-earth containing extruded magnesium alloy. *Materials Science and Engineering A, 575*, 65–73. doi:10.1016/j.msea.2013.03.041

Mirza, F. A., Liu, K., & Chen, X.-G. (2017). Cyclic Stress-Strain Behavior and Low Cycle Fatigue Life of AA6061 Aluminum Alloy. In Minerals, Metals and Materials Series (pp. 447–452). Springer. doi:10.1007/978-3-319-51541-0_56

Pantazopoulos, G. A. (2019). A short review on fracture mechanisms of mechanical components operated under industrial process conditions: Fractographic analysis and selected prevention strategies. *Metals, 9*(2), 148. Advance online publication. doi:10.3390/met9020148

Papadopoulou, S., Vazdirvanidis, A., Toulfatzis, A., Rikos, A., & Pantazopoulos, G. (2020). Failure Investigation of Products and Components in Metal Forming Industry: Root Cause Analysis and Process-Based Approach. *Journal of Failure Analysis and Prevention, 20*(1), 106–114. doi:10.1007/s11668-020-00801-4

Papakyriacou, M., Mayer, H., Pypen, C., Plenk, H. Jr, & Stanzl-Tschegg, S. (2001). Influence of loading frequency on high cycle fatique properties of b.c.c and h.c.p. metals. *Materials Science and Engineering A, 308*(1–2), 143–152. doi:10.1016/S0921-5093(00)01978-X

Payne, A. O. (1976). The fatigue of aircraft structures. *Engineering Fracture Mechanics, 8*(1), 157–203. doi:10.1016/0013-7944(76)90085-0

Piñeiro-Jiménez, A., Villalobos-Gutiérrez, C., Staia, M. H., & Puchi-Cabrera, E. S. (2007). Tensile and fatigue properties of 6063-T6 aluminium alloy coated with electroless Ni–P deposit. *Materials Science and Technology, 23*(3), 253–263. doi:10.1179/174328407X157317

Salerno, G., Magnabosco, R., & Moura Neto, C. (2007). Mean strain influence in low cycle fatigue behavior of AA7175-T1 aluminum alloy. *International Journal of Fatigue, 29*(5), 829–835. doi:10.1016/j.ijfatigue.2006.09.004

Sangid, M. D. (2013). The physics of fatigue crack initiation. *International Journal of Fatigue, 57*, 58–72. doi:10.1016/j.ijfatigue.2012.10.009

Sharma, V. M. J., Rao, G. S., Sharma, S. C., & George, K. M. (2014). Low Cycle Fatigue Behavior of AA2219-T87 at Room Temperature. *Materials Performance and Characterization, 3*(1), 20130092. doi:10.1520/MPC20130092

SreeArravind, M., Ramesh Kumar, S., RaviShankar, B., & Senthil Kumar, S. (2020). Low cycle fatigue behavior of aluminium 6063 alloy under the cyclic frequency of 0.2 Hz. *Materials Today: Proceedings, 27*, 2376–2380. Advance online publication. doi:10.1016/j.matpr.2019.09.133

Srivatsan, T. (2002). An investigation of the cyclic fatigue and fracture behavior of aluminum alloy 7055. *Materials & Design, 23*(2), 141–151. doi:10.1016/S0261-3069(01)00071-1

Srivatsan, T. S. (1991). The low-cycle fatigue and cyclic fracture behaviour of 7150 aluminium alloy. *International Journal of Fatigue, 13*(4), 313–321. doi:10.1016/0142-1123(91)90358-6

Totten, G. E., MacKenzie, D. S., & MacKenzie, D. S. (2003). Handbook of Aluminum. In *Handbook of Aluminum*. CRC Press. doi:10.1201/9780203912591

Yahya, M. M., Mallik, N., & Chakrabarty, I. (2015). Low Cycle Fatigue (LCF) Behavior of AA6063 Aluminium Alloy at Room Temperature. *International Journal of Emerging Technology and Advanced Engineering*, *5*(12).

Chapter 9
Maximization of Tensile Strength of Aluminum 6061 Alloy T6 Grade Friction Welded Joints by Using the Desirability Function

Maneiah Dakkili
https://orcid.org/0000-0002-1462-9470
CMR Technical Campus, India

Debashis Mishra
CMR Technical Campus, India

K. Prahlada Rao
Jawaharlal Nehru Technological University, Anantapuramu, India

K. Brahma Raju
SRKR Engineering College, India

ABSTRACT

Various joining techniques are consistently used in fabrications and maintenance applications of numerous parts in manufacturing industries. Typically, the friction welding technique acquired attention in joining of aluminum and its different alloys for very general structural usages in small to medium to large-scale manufacturing sectors. This is an experimental attempt to weld aluminum 6061 alloy T6 grade of 3mm thickness metal sheets. The hexagonal-shaped steel pin of grade H13 is used. The experiment is performed by using the Taguchi L9 approach, and nine welded specimens are prepared. The chosen factors are rotating speed of the tool, tilting angle, and feed. After the welding, the tensile testing is followed for the measurement of strength of the welded samples. The analysis suggested that the chosen working limits of feed and rotational speed is significant and having impacts on weld strength. The maximum strength is obtained as 212MPa when the ranges of above said factors are 560RPM, 0degree, and 20mm/min.

DOI: 10.4018/978-1-7998-7864-3.ch009

INTRODUCTION OF FRICTION WELDING TECHNIQUE AND LITERATURE STUDIES

The friction welded joints were notably fabricated initially by the insertion of a rotated and not consumable tool in the adjoining interfaces of the two metal sheets (Maneiah & Debashis Mishra, 2020). The welding actions performed were named such as plunging, dwelling, welding by the traverse of the tool, dwelling, and to the end dragging the tool out. Aluminum alloy 6061-T6 grade is a non-ferrous alloy commonly contents aluminium as larger percentages and 0.8 to 1.2% magnesium, copper 0.15 to 0.4%, 0.8% silicon and 0.04 to 0.35% chromium. The wide availability and exceptional mechanical properties like low-density 2.7gm/cc, 310MPa tensile with 12% elongation, and 95BHN hardness make it very functional material for various use as fittings and frames in different products and assemblies. These alloys were also become very useful material because of its strength, lighter in weight, excellent welding, machinability, workability, and corrosion resistive characteristics. Many researchers attempted in friction stir and various issues, concerns about the efficiency and effectiveness of the process, and findings are reported widely. Few from such reports were discussed below to get a thorough understanding of the key concepts of the welding process and experimental investigations done on it. In one report, aluminum alloy AA2219 grade was welded by the friction welding method. Hexagon profile found to be good among all the profiles (Venkata Rao et al., 2015). Aluminum and silicon carbide metal matrix composites were fabricated with the use stir casting process. The maximum strength of the composite was reported as 170MPa with addition of 10% silicon carbide into the aluminum (Mishra & Tulasi, 2020). The factors like tool rotation and weld speed were exceedingly dominant for the welding of aluminum alloys 6061 and 7075 grades (Ugrasen et al., 2018). A mathematical model was formulated to recognize the impact of chosen welding conditions upon the tensile and yield strength of the welded joints (Elatharasan & Kumar Senthil, 2013). The important factors were feed, tool rotational and traverse speed, geometry and tilting of the tool in friction welding and processing. These factors have a major account for heat generation, material plasticization, and its movement from the receding towards the progressing side. Thus, the optimization of various factors should be performed to identify the impressions of chosen factors on the weld and base metal zone (Padhy et al., 2018). It was stated that weld travel speed was a highly influential factor in the friction welding of the aluminum 6061 alloy T6 grade of thickness 4mm metal sheets (Liu et al., 2013). The aluminum AA 2014-T6 grade alloy of 300×80×5mm sheet size was welded with the help of different pin profile geometries like conical, square, pentagonal, triangular, and hexagonal. A regression model was developed at a 95% confidence interval to state the impacts of change in factors on the tensile strength of the welded joints. The hexagon tool profile was produced as an exceptional tensile strength than other profiles (Ramanjaneyulu et al., 2015). A report is presented as aluminum alloy of two different grades A356 and 6061 was welded by changing the factors. The results were shown that the rotational speed was highly desirable to in welding of aluminum alloys (Ghosh et al., 2010). Aluminum alloys AA6061 and AA 5086 grades were welded by the help of different probe profiles like cylindrical, tapered, and threaded cylindrical. The threaded probe produces welds with higher material flow and thus the hardness value was obtained maximum as 83HV and tensile strength as 169MPa (Ilangovan et al., 2015). Aluminum alloys welding that may similar or dissimilar was in ever raising a requirement for various structural uses. One of the studies revealed the dissimilar welding of AA6061-T6 and AA8011-h14 aluminum alloy grades. It was reported that the tensile strength of the welded joints was obtained as 77MPa with the elongation of 21.96%. The effective factors and their values were like 1070RPM tool rotational speed, 2degree tool tilting angle, and 50mm/min traverse speed of the tool (Khanna et al., 2020). In a report, this was mentioned as friction welding can be successfully used to weld soft metals like aluminum, magnesium and its alloys. The factors can be considered as tool rotational speed, feed rate, and welding medium to produce the welding specimens. Aluminium 6061-T6 alloy can be easily welded by this process and produced improved mechanical properties. Friction stir welding generally needs low energy for welding of aluminum and its alloys (Patel Satyam, 2017). Mechanical behavior of the friction welded joints were depending upon the welding process parameters and the optimization of process was desirable to obtain the anticipated properties (Cavaliere, 2013; Mishra & Ma, 2005). Aluminum alloy AA5052-H32 metal sheets was welded. The effect of the selected factors on the joint design have been reported. The tapered square tool was found the most suitable one compared to others. The regression analysis was reported to convey the adequacy of the experimental model (Shanavas & Edwin Raja Dhas, 2017). Magnesium alloy AZ31 grade metal sheet of 5mm thickness was welded followed by optimization of the welding factors. The tool pin profile was the most principal reported factor. The mathematical model was presented for the prediction of the desired tensile properties (Ugender, 2018). Dissimilar welded joints were produced with several aluminum alloys grades and the optimization of welding factors was performed to co-relate the chosen factors and obtained measured results. The effects of various factors were experimentally investigated to suggest ranges of process parameters and suitable probe to produce the high-performance welded joints (Boumerzoug & Helal, 2017; Raj Kumar et al., 2014). The optimization of the manufacturing process was essential to find the optimum condition. The fused deposition modeling factors were optimized and regression expression was developed to predict the tensile strength. The Taguchi L9 approach was used for analysis to measure the effect of process parameters upon the tensile strength of fabricated specimens (Mishra & Das, 2021). In the current experimental investigation aluminum 6061-T6 grade of 3mm thickness was welded. The experiment was planned to formulate a mathematical model to identify the impacts of factors on strength of the welded joints.

MATERIALS AND EXPERIMENTAL METHOD

Aluminum alloy 6061-T6 grade of size 150×150×3mm metal sheets are chosen for the welding. The welding setup is revealed in fig. 1. The hexagonal structured pin profile is designed and prepared using H13 tool steel grade material which is chosen over other profiles like circular, tapered, triangular, and square following the literature, and prior welding trials as shown in fig. 2. Friction welding at the weld angle 60 degree and the welded sheets are presented in fig.3. ASTM E8M/08 standard is followed for the withdrawal of tensile samples from the welded coupons. The complete dimension of tensile specimens is illustrated in fig. 4. The collected samples are undertaken for the tensile testing and all specimens before and after tensile testing are shown in fig. 5. The nine welding settings and tensile tested results are presented in table 1.

REGRESSION ANALYSIS TO FIND OUT THE OPTIMAL SOLUTION

The experimental model is analyzed by the help of Taguchi optimization and analysis of variance (ANOVA) technique. The Minitab software are used to obtain an optimized solution of the impacts of the chosen process parameters in maximization of the tensile strength of the aluminum weld joint at 95%

Table 1. Nine experimental frictions stir welding conditions with the tensile test results.

S. No.	Tool rotational speed (RPM)	Tilting angle of the tool (Degree)	Feed (mm/ min)	Tensile strength (MPa)	Percentage elongation (%)
1	560	0	20	212	10.0
2	560	0.5	63	137	5.4
3	560	1	100	200	9.4
4	900	0	63	192	11.4
5	900	0.5	100	205	9.2
6	900	1	20	187	12.0
7	1400	0	100	181	13.0
8	1400	0.5	20	184	11.4
9	1400	1	63	176	5.2

Figure 1. Friction stir welding setup used for the experiment.

Figure 2. Hexagonal structuredH13 tool steel pin profile.

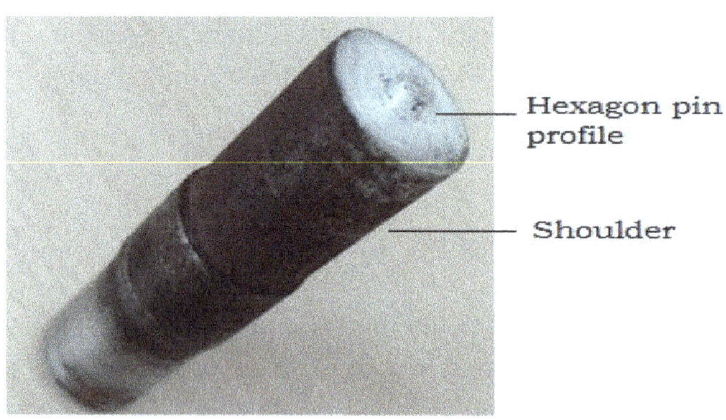

Figure 3. FS welding at 60-degree weld angle and welded sheets of Al 6061-T6 alloy respectively.

Figure 4. Schematic sketch of the extracted tensile samples (ASTM standard E8/E8M-09)
(All dimensions are in mm) (Maneiah & Debashis Mishra, 2020).

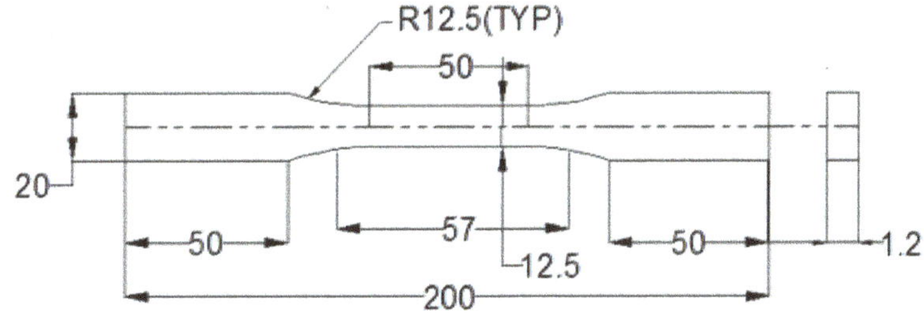

Figure 5. Before and after tensile testing FS welded Al 6061-T6 alloys samples respectively.

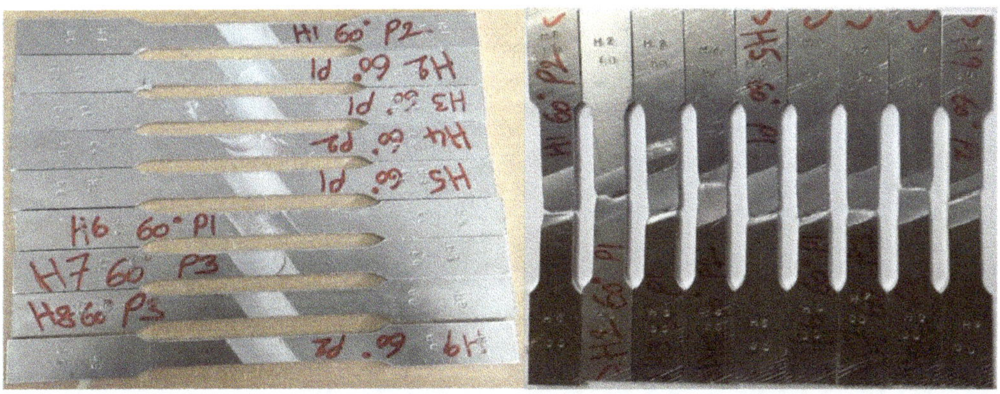

confidence and prediction limit. Then the constraint for the calculation of values of signal to noise ratio (S/N ratio) and the mean of measured strength is the maximization is superior. Table 2 is the complete representations of calculated values. The calculated delta values are such as 0.69, 1.03, and 1.37 for the S/N ratio and 14,3, 19.7, and 27.0 for the mean of tensile strength values respectively. The maximum value is considered as the most effective factor i.e., feed then tilting angle and finally tool rotational speed. The fig. 6 is the graphical representation of welding ranges and respective S/N ratio and mean of the measured strength values. The developed experimental model is evaluated by the use of ANOVA and regression analysis. The analysis suggests that whether the chosen model terms can be considered for the prediction of response values by taking F and P statistics. The assessed F- values of selected factors are like 0.04, 0.11, and 0.00 respectively. Thus, the chosen factors such as tool rotational speed and feed are conveyed to be suggestive factors whereas the tilting angle is not. The regression expression is a valid term as the F-value is obtained as 0.05 and the complete formulations are shown in table 3. The elected measurand are noticed as a suitable term as P-value is obtained as 0.002. The whole interpretations are indicated in table 4. The formulated mathematical expression is given in equation 1.

Table 2. Calculated Signal to noise ratio and mean values of measured tensile strength.

Process parameters/ Level	Tool rotational speed (RPM)		Tilting angle (Degree)		Feed (mm/min)	
	Value of calculated S/N ratio	Mean value of tensile strength	Value of calculated S/N ratio	Mean value of tensile strength	Value of calculated S/N ratio	Mean value of tensile strength
1	45.09	183.0	45.78	195.0	45.75	194.3
2	45.78	194.7	44.76	175.3	44.44	168.3
3	45.12	180.3	45.46	187.7	45.80	195.3
Delta	0.69	14.3	1.03	19.7	1.37	27.0
Effect	3		2		1	

Figure 6. Core effect plot where A- tool rotational speed, B-tilting angle, and C- feed.

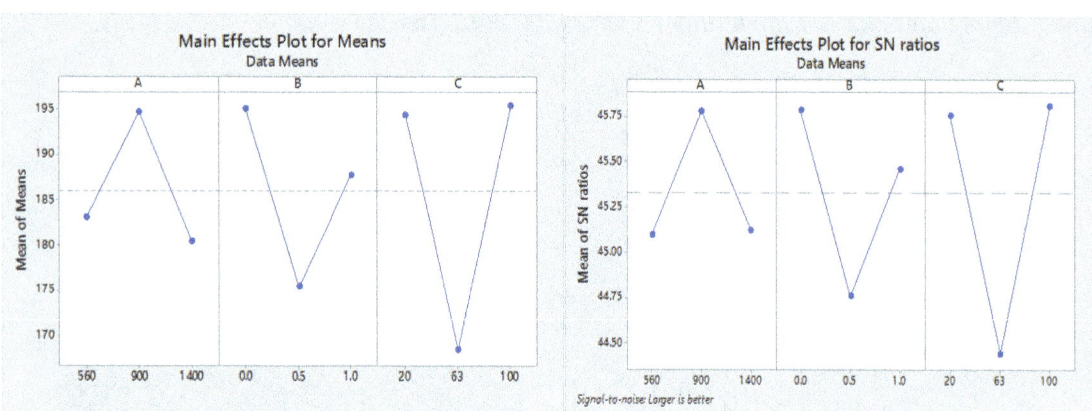

Table 3. Regression analysis of the Taguchi L9 experimental model.

Experimental model phrases	Degrees of freedom	Corrected values of sum of square	Corrected values of mean square	F-Value	Model validation
Regression	3	108.45	36.150	0.05	Yes
Tool rotational speed	1	27.63	27.627	0.04	Yes
Tilting angle	1	80.67	80.667	0.11	no
Feed	1	0.16	0.158	0.00	Yes
Error	5	3691.55	738.310		
Total	8	3800.00			

Table 4. Co-efficient of chosen expressions and experimental model terminologies.

Expressions	Co-efficient	Sum of error in the coefficient	T-Value	P-Value
Measurand	195	33.4	5.83	0.002
Tool rotational speed	-0.0052	0.0263	-0.19	0.854
Tilting angle	-7.29	22.2	-0.33	0.754
Feed	-0.039	0.277	-0.01	0.989

Prediction expression

$$\text{Tensile strength} = 195 - 0.0052 \times \text{rotational speed} - 7.29\times \text{tilt angle} - 0.039\times \text{feed} \quad (1)$$

MAXIMIZATION OF TENSILE STRENGTH BY USING DESIRABILITY AND PREDICTION FUNCTION

The optimizations of performed nine experimental orders are performed. The optimization of the selected factors is conducted to find the maximum strength. The procedure is set at a lower value of tensile strength 137MPa and a higher value of 212MPa and shown in table 5. The estimation of finest condition is achieved composite desirability analysis and all particulars are revealed in fig. 7. The complete assessment is shown in table 6. The optimum specification is achieved as tool rotating speed 560RPM, tilting angle 0 degree, and feed 20mm/min with the desirability of 0.73 i.e., 73%. The tensile strength at the optimum condition is experimentally gained as 212MPa which is assumed to be 192MPa.The assumption error is obtained as 21%. The optimality is assessed by the 95%confidence and prediction period. The lower and higher limit of the strength is predicted to be 137.9 to 245.8MPaand 103.6 to 280.1MPa respectively. The fig. 8, 9 and 10are the representations of the contour plots which is two-dimensional illustration of measured values of strength in light to deep green colored regions against the selected limits of the process parameters in the different axis. Lighter green colored area is representing to low strength. And deep green colored areas are illustrating the higher strength values.

Table 5. Maximization of strength using a response optimization technique

Measured factor	Aim	The lower value of strength	The target value of strength
Tensile strength	Maximization	137MPa	212MPa

Figure 7. Composite desirability function to get an optimum condition.

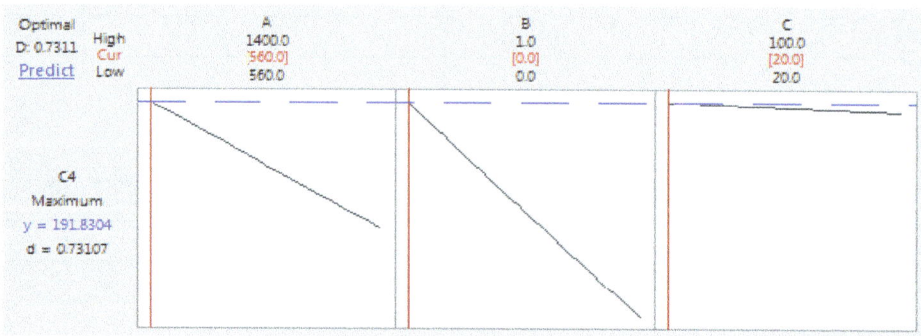

Table 6. Estimation of strength using confidence and prediction interval

Fixing of variable	Tool rotating speed (RPM)	Tilting Angle (Degree)	Feed (mm/min)	Response Tensile strength	Assumed value of strength (MPa)	Experimentally measured strength of the welded joints (MPa)	Error in assumption (%)	95% Assurance Interval (MPa)	95% Prediction Interval (MPa)
	560	0	20		191.8	212	21	(137.9 - 245.8)	(103.6 – 280.1)

Figure 8. Contour plan between tensile strength (C4) versus rotational speed and tilt angle.

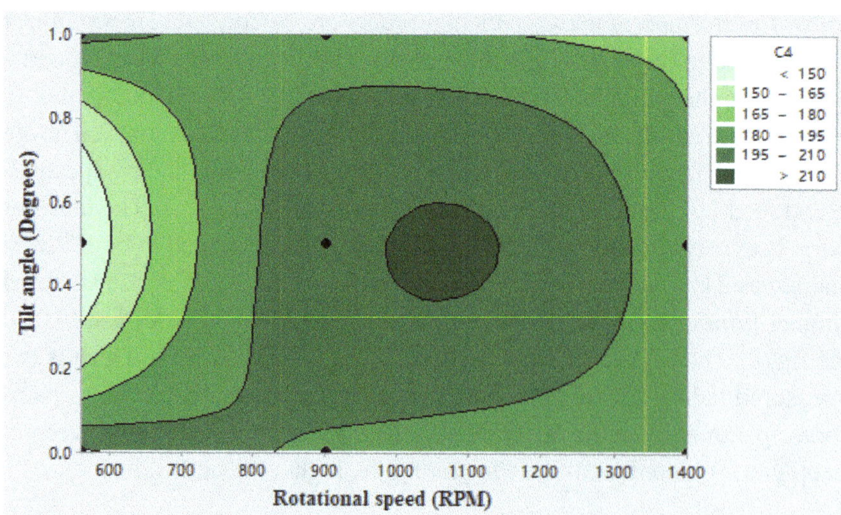

Figure 9. Contour plan between tensile strength (C4) versus rotational speed and feed.

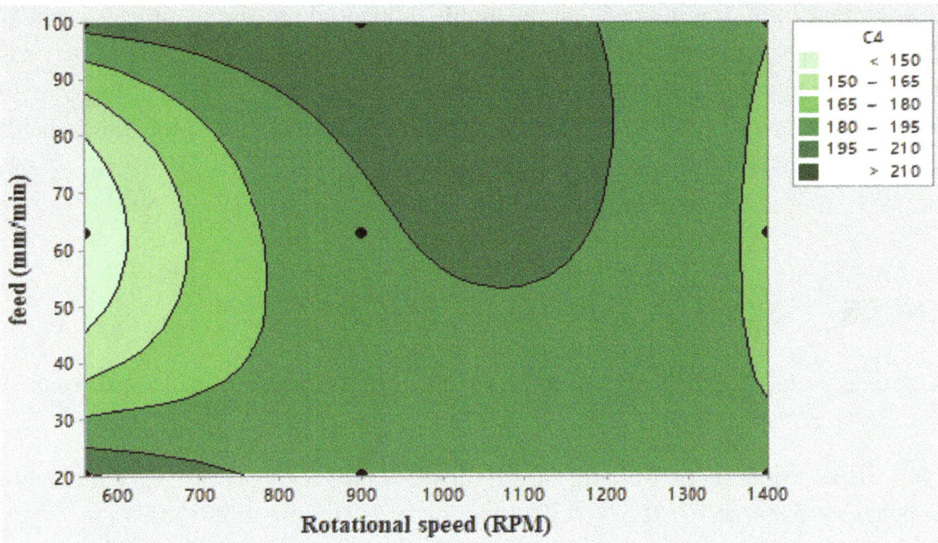

Figure 10. Contour plan between tensile strength (C4) versus tilt angle and feed.

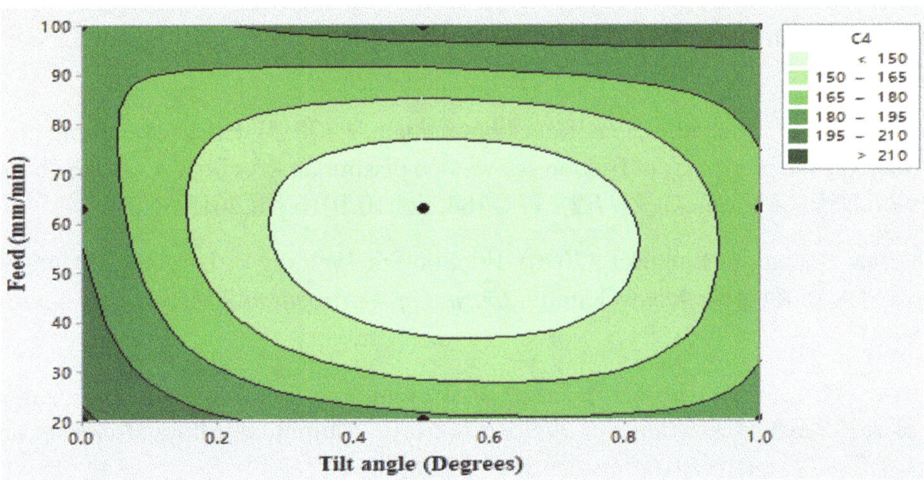

CONCLUSION

- The aluminum welded joints are successfully produced adhering the various experimental standards. The effect of change in factors are analyzed and reported in a significant manner.
- An optimum welding order is succeeded and that is of 560RPM, 0degree, and 20mm/min. The optimum factor is then examined for its desirability to attain the level of maximization and obtained as 0.73 i.e., 73%. The maximum welded joint strength is achieved like 212MPa.Whereas it is predicted to be 192MPa.

- From the regression analysis the experimental model is quantified as a valid forecasting model at a 95% assurance limit. It is noticed that the highly influential process parameter is rate of feed and then rotational speed, and finally tilting angle.
- The response prediction is assessed and the 95% confidence and prediction range are obtained as 137.9 to 245.8MPa and 103.6 to 280.1MPa respectively. The contour plot shows the distribution of measured strength to the considered limits of factors.

REFERENCES

Boumerzoug, Z., & Helal, Y. (2017). *Friction stir welding of dissimilar materials Aluminum Al 6061-T6 to Ultra Low Carbon Steel.* MDPI, Metals.

Cavaliere, P. (2013). Friction stir welding of Al alloys: Analysis of processing parameters affecting mechanical behavior. *Procedia CIRP*, *11*, 139–144. doi:10.1016/j.procir.2013.07.039

Elatharasan & Kumar Senthil. (2013). An experimental analysis and optimization of process parameters on friction stir welding of AA6061-T6 aluminum alloys using RSM. *Procedia Engineering, 64*, 1227-1234.

Ghosh, M., Kumar, K., Kailas, S. V., & Ray, A. K. (2010). Optimization of friction stir welding parameters for dissimilar aluminum alloys. Elsevier. *Materials & Design*, *31*(6), 3033–3037. doi:10.1016/j.matdes.2010.01.028

Ilangovan, M., Rajendra Boopathy, S., & Balasubramanian, V. (2015). Effect of tool pin profile on microstructure and tensile properties of friction stir welded dissimilar AA 6061eAA 5086 aluminum alloy joints. Elsevier. *Defence Technology*, *11*(2), 174–184. doi:10.1016/j.dt.2015.01.004

Khanna, Sharma, Bharati, & Badheka. (2020). Friction Stir Welding of Dissimilar Aluminum Alloys AA6061-T6 and AA8011-h14: A Novel Study. *Journal of the Brazilian Society of Mechanical Sciences and Engineering.*

Liu, H. J., Hou, J. C., & Guo, H. (2013). Effect of Welding Speed on Microstructure and Mechanical Properties of self-Reacting Friction Stir Welded 6061-T6 Aluminum Alloy. *Materials and Design, Elsevier*, *50*, 872–878. doi:10.1016/j.matdes.2013.03.105

Maneiah, D., & Debashis Mishra, K. (2020). Process parameters optimization of friction stir welding for optimum tensile strength in Al 6061-T6 alloy butt welded joints. *Materials Today: Proceedings*, *27*, 904–908. doi:10.1016/j.matpr.2020.01.215

Mishra, D., & Das, A. K. (2021). Linear model analysis of fused deposition modeling process parameters for obtaining the maximum tensile strength in acrylonitrile butadiene styrene (ABS) and carbon fiber polylactic acid (PLA) materials. Multidiscipline Modeling in Materials and Structures. doi:10.1108/MMMS-09-2020-0239

Mishra, D., & Tulasi, T. (2020). *Experimental Investigation on Stir Casting Processing and Properties of Al 6082/SiC Metal Matrix Composites. In Lecture Notes in Mechanical Engineering*. Springer.

Mishra, R. S., & Ma, Z. Y. (2005). Friction stir welding and processing. Elsevier. *Materials Science and Engineering R Reports*, *50*(1-2), 1–78. doi:10.1016/j.mser.2005.07.001

Padhy, Wu, & Gao. (2018). Friction Stir Based Welding and Processing Technologies -Processes, Parameters, Microstructures and Applications: A review. *Journal of Materials Science and Technology.*

Patel Satyam. (2017). A Review on an Investigation on the Effect of Immersed. Medium in Friction Stir Welding of Aluminium Alloys (AA 6061-T6). *Journal of Material & Metallurgical Engineering, Volume*, *07*(1).

Raj Kumar, Venkatesh Kannan, Sadeesh, Arivazhagan, & Devendranath Ramkumar. (2014). Studies on the effect of tool design and welding parameters on the friction stir welding of dissimilar aluminum alloys AA 5052 – AA 6061. *Procedia Engineering, 75*, 93-97.

Ramanjaneyulu, Madhusudhan Reddy, & Gokhale. (2015). Optimization of process parameters of aluminum alloy AA2014 - T6 friction stir welds by response surface methodology. *Defence Technology, 11*, 209 - 212.

Shanavas, S., & Edwin Raja Dhas, J. (2017). Parametric Optimization of Friction Stir Welding Parameters of Marine Grade Aluminium Alloy using Response Surface Methodology. *Transactions of Nonferrous Metals Society of China, Volume*, *27*(11), 2334–2344. doi:10.1016/S1003-6326(17)60259-0

Ugender. (2018). Influence of tool pin profile and rotational speed on the formation of friction stir welding zone in AZ31 magnesium alloy. *Journal of Magnesium and Alloys, 6*, 205–213.

Ugrasen, G., Bharath, G., Kishor Kumar, G., Sagar, R., Shivu, P. R., & Keshavamurthy, R. (2018). Optimization of Process Parameters for Al6061-Al7075 alloys in Friction Stir Welding using Taguchi's Technique. *Materials Today: Proceedings*, *5*(1), 3027–3035. doi:10.1016/j.matpr.2018.01.103

Venkata Rao, Ch., Madhusudhan Reddy, G., & Srinivasa Rao, K. (2015). Influence of Tool Pin Profile on Microstructure and Corrosion Behaviour. Elsevier. *Defence Technology*, *11*, 197–208. doi:10.1016/j.dt.2015.04.004

Chapter 10
Mechanical and Corrosion Behavior of Friction Stir Welded AA 6063 Alloy

Radha R.
https://orcid.org/0000-0002-4819-8648
VIT University, Chennai, India

Sreekanth D.
VIT University, Chennai, India

Tushar Bohra
VIT University, Chennai, India

Surya Bhan Pratap Singh
VIT University, Chennai, India

ABSTRACT

Friction stir welding (FSW) is considered to be the most significant development in solid state metal joining processes. This joining technique is energy efficient, environmentally friendly, and versatile. In particular, it can be used to join high-strength aerospace aluminum alloys and other metallic alloys that are hard to weld by conventional fusion welding. The project aims to join Aluminum 6063 alloy plates by FSW and emphasize the (1) mechanisms responsible for the formation of welds without any defects, microstructural refinement, and (2) effects of FSW parameters on resultant microstructure, mechanical, and corrosion properties.

INTRODUCTION

Friction stir welding (FSW) is a solid-state joining process that uses a non-consumable tool to join two facing work pieces without melting the work piece material (Salem, 2003; Sharma et al., 2017; Sivashanmugam et al., 2010). The heat is generated by friction between the rotating tool and the work

DOI: 10.4018/978-1-7998-7864-3.ch010

piece, which leads to a softened region near the FSW tool. When the tool is traversed along the joint line, it mechanically helps to bond the two pieces of metal with the application of pressure. A rotating cylindrical tool with a profiled probe is fed into a butt joint between two clamped work pieces, until the shoulder which has a larger diameter than the pin touches the surface of the work pieces (Desai & Kapopara, 2009). The probe is slightly shorter than the weld depth required and the tool shoulder riding on top of the work surface. After a short dwell time, the tool is moved forward along the joint line at the pre-set welding speed. The heat due to friction is generated between the wear-resistant tool and the work pieces. Along with that heat generated by the mechanical mixing process and the adiabatic heat within the material cause the stirred materials to soften without melting (Ating et al., 2010; Lueth & Hale, 1970; Obi-Egbedi et al., 2012). As the tool moves forward, a special profile on the probe forces plasticized material from the leading face to the rear, where the high forces assist in a forged consolidation of the weld. This process of the tool traversing along the weld line in a plasticized tubular shaft of metal results in severe solid-state deformation involving dynamic recrystallization of the base material. Jingming and Yifu (2017) investigated the effects of preheating treatment on temperature distribution and material flow of friction stir welded aluminum alloy and steel (Tang & Shen, 2017). Ning and Yunlong (2017) found that better mechanical properties was attained with the rotating speed of tool as 610 rpm and feed 60mm/min (Guo et al., 2017). The hardness of friction stir welded alumnium 7449 alloy was improved with the speed of 1500 rpm and feed 8mm/min (Martinez et al., 2017). The dissimilar friction stir welds were obtained by reinforcing SiC in AA6082-T6 and AA5083-H111 aluminum alloys and the hardness was observed to be higher in the weld nugget (Pantelis et al., 2016). The optimum tensile strength of Al 1100 and St37 steel lap welds was achieved at minimum welding speeds and maximum speed of the tool (Pourali et al., 2017). Yu Chen et al.(2015) found that the welding heat input has a very significant effect on the hardness of the welds (Yu et al., 2015). The ductile mode fracture was observed in the friction stir welded aluminium alloy (Kumbhar et al., 2011).The microstructures of weld varied depending on length of the probe, rotational speed of the tool and its holding time and it was observed that the tensile strength was improved with increasing in probe length (Tozaki et al., 2007). Elangovan (2007) investigated the properties of friction stir welds by different tool geometries and they found square pin tool yield defect free welds with better mechanical properties (Elangovan & Balasubramanian, 2007). The objective of the present work is to investigate the effect of tool geometry on the mechanical and corrosion properties of friction stir welded A6063-O aluminium alloy. The corrosion properties of welded lap joints of AA6061-T6 aluminum alloy produced by FSW process has been investigated and it was observed that intermetallic particles distributed in the weld region causes galvanic coupling leads to accelerate the corrosion (Gharavi et al., 2015). The resistance to pitting corrosion was observed owing to the grain refinement, heat input, and passivation layer at stir region of friction stir welded joints of aluminum alloy AA5052-H32 using a tool with featured shoulder and threaded pin (Soto-Díaz et al., 2021). The corrosion behaviour of friction stir welded AA6061-T3 alloy joints were investigated and it was observed that the conical pin profile of the tool rotated at 850 rpm exhibited better corrosion resistance than the straight pin profile (Abbas et al., 2021). The corrosion study of the dissimilar (between AA7075-T651 and AA2014-T6) friction stir welds was carried out and it was found that the weld nugget stir zone experienced the inter-granular corrosion, while heat affected zone on advancing side experienced pitting corrosion (Raturi & Bhattacharya, 2021).

MATERIALS AND METHODS

Workpiece and Tool Material

The base material used in the present study was 6 mm thick 6063-O (Unheated-Un tempered) with the chemical composition listed in Table 1. The plates were cut and machined into rectangular welding specimens of cross-section 75x36 mm^2 and 90x63 mm^2. The specimens were longitudinally friction stir butt welded with two different tool profiles (Square and Triangular) and the chemical composition of tool is shown in table 2. The design of the tool is depicted in Figure 1 and the welding parameters are listed in table 3 respectively.

Table 1. Chemical composition of AA 6063 (wt. %) of base metal

Material	Si	Fe	Cu	Mn	Mg	Cr	Zn	Ti	Al
AA 6063	0.2-0.6	^{3} 0.35	^{3} 0.1	^{3} 0.1	0.45-0.9	^{3} 0.1	^{3} 0.1	^{3} 0.1	Bal.

Table 2. Chemical composition of tool (wt. %)

Material	C	Mn	Si	Cr	Mo	V
HCHC- D2	1.5	0.3	0.3	12.0	0.75	0.9

Microstructural Characterization

Three samples one with base metal and other two with square and triangular pin are prepared for microstructure study as finer grain structure was expected after friction stir welding. First all the samples are rubbed against emery papers of grades 1- 4 to remove scratches if present on samples followed by diamond polishing on the polishing machine. After this samples are washed with ethanol followed by treatment with etchant. Here in this microstructure study, weck's reagent is used as etchant. The composition of weck's reagent is 100 ml distilled water, 4g KMnO4 and 1g NaOH. The samples were etched for 3 minutes in order to get clear image. Once etching is done, samples are placed on optical microscope with 200x magnification.

Hardness Measurement

Hardness is measured along the weld length in welded zone, HAZ, and base material through micro hardness tester. The load applied was 150 kgf with standard equipment. Intender used for testing is 1/8" inch ball type. Dwelling time is 30 sec just to ensure proper indentation. ASTM standard E18 was followed for the preparation of sample and procedure for hardness testing.

Figure 1. Tool geometry (a) square pin tool (b) triangular pin tool (c) CAD of Tool (d) friction stir welding tool

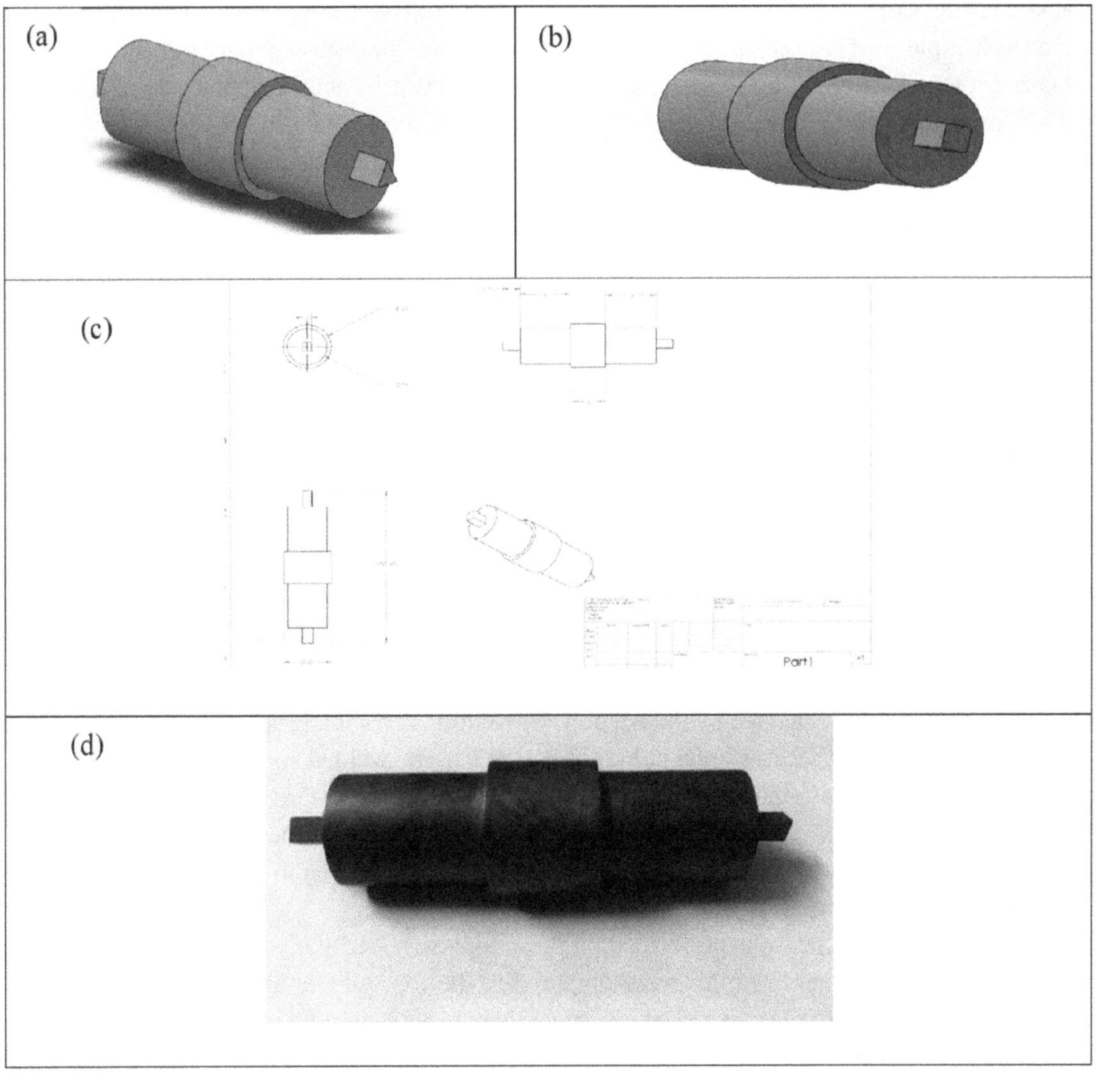

Table 3. Friction stir welding parameters

Welding parameters	Values
Spindle Speed	1000 rpm
Feed rate or transverse speed	0.2 mm/s
Piercing depth	5.6 mm
Piercing Rate	0.04 mm/s
Weld Length	80 mm

Tensile Properties

The tensile testing samples are prepared according to ASTM E8/E8M standard. CNC milling is used to machine the sample. Just to confirm the fact obtained from literature survey that Un-heat-treated 6063 has maximum tensile strength no more than 130 MPa (19,000 psi), and no specified maximum yield strength. The material has elongation (stretch before ultimate failure) of 18%. Samples were tested on universal testing machine for tensile test and ultimate tensile strength was measured.

Corrosion Measurement

The potentiodynamic polarization test was conducted on both welded zone and the base metal in 3.5 wt.% NaCl medium using a potentiostat (Interface 1010, Gamry Instruments) by exposing 0.5 cm^2 area of the sample. The corrosion cell consisted of a saturated calomel electrode (SCE) and a platinum wire, as reference and counter electrodes respectively. Tafel plots were generated by polarizing the specimen about 0.2 V anodically and cathodically with reference to open circuit potential (OCP) at a scan rate of 0.5 mV s^{-1} after an initial delay of 300 seconds. After the measurements were obtained, the Tafel data was analysed by curve fitting and equivalent circuit modelling using Gamry Echem Analyst software.

RESULTS AND DISCUSSION

The defect free welds were obtained by the friction stir welding at a spindle speed of 1000 rpm and feed rate between 0.2 to 0.4 mm/sec as shown in Figure 2(a) & (b) The welded zones have well refined grain structure as compared to base metal's grain structure or in other word more grain refinement has been observed at the weld zone as depicted in Figure 2 (c) & (d). The results of the mechanical properties are given in table 4. And it was observed that specimens welded with square pin have more tensile strength as compared to specimens welded with triangular pin.

The Rockwell hardness test was performed on the samples welded with different pin profile across welded zone and the hardness profiles were obtained at the top, middle and bottom portions of the cross- section of the samples. The increase in hardness and tensile strength of the samples welded with square pin was attributed to the grain refinements which further prevents the dislocation movement. The higher rate of pulsating action employed in the stir zone of square pin profiled tool produces well refined grain structure and in turn yields better tensile strength and higher hardness. The corrosion test results revealed that base metal has the least corrosion resistance ability. The sample welded with triangular pin has the maximum corrosion resistance than the square pin welded sample and the corrosion results obtained from Tafel extrapolation are shown in Table 5. Among all the samples the samples welded with triangular pin has shown the less corrosion current density (i_{corr}) and it was observed from Figure 3., the anodic branch of the sample welded with triangular pin shows the passivation. The protective layer formed on the surface of the sample prevents from corrosion.

Figure 2. (a) Weld joint using square pin, (b) weld joint using triangular pin, (c) optical micrograph of weld zone using square pin, (d) optical micrograph of weld zone using triangular pin

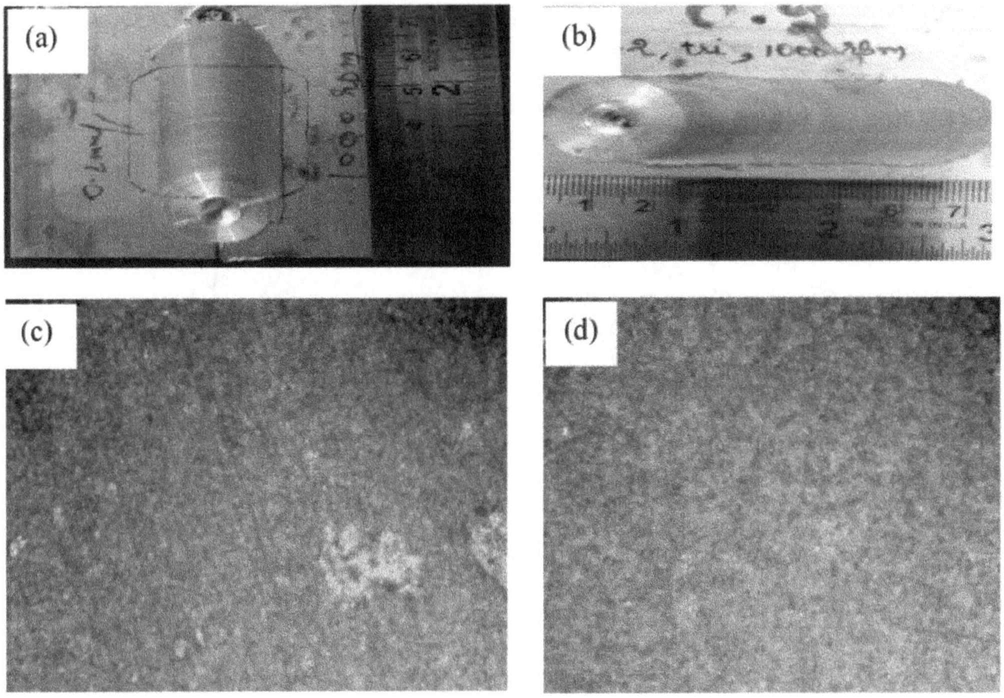

Table 4. Mechanical properties

Values	Square Pin	Triangular Pin
Max. Force (N)	5600.79	5421.88
Max. Stress (N/mm^2)	90.20	87.31
Max. Stroke Strain %	20.6624	18.23
Hardness in the weld zone (HRB)	57	52

Table 5. Tafel results of base metal, square pin and triangular pin.

Parameters	Value (base metal)	Value (square pin)	Value (triangular pin)
β_A	1.219 V/decade	672.2e-3 V/decade	541.5e-3 V/decade
β_C	263.3e-3 V/decade	397.7e-3 V/decade	164.4e-3 V/decade
I_{corr}	3.470 µA	1.700 µA	867.0 nA
E_{corr}	-1.300 V	-901.0 mV	-1.220 V
Corrosion rate	8.936 mpy	4.382 mpy	2.240 mpy

Figure 3. Tafel curves

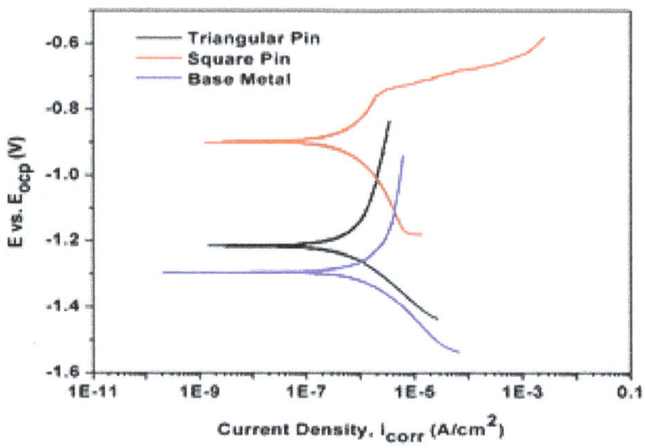

CONCLUSION

Aluminum alloy AA6063-O plates were successfully joined using the friction stir welding technique by using tool with varying profiles such as square and triangular pin. The joined plates were subjected to hardness test, tensile test and corrosion test. The results are as follows:

Friction stir welding is an effective methodology for joining aluminum alloy AA6063-O plates and properties are noticeably affected by varying the tool profile. Friction stir welding of a specimen with square pin resulted in the maximum ultimate tensile strength of 90.20 MPa with the high strain rate of 20.67% and hardness of 57 HRB. The grain refinement with equiaxed structure was observed in the friction stir welded AA6063-O plates with square pin and triangular pin tool. The samples welded with triangular pin profile of the tool exhibits the higher corrosion resistance compared to the square pin tool profile. This may be due to the coarsening of precipitates in the thermomechanical affected zone and heat affected zone.

REFERENCES

Abbas, W. S., Alali, M., & Salim, J. A. (2021, March). Effect of rotational speed and tool pin profile on the corrosion rate of friction stir welded AA6061-T3. *IOP Conference Series. Materials Science and Engineering*, *1090*(1), 012124. doi:10.1088/1757-899X/1090/1/012124

Ating, E. I., Umoren, S. A., Udousoro, I. I., Ebenso, E. E., & Udoh, A. P. (2010). Leaves extract of Anan-assativum as green corrosion inhibitor for aluminium in hydrochloric acid solutions. *Green Chemistry Letters and Reviews*, *3*(2), 61–68. doi:10.1080/17518250903505253

Desai, P.S., & Kapopara, S.M. (2009). *Inhibiting effect of anisidines on corrosion of aluminium in hydrochloric acid.* Academic Press.

Elangovan, K., & Balasubramanian, V. (2007). Influences of pin profile and rotational speed of the tool on the formation of friction stir processing zone in AA2219 aluminium alloy. *Materials Science and Engineering A*, *459*(1-2), 7–18. doi:10.1016/j.msea.2006.12.124

Gharavi, F., Matori, K. A., Yunus, R., Othman, N. K., & Fadaeifard, F. (2015). Corrosion behavior of Al6061 alloy weldment produced by friction stir welding process. *Journal of Materials Research and Technology*, *4*(3), 314–322. doi:10.1016/j.jmrt.2015.01.007

Guo, N., Fu, Y., Wang, Y., Meng, Q., & Zhu, Y. (2017). Microstructure and mechanical properties in friction stir welded 5A06 aluminum alloy thick plate. *Materials & Design*, *113*, 273–283. doi:10.1016/j.matdes.2016.10.030

Kumbhar, N. T., Sahoo, S. K., Samajdar, I., Dey, G. K., & Bhanumurthy, K. (2011). Microstructure and microtextural studies of friction stir welded aluminium alloy 5052. *Materials & Design*, *32*(3), 1657–1666. doi:10.1016/j.matdes.2010.10.010

Lueth, R. C., & Hale, T. E. (1970). Compressive Strength of Cemented Carbides—Failure Mechanics and Testing Methods. *Materials Research and Standards*, *10*(2), 23–28.

Martinez, N., Kumar, N., Mishra, R. S., & Doherty, K. J. (2017). Microstructural variation due to heat gradient of a thick friction stir welded aluminum 7449 alloy. *Journal of Alloys and Compounds*, *713*, 51–63. doi:10.1016/j.jallcom.2017.04.147

Obi-Egbedi, N. O., Obot, I. B., & Umoren, S. A. (2012). Spondias mombin L. as a green corrosion inhibitor for aluminium in sulphuric acid: Correlation between inhibitive effect and electronic properties of extracts major constituents using density functional theory. *Arabian Journal of Chemistry*, *5*(3), 361–373. doi:10.1016/j.arabjc.2010.09.002

Pantelis, D. I., Karakizis, P. N., Daniolos, N. M., Charitidis, C. A., Koumoulos, E. P., & Dragatogiannis, D. A. (2016). Microstructural study and mechanical properties of dissimilar friction stir welded AA5083-H111 and AA6082-T6 reinforced with SiC nanoparticles. *Materials and Manufacturing Processes*, *31*(3), 264–274. doi:10.1080/10426914.2015.1019095

Pourali, M., Abdollah-Zadeh, A., Saeid, T., & Kargar, F. (2017). Influence of welding parameters on intermetallic compounds formation in dissimilar steel/aluminum friction stir welds. *Journal of Alloys and Compounds*, *715*, 1–8. doi:10.1016/j.jallcom.2017.04.272

Raturi, M., & Bhattacharya, A. (2021). Mechanical strength and corrosion behavior of dissimilar friction stir welded AA7075-AA2014 joints. *Materials Chemistry and Physics*, *262*, 124338. doi:10.1016/j.matchemphys.2021.124338

Salem, H. G. (2003). Friction stir weld evolution of dynamically recrystallized AA 2095 weldments. *Scripta Materialia*, *49*(11), 1103–1110. doi:10.1016/j.scriptamat.2003.08.010

Sharma, N., Khan, Z. A., & Siddiquee, A. N. (2017). Friction stir welding of aluminum to copper—An overview. *Transactions of Nonferrous Metals Society of China*, *27*(10), 2113–2136. doi:10.1016/S1003-6326(17)60238-3

Sivashanmugam, M., Kumar, T., Ravikumar, S., Rao, V. S., & Muruganandam, D. (2010, November). A review on friction stir welding for aluminium alloys. *Frontiers in Automobile and Mechanical Engineering, 2010*, 216–221. doi:10.1109/FAME.2010.5714839

Soto-Díaz, R., Sandoval-Amador, A., & Unfried-Silgado, J. (2021). Experimental evaluation of rotational and traverse speeds effects on corrosion behavior of friction stir welded joints of aluminum alloy AA5052-H32. *International Journal of Advanced Manufacturing Technology*, 1–11.

Tang, J., & Shen, Y. (2017). Effects of preheating treatment on temperature distribution and material flow of aluminum alloy and steel friction stir welds. *Journal of Manufacturing Processes, 29*, 29–40. doi:10.1016/j.jmapro.2017.07.005

Tozaki, Y., Uematsu, Y., & Tokaji, K. (2007). Effect of tool geometry on microstructure and static strength in friction stir spot welded aluminium alloys. *International Journal of Machine Tools & Manufacture, 47*(15), 2230–2236. doi:10.1016/j.ijmachtools.2007.07.005

Yu, C., Hua, D., Li, J. Z., Zhao, J. W., Fu, M. J., & Li, X. H. (2015). Effect of welding heat input and post-welded heat treatment on hardness of stir zone for friction stir-welded 2024-T3 aluminum alloy. *Transactions of Nonferrous Metals Society of China, 25*(8), 2524–2532. doi:10.1016/S1003-6326(15)63871-7

Section 3
Applications

Chapter 11
Fabrication of Tailor-Made Metallic Structures for Lightweight Applications and Mechanical Behaviour

R. Ganesh Narayanan
Department of Mechanical Engineering, Indian Institute of Technology, Guwahati, India

Perumalla Janaki Ramulu
School of Mechanical, Chemical and Materials Engineering, Adama Science and Technology University, Adama, Ethiopia

Satheeshkumar V.
Department of Production Engineering, National Institute of Technology Tiruchirappalli, India

Arvind K. Agrawal
https://orcid.org/0000-0003-2151-6404
Madanapalle Institute of Technology and Science, India

Sumitesh Das
Research and Development, Tata Steel Limited, Jamshedpur, India

Ajay Kumar P.
Department of Mechanical Engineering, Indian Institute of Technology Tirupati, India

V. Vishnu Namboodiri
National Institute of Construction Management and Research, Hyderabad, India

ABSTRACT

Tailor-made metallic structures are fabricated by welding, adhesive bonding, and mechanical joining methods. Here the aim is not only to fabricate lightweight structures, but also to develop novel methods of joining. Lightweight structures are advantageous in several ways including reduction of fuel consumption and vehicle emissions. Developing novel methods of joining is advantageous due to the possibility of joining of dissimilar materials, improved mechanical performance, and microstructures. In the chapter, initially, tailor-welded blanks (TWB) are introduced, and after that, fabrication of TWBs by laser welding, friction stir welding, and friction stir additive manufacturing are elaborately discussed. Some critical issues in modeling the deformation during fabrication of TWBs is also discussed. A brief account of mechanical behavior of adhesive bonded sheets and mechanical joining are presented in the later part.

DOI: 10.4018/978-1-7998-7864-3.ch011

INTRODUCTION TO TWB AND FABRICATION BY LASER WELDING

Introducing TWB

Tailor Welded Blank (TWB) consists of metallic sheets with similar or dissimilar thickness, materials, coatings etc. welded in a single plane before forming. TWBs are formed like un-welded blanks to manufacture automotive components. By using this technology, it is possible to produce parts with varying mechanical properties, leading to component optimization (Pallett & Lark, 2001). Among all the welding methods, laser welding is used predominantly to TWBs as the process is quick and produces smaller welds. The main catalyst behind the sudden growth of TWB is to maintain the market share of steel material in the face of competition generated by aluminium alloys towards weight reduction (Pallett & Lark, 2001).

Some of the advantages of using TWBs in the automotive sector are, (1) scrap materials from stamping industries can be reused to have new stamped products, (2) by distributing material thickness and properties, part consolidation is possible resulting in cost reduction and better quality, stiffness and tolerances, (3) provides greater flexibility for component designers, (4) weight reduction of the product can be achieved and hence fuel consumption is lowered, and (5) improved corrosion resistance and product quality. Research shows that a 1% reduction in vehicle weight can result in a reduction of fuel consumption to 0.6 – 1% (Pallett & Lark, 2001). With these advantages, the potential of TWBs was soon recognized by steel industries and a consortium called Ultra Light Steel Auto Body (ULSAB) concept was then developed (WorldAutoSteel, 2020)

Some of the applications of TWB include center pillar, bumper, front door inner, and rear door inner (Irving, 1995; Prange et al., 1994). Nevertheless, the applications are not restricted to automotive sector, as construction industries also show encouraging pathway in using TWB as part of buildings (Pallett & Lark, 2001). Some applications of laser-welded TWBs are listed in Table 2.1(WorldAutoSteel, 2020). In the case of aluminium alloys, 5xxx for automotive inner body panels, and 6xxx for outer body panels are mainly used. Specifically, these include 6111-T4, 5754-O, 6061, 5182, 5052 and 5454. Some of the applications of aluminium TWBs made by laser welding, include deck lids, hoods, floor and door inner panels, side frame rails etc. (Das, 2000). Several applications in European and USA market, variety of welded sheets in the form of three piece blanks, non-linear weld lines, details on noise separation, and introduction of multi-piece blank lines can be noted (Rooks, 2001).

In laser welding, CO_2 and ND:YAG (Neodymium:Yttrium-Aluminum-Garnet) laser welding techniques are widely used to produce TWBs. A typical experimental setup is shown in Figure 1 (a) in which two steel sheets are clamped for CO_2 laser welding. Initially, a pointer is displaced at the interface of sheets to check the alignment of sheets with laser source. Figure 1 (b) shows the laser-welded TWB made of steel sheets. Edge quality is crucial and generally, machined edges are preferred for laser based TWB fabrication (Ono et al., 2004). Some of the problems encountered while welding Aluminium sheets are (i) excessive porosity and hot cracking of the fusion zone, (ii) poor coupling during laser welding because of high reflectivity of the metal, (iii) degradation of weld zone properties, and (iv) loss of alloying elements in fusion zone (Das, 2000). Of the two techniques, CO_2 based is accepted widely, safer and less expensive. However, it is subjected to high reflectivity of laser beam on the Al solid surface for a particular wavelength. ND:YAG laser welding has the advantage of uniform energy distribution than CO_2 laser beam over a wide area. It provides the ability to span over large gaps, while maintaining weld speed and quality.

Table 1. Steel TWB products with part specifications (https://www.worldautosteel.org/)

TWB product	Thickness combination (in mm)	Material combination *	Yield strength combination (in MPa)
Body side outer	1.5/1.0/0.7	DP/DP/BH	700/700/250
Wheel inner house	1.4/0.6/1.1/0.6	DP/DP/DP/DP	700/500/700/500
Floor rear	0.7/0.6/1.1/1.1	DP/BH/DP/DP	700/210/350/350
Rear rail	1.8/1.1	DP/DP	700/500
Rocker inner	1.5/0.7	DP/DP	700/700
Reinforcement B-pillar rocker rear	1.2/1.4	DP/DP	700/700
Body side outer	0.7/1.5/1.0/1.2/0.7	BH/DP/DP/DP/BH	260/700/700/700/260
Rail rear outer floor extension	0.6/1.0	BH/DP	210/500
Front door inner front	1.0/1.2	MS/MS	140/140
Rear door inner front	1.0/1.2	MS/MS	140/140

[***Material combination:** DP – Dual Phase steel; BH – Bake Hardenable steel; MS –Mild Steel]

Figure 1. (a) Laser welding set up for fabricating tailor-welded sheets, (b) Laser welded tailored-sheets deformed until fracture (Narayanan, 2007)

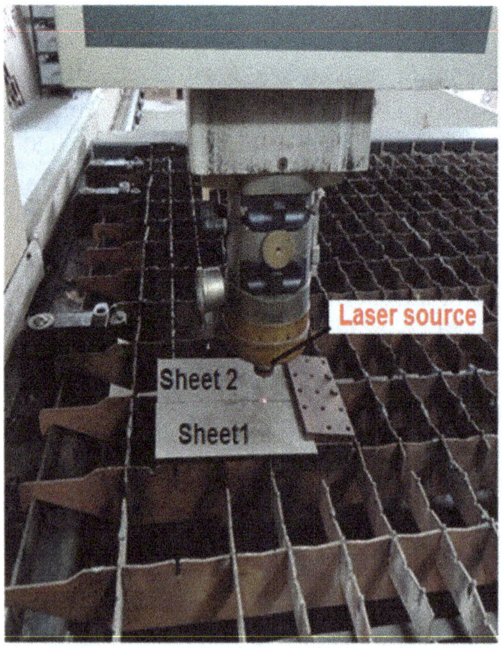

(a)

(b)

In the following sections, the effect of various laser based TWB parameters on the forming performance are summarized.

Influence of Thickness Ratio and Strength Ratio

In addition to the basic material properties, the forming behavior of TWB is affected by thickness ratio, strength ratio of the sheets welded, weld conditions such as weld orientation, weld location, weld properties, weld microstructure etc. Thickness ratio is the ratio of one blank thickness to other blank thickness, which constitute TWB.

Forming limit is found to decrease with increasing thickness ratio (thick/thin). In addition, as the thickness ratio reduces, the forming limit of TWB approaches to that of un-welded blank. Chan et al. (2003) studied the effect of thickness ratio by performing tensile test and swift cup test. The thickness ratio selected were 2.0 (1.0/0.5), 1.67 (1.0/0.6), 1.25 (1.0/0.8) and SPCC material was chosen for the analysis. It was found from the tensile tests that the uniform elongation of TWB is lower than that of un-welded blanks. In addition, with increase in thickness ratio, the uniform elongation decreases. TWBs with thickness ratio of 2.0 and 1.67 exhibit about 20% uniform elongation, while TWB with 1.25 has 23% uniform elongation. Swift cup test revealed similar behavior with forming limit of TWB lower than that of un-welded blank. The same research group presented more data on formability (Chan et al., 2005) and results reveal the same. Others present similar results as well in Aluminum alloys and carbon steel (Shakeri et al., 2002; Buste et al., 1999; Friedman & Kridli, 2000; Ghoo et al., 2001). The reason behind reduction in forming limit is because of the non-uniform plastic deformation exhibited by the thinner and thicker blanks in TWB. Longitudinal and 45° weld orientations show lesser forming limit strains in comparison to that of un-welded blank, but higher than transverse weld TWB for varied thickness ratios (Buste et al., 1999), which is due to the alignment of weld with major straining direction changing the strain path from plane strain to bi-axial stretching. The dome height at failure of TWB decreases when compared with un-welded blank, and it decreases with increase in thickness ratio (Bhaskar et al., 2004; Nakagawa et al., 1993; Raymond et al., 2004; Siegert & Knabe, 1995; Baysore et al., 1995).

In deep drawing, the drawability parameters monitored are limit draw ratio (LDR) (Kampus & Balic, 2003; Tusek et al., 2001; Reis et al., 2004), earing characteristics, and predominantly weld zone movement and control (Choi et al., 2000; Heo et al., 2001; Ahmetoglu et al., 1995; Kinsey et al., 2000; Meinders et al., 2000; Saunders & Wagoner, 1996; Bhagwan et al., 2002; Pepelnjak et al., 1997). LDR of 1.67 to 2.0 is achieved by Ahmetoglu et al. (1995) and 2.25 to 2.50 (Kampus & Balic, 2003) for various thickness and material combination. Increase in thickness ratio increases the weld line shift. It is demonstrated by Choi et al. (2000) that the weld line shift in a steel grade TWB is higher in the case of 0.8×1.6 mm thickness combination for different weld line positions. The weld line movement is due to the heterogeneity in plastic deformation achieved by the thicker (or stronger) and thinner (or weaker) blanks during forming. As a result, the thinner (or weaker) blank deforms largely allowing the weld zone to move in the cup wall region. With increase in thickness difference, weld zone shift increases. The weld zone not only moves but also rotates slightly (about 5°) with respect to vertical axis during drawing (Ahmetoglu et al., 1995; Pepelnjak et al., 1997).

Strength ratio is defined as the ratio of yield strength of one material to that of another material constituting TWB. Generally, increase in strength difference between the materials welded deteriorates the forming behavior of TWB. Therefore, with increase in strength ratio, forming limit strain, dome height, limiting draw ratio, and cup height should decrease and weld line movement should increase (Nakagawa et al., 1993; Kampus & Balic, 2003; Saunders & Wagoner, 1996; Kusuda et al., 1997; Merklein et al., 2004; Hayashi et al., 1998; Koruk et al., 2001).

To summarize, increase in thickness ratio and strength ratio, deteriorates the forming behavior of laser-welded TWB significantly. Weld line shift increases with increasing thickness or strength ratio. However, the effect depends on weld location and orientation as well.

Influence of Weld Orientation and Weld Location

In the case of laser-welded TWBs, the weld orientations mostly studied are 0° (weld parallel to major straining direction), 90° (weld normal to major straining direction) and 45° (weld at 45° to major straining direction). In Cheng et al. (2005) work on TWB made of 1 mm thick 5754-O sheets, transverse weld samples exhibit tensile and forming limit equivalent to that of un-welded blank. In longitudinal weld orientation, tensile and forming limit of TWBs are drastically reduced. This is due to the difference in volume fraction of weld zone in the two orientations during tensile tests. Failure is away from the weld region in transverse weld orientation, while it is normal to the weld in longitudinal weld orientation. In Buste et al. (1999) work, formability of AA5754 alloy is studied for thickness ratios, 1:2, 1:1.6, in plane strain condition. Here, TWB with longitudinal weld orientation has limit strain closer to that of un-welded blank, while TWB with transverse weld shows reduction. This is due to the change in strain path from plane strain to bi-axial stretching in TWB with longitudinal weld.

It should be noted that formability of laser-welded TWBs made of AA5182-O material (thickness: 0.84 and 1.55 mm) with 45° weld is superior to formability of TWBS with 90º weld orientation (Friedman & Kridli, 2000). Shakeri et al. (2001) presented a similar forming behavior.

The effect of weld location in TWB is also equally important like weld orientation. Choi et al. (2000) studied the effect of initial weld line location on the weld line movement during deep drawing of laser-welded TWB. TWB with weld at 24 mm offset shows larger weld line movement as compared to weld line at center. Weld line movement is larger in the case of TWB with weld at 24 mm offset, i.e., when it is placed at a non-symmetric location of the cup, during deep drawing of TWB with drawbeads on thinner sheet (Heo et al., 2001). Hayashi et al. (1998) fabricated laser-welded TWB made of EDDQ steel and high strength steel of 0.7 mm thickness and found two compounding mechanisms of weld line movement. Primary weld line movement, in which weld line moves because of the geometric change in shape and size of the blank, and secondary weld line movement, which depends on the heterogeneity of properties in the base metal and weld region i.e., hardening degree of weld region. These mechanisms are analyzed by monitoring movement of scratched line and weld region during deep drawing. Similar studies are carried out by Scriven et al. (1997).

Finite element simulation of tensile tests with different transverse weld locations reveals that forming limit strain is larger when weld is located at the center, while it decreases when placed away from the center, and then increases to that of un-welded sheet (Narayanan & Narasimhan, 2005). Narayanan and Narasimhan (2008a) showed that forming limit of laser-welded Interstitial Free (IF) steel sheet could be improved by changing the weld locations and orientations. Offset weld locations deteriorated the forming limit evaluated by formability tests (Figure 2). Forming limit strains were also predicted by various limit strain criteria (Narayanan and Narasimhan, 2008b; Naik et al, 2010). To summarize, the initial weld line placement and orientation can affect the forming behavior of TWBs such as weld line movement, dome height at failure, forming limit strains, and strain distribution considerably. Under specific locations and orientations, formability can be improved. Post-weld heat treatment, forming temperature and speed of deformation can also change failure pattern and forming limit of laser-welded tailored blanks as proposed by Liu et al. (2018).

Figure 2. Weld location and weld orientation effect on the forming limit curves of TWBs made of IF steel (Narayanan, 2007)

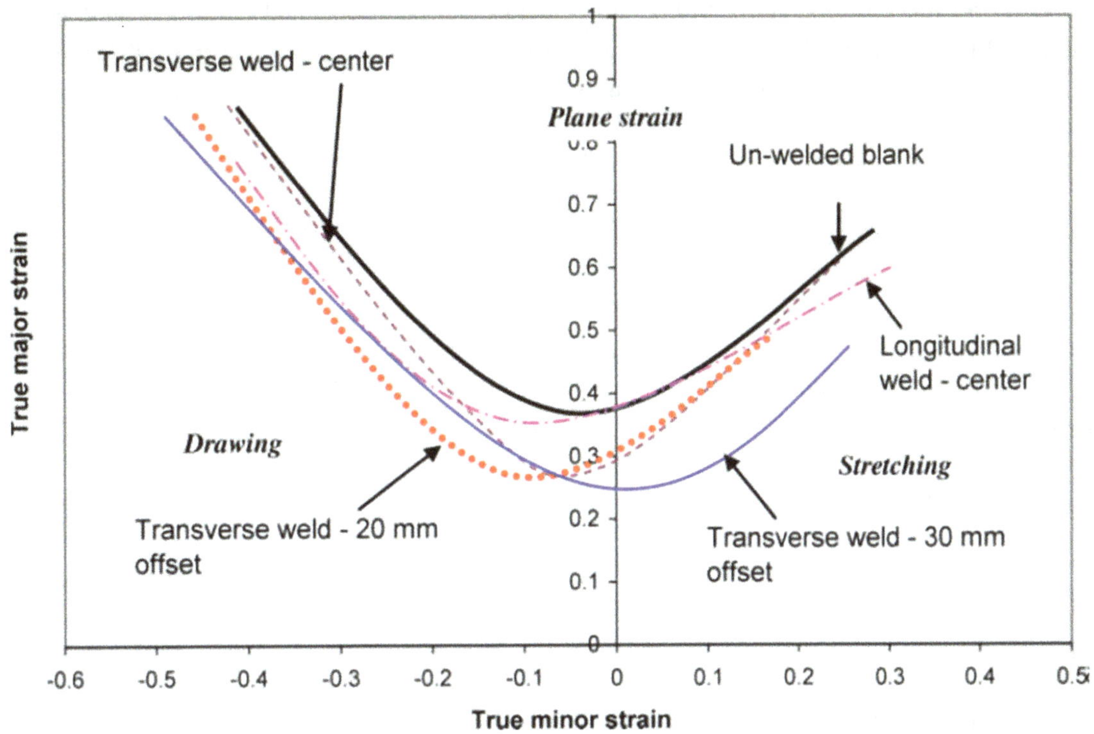

Fiber laser welding is also used to fabricate laser-based TWBs. Dissimilar joints of DP600-DP980 were made by fiber laser welding with a focused and defocused beam (Jia et al., 2016). Fusion zone showed higher hardness (about 456 Hv) across the joint. DP980 side experienced severe HAZ softening (282 Hv), whereas on the DP600 side, an extremely small hardness drop (5 Hv) was found. TWBs made by focused and defocused beam showed almost same tensile formability and strength requirements. Another example sighting the application of fiber laser welding for DP steel TWB fabrication is noteworthy (Bandyopadhyay et al., 2014). Productivity, costs, weld quality, mechanical properties of TWBs made by laser welding with CO_2, Nd/YAG, and fiber laser sources are compared (Assunção et al., 2010). Results emphasized efficient performance of fiber lasers when compared to Nd:YAG and CO_2 lasers. Moreover, highest welding speeds with lowest cost/hour was also witnessed in case of fiber laser showing higher productivities than its competitors. Formability tests showed good performance of fiber laser welds with failure outside the weld and in thinner material. Recently, Diode laser source is utilized to fabricate TWBs made of SS304 with thickness combination of 2 mm and 1 mm (Zhu et al., 2020; Lin et al, 2021). Experimental results demonstrate fabrication of a defect free part with failure occurring in the thinner sheet away from weld zone.

Effect of Laser Welding Parameters

The mechanical properties and microstructure of weld zone in TWB are decided not only by the materials that are welded, but also by the welding techniques and welding parameters. Some of the laser welding parameters affecting the forming behavior of laser-welded TWBs are welding speed, power, shielding gas, shielding gas flow rate, beam misalignment and gap (Stasik & Wagoner, 1998; Stasik & Wagoner, 1996; Eisenmenger et al., 1995). The effect of welding speed and power are predominantly studied. At 3 kW power (in CO_2 laser welding), with increase in welding speed from 95 mm/s to 148 mm/s, the ductility of Al 6111-T4 TWB is found to decrease from 15% to 5%. Similar results are seen in Al 5754-O as well. Increasing laser power from 3 kW to 5 kW for 127 mm/s speed increases the total elongation of Al 5754-O TWB with longitudinal weld.

Weld Line Movement and Controlling Methods

Weld line movement during forming of TWB is because of the heterogeneity in plastic deformation attained by the thinner or weaker material and thicker or stronger material during forming. Stamping behavior deteriorates because of this. The weld line movement can be controlled by various methods as discussed below.

Differential Blank Holding Force (BHF) Technique

If the BHF acting on both base metals is the same, the thinner material deforms largely than the thicker material. By applying differential BHF, higher BHF holds the thinner material and the lower BHF allows the thicker material to deform, compensating stretching of the thinner material, reducing weld line movement.

Ahmetoglu et al. (1995) demonstrated the differential sheet holding technique during deep drawing of laser-welded TWB made of 0.8 and 1.8 mm thick AKDQ steel with weld at centre and at some offset from the center. In order to compensate for different pressure requirements at different positions on the blank, a multi-point pressure-control technique was employed to vary the pressure around the periphery of the blank holder. Initially, a constant BHF of 5 tons was applied on the thinner and thicker blanks. A cup with weld line at a critical location was obtained. Later, a differential BHF of 10 tons was applied on the thinner side and 1 ton was applied on the thicker side, reducing the weld line movement. A cup with weld line at the center was obtained. In He et al. (2003) work, weld line movement in strip drawing and box drawing was minimized with differential BHF of 1.2 ton and 1.4 ton in strip drawing, and 2.4 ton and 7.2 ton in box drawing. Detailed description on segmented BH system can be obtained from existing work (Hassan et al., 2003; Hassan et al., 2005; Gunnarsson et al., 2001).

Using Stepped Die, Blank Holder or Punch to Compensate Thickness Difference

In the case of TWB with dissimilar thickness sheets, uniform blank holder pressure cannot be applied, as uniform contact between the thinner blank and the tools is absent. In this case, the thickness difference should be compensated without which wrinkling of thinner sheet, tearing near the weld region, weld line movement will occur deteriorating the formability of TWB. Compensating thickness difference can be done few ways – (1) by providing shim, and (2) by using stepped die or blank holder (or lower, upper binder). Shim is nothing but a piece of sheet inserted below or above the thinner blank. In the second case, die or blank holder is segmented into different regions, which has step near the weld zone.

Using Drawbeads at Different Locations of TWB

Weld line movement can be reduced by providing drawbeads or by applying drawbead restraining forces at various locations of TWB as suggested by Heo et al. (2001). By increasing the radius and height of drawbead placed on the thinner sheet, weld line movement is significantly minimized.

Optimum Initial Weld Line Location

Weld line movement can also be reduced by locating weld zone at an optimum position such that the weld line is at the same position, or it will move to a desired position after forming. In Choi et al. (2000) work, it is shown that when the weld line is at symmetric position, i.e., at the center of TWB, weld line movement is minimized during deep drawing. Movement is considerable when it is located at some offset. The desired position after deformation can be predicted by following available criteria (Kinsey & Cao, 2003).

Specially Designed Local Adaptive Controllers

Local adaptive controllers can also be used to control weld line movement (Kinsey et al., 2000). The method involves restricting weld line movement by subjecting neighboring areas of weld region to local hydraulic pressure or clamping force. Hydraulic pressure can be applied at one or multiple locations or along the entire weld line by designing multi local hydraulic pressure system. They presented a case study of door inner part. By using the clamping mechanism at suitable locations, the weld line movement is reduced.

FABRICATION OF TWBS BY FRICTION STIR WELDING

FSW Tailored Blanks Fabrication

In Friction Stir Welding (FSW), a rotating rigid metallic tool is plunged into the sheets or plates that are clamped rigidly in a milling machine or a special purpose machine, and traversed along the interface edges of the sheets. Due to intense plastic stirring, material mixing occurs below the tool, and welding is accomplished (Mishra & Ma, 2005; Ma, 2008). The tool has two regions, cylindrical shoulder and pin, that are helpful in accomplishing the stirring action. Figure 3 shows the stages of fabrication of FSW tailored blanks with tool details. FSW parameters generally varied are rotational speed, transverse speed, plunge depth, plunge speed, shoulder diameter, pin profile and dimensions, shoulder contact surface, tilt angle, and lubricants. Initially, FSW was used to fabricate Aluminium structures. Recently Steel, Titanium and Magnesium alloys are also welded. There are several variants of FSW in practice now. In the following section, a summary of influence of various parameters on the fabrication of FSW tailored sheets and their mechanical behavior are presented.

Figure 3. Fabrication of FSW tailored blanks (a) at the start of welding, (b) during FSW, (c) FSW tool

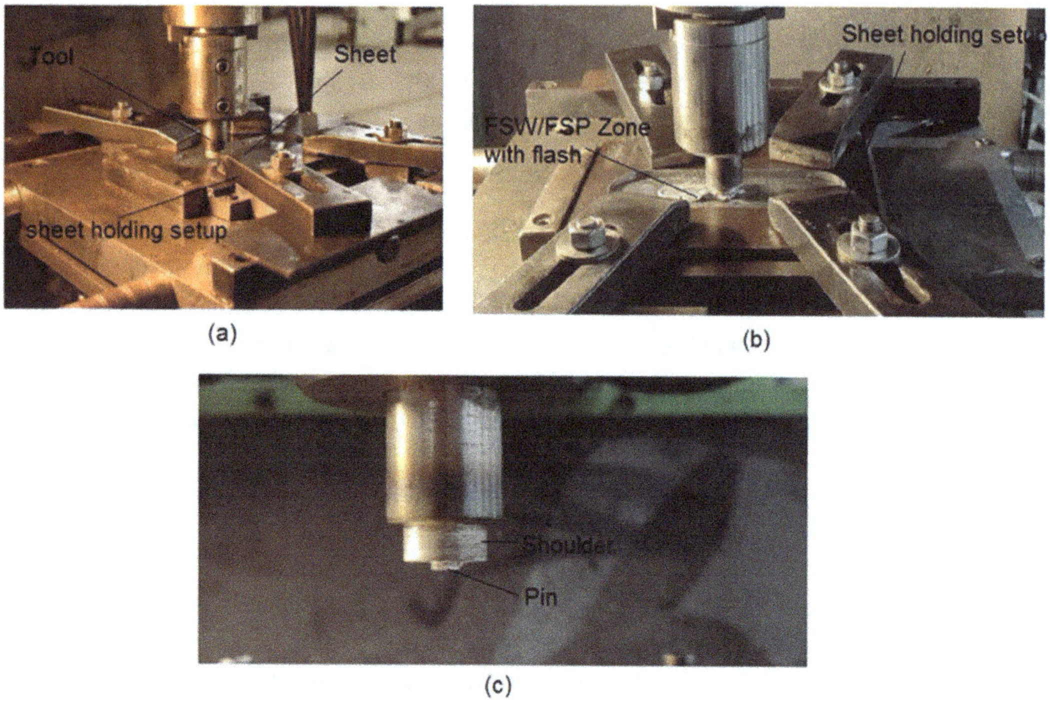

Defects in FSW

Formation of defects affects the joint formation and joint strength of the FSW tailored-sheets. Chen et al. (2006) presented defects such as groove, void, channel defect, kissing bond, lazy S, while welding 4 mm thick Al 5456 plates. It is suggested that

- When the plunge depth is larger, the plasticized material near the pin extrudes and weld flash is formed. At lower pin plunge depth, groove defect is generated at the advancing side. Groove defect is formed when the toll tilt angle is small (= 1°) (Chen et al., 2006).
- With increase in tool tilt angle say to 1.5°, groove type defect was not seen, however, void is observed in the weld surface and in advanced side.
- When the tilt angle is increased further (say about 3.5°), sound joint occurs with onion ring formation, and further increase (larger than 4.5°) resulted in weld flash in the retreating side, and partial fill up of nugget cavity is seen resulting in channel defect. The case at 3.5° tilt angle characterized by onion ring showed larger elongation among all the cases (Chen et al., 2006).
- Kissing bond defect is formed due to insufficient material deformation because of insufficient pressure and stirring action. It is characterized by improper bonding due to oxide layer on the Aluminium surface. It is formed at the root of the weld and/or at the interface, and is difficult to identify this by non-destructive testing (Chen et al., 2006; Khan et al., 2015). This defect can deteriorate the quality of the FSW TWB.

- Lazy S type defect originates from the oxide layer that is broken, extruded and deformed during FSW. It is characterized by chaotic mixing of oxide layer on the retreating or advancing side (Chen et al., 2006).

Identification of defects is a crucial task. Ramulu et al. (2013a) developed a simple criterion (equation 1) to predict the internal defect formation during shop floor trials of FSW of 6xxx and 5xxx sheets.

$$\left(\frac{\partial \tau}{\partial p}\right)_{defective} > \left(\frac{\partial \tau}{\partial p}\right)_{defect\ free} \quad and \qquad (1)$$

$$\left(\frac{\partial F}{\partial p}\right)_{defective} > \left(\frac{\partial F}{\partial p}\right)_{defect\ free}$$

where ¶p is the change in parameters such as welding speed, rotation speed or plunge depth.

By using the criterion, Ramulu et al. (2013a) optimized the FSW parameters such as shoulder diameter, plunge depth, rotational speed, and welding speed as shown in Table 2. From the optimized range, FSW was performed to fabricate tailored sheets made of 6061-T6 and their formability was studied.

Table 2. Optimized FSW parameter range for fabricating FSW sheets

Shoulder diameter (mm)	Plunge depth (mm)	Rotation speed (rpm)	Welding speed (mm/min)
12, 15, 18	1.85, 1.9, 1.95 mm	1300, 1400, 1500	90, 100, 120

Other methods useful for identification of FSW defects during FSW tailored-blank fabrication are (i) machine learning methods with non-destructive testing based data set generation (Nadeau et al., 2020), (ii) artificial neural network based surface quality prediction (Hartl et al., 2020), (iii) measurement of force signals (frequency, time data) and correlating to the occurrence and size of discontinuities such as subsurface voids, volume defects during FSW (Shrivastava et al., 2017), and (iv) monitoring real time torque signals during FSW (Das et al., 2017).

Failure Pattern and Formability of FSW TWB

Using optimized parameters in Table 1, FSW tailored-blanks were fabricated and their formability was evaluated through limiting dome height tests and developing forming limit curves (FLC) on the stretching side of forming limit diagram (FLD). Some of the formed un-welded and FSW tailored-sheets are shown in Figure 4 and 5 (Ramulu, 2013). In the case of FSW blanks, failure pattern in 100 mm × 200 mm strain path is normal to the weld like in un-welded blanks, whereas in 125 mm × 200 mm strain

Figure 4. Un-welded sheet formed at different lubrication conditions (Ramulu, 2013)

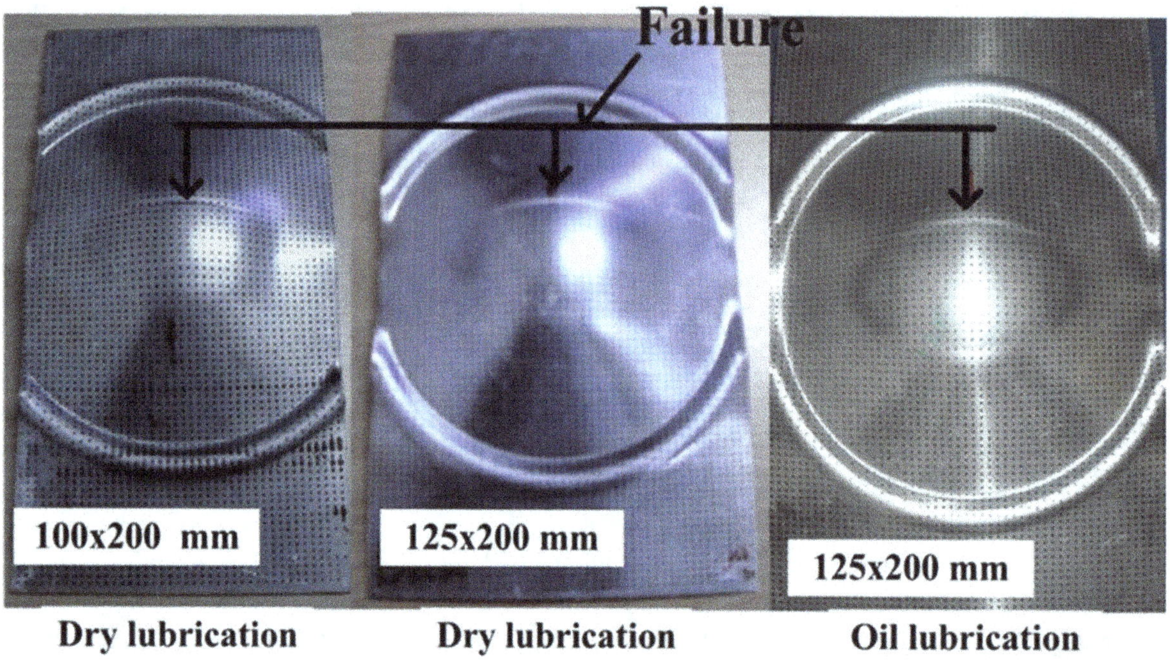

path failure is seen at interface of weld (Figure 4 and 5). Other stretching strain paths exhibit similar failure pattern (Ramulu et al., 2013b).

Weld location and orientation play a significant role in deciding the formability of FSW sheets like in the case of laser-welded TWBs (Ramulu et al., 2015). In the case of longitudinal welds, the weld zone was located at three different positions including one at the geometric center, and at 5 mm and 10 mm offsets. The failure pattern in 100 mm × 200 mm strain path (plane-strain condition) is perpendicular to the weld irrespective of weld locations. However, in 125 mm × 200 mm strain path, failure is seen at the interface of weld in all the locations (Figure 6). While comparing weld orientations, weld at 0° orientation encountered failure normal to the weld in 100 mm × 200 mm strain path. The strain path of 125 mm × 200 mm in 0° weld orientation failed along the weld interface. In 45° orientation, failure is observed at the weld interface, independent of the strain paths deformed. In the case of transverse weld, failure is observed at the weld interface like in 45° weld orientation (Figure 7) (Ramulu, 2013; Ramulu et al., 2015).

Developing FLD provides useful insight about the formability of FSW tailored-sheets and selection of material and process conditions for successful fabrication. As an example, Figure 8 shows the master FLD presenting the effect of tool shoulder diameters (12 mm and 18 mm) on the FLCs of FSW tailored-blanks (Ramulu, 2013; Ramulu et al., 2013b). Forming limit of the un-welded parent sheet and FSW sheets are almost same in near plane-strain strain path, and formability of FSW sheet increases significantly in stretching strain paths improving its overall formability. Ramulu et al. work (Ramulu et al., 2013c) revealed the influence of rotational speed and feed rate on the forming limit of FSW sheets made of AA6061-T6. With increase in the tool rotation speed, from 1300 to 1400 rpm, for a constant feed rate, the forming limit of FSW sheet has improved and with increase in feed rate, from 90 to 100 mm/min, for a constant tool rotation speed, it has decreased (Ramulu et al., 2013c). Forming limit data

Figure 5. FSW sheets made at 1.9 mm plunge depth, 1300 rpm rotational speed, 90 mm/min welding speed using tool with 12 mm shoulder diameter (Ramulu, 2013)

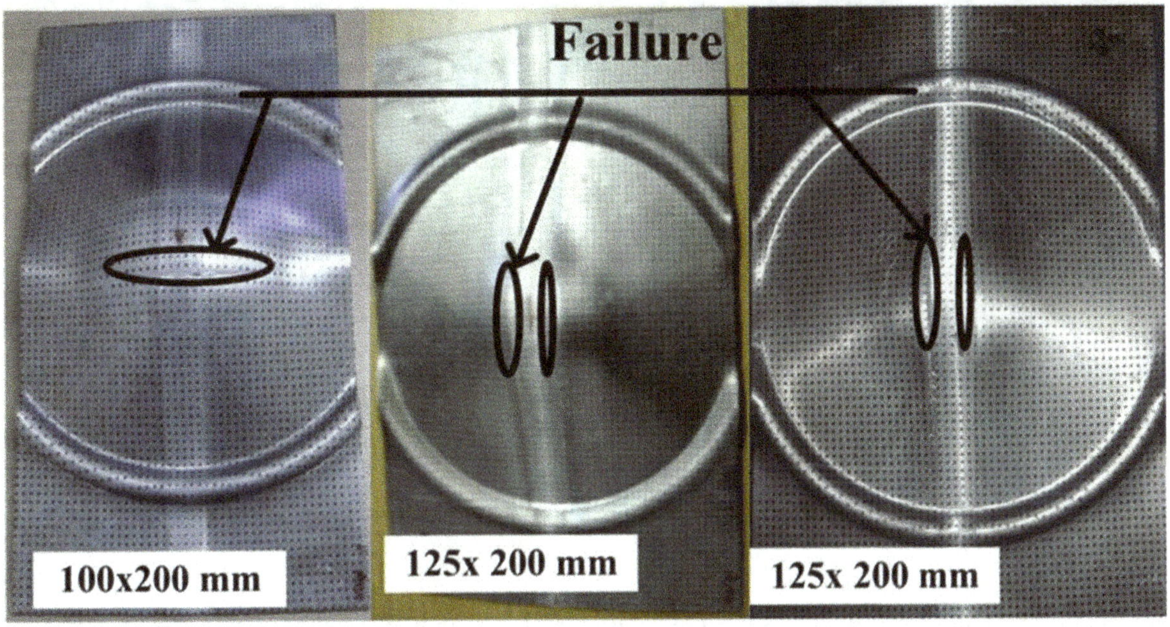

Dry lubrication Dry lubrication Oil lubrication

Figure 6. Failure pattern in FSW TWB with longitudinal weld orientation (Ramulu, 2013)

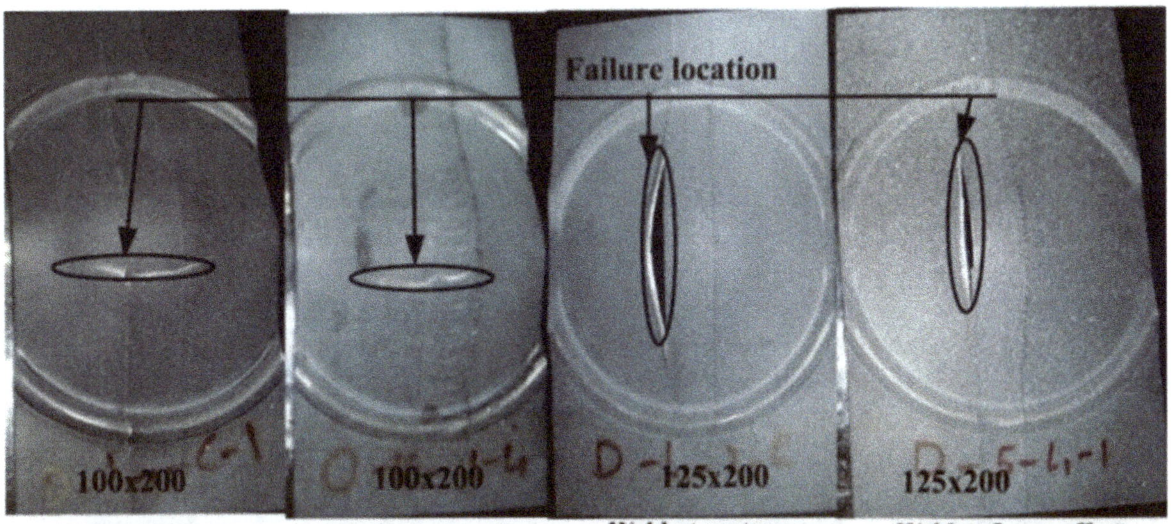

Weld at centre
Dry lubrication

Weld at 10 mm offset
Oil lubrication

Weld at centre
Dry lubrication

Weld at 5 mm offset
Dry lubrication

Figure 7. Failure pattern in FSW TWB with weld at 0°, 45° and 90° weld orientations (Ramulu, 2013)

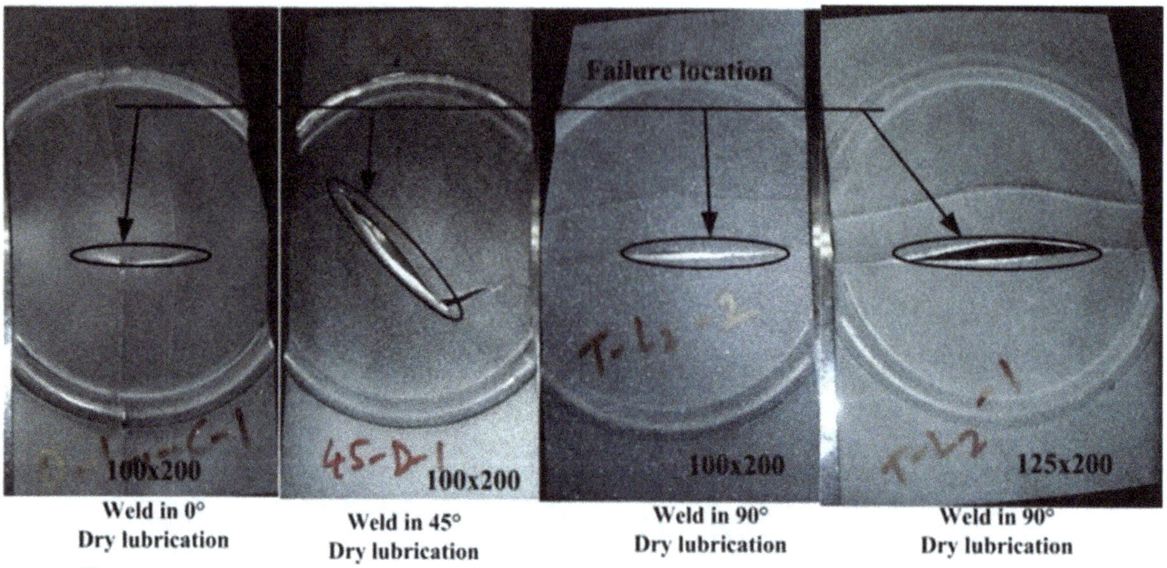

at different weld locations indicates weld at center and a longitudinal weld are preferable (Figure 9 a, b) (Ramulu et al., 2015). The experimental data of Miles et al. (2004, 2005) reveal decreasing or same forming limit in bi-axial stretching side of FLD as compared to plane strain in case of FSW tailored-sheets made of 5182-O, 5754-O, and 6022-T4 sheets for similar and dissimilar combinations. They had equivalent or better formability when compared to that of base sheet depending on the quality of base sheet. Kesharwani et al. (2015) proposed an optimization procedure to enhance the strength and ductility, while reducing the surface roughness and energy consumption of FSW tailored-sheet made of AA5754-H22 and AA5052-H32 materials. They have also proposed to use modified conical tractrix die to improve the deep drawability of FSW sheets by 27%.

The effect of external weld flash on the plane strain formability of FSW sheets made of AA6061T6 and AA5052H32 is negligible (Kumar et al., 2013). They also presented that increasing the traverse speed or decreasing the plunge depth decreases the flash height, while increase in tool rotational speed or shoulder diameter increases the flash height. Aktarer et al. (2019) presented formability data of tailored-blanks made by FSW of IF-steel and AA6061 sheets. They found a drastic decrease in elongation of joint to about 4% as compared to parent sheets due to the formation of intermetallic compound. Erichsen index of the joint is in between that of parent sheets. Reduction in forming limit was also identified for FSW sheets made of low carbon steel with dissimilar thicknesses (Habibi et al., 2018). Miles et al. (2006) showed FSW tailored sheets made of Dual Phase (DP 590) steel with 1.6 and 1 mm thicknesses had better formability as compared to laser-welded sheets by about 20% indicating the possibility of using it in fabricating TWB. Later, Miles et al. (2009) showed the possibility of joining DP 590 steel, Transformation Induced Plasticity (TRIP) 590 steel and DP 980 steels by FSW at various process conditions. Kim et al. (2010) presented that (i) strength and ductility of FSW TWBs made of 6111-T4 sheets are lower than that of base sheets, and (ii) strength and ductility improved slightly in the order of RD||RD, TD||RD and TD||TD. Deep drawability of FSW tailored-sheets made of AA5182-H111 and AA6016-T4 indicated the importance of sheet quality (similar or dissimilar combinations),

Figure 8. Influence of shoulder diameter on FLC of FSW tailored sheets (Ramulu, 2013)

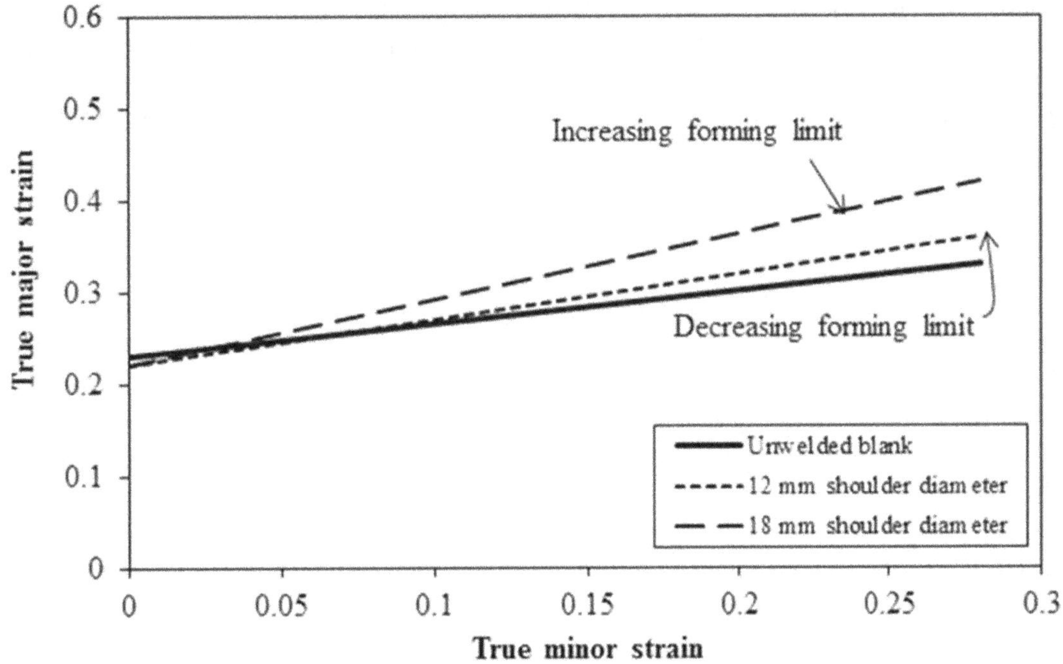

matching of rolling directions (LD vs TD), and initial size of the sheet (Leitao et al., 2009). FLC data of FSW TWBs made of 1.2 mm and 1.4 mm thick stainless steel sheets reveal that (i) transverse weld orientation is the best in uniaxial tension mode; while in plane strain mode of deformation, longitudinal weld orientation is acceptable. In biaxial tension, changing the weld orientation has no effect on formability (Moayedi et al., 2020). Spinning of FSW welded Al alloy tubes for hydroforming applications was attempted by Wang et al. (2014). Formability of spun FSW tubes by hydraulic bulge tests reveal the

Figure 9. Forming limit comparison of FSW blanks at (a) different weld locations, (b) different weld orientations (Ramulu, 2013)

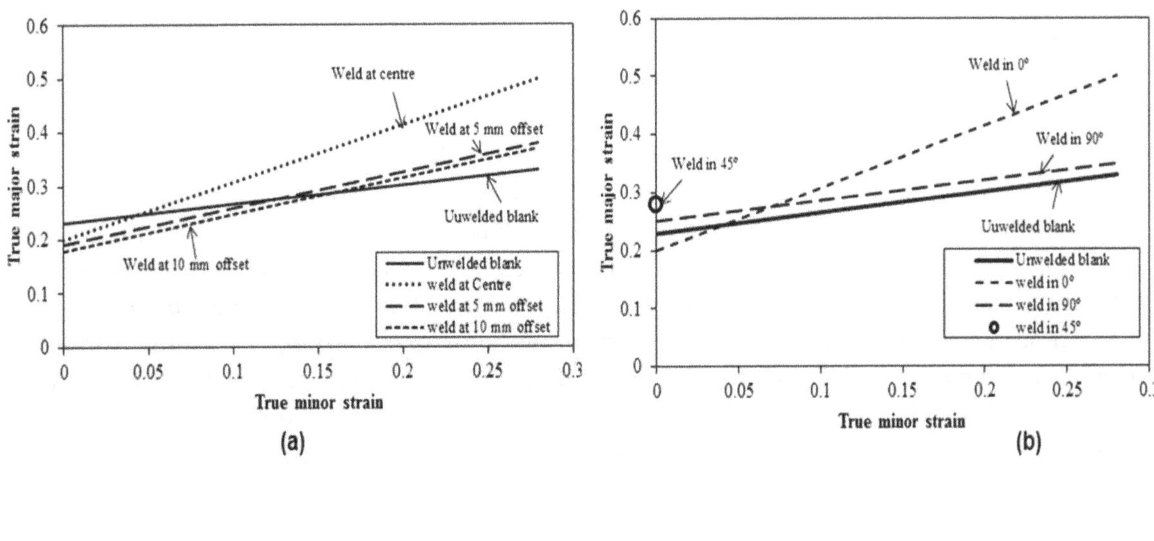

grains of the FSW tube decrease with increasing thickness reduction, and the effect of grain refinement is more prominent for base metal compared to that of weld. Yuan et al. (2012) fabricated a circular tubes by coiling a rolling sheet and FSW, and hydraulic bulge formability tests revealed that the tube exhibits high formability and shows almost two times higher elongation compared to extruded tubes. Instead of coiling and FSW, in the past few years, Agrawal et al. (2017, 2019, 2021) and Agrawal and Narayanan (2020a) attempted series of FSP trials on Aluminum tubes for end forming applications. There are several merits of using FSW Al tubes in place of raw tubes depending on the end forming operations and FSP parameters. After FSW/FSP, solution treatment of joints made of 2024-O alloys showed better formability compared to the as-welded joints, with former showing 1.5 times bulge height as compared to that of as-welded (Pang et al., 2019). Preheating treatments and temperatures also improve formability of FSW sheets made of AA2219 alloy (Liu et al., 2018). Studies on single point incremental forming of FSW sheets made of dissimilar alloys provide crucial insights into formability as compared to base sheets (Tayebi et al., 2018). Gas tungsten inert gas welding and friction stir vibration welding are other methods of fabricating TWBs, and their mechanical performance is inferior when compared with TWBs made by conventional FSW (Bagheri et al., 2021; Karami et al., 2021). Post-FSW forming operations such as cold rolling deteriorates the formability of FSW sheets made of AA5754-H114 as compared to base sheets, and formability decreases significantly with increase in thickness reduction during rolling (Casalino et al., 2018).

OTHER WELDING METHODS FOR FABRICATING TWBS

Mash seam welding, upset welding, flash welding, plasma arc welding are other methods of fabricating TWB, however, implemented restrictedly. Mash seam welding is a resistance welding process in which rotating electrodes are used to crush/mash the ends of the sheets during joining. Merits include best productivity (in terms of welding speed), lower equipment costs compared to laser welding, and no need of high-precision cutting of sheets before welding as observed in laser welding. Demerits include poor weld quality, and inferior weld performance. Plasma arc welding has the advantage of low-cost welding system, however, faces the problem of low welding speed, high heat input, and poor weld performance (Uchihara and Fukui, 2006). Laser welding, mash seam welding and plasma arc welding of series of high–strength steels indicate superiority of laser welding over other methods due to presence of narrow weld, and small HAZ softening width. As a result, the formability is better in laser-welded sheets (Uchihara and Fukui, 2006). Uchihara and Fukui (2002) also concluded that laser welding is the best when formability is considered because of its small heat input, and it is the only welding process that can be adapted to 780 MPa high strength steels. They are reported unacceptable fatigue performance of mash seam welded tailored-blanks as compared to laser welded and plasma arc welded sheets.

Mash seam welding has the merit of generating low hardness in weld region as compared to laser welding (Min and Kang, 2000). Flash welding, which is another resistance welding process, is also used to fabricate TWB of steel sheets, but with broad weld zone (Min and Kang, 2000). Upset welding is similar to the flash welding when device is concerned. It is meant for a joint of small cross-section, acceptable process in respect of the working environment, and repairing facility as there is little sparking. It is an economical method too. The best quality of upset welding is its productivity as the welding time is about 1.0±1.5 s only (Min et al., 2000). Ono et al. (2004) indicates loss of ductility of mash seam welds as compared to laser welds for two different high-strength steel sheets. The fatigue strength of the mash seam joints is also lower than that of the laser-welded joints due to stress concentration in the overlap region. Wang et al. (1995) performed axial impact tests and bending tests on laser welded, mash seam welding and resistance spot welded sheets made of aluminum-killed drawing-quality (AKDQ) and high strength low-alloy (HSLA) steels. Results reveal that both laser welded and mash seam welded sheets perform equally in terms of the maximum energy absorption, peak load, and maximum crush distance. However, performance is not acceptable in case of resistance spot welded sheets. There are mash seam welding parameters such as weld current, sheet overlap, weld force, with or without planishing that can be optimized for better formability of mash seam welded parts, and at par with laser welded sheets (Bhatt et al., 1995). Use of mash seam welded automotive bumper beam as TWB reduced weight and cost by about 10.5% (Shin et al., 2005). Micro plasma transfer arc is another method by which TWBs can be made (Dwibedi et al., 2018).

Since light-weight metal alloys (LWMA) such as Aluminium, Magnesium have found a place as substitute for steel, further research is need to lower the cost of LWMA presently used for making body parts or new development methods are required for using less expensive alloys. Maintaining the quality of joining, the joining methods also need improvement. Disposal after service of LWMA is of great concern as it accounts for more than half of scrap material value. In the recycling of both aluminum and magnesium, recycling processing costs are a small portion of the prime reduction costs. The energy consumption per ton of recycled aluminum ingot is about 2 kWh/kg, which is about 5% of the cost of mining, alumina refining, and aluminum reduction. The capital costs of a secondary smelter and upstream scrap processing are about 500 USD per annual ton of aluminum, which is again 5% of the costs of a prime smelter, power plant, and alumina refinery. The cost of re-melting scrap aluminum or magnesium is lower than that of electric-arc-furnace production of steel from scrap. This is mainly due to the lower melting temperatures of the light metals than steel. This comparison is especially attractive for light metals on a per-volume basis (Gesing, 2004; Padmanabhan et al., 2011).

MODELLING DEFORMATION DURING FABRICATION AND FORMING OF TWB

Understanding heat transfer and material flow during FSW is inevitable, especially in process design and selection of materials. The temperature evolution in the work piece will help to understand the material flow in the work piece. This can be achieved through development of suitable numerical simulation strategies and subsequent experimental validation. The following are some examples in this context.

Song and Kovacevic (2003) successfully developed an integrated transient 3D heat transfer numerical model with moving coordinate systems for FSW process. The FSW process initially divided into three various process periods, (1) penetration period, (2) weld period and (3) tool pulling out period. One of the assumptions in the model was no heat flow into the work piece if the local temperature reaches the material melting point. A heat transfer control equation is developed with moving coordinate systems. The heat generation model consists of (1) heat input from tool shoulder, (2) heat input from tool pin. Initial boundary conditions were established for tool shoulder/work piece interface, tool pin/ work piece interface, convection boundary conditions, symmetric boundary conditions, tool penetration and extraction and initial condition to start with. The finite difference method was used to numerically solve heat transfer control equations supported with the boundary conditions. A non-uniform mesh has been considered for the evaluation. The explicit central difference scheme was used to discretize the heat transfer control equations. The heat conduction coefficients were assumed. AA6061T6 material properties were considered for the simulations. The numerically obtained temperature profiles were validated with experimental data from thermocouple. The numerically obtained maximum temperature distribution of weld was compared with microstructural morphology. The temperature contour was obtained for penetration period, and weld period. The local thermal cycles influence in the microstructural morphologies in the weld zone. This can be classified as nugget zone, TMAZ, HAZ and base metal zone. The calculated peak temperature in nugget zone is about 820 K, 751-612 K for TMAZ and 612-439 K for HAZ. The numerically obtained temperature profile shows good agreement with experimental data. Similar numerical attempts were made by Schmidt and Hattel (2006), Nandan et al. (2006), Zhang and Zhang (2007), Nandan et al. (2007), Su et al. (2013), Pandian and Kannan (2020).

Selection of constitutive laws during FSW affects the overall prediction accuracy (Kuykendall et al., 2013). The authors have found that changing constitutive laws in FSW can provide variation of about 21% in peak temperature. About 130% and 160% difference in peak strains and strain rates are observed by changing the constitutive laws. Sellars and Tegart and Kocks and Mecking models predict saturation at high strains in the nugget zone near the pin, while Johnson–Cook and Kocks and Mecking models capture strain hardening occurring in TMAZ at a location in the vicinity of pin.

Cho et al. (2005, 2008) to address the accuracy in describing the stress-strain constitutive relationship for stainless steel during FSW introduced a Voce-like saturation limit for state variable (K) in Hart's model. They proposed the evolution of K to follow

$$\frac{\mathrm{D}}{\mathrm{D}t}(K) = h_0\left(1-\frac{K}{K^{sat}}\right)^{n0}\bar{D} \tag{2}$$

where h_0 and n_0 are determined from experiments.

Such a constitutive equation predicted FSW outputs accurately when experiments and simulation data are compared.

Modelling friction during FSW is a crucial task. Several attempts are made in the existing literature. For instance, Zhang (2008) modelled friction during FSW using classical Coulomb friction law and modified Coulomb friction law under lower and higher angular velocity conditions. Coulomb friction law is given by

$$\tau c_{rit} = \mu \mathrm{p} \tag{3}$$

where μ is coefficient of friction and p is the contact pressure.

When $\tau_{crit} = \tau_{eq} = \sqrt{\tau_1^2 + \tau_2^2}$ slip occurs.

In a modified Coulomb friction law, the critical friction stress is evaluated as

$$\tau_{crit} = \min(\mu p, \sigma_s / \sqrt{3}) \tag{4}$$

Zhang observed that in lower angular velocities, both the contact models show negligible difference in the temperature and material flow predictions during simulation of FSW. However, in higher angular velocities, the classical Coulomb friction model fails to complete the simulation because of the increase of the dynamic effect of the welding tool. Recently, Dialami et al. (2017) modeled interface friction during FSW by modified Norton's friction law and compared the thermal and mechanical results with the models using Coulomb friction model. In modified Norton's friction law, both sliding velocity and pressure distribution with respect to the rotation axis are considered. The modified model showed realistic predictions when compared with experimental data.

The term Cellular Automata consists of two terms – Cellular & Automata. Cellular means "cells" be it in any form or size, in multi-dimensional space. Von Neumann was one of the pioneers who represented space with this powerful concept of repetitive finite domains called cells. Cells could potentially take up any form – however from a visualization and computational angle – the square shaped cells are usually preferred. The cells embed within them the length scale aspect. In other words, the overall dimension of the cell encapsulates the underlying physical phenomena at that length scale. This is a very vital and powerful thought. Similar cell like approach in treatment of space is seen in complementary computational methods also viz. Finite Element, Finite Difference.

The second term is the "automata" part. Automation refers to the engine that drives particular physical phenomena to be digitally represented and encapsulated through numerical formulations within the cell and its neighborhood. In other words, the "automation" rules govern the state of the cell (e.g. living, dead) in relation to its neighboring cells.

The earliest applications of the Cellular Automata methodology in materials modelling were by Davies (1995) in recrystallization. Pioneering work done by Gandin and Rappaz (1994) and Rollett and Rabbe (2001) demonstrate the flexibility and adaptability of this technique. Das et al. (2003) addressed the next frontier in materials modelling using Cellular Automata. This was in the area of thermo-mechanical processes, specifically in metal forging and rolling at high temperatures. Shterenlikht and Howard (2004) used a coupled cellular Automata approach to capture the transitional ductile-brittle fracture in steels. In this particular case, the linkage of the cellular automation is done using the material model (UMAT) in the finite element package, ABAQUS. Das (2010) outlined the concept of a designer microstructure using multi-scale Cellular Automata Finite Element framework. The explicit example was to use images of single and dual phase microstructures in steel and convert these into cellular structures. Rauch and Madej (2010) also present a similar concept. They present a digital representation of the microstructure using automatic image processing. Here algorithms extract the underlying information of grain size and volume fractions to develop the cells. This effort of creating a digital framework is gaining traction in the recent years as seen in the efforts by Madej et al. (2013, 2007) and Spytkowski et al. (2009).

Patel et al. (2012) and Saluja et al. (2012) applied the digital representation for the first time in FSW. Here in the evolution of the welded zone and its properties are captured at the cellular level while the macro parameters of the friction process are evaluated at the finite element level. They have predicted the grain size during FSW (Figure 10) and presented a strategy to predict tensile stress-strain curve of FSW sheets with defects in weld zone using the CAFÉ model (Figure 11) (Saluja et al., 2012). The technique also makes use of fuzzy rules to guide the evolution of the cellular states. Later, Valvi et al. (2016) extended the CA model to predict the dislocation density in the stir zone from grain size, and tool selection among various pin profiles was done by Rajpoot et al. (2018) using the same CA model. Samanta et al (2018) extended this technique to study the microstructure evolution for Friction Stir Blind Riveting. Here, the dynamic recrystallization process is captured on the cellular level while the macro thermos-mechanical loading conditions are obtained using a mesh-free Lagrangian particle based smooth particle hydrodynamics method.

Modeling of forming of TWB is a critical issue because of the presence of weld region, which has relatively different mechanical properties as that of base sheets. Forming behavior of TWB can be mod-

Figure 10. Grain size predictions from café model and experimental data comparison for various FSW cases (cases (a) to (d)) (with permission from Saluja et al., 2012; Elsevier)

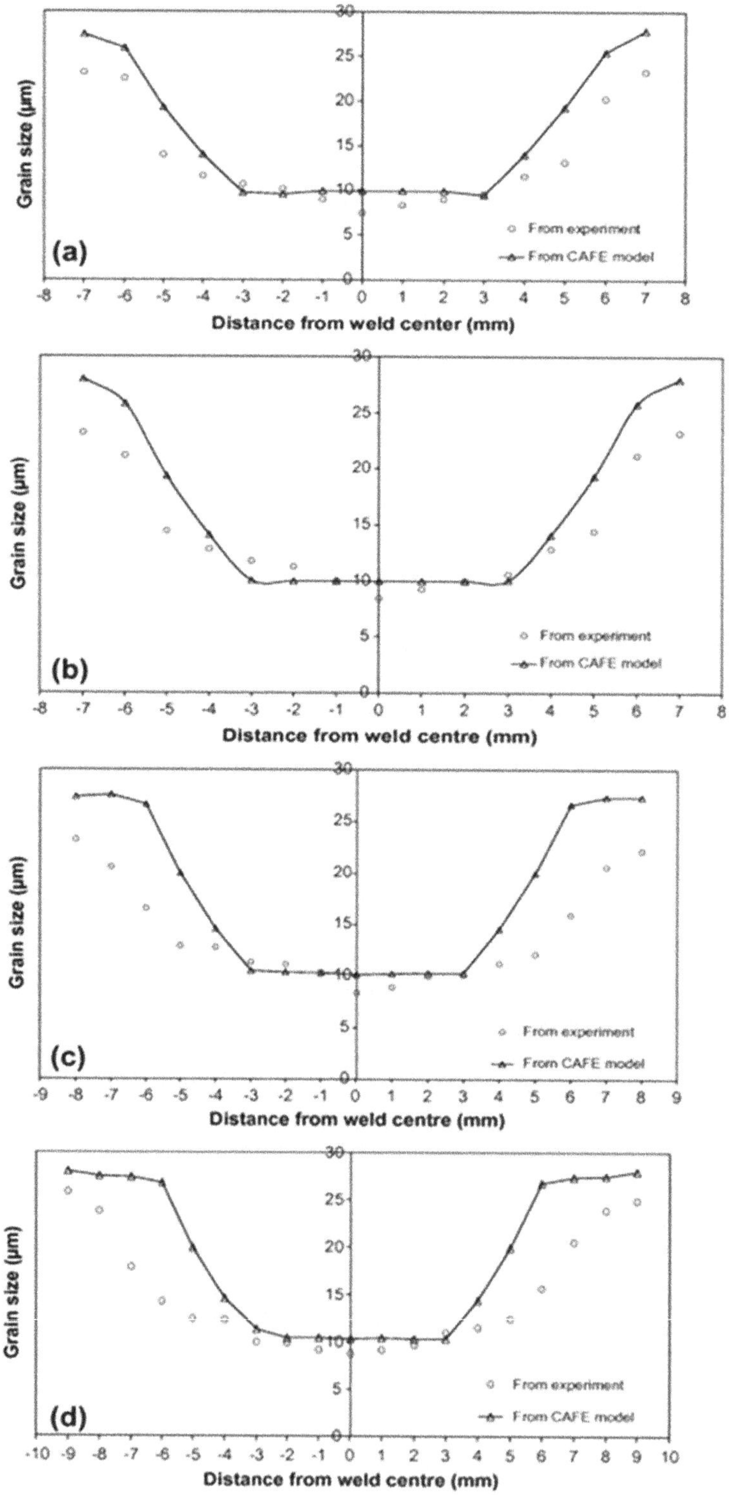

Figure 11. (a) Schematic description of cafe model used for tensile test simulation of FSW sheet, (b) comparison of stress–strain behavior from FE simulation and from simulation with CA cells containing defects and without defects (with permission from Saluja et al., 2012; Elsevier)

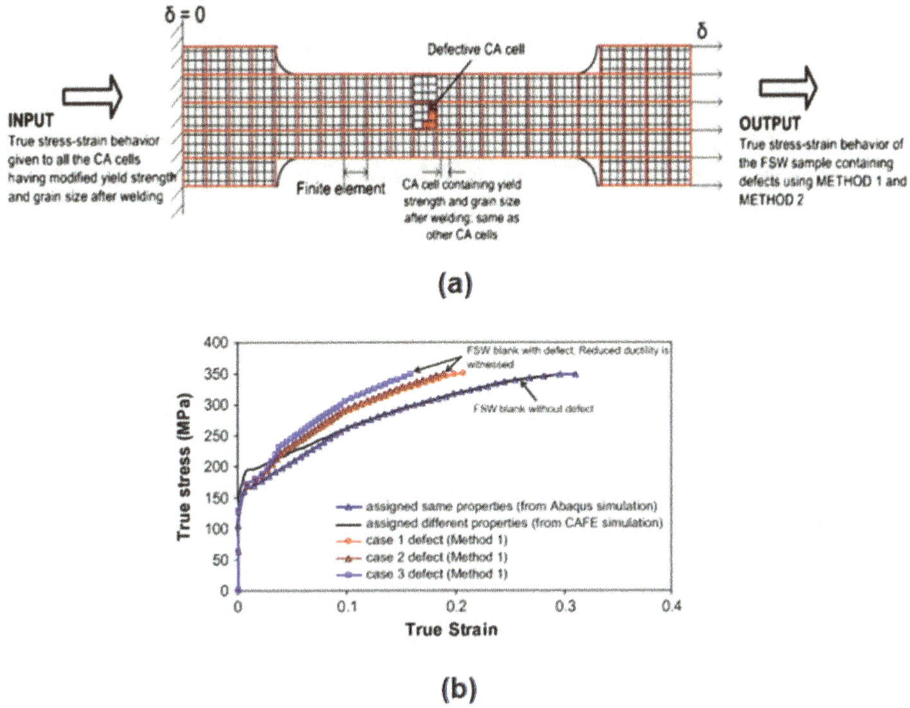

elled by two methods – (i) by incorporating the mechanical properties of the weld zone separately like base metals, (ii) by assuming weld region as weld line without considering properties. The constitutive behavior of weld zone is obtained by conducting tensile tests on sub-size specimens with longitudinal weld by following rule of mixtures method. These assumptions (weld line or zone) followed in modeling TWB play a vital role in predicting the forming behavior. The following section highlights the modeling aspects of forming of TWB in this direction. In addition, the methodology to evaluate weld mechanical properties is also described.

Several research groups focused on modeling TWBs without incorporating weld properties, i.e., weld line assumption is followed. The reasons behind this assumption are, (1) evaluating weld properties is difficult as laser welded blanks, which is widely used, have weld width of the order of 2 mm, and (2) weld line assumption can predict the forming behavior of TWB with acceptable accuracy, as the weld region occupies lesser area in the entire TWB part produced by laser welding such as door inner, fender etc. Only few weld properties are evaluated, while others are assumed. Though the weld line assumption can predict forming behavior acceptably, the relative change in mechanical properties of the weld region with that of base material, say softer or harder weld zone, will affect the failure location and formability prediction considerably. This issue is not considered if the weld line assumption is followed.

The weld line assumption is followed in available literature (Meinders et al., 2000; Heo et al., 2001) to study the forming behavior of TWB like strain distribution, weld line movement control, deep drawability etc. Some incorporated weld properties during simulations (Dry et al., 2001; Ghoo et al., 1998;

Figure 12. Schematic of load sharing by base metals and weld region in TWB with longitudinal weld

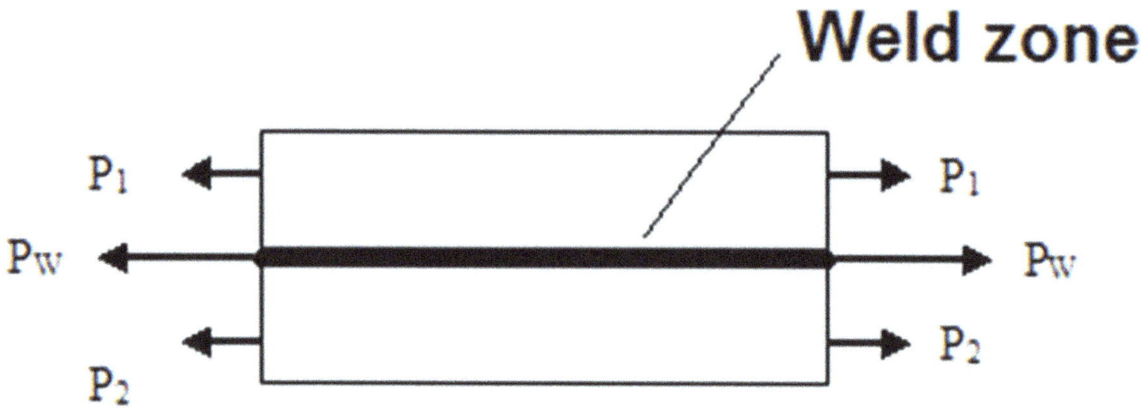

Zhao et al., 2001; Narayanan and Narasimhan, 2008b). For instance, the weld region is incorporated with homogeneous properties to model the forming behavior of TWB (Narayanan & Narasimhan, 2005; Narayanan et al., 2004; Narayanan & Narasimhan, 2006; Narayanan & Narasimhan, 2007). In a few other work including Ghoo et al. (1998), weld properties are included in modeling weld region with HAZ properties separately. Here, HAZ properties are given as triangular variation in the 2-D FE model, which contains linear element, and gradual stepped variation inside the HAZ 3-D model containing rectangular elements. The door inner panel and hemispherical dome stretching are simulated to predict the weld line movement. Recently, Aminzadeh et al. (2020) predicted the deep drawing behavior of laser-made TWB by FE simulations and optimization. In this, initially laser welding simulation was performed in Simufact Welding, and later the model is transferred to Abaqus for deep drawing simulation. In this way, weld zone properties are considered during deep drawing modeling. Dual beam laser welding for TWB applications was numerically modelled by Zhang et al. (2019) and provided new insights into the weld zone characteristics useful for TWB fabrication.

Evaluating the constitutive behavior of weld region is crucial and was performed even before the application of TWBs. However, several applications used arc-welding in which weld region (weld zone + HAZ) is of the order of 15 mm and the weld properties were obtained accurately. In the case of laser-welded TWB, obtaining the properties of a smaller weld zone having less than 2 mm width accurately is relatively difficult.

Several existing work deal with the methodology to evaluate the weld zone properties (Saunders & Wagoner, 1996; Abdullah et al., 2001; Ghoo et al., 2001; Cheng et al., 2005; Liu & Chao, 2004; Narasimhan & Narayanan, 2011). Abdullah et al. (2001) used the rule of mixtures technique with TWB having longitudinal weld (Figure 12) to evaluate the stress-strain behavior of the TWB and the weld region using equation 5.

$$\sigma_W = \frac{P - \left(K_1 \varepsilon_w^{\ n1}\right) A_1 - \left(K_2 \varepsilon_w^{\ n2}\right) A_2}{A_W} \tag{5}$$

Equation (5) defines the stress-strain relation in the weld region. Here, except σ_w and ε_w all other values can be obtained from tensile tests of base metal using sub-size samples. Area of the weld region 'A_w' can be calculated by knowing the area of cross sections of base metals 1 and 2 or from weld microstructure. Results from Abdullah et al. (2001) indicates longitudinal TWB show higher yield strength and lower ductility than that of parent sheet. In Ghoo et al. (2001), it has been shown that effect of tensile specimen sizes and offset weld positions on the stress-strain behavior of weld region is negligible.

The global stress-strain behavior of the welded plate, behavior of individual regions like HAZ, weld nugget can be obtained using iso-stress condition in transverse welded plate (Liu & Chao, 2004) (Figure 13).

Figure 13. Schematic view of transverse welded specimen under tensile loading

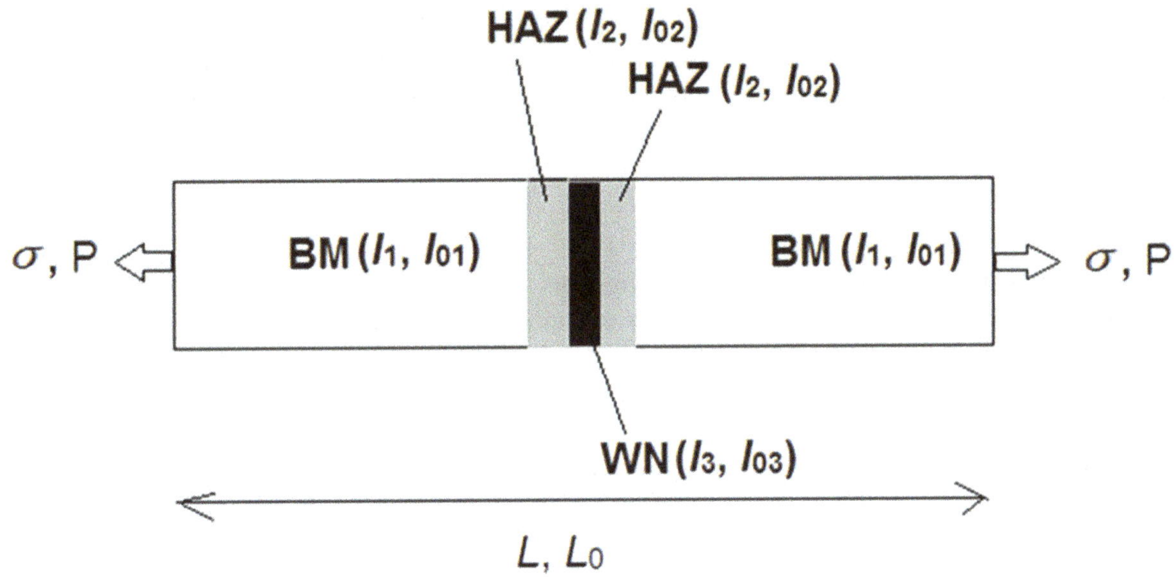

The following relationship describes the global stress-strain behavior of transverse welded plate.

$$\varepsilon = ln\left(\frac{2e^{\left(\frac{\sigma}{K1}\right)^{1/n1}} l_{01} + 2e^{\left(\frac{\sigma}{K2}\right)^{1/n2}} l_{02} + 2e^{\left(\frac{\sigma}{K3}\right)^{1/n3}} l_{03}}{2l_{01} + 2l_{02} + 2l_{03}} \right) \tag{6}$$

The stress-strain response from this model is compared to that of experimental stress-strain behavior of Al alloy 2024-T351 (Liu & Chao, 2004) and stress-strain data and mechanical properties obtained from the model and experiments correlate well. The properties evaluated from TWB with longitudinal weld also match with those from transverse weld.

FRICTION STIR ADDITIVE MANUFACTURING: A ROUTE TO FABRICATE TAILOR-MADE STRUCTURES

Additive manufacturing (AM) is a rapidly growing route to manufacture tailor-made structures. Addition of structures is possible through various sources such as laser based, electron beam based, arc based, deformation based, and other hybrid varieties. AM methods can also be classified based on raw materials such as powder or wire or other raw materials. In AM, layer-by-layer addition of metallic or non-metallic materials results in a three-dimensional structure. There are several review articles in a variety of topics of AM (Wong and Hernandez, 2012; Vaezi et al., 2013; Frazier, 2014; Uriondo et al., 2015; Lewandowski and Seifi, 2016; Bikas et al., 2016; Lee et al., 2017; Lindgren and Lundbäck, 2018; Jiang et al., 2018; Stavropoulos and Foteinopoulos, 2018; Singh et al., 2019; Li et al., 2019; du Plessis et al., 2019).

Deformation based AM such as Friction stir additive manufacturing (FSAM) is a solid-state additive manufacturing method in which the raw materials in the form of sheet or plate or rod are lap joined one above another using friction stir processing. There are several advantages of FSAM as compared to fusion welding and ultrasonic based AM routes (Palanivel et al., 2015a), and certain demerits as well. Some of the advantages are,

- Potential to fabricate large sized components
- Contamination of faying surface is not as vigorous as in ultrasonic based AM
- Potential to join a variety of materials that are difficult by other methods such as fusion welding and ultrasonic based AM
- Process forces and tool geometry (such as shoulder diameter, pin geometry and size) govern formation of joint
- Encouraging reproducibility and metallurgical properties in the joint region
- Tailor-made microstructures and mechanical properties for typical applications

Palanivel et al. (2015a) fabricated metallic structure via FSAM of a Magnesium alloy, whose strength is about 400 MPa and ductility of 17%. Tailor-made properties of the structure were achieved by four layers of sheets, each of 1.7 mm thickness. However, there was material cracking due to thermal residual stresses. Later, Palanivel et al. (2015b) extended FSAM for AA5083 sheets as well. There are other examples highlighting the importance of FSAM for fabricating tailor-made structures of a variety of materials. Some examples include

- AA7075 structure containing nine layers of 5 mm thick plates fabricated by Yuqing et al. (2016)
- Cylindrical structure fabricated by Dilip et al. (2012) with 19 mm diameter and 78 mm height using friction welding, and later used friction deposition to deposit AISI 304 to Mild steel rod
- Manufacture of Ti-6Al-4V structure by TWI (Russell et al., 2008)
- Deposition of Ni–Cr–Mo superalloy onto Ni–Cr–Fe alloy plate by gas tungsten arc process and subsequently FSPed demonstrating the importance of hybrid AM process (Rodelas and Lippold, 2013)
- A recent attempt by Hang et al. (2018) in which a hollow shoulder, through which the depositing materials in the form of solid rod or powder are fed, was used to construct layers through intense stirring

In all the above examples, excellent microstructures and mechanical properties are achieved. Recent reviews on friction stir welding by Padhy et al. (2018) and Bhardwaj et al. (2019) on various aspects are noteworthy.

FABRICATION OF ADHESIVE BONDED SANDWICH STRUCTURES

Adhesive bonding is a suitable technique for joining variety of materials like metals, polymers, ceramics, and combination of these materials. Adhesive bonding technology has substantial features and advantages like flexibility in multi-material component design, energy saving with reduced weight components, cost reduction through repair ability and recyclability in manufacturing industry. In this section, fabrication of adhesive bonded structures with modification of adhesive properties, surface treatments of adherends, environmental concerns, and their mechanical behaviour are discussed.

Adhesive bonded sheets are fabricated by structural adhesives made of hardener and resin at suitable mixing ratio. Maintaining uniform thickness of adhesive layer is a major concern in the fabrication of adhesive bonded sheets for avoiding defects and better performance. At lab scale, adhesive bonded sheets are fabricated by using suitable molds for maintaining uniform thickness as shown in Figure 14. In the case of large-scale fabrication, sophisticated machines are used for the purpose. The adhesive bonded sheets are kept in hot air oven or autoclave for maintaining uniform temperature and pressure for appropriate curing of adhesive. Apart from two parts adhesive, other adhesives like hot melt adhesives, adhesive tapes, and pressure sensitive adhesives are also used for bonding.

Figure 14. Schematic of lab scale fabrication of adhesive bonded sheets

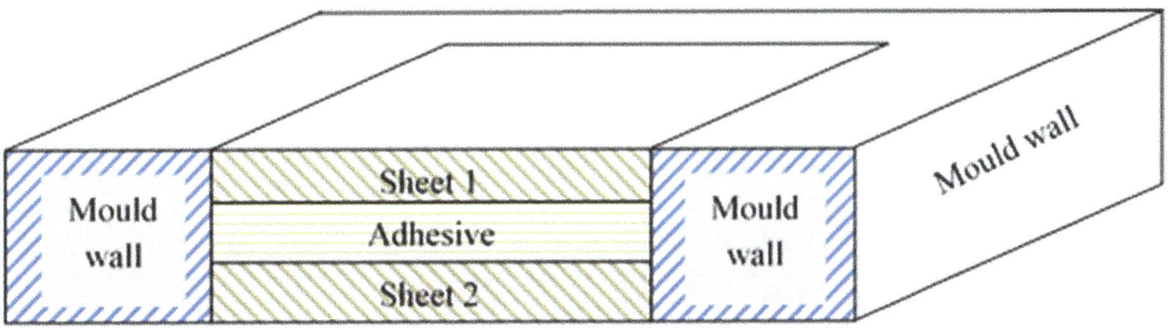

In the fabrication of adhesive bonded blanks, adhesion is established by phenomena like various principles of rheology, mechanical interlocking, electron transfer, electrostatics, adsorption, and diffusion at the interface (Schreiber and Ouhlal, 2003). Annealing above glass transition temperature of assemblies made of polypropylene/linear low-density polyethylene, polystyrene/polyvinyl chloride, polystyrene/polymethyl methacrylate, and polyvinyl chloride/polyvinylidene fluoride exhibits diffusion bonding at the interface and better strength because of acid-base interaction (Schreiber and Ouhlal, 2003). In an investigation carried out by He et al. (2020), it was found that the failure mode of adhesive bonded blanks depends on strength of the adhesive at room temperature. Increase in temperature above glass transition temperature results adhesive-steel interface failure, irrespective of the mechanical properties of the adhesive. In the case of double containment corner joint, increase in support length and the slot depth have a significant effect on reduction in the peak adhesive stresses at the critical points and increase in the overall stiffness of the corner joint (Apalak et al., 1995). Kanani et al. (2020) demonstrated a simple strategy of fabricating adhesive bonded blanks in order to improve lap shear strength. Notches with 2 mm width and 0.5 mm depth have been created on the surface of aluminium alloy and Polyphthalamide plates constituting adhesive bonded single lap joint. The results showed about 13.26% increase in lap shear strength with increase in number of notches in the bond length (3 notches in the bond length of 29 mm). The surface treatments of substrates like mechanical (abrasion and grit blasting), chemical (acid etching), electrochemical (anodizing), coupling agent (silane and sol-gel), and dry surface treatments (laser texturing, plasma-sprayed, and ion beam enhanced deposition of Alumina) (Park et al., 2010) have substantial effect on mechanical performance of adhesive bonded blanks. Alumina blasting method produces better bonding strength regardless of adhesive type (Shakal and Pfeiffer, 1998). Using primers improves adhesion strength and the environmental durability of epoxy adhesive bonds on grit-blasted aluminium substrates (Chabot and Brescia, 1993).

In the case of surface treatments of metallic materials, there is a wide variety of techniques to modify their physiochemical properties, to improve tensile strength, bonding strength, durability corrosion resistance of adhesive-bonded joints. Laser-ablated adhesively bonded joints exhibit a significant improvement in mechanical performance. Table 3 shows comparison of different surface treatments suitable for metallic materials used for fabricating adhesive bonded joints and their performance (Min et al., 2020).

Table 3. Comparison of surface treatments for adhesive bonding (Min et al., 2020)

Pretreatment methods	Substrates	Nature of treatment	Surface roughness	Bonding strength	Durability	Efficiency
Abrasion and solvent wipe	C45 steel, AA6061-T6	Contamination removal	Macro	Adequate	Poor	Low
Grit blasting	Mild steel	Contamination removal	Macro	High	-	High
Acid etch	AA1050	Etch	Micro	Low	-	Low
Alkaline etch	AA1050	Etch	Micro	Adequate	-	Low
Boiling water immersion	AA6061-T6	Hydration	Micro	High	Excellent	Low
Silane gel	AZ31 AA1050-H14	Coupling and oxidation	None	High	Excellent	Low
Plasma treatment	AA6061-T6 AA7075	Ablation and activation	Micro	High	Excellent	High
Laser ablation treatment	AA6061-T6, AA6016, DP500, AISI304	Ablation and oxidation	Macro/Micro	High	Excellent	High

Reinforcing with wires and fibres, and filling with different powder particles in the adhesive have significant influence in improving mechanical performance of adhesive bonded blanks. Research work performed by Khalili et al. (2008), Zhang et al. (2011) in case of flat structures, and work done by Albiez et al. (2019) and Braiek et al. (2018) on metallic tubes are worth referring.

Formability of adhesive bonded sandwich sheets is presented in the section. During forming of aluminium/polypropylene/aluminium sandwich sheets, springback occurs and this could be reduced by forming at elevated temperature. The elevated temperature of core material decreases strength resulting in wrinkling (Weiss et al., 2007). Aghchai et al. (2008) found that the formability of a low formable constituent improves when bonded with high formable material. Varying core thickness of sandwich sheets (Al/PP/Al) decreases limiting drawing ratio (Parsa et al., 2010), and forming limit of Al/PE/Al sandwich structure improves with increase in interfacial strength (Liu et al., 2012). During deep drawing, wrinkling could be reduced by increasing blank holding force (Morovvati et al., 2010, 2011). Kee et al. (2009) demonstrated the fabrication of automotive hood part made of sandwich sheets, which is 65% lighter than the steel sheet and 30% lighter than an aluminum alloy without compromising the mechanical performance. Similarly, in the formability evaluation of AA/PP/AA sandwich sheets for automotive hood, it was suggested that the sandwich sheet could reduce the weight and maintain the flexural rigidity as compared to the steel sheet (Kim et al., (2010)). D'Urso and Maccarini (2012) found that foam density influences formability significantly and suggested to use low-density foam. Formability changes are recorded in other articles as well (Liu and Xue, 2013; Sokolovan et al., 2012; Oya et al., 2010). Satheeshkumar and Ganesh Narayanan (2014a, 2014b, 2015a, 2015b, 2016) demonstrated the strategy of improving formability of adhesive bonded steel sheets by varying hardener to resin (H/R) ratio, by filling carbon black particle in adhesive, and by reinforcing metal wire reinforcement like Cu and Al in the epoxy adhesive. It was suggested that the forming limit strain and deep drawability of adhesive bonded sheets could be improved with increase in (H/R) ratio of adhesive, by filling suitable amount of fillers in the adhesive and by reinforcing suitable level of high ductile metal wire in the low ductile adhesive. Satheeshkumar et al. (2015) demonstrated that springback of adhesive bonded steel sheets can be controlled by modifying hardener to resin (H/R) ratios of adhesive.

Adhesive bonding technology and bonded blanks play a vital role in green and sustainable manufacturing processes. The features of adhesive bonded blanks like flexibility in multi-material design, lightweight, repair ability, recyclability, reusability, design for manufacturing assembly and disassembly help sustainable production and cost reduction in the shop floor. Guo et al. (2019) developed a green adhesive based on renewable copolymerisation of β-myrcene with SO_2 at first time. This green adhesive exhibits higher adhesion even on wet wood surfaces than other materials. Oliveira et al. (2020) developed a bio-polyurethane (bio-PU) adhesive made from castor oil plant to fabricate eco-friendly sandwich panels made from recycled plastic waste. Nakamura (2009) showed that the fracture strain of green composites at high temperature increases significantly as compared to that of room temperature. Satheeshkumar and Ganesh Narayanan (2014a) suggested a green concept with adhesive bonded sheets by varying adhesive properties. The resin rich formulation may be used when strength is the desired quantity, and hardener rich formulation may be used when formability is the requirement. It was also suggested to use ductile adhesives between two metal sheets to improve formability and to utilize the advantages of adhesive. Later, adhesive can be removed for using the remaining single sheet components for automotive applications. Shailesh et al. (2019) fabricated spur gear made of epoxy adhesive with different hardener to resin ratio with aim of checking repair ability and reusability and demonstrated the performance. It was found that there is not much difference in performance of actual and repaired gears.

Figure 15. Schematic of (a) Self-pierce riveting, and (b) Mechanical clinching (not to scale)

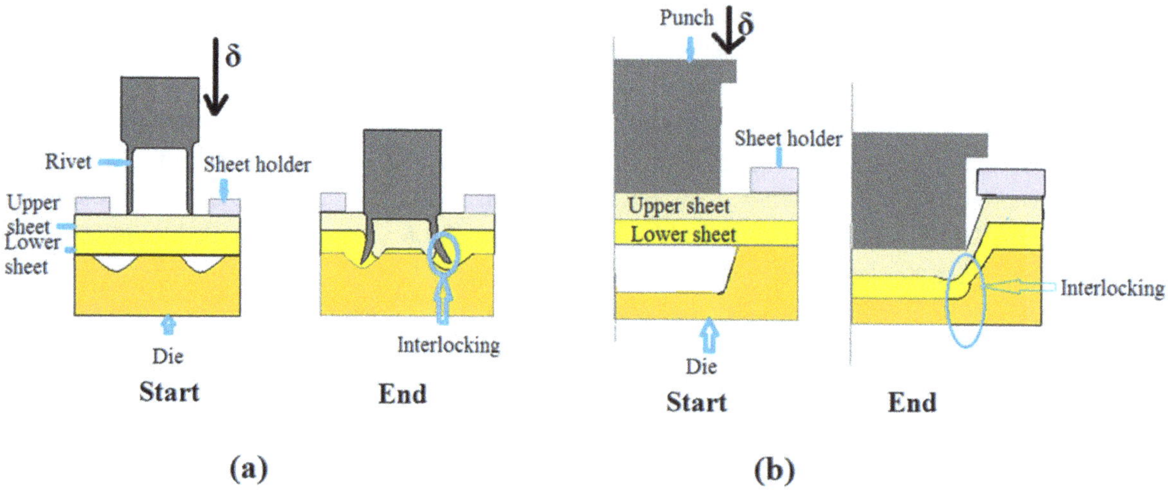

FABRICATION OF TAILOR-MADE STRUCTURES BY MECHANICAL JOINING

Fabrication of metallic structures by mechanical joining is done by plastic deformation. Here, interfacial pressure and interlocking are accomplished by plastic deformation without thermal effects during joining. Joining of a tube to a sheet and a tube to a tube (Agrawal and Narayanan, 2017; Agrawal and Narayanan, 2020b), self-pierce riveting (SPR), mechanical clinching, electromagnetic joining etc. are some examples. These are used predominantly in mechanical and civil engineering structures. A brief account of such processes is presented here.

Joining dissimilar materials such as steel and aluminium having different melting points can be performed successfully with processes such as SPR and mechanical clinching. In these processes, the sheets are mechanically interlocked without metallurgical bonding by controlled plastic deformation. In SPR, the tubular rivet is driven through the upper sheet and the rivet end is flared in the lower sheet to create an interlock (Figure 15a). Here the rivet must have sufficient strength to penetrate the upper sheet, and sufficient ductility for flaring. This process is mainly used to join aluminum sheets, but is also applied to coated steel, copper, sandwich materials, and plastics. The interlock in SPR depends on the thickness of the sheets, flow stress and ductility of sheets and the rivet, and shape of tools. The die bottom has a ring groove that helps in rivet flaring at its end. Rivets are made from high strength boron steel, with a hardness between 400 and 530 VHN, often plated with zinc to prevent corrosion. Lennon et al. (1999) carried out shear tests on four types of mechanical connections such as SPR, press joining, pop riveting and self-tapping screws, with sheet thicknesses of 1.0, 1.2, 1.6 and 2.0 mm. The joints made by SPR possess higher peak load and better ductility when compared to others.

In mechanical clinching operation (Figure 15b), the sheets are bulged with the punch, and then compressed with the punch and the die. An interlock is formed between the upper and lower sheets by controlled plastic deformation of the sheets, which creates a round bottom. The ring groove like structure in the die assists the interlock formation. Here thinning of sheets, which finally results in fracture, should be avoided. Mechanical clinching is useful for joining zinc-coated steel sheets (Varis and Lepisto,

Figure 16. Schematic of joining of metallic structures using tube end forming (a) joining of a tube to a sheet, and (b) joining of a tube to a tube

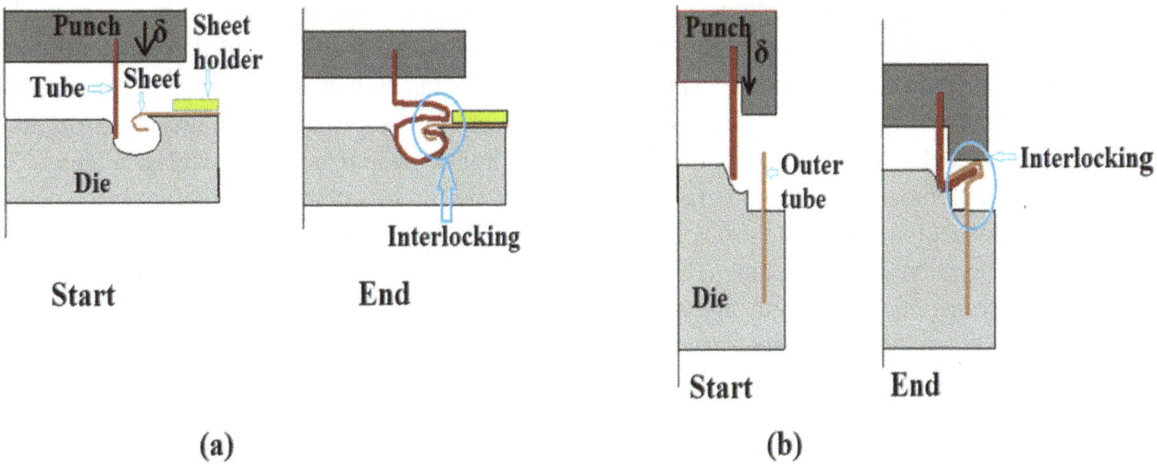

2003), which cause problems if resistance spot welding is used. Round tools are generally employed for mechanical clinching to obtain stronger joints.

Alves et al. (2011) presents flexible joining of sheet panels to thin-walled tubular profiles by means of a two-stage tube end forming processes. They compared the joint strength between the lever and fulcrum of a safety part from an automotive braking system, which is made through traditional MIG/MAG welding and through the method proposed by the authors. Torsional test has been conducted on both joints and identical performance was seen. Tube end forming has been utilized to join a tube to a sheet and a tube to a tube (Agrawal and Narayanan, 2017; Agrawal and Narayanan, 2020b). A specially designed die-punch set-up has been used for fabrication of the joints. The start and end of the processes has been shown in Figure 16 (a, b). In the tube-sheet joint, the tube travels through a die groove and

Figure 17. Experimental load evolution curves for the joints made using different joining technologies

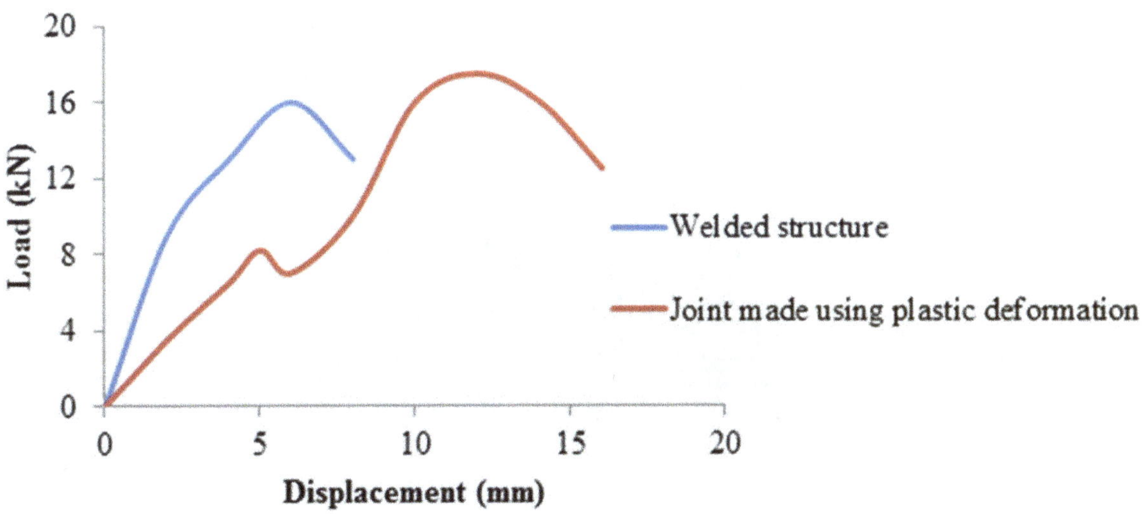

enters into the bent part of the sheet. Eventually a neck is formed just above the sheet and a tube-sheet joint is obtained. The interlocked region can be seen in Figure 16a. In the tube-tube joint, the inner tube travels through the die circular arc radius, while the outer tube touches the projected height of the punch. An outward bulge is formed in the unsupported region of the outer tube. Upon further displacement, the inner tube enters into the bulged region of the outer tube. When both inner and outer tubes are compressed simultaneously, a tube-tube joint is obtained by interlocking mechanism (Figure 16b). Agrawal and Narayanan (2018) evaluated the pull-out strength of joints made using end forming (Figure 17) and compared with tube-sheet joint obtained through gas welding. It is observed that joint made using mechanical deformation is better in terms of peak load and displacement to failure.

Joining by electromagnetic forming (Psyk et al., 2011), die-less forming, mechanical and hydraulic crimping and by incremental forming are few other in the category. In hydraulic crimping, an elastomer has been used for application of necessary fluid pressure in such a way that the tube deforms inside the groove and a joint is obtained between tube and mandrel.

Recently, Narayanan (2018a) proposed a mechanical joining method called as joining by accumulation, which can be used to join metals and non-metals. The process can be accomplished in a lathe machine. Figure 18 shows various stages of joint formation. Here there is no need to deform the sheet, only rod gets plastically deformed. Joining steel sheet and aluminium rod (for example as attempted in Figure 18) is not possible by conventional welding methods without severe distortion. Figure 18f shows joining performed between a Nylon rod and Steel sheet by this method (Narayanan, 2018b).

Figure 18. Stages in joining by material accumulation (Narayanan, 2018a; Narayanan, 2018b)

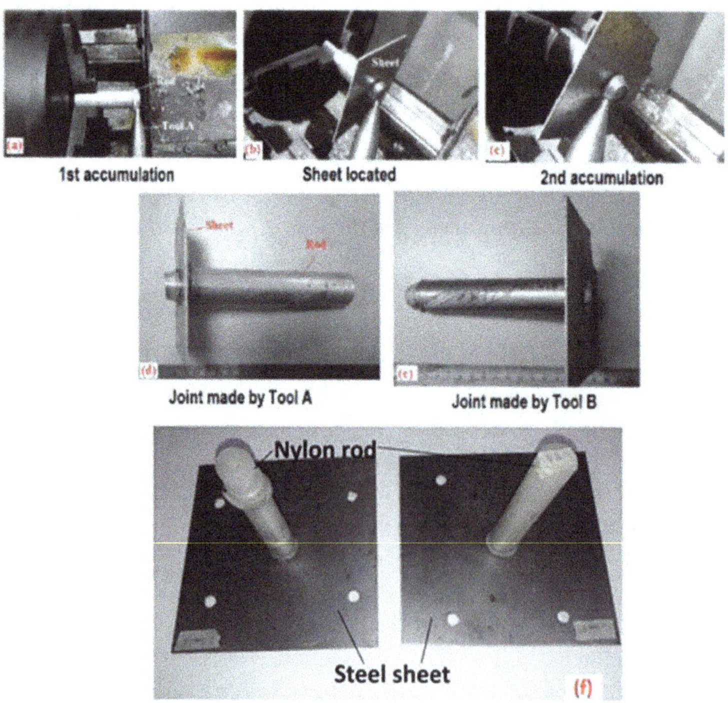

To summarize the section, joints made using plastic deformation retain strength at par with the traditionally made joints. Intelligent application of plastic deformation in joining provides useful solution for meeting high productivity and low cost requirements. In this context, methods of joining using plastic deformation should be made more economical, versatile and environmental friendly. Further, joining using plastic deformation require certain amount of ductility for the materials being joined. However, the joining of newer high strength materials having reduced ductility is also mandatory. In this direction, laser heating of sheets prior to joining operations can be attempted.

SUMMARY

In the chapter, fabrication of tailor-made metallic structures through laser welding, friction stir welding, friction stir additive manufacturing, adhesive bonding and joining by plastic deformation is described with research case studies. A brief account of modeling of deformation during the fabrication of tailor-welded blanks is presented. In the case of laser-welded and friction stir welded tailored blanks, optimizing weld conditions is crucial to enhance the mechanical performance. In case of friction stir additive manufacturing, clamping system and adding strategy need appropriate design methodology. Adhesive bonding to fabricate tailor-made structures including sandwich structures provide flexibility in terms of material processing and strength requirements. An important advantage of joining by plastic deformation is material hardening during joining occurs improving the overall joint strength. In addition to this, dissimilar materials are joined without heating and fume generation maintaining sustainability. A few methods are discussed briefly to present their importance. Finally, there is lot of scope for modeling of deformation during fabrication in terms of process and materials design.

Finally, selection of joining methods depends on applications and material response. Moreover, welds should satisfy criteria such as uniform weld configuration and closer to that of base materials, superior weld strength as compared to the base metal, and no loss in the press formability as compared to the base metal formability. Laser welding and FSW are attempted when permanent joints are required during the manufacture of TMB, which is the case always. In this, laser welding is preferred for Steel sheets, and when the application necessitates reduced weld size. Fiber laser welding is found to produce TWBs that are better than those produced by other laser sources. FSW is advantageous for non-ferrous sheets, and it also enhances the ductility and formability of sheets after joining. Adhesive bonding and joining by plastic deformation do not introduce metallurgical bonding, and the strength depends only on adhesive quality and strain hardening during plastic deformation. Adhesive bonding is to some extent repairable, however, this is not the case in joining by plastic deformation. When functionally grades materials are required, FSAM and other additive manufacturing methods are implemented.

REFERENCES

Abd El Salam Shakal, M., & Pfeiffer, P. (1998). Effect of different surface preparations on the tensile strength of adhesively bonded metal joints. *Journal of Adhesion Science and Technology*, *12*(3), 349–361. doi:10.1163/156856198X00920

Abdullah, K., Wild, P. M., Jeswiet, J. J., & Ghasempoor, A. (2001). Tensile testing for weld deformation properties in similar gage tailor welded blanks using the rule of mixtures. *Journal of Materials Processing Technology*, *112*(1), 91–97. doi:10.1016/S0924-0136(01)00555-6

Aghchai, A. J., Shakeri, M., & Mollaei-Dariani, B. (2008). Theoretical and experimental formability study of two-layer metallic sheet (Al1100/St12). *Proceedings of the Institution of Mechanical Engineers. Part B, Journal of Engineering Manufacture*, *222*(9), 1131–1138. doi:10.1243/09544054JEM1140

Agrawal, A. K., & Narayanan, R. G. (2017). Joining of a tube to a sheet through end curling. *Journal of Materials Processing Technology*, *246*, 291–304. doi:10.1016/j.jmatprotec.2017.03.035

Agrawal, A. K., & Narayanan, R. G. (2018). Pull-out tests on tube to sheet joints fabricated by endforming. *Journal of Constructional Steel Research*, *144*, 186–197. doi:10.1016/j.jcsr.2018.01.027

Agrawal, A. K., & Narayanan, R. G. (2020a). Friction stir processing of AA6063-T6 tubes and end forming characterization at varying tool pin profiles. *Materials Performance and Characterization*, *9*(1), 497–517. doi:10.1520/MPC20190248

Agrawal, A. K., & Narayanan, R. G. (2020b). Experimental and numerical studies on joining steel tubes by end forming. *Journal of Constructional Steel Research*, *167*, 105792. doi:10.1016/j.jcsr.2019.105792

Agrawal, A. K., Narayanan, R. G., & Kailas, S. V. (2017). End forming behaviour of friction stir processed Al6063-T6 tubes at different tool rotational speeds. *Journal of Strain Analysis for Engineering Design*, *52*(7), 434–449. doi:10.1177/0309324717724662

Agrawal, A. K., Narayanan, R. G., & Kailas, S. V. (2019). Experimental evaluation and prediction of end forming behaviour of friction stir processed Al6063T6 tubes at different tool traverse speeds. *International Journal of Advanced Manufacturing Technology*, *104*(9-12), 3607–3627. doi:10.100700170-019-04068-4

Agrawal, A. K., Narayanan, R. G., & Kailas, S. V. (2021). Formability and instability evaluation of friction stir processed AA6063-T6 tubes during end forming and numerical prediction. *Journal of Materials Engineering and Performance*, *30*(2), 973–993. doi:10.100711665-020-05400-w

Ahmetoglu, M. A., Brouwers, D., Shulkin, L., Taupin, L., Kinzel, G. L., & Altan, T. (1995). Deep drawing of round cups from tailor-welded blanks. *Journal of Materials Processing Technology*, *53*(3-4), 684–696. doi:10.1016/0924-0136(94)01767-U

Aktarer, S. M., Sekban, D. M., Kucukomeroglu, T., & Purcek, G. (2019). Microstructure, mechanical properties and formability of friction stir welded dissimilar materials of IF-steel and 6061 Al alloy. *International Journal of Minerals Metallurgy and Materials*, *26*(6), 722–731. doi:10.100712613-019-1783-z

Albiez, M., Vallée, T., Fricke, H., & Ummenhofer, T. (2019). Adhesively bonded steel tubes—Part I: Experimental investigations. *International Journal of Adhesion and Adhesives*, *90*, 199–210. doi:10.1016/j.ijadhadh.2018.02.005

Alves, L. M., Dias, E. J., & Martins, P. A. (2011). Joining sheet panels to thin-walled tubular profiles by tube end forming. *Journal of Cleaner Production*, *19*(6-7), 712–719. doi:10.1016/j.jclepro.2010.12.014

Aminzadeh, A., Parvizi, A., & Moradi, M. (2020). Multi-objective topology optimization of deep drawing dissimilar tailor laser welded blanks; experimental and finite element investigation. *Optics & Laser Technology*, *125*, 106029. doi:10.1016/j.optlastec.2019.106029

Apalak, M. K., Davies, R., & Apalak, Z. G. (1995). Analysis and design of adhesively bonded double-containment corner joints. *Journal of Adhesion Science and Technology*, *9*(2), 267–293. doi:10.1163/156856195X01166

Assunção, E., Quintino, L., & Miranda, R. (2010). Comparative study of laser welding in tailor blanks for the automotive industry. *International Journal of Advanced Manufacturing Technology*, *49*(1-4), 123–131. doi:10.100700170-009-2385-0

Bagheri, B., Abbasi, M., & Hamzeloo, R. (2021). Comparison of different welding methods on mechanical properties and formability behaviors of tailor welded blanks (TWB) made from AA6061 alloys. Proc. of IMechE, Part C. *Journal of Mechanical Engineering Science*, *235*(12), 2225–2237. doi:10.1177/0954406220952504

Bandyopadhyay, K., Panda, S. K., & Saha, P. (2014). Investigations Into the Influence of Weld Zone on Formability of Fiber Laser-Welded Advanced High Strength Steel. *Journal of Materials Engineering and Performance*, *23*(4), 1465–1479. doi:10.100711665-014-0881-3

Baysore, J. K., Williamson, M. S., Adonyi, Y., & Milian, J. L. (1995). Laser beam welding and formability of tailored blanks. *Welding Research Supplement*, 345-351.

Bhagwan, A. V., Kridli, G. T., & Friedman, P. A. (2002). *Influence of weld characteristics on numerically predicted deformation behavior of aluminum tailor welded blanks* (No. 2002-01-0386). SAE Technical Paper.

Bhardwaj, N., Narayanan, R. G., Dixit, U. S., & Hashmi, M. S. J. (2019). Recent developments in friction stir welding and resulting industrial practices. *Advances in Materials and Processing Technologies*, *5*(3), 461–496. doi:10.1080/2374068X.2019.1631065

Bhaskar, V. V., Narayanan, R. G., & Narasimhan, K. (2004, June). Effect of thickness ratio on formability of tailor welded blanks (TWB). In. AIP Conference Proceedings: Vol. 712. *No. 1* (pp. 863–868). American Institute of Physics. doi:10.1063/1.1766635

Bhatt, K., Eisenmenger, M., & Shi, M. (1995). Formability of Mash Seam Welded Blanks: Effects of Welding Set-Up Conditions. *SAE Transactions*, *104*, 889–895. doi:10.4271/950923

Bikas, H., Stavropoulos, P., & Chryssolouris, G. (2016). Additive manufacturing methods and modelling approaches: A critical review. *International Journal of Advanced Manufacturing Technology*, *83*(1-4), 389–405. doi:10.100700170-015-7576-2

Braiek, S., Khalifa, A. B., Zitoune, R., & Zidi, M. (2018). Experimental and numerical investigation of adhesively bonded±55° filament wound tubular specimens under internal pressure. *Engineering Fracture Mechanics*, *199*, 461–475. doi:10.1016/j.engfracmech.2018.06.009

Buste, A., Lalbin, X., Worswick, M. J., Clarke, J. A., Altshuller, B., Finn, M., & Jain, M. (1999). Prediction of strain distribution in aluminum tailor welded. *Proceedings of the 4th International conference and workshop on Numerical simulation of 3D sheet forming process, NUMISHEET'99.*

Casalino, G., El Mehtedi, M., Forcellese, A., & Simoncini, M. (2018). Effect of Cold Rolling on the Mechanical Properties and Formability of FSWed Sheets in AA5754-H114. *Metals*, *8*(4), 223. doi:10.3390/met8040223

Chabot, K. A., & Brescia, J. A. (1993). Evaluation of primers for adhesively-bonded aircraft repair. *Journal of Adhesion Science and Technology*, *7*(11), 1183–1194. doi:10.1163/156856193X00042

Chan, L. C., Cheng, C. H., Chan, S. M., Lee, T. C., & Chow, C. L. (2005). Formability analysis of tailor-welded blanks of different thickness ratios. *Journal of Manufacturing Science and Engineering*, *127*(4), 743–751. doi:10.1115/1.2034518

Chan, S. M., Chan, L. C., & Lee, T. C. (2003). Tailor-welded blanks of different thickness ratios effects on forming limit diagrams. *Journal of Materials Processing Technology*, *132*(1-3), 95–101. doi:10.1016/S0924-0136(02)00407-7

Chen, H. B., Yan, K., Lin, T., Chen, S. B., Jiang, C. Y., & Zhao, Y. (2006). The investigation of typical welding defects for 5456 aluminum alloy friction stir welds. *Materials Science and Engineering A*, *433*(1-2), 64–69. doi:10.1016/j.msea.2006.06.056

Cheng, C. H., Chan, L. C., Chow, C. L., & Lee, T. C. (2005). Experimental investigation on the weldability and forming behavior of aluminum alloy tailor-welded blanks. *Journal of Laser Applications*, *17*(2), 81–88. doi:10.2351/1.1848521

Cheng, C. H., Chan, L. C., Tang, C. Y., & Chow, C. L. (2005). Determination of true stress-strain curve for the weldment of aluminum laser-welded blanks. *Journal of Laser Applications*, *17*(3), 159–170. doi:10.2351/1.1961652

Cho, J. H., Boyce, D. E., & Dawson, P. R. (2005). Modeling strain hardening and texture evolution in friction stir welding of stainless steel. *Materials Science and Engineering A*, *398*(1-2), 146–163. doi:10.1016/j.msea.2005.03.002

Cho, J. H., Kang, S. H., Han, H. N., & Oh, K. H. (2008). Modeling friction stir welding process of aluminum alloys. *Metals and Materials International*, *14*(2), 247–258. doi:10.3365/met.mat.2008.04.247

Choi, Y., Heo, Y., Kim, H. Y., & Seo, D. (2000). Investigations of weld-line movements for the deep drawing process of tailor welded blanks. *Journal of Materials Processing Technology*, *108*(1), 1–7. doi:10.1016/S0924-0136(00)00536-7

D'Urso, G., & Maccarini, G. (2012). The formability of aluminum foam sandwich panels. *International Journal of Material Forming*, *5*(3), 243–257. doi:10.100712289-011-1036-9

Das, B., Pal, S., & Bag, S. (2017). Torque based defect detection and weld quality modelling in friction stir welding process. *Journal of Manufacturing Processes*, *27*, 8–17. doi:10.1016/j.jmapro.2017.03.012

Das, S. (2000). Aluminium tailor welded blanks. *Advanced Materials & Processes*, 41–49.

Das, S. (2010). Modeling mixed microstructures using a multi-level cellular automata finite element framework. *Computational Materials Science*, *47*(3), 705–711. doi:10.1016/j.commatsci.2009.10.012

Das, S., Palmiere, E. J., & Howard, I. C. (2003). CAFE: a potential tool for modeling thermomechanical processes. *Proceedings of International Conference on Thermomechanical Processing: Mechanics, Microstructure and Control.*

Davies, C. H. J. (1995). The effect of neighbourhood on the kinetics of a cellular automaton recrystallisation model. *Scripta Metallurgica et Materialia*, *33*(7), 1139–1143. doi:10.1016/0956-716X(95)00335-S

Dialami, N., Chiumenti, M., Cervera, M., Segatori, A., & Osikowicz, W. (2017). Enhanced friction model for Friction Stir Welding (FSW) analysis: Simulation and experimental validation. *International Journal of Mechanical Sciences*, *133*, 555–567. doi:10.1016/j.ijmecsci.2017.09.022

Dilip, J. J. S., Janaki Ram, G. D., & Stucker, B. E. (2012). Additive manufacturing with friction welding and friction deposition processes. *International Journal of Rapid Manufacturing*, *3*(1), 56–69. doi:10.1504/IJRAPIDM.2012.046574

Dry, D., Hughes, D., & Owen, R. (2001). Methods of assessing influence of weld properties on formability of laser welded tailored blanks. *Ironmaking & Steelmaking*, *28*(2), 89–95. doi:10.1179/030192301677984

du Plessis, A., Broeckhoven, C., Yadroitsava, I., Yadroitsev, I., Hands, C. H., Kunju, R., & Bhate, D. (2019). Beautiful and functional: A review of biomimetic design in additive manufacturing. *Additive Manufacturing*, *27*, 408–427. doi:10.1016/j.addma.2019.03.033

Dwibedi, S., Jain, N. K., & Pathak, S. (2018). Investigations on joining of stainless steel tailored blanks by μ-PTA process. *Materials and Manufacturing Processes*, *33*(16), 1851–1863. doi:10.1080/10426914.2018.1476766

Eisenmenger, M., Bhatt, K. K., & Shi, M. F. (1995). Influence of laser welding parameters on formability and robustness of blank manufacturing: An application to a body side frame. *SAE Transactions*, 877–888. doi:10.4271/950922

Frazier, W. E. (2014). Metal additive manufacturing: A review. *Journal of Materials Engineering and Performance*, *23*(6), 1917–1928. doi:10.100711665-014-0958-z

Friedman, P. A., & Kridli, G. T. (2000). Microstructural and mechanical investigation of aluminum tailor-welded blanks. *Journal of Materials Engineering and Performance*, *9*(5), 541–551. doi:10.1361/105994900770345674

Gandin, C. A., & Rappaz, M. (1994). A coupled finite element-cellular automaton model for the prediction of dendritic grain structures in solidification processes. *Acta Metallurgica et Materialia*, *42*(7), 2233–2246. doi:10.1016/0956-7151(94)90302-6

Gesing, A. (2004). Assuring the continued recycling of light metals in end-of-life vehicles: A global perspective. *JOM*, *56*(8), 18–27. doi:10.100711837-004-0176-5

Ghoo, B. Y., Back, S. J., Keum, Y. T., & Kang, S. Y. (1998). Finite element analysis of tailored sheet forming processes considering laser welding zone. *Metals and Materials*, *4*(4), 862–870. doi:10.1007/BF03026414

Ghoo, B. Y., Keum, Y. T., & Kim, Y. S. (2001). Evaluation of the mechanical properties of welded metal in tailored steel sheet welded by CO_2 laser. *Journal of Materials Processing Technology, 113*(1-3), 692–698. doi:10.1016/S0924-0136(01)00674-4

Ghoo, B. Y., Park, S. W., & Keum, Y. T. (2001). New forming limit diagram of laser tailored blank. *Journal of Strain Analysis for Engineering Design, 36*(2), 143–152. doi:10.1243/0309324011512694

Gunnarsson, L., & Schedin, E. (2001). Improving the properties of exterior body panels in automobiles using variable blank holder force. *Journal of Materials Processing Technology, 114*(2), 168–173. doi:10.1016/S0924-0136(01)00727-0

Guo, X., Liu, X., Shan, S., Zhao, W., Su, H., & Jia, Q. (2019). Green approach toward sustainable adhesive: Synthesis and characterization of poly (myrcene sulfone). *Journal of the Taiwan Institute of Chemical Engineers, 95*, 208–216. doi:10.1016/j.jtice.2018.07.004

Habibi, M., Hashemi, R., Fallah Tafti, M., & Assempour, A. (2018). Experimental investigation of mechanical properties, formability and forming limit diagrams for tailor-welded blanks produced by friction stir welding. *Journal of Manufacturing Processes, 31*, 310–323. doi:10.1016/j.jmapro.2017.11.009

Hang, Z. Y., Jones, M. E., Brady, G. W., Griffiths, R. J., Garcia, D., Rauch, H. A., Cox, C. D., & Hardwick, N. (2018). Non-beam-based metal additive manufacturing enabled by additive friction stir deposition. *Scripta Materialia, 153*, 122–130. doi:10.1016/j.scriptamat.2018.03.025

Hartl, R., Praehofer, B., & Zaeh, M. F. (2020). Prediction of the surface quality of friction stir welds by the analysis of process data using Artificial Neural Networks. *Proceedings of the Institution of Mechanical Engineers, Part L: Journal of Materials: Design and Applications, 234*(5), 732-751. 10.1177/1464420719899685

Hassan, M. A., Suenaga, R., Takakura, N., & Yamaguchi, K. (2005). A novel process on friction aided deep drawing using tapered blank holder divided into four segments. *Journal of Materials Processing Technology, 159*(3), 418–425. doi:10.1016/j.jmatprotec.2004.06.006

Hassan, M. A., Takakura, N., & Yamaguchi, K. (2003). A novel technique of friction aided deep drawing using a blank-holder divided into four segments. *Journal of Materials Processing Technology, 139*(1-3), 408–413. doi:10.1016/S0924-0136(03)00514-4

Hayashi, H., Pak, S. W., & Usuda, M. (1998). Metal flow of a tailored blank in square cup deep drawing. *SAE Transactions*, 273–279. doi:10.4271/980447

He, J., Xian, G., & Zhang, Y. X. (2020). Effect of moderately elevated temperatures on bond behaviour of CFRP-to-steel bonded joints using different adhesives. *Construction & Building Materials, 241*, 118057. doi:10.1016/j.conbuildmat.2020.118057

He, S., Wu, X., & Hu, S. J. (2003). Formability enhancement for tailor-welded blanks using blank holding force control. *Journal of Manufacturing Science and Engineering, 125*(3), 461–467. doi:10.1115/1.1580853

Heo, Y., Choi, Y., Kim, H. Y., & Seo, D. (2001). Characteristics of weld line movements for the deep drawing with drawbeads of tailor-welded blanks. *Journal of Materials Processing Technology, 111*(1-3), 164–169. doi:10.1016/S0924-0136(01)00503-9

Irving, B. (1995). Welding tailored blanks is hot issue for automakers. *Welding Journal, 74*(8), 49–52.

Jia, Q., Guo, W., Li, W., Zhu, Y., Peng, P., & Zou, G. (2016). Microstructure and tensile behavior of fiber laser-welded blanks of DP600 and DP980 steels. *Journal of Materials Processing Technology, 236*, 73–83. doi:10.1016/j.jmatprotec.2016.05.011

Jiang, J., Xu, X., & Stringer, J. (2018). Support structures for additive manufacturing: A review. *Journal of Manufacturing and Materials Processing, 2*(4), 64. doi:10.3390/jmmp2040064

Kampuš, Z., & Balič, J. (2003). Deep drawing of tailored blanks without a blankholder. *Journal of Materials Processing Technology, 133*(1-2), 128–133. doi:10.1016/S0924-0136(02)00215-7

Kanani, A. Y., Hou, X., & Ye, J. (2020). The influence of notching and mixed-adhesives at the bonding area on the strength and stress distribution of dissimilar single-lap joints. *Composite Structures, 241*, 112136. doi:10.1016/j.compstruct.2020.112136

Karami, V., Dariani, B. M., & Hashemi, R. (2021). Investigation of forming limit curves and mechanical properties of 316 stainless steel/St37 steel tailor-welded blanks produced by tungsten inert gas and friction stir welding method. *CIRP Journal of Manufacturing Science and Technology, 32*, 437–446. doi:10.1016/j.cirpj.2021.02.002

Kee, K. J., Rhee, M. H., Choi, B. I., Kim, C. W., Sung, C. W., Han, C. P., Kang, K. W., & Won, S. T. (2009). Development of application technique of aluminum sandwich sheets for automotive hood. *International Journal of Precision Engineering and Manufacturing, 10*(4), 71–75. doi:10.100712541-009-0073-5

Kesharwani, R. K., Panda, S. K., & Pal, S. K. (2015). Experimental investigations on formability of aluminum tailor friction stir welded blanks in deep drawing process. *Journal of Materials Engineering and Performance, 24*(2), 1038–1049. doi:10.100711665-014-1361-5

Khalili, S. M. R., Shokuhfar, A., Hoseini, S. D., Bidkhori, M., Khalili, S., & Mittal, R. K. (2008). Experimental study of the influence of adhesive reinforcement in lap joints for composite structures subjected to mechanical loads. *International Journal of Adhesion and Adhesives, 28*(8), 436–444. doi:10.1016/j.ijadhadh.2008.04.009

Khan, N. Z., Siddiquee, A. N., Khan, Z. A., & Shihab, S. K. (2015). Investigations on tunneling and kissing bond defects in FSW joints for dissimilar aluminum alloys. *Journal of Alloys and Compounds, 648*, 360–367. doi:10.1016/j.jallcom.2015.06.246

Kim, D., Lee, W., Kim, J., Kim, C., & Chung, K. (2010). Formability evaluation of friction stir welded 6111-T4 sheet with respect to joining material direction. *International Journal of Mechanical Sciences, 52*(4), 612–625. doi:10.1016/j.ijmecsci.2010.01.001

Kim, K. J., Rhee, M. H., Choi, B., Kim, C. W., Sung, C. W., Han, C. P., Kang, K. W., & Won, S. T. (2010). Development of application technique of aluminum sandwich sheets for automotive hood. *International Journal of Precision Engineering and Manufacturing, 10*(4), 71–75. doi:10.100712541-009-0073-5

Kinsey, B., Liu, Z., & Cao, J. (2000). A novel forming technology for tailor-welded blanks. *Journal of Materials Processing Technology, 99*(1-3), 145–153. doi:10.1016/S0924-0136(99)00412-4

Kinsey, B. L., & Cao, J. (2003). An analytical model for tailor welded blank forming. *Journal of Manufacturing Science and Engineering*, *125*(2), 344–351. doi:10.1115/1.1537261

Koruk, A. N., Cretteur, L., & Corostola, J. R. (2001). Laser welded tailored blanks made of stainless steel/carbon steel. Proceedings of Laser Assisted Net Shape Engineering 01, LANE 2001.

Kumar, M., Kailas, S. V., & Narayanan, R. G. (2013). Influence of external weld flash on the in-plane plane-strain formability of friction stir welded sheets. *Journal of Strain Analysis for Engineering Design*, *48*(6), 376–385. doi:10.1177/0309324713488884

Kusuda, H., Takasago, T., & Natsumi, F. (1997). Formability of tailored blanks. *Journal of Materials Processing Technology*, *71*(1), 134–140. doi:10.1016/S0924-0136(97)00159-3

Kuykendall, K., Nelson, T., & Sorensen, C. (2013). On the selection of constitutive laws used in modeling friction stir welding. *International Journal of Machine Tools & Manufacture*, *74*, 74–85. doi:10.1016/j.ijmachtools.2013.07.004

Lee, H., Lim, C. H. J., Low, M. J., Tham, N., Murukeshan, V. M., & Kim, Y. J. (2017). Lasers in additive manufacturing: A review. *International Journal of Precision Engineering and Manufacturing-Green Technology*, *4*(3), 307–322. doi:10.100740684-017-0037-7

Leitao, C., Emílio, B., Chaparro, B. M., & Rodrigues, D. M. (2009). Formability of similar and dissimilar friction stir welded AA 5182-H111 and AA 6016-T4 tailored blanks. *Materials & Design*, *30*(8), 3235–3242. doi:10.1016/j.matdes.2008.12.005

Lennon, R., Pedreschi, R., & Sinha, B. P. (1999). Comparative study of some mechanical connections in cold formed steel. *Construction & Building Materials*, *13*(3), 109–116. doi:10.1016/S0950-0618(99)00018-5

Lewandowski, J. J., & Seifi, M. (2016). Metal additive manufacturing: A review of mechanical properties. *Annual Review of Materials Research*, *46*(1), 151–186. doi:10.1146/annurev-matsci-070115-032024

Li, N., Huang, S., Zhang, G., Qin, R., Liu, W., Xiong, H., Shi, G., & Blackburn, J. (2019). Progress in additive manufacturing on new materials: A review. *Journal of Materials Science and Technology*, *35*(2), 242–269. doi:10.1016/j.jmst.2018.09.002

Lin, X., Wang, P., Zhu, H., Zhang, Y., Ning, Y., & Wang, L. (2021). A novel method for the welding of tailor-welded blanks with different thicknesses based on the diode laser source. *Optics & Laser Technology*, *141*, 107100. doi:10.1016/j.optlastec.2021.107100

Lindgren, L. E., & Lundbäck, A. (2018). Approaches in computational welding mechanics applied to additive manufacturing: Review and outlook. *Comptes Rendus. Mécanique*, *346*(11), 1033–1042. doi:10.1016/j.crme.2018.08.004

Liu, J., Wang, A., Gao, H., Gandra, J., Beamish, K., Zhan, L., & Wang, L. (2018). Transition of failure mode in hot stamping of AA6082 tailor welded blanks. *Journal of Materials Processing Technology*, *257*, 33–44. doi:10.1016/j.jmatprotec.2018.02.028

Liu, J., & Xue, W. (2013). Formability of AA5052/polyethylene/AA5052 sandwich sheets. *Transactions of Nonferrous Metals Society of China*, *23*(4), 964–969. doi:10.1016/S1003-6326(13)62553-4

Liu, J. G., Wei, L., & Wang, J. X. (2012). Influence of interfacial adhesion strength on formability of AA5052/polyethylene/AA5052 sandwich sheet. *Transactions of Nonferrous Metals Society of China, 22*, s395–s401. doi:10.1016/S1003-6326(12)61737-3

Liu, S., & Chao, Y. J. (2004). Determination of global mechanical response of friction stir welded plates using local constitutive properties. *Modelling and Simulation in Materials Science and Engineering, 13*(1), 1–15. doi:10.1088/0965-0393/13/1/001

Liu, W., Cheng, W., Xu, Y., & Yuan, S. (2018). Enhancing Formability of AA2219 Aluminum Alloy Friction Stir Welded Blanks with Preheating Treatment. *Journal of Materials Engineering and Performance, 27*(9), 4819–4828. doi:10.100711665-018-3544-y

Ma, Z. Y. (2008). Friction stir processing technology: A review. *Metallurgical and Materials Transactions. A, Physical Metallurgy and Materials Science, 39*(3), 642–658. doi:10.100711661-007-9459-0

Madej, L., Hodgson, P. D., & Pietrzyk, M. (2007). The validation of a multiscale rheological model of discontinuous phenomena during metal rolling. *Computational Materials Science, 41*(2), 236–241. doi:10.1016/j.commatsci.2007.04.002

Madej, L., Sieradzki, L., Sitko, M., Perzynski, K., Radwanski, K., & Kuziak, R. (2013). Multi scale cellular automata and finite element based model for cold deformation and annealing of a ferritic–pearlitic microstructure. *Computational Materials Science, 77*, 172–181. doi:10.1016/j.commatsci.2013.04.020

Meinders, T., van den Berg, A., & Huetink, J. (2000). Deep drawing simulations of tailored blanks and experimental verification. *Journal of Materials Processing Technology, 103*(1), 65–73. doi:10.1016/S0924-0136(00)00420-9

Merklein, M., Giera, A., & Geiger, M. (2004). Deep drawing of friction stir welded aluminium tailored blanks. *Proceedings of International Deep Drawing Research Group conference (IDDRG 2004).*

Miles, M. P., Nelson, T. W., & Decker, B. J. (2004). Formability and strength of friction-stir-welded aluminum sheets. *Metallurgical and Materials Transactions. A, Physical Metallurgy and Materials Science, 35*(11), 3461–3468. doi:10.100711661-004-0183-8

Miles, M. P., Nelson, T. W., & Melton, D. W. (2005). Formability of friction-stir-welded dissimilar-aluminum-alloy sheets. *Metallurgical and Materials Transactions. A, Physical Metallurgy and Materials Science, 36*(12), 3335–3342. doi:10.100711661-005-0008-4

Miles, M. P., Nelson, T. W., Steel, R., Olsen, E., & Gallagher, M. (2009). Effect of friction stir welding conditions on properties and microstructures of high strength automotive steel. *Science and Technology of Welding and Joining, 14*(3), 228–232. doi:10.1179/136217108X388633

Miles, M. P., Pew, J., Nelson, T. W., & Li, M. (2006). Comparison of formability of friction stir welded and laser welded dual phase 590 steel sheets. *Science and Technology of Welding and Joining, 11*(4), 384–388. doi:10.1179/174329306X107737

Min, J., Wan, H., Carlson, B. E., Lin, J., & Sun, C. (2020). Application of laser ablation in adhesive bonding of metallic materials: A review. *Optics & Laser Technology, 128*, 106188. doi:10.1016/j.optlastec.2020.106188

Min, K. B., & Kang, S. S. (2000). A study on resistance welding in steel sheets for tailor welded blank: Evaluation of flash weldability and formability (2nd Report). *Journal of Materials Processing Technology, 103*(2), 218–224. doi:10.1016/S0924-0136(00)00494-5

Min, K. B., Kim, K. S., & Kang, S. S. (2000). A study on resistance welding in steel sheets using a tailor-welded blank (1st report): Evaluation of upset weldability and formability. *Journal of Materials Processing Technology, 101*(1–3), 186–192. doi:10.1016/S0924-0136(00)00476-3

Mishra, R. S., & Ma, Z. Y. (2005). Friction stir welding and processing. *Materials Science and Engineering R Reports, 50*(1-2), 1–78. doi:10.1016/j.mser.2005.07.001

Moayedi, H., Darabi, R., Ghabussi, A., Habibi, M., & Foong, L. K. (2020). Weld orientation effects on the formability of tailor welded thin steel sheets. *Thin-walled Structures, 149*, 106669. doi:10.1016/j.tws.2020.106669

Morovvati, M. R., Fatemi, A., & Sadighi, M. (2011). Experimental and finite element investigation on wrinkling of circular single layer and two-layer sheet metals in deep drawing process. *International Journal of Advanced Manufacturing Technology, 54*(1), 113–121. doi:10.100700170-010-2931-9

Morovvati, M. R., Mollaei-Dariani, B., & Asadian-Ardakani, M. H. (2010). A theoretical, numerical, and experimental investigation of plastic wrinkling of circular two-layer sheet metal in the deep drawing. *Journal of Materials Processing Technology, 210*(13), 1738–1747. doi:10.1016/j.jmatprotec.2010.06.004

Nadeau, F., Thériault, B., & Gagné, M. O. (2020). Machine learning models applied to friction stir welding defect index using multiple joint configurations and alloys. *Proceedings of the Institution of Mechanical Engineers, Part L: Journal of Materials: Design and Applications, 234*(5), 752-765. 10.1177/1464420720917415

Naik, S. B., Janaki Ramulu, P., & Ganesh Narayanan, R. (2010). Application of a few necking criteria in predicting the forming limit of unwelded and tailor-welded blanks. *Journal of Strain Analysis for Engineering Design, 45*(2), 79–96. doi:10.1243/03093247JSA562

Nakagawa, N., Ikura, S., Natsumi, F., & Iwata, N. (1993). Finite element simulation of stamping a laser-welded blank (No. 930522). *SAE Technical Paper*, 734-745.

Nakamura, R., Goda, K., Noda, J., & Ohgi, J. (2009). High temperature tensile properties and deep drawing of fully green composites. *Express Polymer Letters, 3*(1), 19–24. doi:10.3144/expresspolymlett.2009.4

Nandan, R., Roy, G. G., Lienert, T. J., & DebRoy, T. (2006). Numerical modelling of 3D plastic flow and heat transfer during friction stir welding of stainless steel. *Science and Technology of Welding and Joining, 11*(5), 526–537. doi:10.1179/174329306X107692

Nandan, R., Roy, G. G., Lienert, T. J., & Debroy, T. (2007). Three-dimensional heat and material flow during friction stir welding of mild steel. *Acta Materialia, 55*(3), 883–895. doi:10.1016/j.actamat.2006.09.009

Narasimhan, K., & Narayanan, R. G. (2011). 2 - Deformation of tailor welded blanks during forming. In B. L. Kinsey & X. Wu (Eds.), *Woodhead Publishing Series in Welding and Other Joining Technologies, Tailor Welded Blanks for Advanced Manufacturing* (pp. 24–47). Woodhead Publishing. doi:10.1533/9780857093851.1.24

Narayanan, R. G. (2007). *Forming behavior of Tailor Welded Blanks* (Ph.D. thesis). IIT Bombay.

Narayanan, R. G. (2018a). A novel method of joining a rod to a sheet by end deformation: A preliminary experimental study. *International Journal of Precision Engineering and Manufacturing*, *19*(5), 773–779. doi:10.100712541-018-0093-0

Narayanan, R. G. (2018b). Sustainable Joining of Metallic Structures by End Forming. In Reference Module in Materials Science and Materials Engineering. Elsevier.

Narayanan, R. G., Bhaskar, V. V., & Narasimhan, K. (2004). Effect of weld conditions on the deformation behavior of Tailor Welded Blanks (TWB). *Proceedings of the 8th International conference on Numerical Methods in Industrial Forming Processes, NUMIFORM '04*.

Narayanan, R. G., & Narasimhan, K. (2005). Effect of weld conditions on the forming limit strains of Tailor Welded Blanks (TWB) – Part I & II. *Proceedings of the 8th ESAFORM conference on material forming, ESAFORM 2005*.

Narayanan, R. G., & Narasimhan, K. (2006). Weld region representation during the simulation of TWB forming behavior. *International Journal of Forming Processes*, *9*(4), 491–518. doi:10.3166/ijfp.9.491-518

Narayanan, R. G., & Narasimhan, K. (2007). Relative Effect of Material and Geometric Parameters on the Forming Behaviour of Tailor Welded Blanks. *International Journal of Forming Processes*, *10*(2), 145–178. doi:10.3166/ijfp.10.145-178

Narayanan, R. G., & Narasimhan, K. (2008a). Influence of the weld conditions on the forming-limit strains of tailor-welded blanks. *Journal of Strain Analysis for Engineering Design*, *43*(4), 217–227. doi:10.1243/03093247JSA344

Narayanan, R. G., & Narasimhan, K. (2008b). Predicting the forming limit strains of tailor-welded blanks. *Journal of Strain Analysis for Engineering Design*, *43*(7), 551–563. doi:10.1243/03093247JSA445

Oliveira, P. R., May, M., Panzera, T. H., Scarpa, F., & Hiermaier, S. (2020). Reinforced biobased adhesive for eco-friendly sandwich panels. *International Journal of Adhesion and Adhesives*, *98*, 102550. doi:10.1016/j.ijadhadh.2020.102550

Ono, M., Yoshitake, A., & Omura, M. (2004). Laser weldability of high-strength steel sheets in fabrication of tailor welded blanks. *Welding International*, *18*(10), 777–784. doi:10.1533/wint.2004.3321

Oya, T., Tiesler, N., Kawanishi, S., Yanagimoto, J., & Koseki, T. (2010). Experimental and numerical analysis of multilayered steel sheets upon bending. *Journal of Materials Processing Technology*, *210*(14), 1926–1933. doi:10.1016/j.jmatprotec.2010.07.003

Padhy, G. K., Wu, C. S., & Gao, S. (2018). Friction stir based welding and processing technologies-processes, parameters, microstructures and applications: A review. *Journal of Materials Science and Technology*, *34*(1), 1–38. doi:10.1016/j.jmst.2017.11.029

Padmanabhan, R., Oliveira, M. C., & Menezes, L. F. (2011). 5 - Lightweight metal alloy tailor welded blanks. In B. L. Kinsey & X. Wu (Eds.), *Woodhead Publishing Series in Welding and Other Joining Technologies, Tailor Welded Blanks for Advanced Manufacturing* (pp. 97–117). Woodhead Publishing. doi:10.1533/9780857093851.2.97

Palanivel, S., Nelaturu, P., Glass, B., & Mishra, R. S. (2015a). Friction stir additive manufacturing for high structural performance through microstructural control in an Mg based WE43 alloy. *Materials & Design, 65*, 934-952.

Palanivel, S., Sidhar, H., & Mishra, R. S. (2015b). Friction stir additive manufacturing: Route to high structural performance. *JOM, 67*(3), 616–621. doi:10.100711837-014-1271-x

Pallett, R. J., & Lark, R. J. (2001). The use of tailored blanks in the manufacture of construction components. *Journal of Materials Processing Technology, 117*(1-2), 249–254. doi:10.1016/S0924-0136(01)01124-4

Pandian, V., & Kannan, S. (2020). Numerical prediction and experimental investigation of aerospace-grade dissimilar aluminium alloy by friction stir welding. *Journal of Manufacturing Processes, 54*, 99–108. doi:10.1016/j.jmapro.2020.03.001

Pang, Q., Zhang, J. H., Huq, M. J., & Hu, Z. L. (2019). Characterization of microstructure, mechanical properties and formability for thermomechanical treatment of friction stir welded 2024-O alloys. *Materials Science and Engineering A, 765*, 138303. doi:10.1016/j.msea.2019.138303

Park, S. Y., Choi, W. J., Choi, H. S., Kwon, H., & Kim, S. H. (2010). Recent trends in surface treatment technologies for airframe adhesive bonding processing: A review (1995–2008). *The Journal of Adhesion, 86*(2), 192–221. doi:10.1080/00218460903418345

Parsa, M. H., Ettehad, M., Matin, P. H., Ahkami, A., & Nasher, S. (2010). Experimental and numerical determination of limiting drawing ratio of Al3105-polypropylene-Al3105 sandwich sheets. *Journal of Engineering Materials and Technology, 132*(3), 031004-1, 031004–031011. doi:10.1115/1.4001264

Patel, C., Das, S., & Narayanan, R. G. (2013). CAFE modeling, neural network modeling, and experimental investigation of friction stir welding. *Proceedings of the Institution of Mechanical Engineers. Part C, Journal of Mechanical Engineering Science, 227*(6), 1164–1176. doi:10.1177/0954406212459150

Pepelnjak, T., Jurkosek, B., Kuzman, K., & Jesenicnik, B. (1997). Research on the forming of tailored blanks in conventional tools. Proceedings of Laser Assisted Net Shape Engineering 97, LANE '97.

Prange, W., Schneider, C., & Jaroni, U. (1994). Tailored Blanks-Contributions to an Optimized Steel Body Shell. In *Laser Assisted Net Shape Engineering, proceedings of the LANE* (Vol. 94, pp. 145–165). Meisenbach Bamberg.

Psyk, V., Risch, D., Kinsey, B. L., Tekkaya, A. E., & Kleiner, M. (2011). Electromagnetic forming—A review. *Journal of Materials Processing Technology, 211*(5), 787–829. doi:10.1016/j.jmatprotec.2010.12.012

Rajpoot, Y. S., Narayanan, R. G., & Das, S. (2018). Predicting the effect of tool configuration during friction stir welding by cellular automata finite element analyses. *International Journal of Manufacturing Research, 13*(4), 359–381. doi:10.1504/IJMR.2018.095377

Ramulu, P. J. (2013). *Forming Behavior of Friction Stir Welded Sheets* (PhD thesis). IIT Guwahati.

Ramulu, P. J., Kailas, S. V., & Narayanan, R. G. (2013c). Influence of tool rotation speed and feed rate on the forming limit of friction stir welded AA6061-T6 sheets. *Proceedings of the Institution of Mechanical Engineers. Part C, Journal of Mechanical Engineering Science, 227*(3), 520–541. doi:10.1177/0954406212463996

Ramulu, P. J., Kailas, S. V., & Narayanan, R. G. (2015). Influence of different weld location and orientation on forming limit of friction stir welded sheets. *International Journal of Materials & Product Technology*, *50*, 147–160. doi:10.1504/IJMPT.2015.067833

Ramulu, P. J., Narayanan, R. G., & Kailas, S. V. (2013b). Forming limit investigation of friction stir welded sheets: Influence of shoulder diameter and plunge depth. *International Journal of Advanced Manufacturing Technology*, *69*(9), 2757–2772. doi:10.100700170-013-5245-x

Ramulu, P. J., Narayanan, R. G., Kailas, S. V., & Reddy, J. (2013a). Internal defect and process parameter analysis during friction stir welding of Al 6061 sheets. *International Journal of Advanced Manufacturing Technology*, *65*(9-12), 1515–1528. doi:10.100700170-012-4276-z

Rauch, L., & Madej, L. (2010). Application of the automatic image processing in modeling of the deformation mechanisms based on the digital representation of microstructure. *International Journal for Multiscale Computational Engineering*, *8*(3), 343–356. doi:10.1615/IntJMultCompEng.v8.i3.90

Raymond, S. D., Wild, P. M., & Bayley, C. J. (2004). On modeling of the weld line in finite element analyses of tailor-welded blank forming operations. *Journal of Materials Processing Technology*, *147*(1), 28–37. doi:10.1016/j.jmatprotec.2003.09.005

Reis, A., Teixeira, P., Duarte, J. F., Santos, A., Da Rocha, A. B., & Fernandes, A. A. (2004). Tailored welded blanks—an experimental and numerical study in sheet metal forming on the effect of welding. *Computers & Structures*, *82*(17-19), 1435–1442. doi:10.1016/j.compstruc.2004.03.039

Rodelas, J., & Lippold, J. (2013). Characterization of engineered nickel-base alloy surface layers produced by additive friction stir processing. *Metallography, Microstructure, and Analysis*, *2*(1), 1–12. doi:10.100713632-012-0056-2

Rollett, A. D., & Raabe, D. (2001). A hybrid model for mesoscopic simulation of recrystallization. *Computational Materials Science*, *21*(1), 69–78. doi:10.1016/S0927-0256(00)00216-0

Rooks, B. (2001). Tailor-welded blanks bring multiple benefits to car design. *Assembly Automation*, *21*(4), 323–329. doi:10.1108/EUM0000000006014

Russell, M. J., Blignault, C., Horrex, N. L., & Wiesner, C. S. (2008). Recent developments in the friction stir welding of titanium alloys. *Welding in the World*, *52*(9), 12–15. doi:10.1007/BF03266662

Saluja, R. S., Narayanan, R. G., & Das, S. (2012). Cellular automata finite element (CAFE) model to predict the forming of friction stir welded blanks. *Computational Materials Science*, *58*, 87–100. doi:10.1016/j.commatsci.2012.01.036

Samanta, A., Shen, N., Ji, H., Wang, W., Li, J., & Ding, H. (2018). Cellular automaton simulation of microstructure evolution for friction stir blind riveting. *Journal of Manufacturing Science and Engineering*, *140*(3), 031016–031026. doi:10.1115/1.4038576

Satheeshkumar, V., & Ganesh Narayanan, R. (2014a). Formability of adhesive bonded steel sheets with artificial finite adhesive defects. *Journal of Strain Analysis for Engineering Design*, *49*(5), 286–300. doi:10.1177/0309324713517380

Satheeshkumar, V., & Ganesh Narayanan, R. (2014b). Investigation on the influence of adhesive properties on the formability of adhesive bonded steel sheets. In Proc IMechE Part C. *Journal of Mechanical Engineering Science*, *228*(3), 405–425. doi:10.1177/0954406213488727

Satheeshkumar, V., & Ganesh Narayanan, R. (2015a). Forming performance of adhesive bonded steel sheets reinforced with metallic wires. *Welding in the World - The International Journal of Materials Joining, 59*(6), 883–900.

Satheeshkumar, V., & Ganesh Narayanan, R. (2015b). Experimental evaluation and prediction of deep drawability of adhesive bonded blanks. *Materialwissenschaft und Werkstofftechnik*, *46*(7), 713–735. doi:10.1002/mawe.201500435

Satheeshkumar, V., & Ganesh Narayanan, R. (2016). Experimental evaluation and prediction of formability of adhesive bonded steel sheets at different adhesive properties. *Journal of Testing and Evaluation*, *44*(3), 1–13. doi:10.1520/JTE20140239

Satheeshkumar, V., Yadav, A. K., & Ganesh Narayanan, R. (2015). Chapter 4, Formability Prediction and Springback Evaluation of Adhesive-Bonded Steel Sheets. Advances in Material Forming and Joining, 61–80.

Saunders, F. I., & Wagoner, R. H. (1996). Forming of tailor-welded blanks. *Metallurgical and Materials Transactions. A, Physical Metallurgy and Materials Science*, *27*(9), 2605–2616. doi:10.1007/BF02652354

Schmidt, H., & Hattel, J. (2006). Analysis of the velocity field in the shear layer in FSW: experimental and numerical modelling. In *Proceedings of 6th International Friction Stir Welding Symposium*. TWI.

Schreiber, H. P., & Ouhlal, A. (2003). Polymer diffusion and the evolution of adhesive bond strength. *The Journal of Adhesion*, *79*(2), 141–153. doi:10.1080/00218460309572

Scriven, P. J., Brandon, J. A., & Williams, N. T. (1997). Relative influence of sheet rolling direction and weld orientation on formability of laser welded steel sheet. *Ironmaking & Steelmaking*, *24*(1), 79–85.

Shailesh, R., Ramu, M., Govindaraju, M., Karthikeyan, K., & Satheeshkumar, V. (2019). Performance Evaluation of Adhesive Spur Gear with the Influence of Properties of Adhesive. In R. Narayanan, S. Joshi, & U. S. Dixit (Eds.), *Advances in Computational Methods in Manufacturing. Lecture Notes on Multidisciplinary Industrial Engineering* (pp. 923–931). Springer. doi:10.1007/978-981-32-9072-3_77

Shakeri, H. R., Buste, A., Worswick, M. J., Clarke, J. A., Feng, F., Jain, M., & Finn, M. (2002). Study of damage initiation and fracture in aluminum tailor welded blanks made via different welding techniques. *Journal of Light Metals*, *2*(2), 95–110. doi:10.1016/S1471-5317(02)00028-7

Shakeri, H. R., Lee, Y., Worswick, M. J., Feng, F., Christy, W., & Clarke, J. A. (2001). Weld failure in formability testing of aluminum tailor welded blanks. *SAE Transactions*, 101–110. doi:10.4271/2001-01-0090

Shin, W. G., Lee, S. H., & Kim, E. S. (2005). The study for the forming technology of Automobile Bumper beam using the Tailored Blank of Mash Seam Welding. *Proceedings of the Korean Society of Precision Engineering Conference*, 1376-1380.

Shrivastava, A., Zinn, M., Duffie, N. A., Ferrier, N. J., Smith, C. B., & Pfefferkorn, F. E. (2017). Force measurement-based discontinuity detection during friction stir welding. *Journal of Manufacturing Processes, 26*, 113–121. doi:10.1016/j.jmapro.2017.01.007

Shterenlikht, A., & Howard, I. C. (2006). The CAFE model of fracture—Application to a TMCR steel. *Fatigue & Fracture of Engineering Materials & Structures, 29*(9-10), 770–787. doi:10.1111/j.1460-2695.2006.01031.x

Siegert, K., & Knabe, E. (1995). Fundamental research and draw die concepts for deep drawing of tailored blanks. *SAE Transactions*, 866–876. doi:10.4271/950921

Singh, S., Ramakrishna, S., & Singh, R. (2017). Material issues in additive manufacturing: A review. *Journal of Manufacturing Processes, 25*, 185–200. doi:10.1016/j.jmapro.2016.11.006

Sokolovan, O. A., Kuhn, M., & Palkowski, H. (2012). Deep drawing properties of lightweight steel/polymer/steel sandwich composites. *Archives of Civil and Mechanical Engineering, 12*(2), 105–112. doi:10.1016/j.acme.2012.05.001

Song, M., & Kovacevic, R. (2003). Thermal modeling of friction stir welding in a moving coordinate system and its validation. *International Journal of Machine Tools & Manufacture, 43*(6), 605–615. doi:10.1016/S0890-6955(03)00022-1

Spytkowski, P., & Klimek, T. (2009). Implementation of cellular automata framework dedicated to digital material representation. *Computer Methods in Materials Science, 9*(2), 283–288.

Stasik, M. C., & Wagoner, R. H. (1996). Forming of tailor welded aluminum blanks, Aluminum and Magnesium for Automotive Applications. The Minerals, Metals & Materials Society, 69-82.

Stasik, M. C., & Wagoner, R. H. (1998). Forming of tailor welded aluminum blanks. *International Journal of Forming Processes, 1*, 9–22.

Stavropoulos, P., & Foteinopoulos, P. (2018). Modelling of additive manufacturing processes: A review and classification. *Manufacturing Review, 5*, 2. doi:10.1051/mfreview/2017014

Su, H., Wu, C. S., Pittner, A., & Rethmeier, M. (2013). Simultaneous measurement of tool torque, traverse force and axial force in friction stir welding. *Journal of Manufacturing Processes, 15*(4), 495–500. doi:10.1016/j.jmapro.2013.09.001

Tayebi, P., Fazli, A., Asadi, P., & Soltanpour, M. (2019). Formability analysis of dissimilar friction stir welded AA 6061 and AA 5083 blanks by SPIF process. *CIRP Journal of Manufacturing Science and Technology, 25*, 50–68. doi:10.1016/j.cirpj.2019.02.002

Tušek, J., Kampuš, Z., & Suban, M. (2001). Welding of tailored blanks of different materials. *Journal of Materials Processing Technology, 119*(1-3), 180–184. doi:10.1016/S0924-0136(01)00937-2

Uchihara, M., & Fukui, K. (2002). Tailored Blanks of High Strength Steels— Comparison of Welding Processes. *Welding in the World, 46*(7-8), 41–48. doi:10.1007/BF03263389

Uchihara, M., & Fukui, K. (2006). Formability of tailor welded blanks fabricated by different welding processes: Study of tailor welded blanks using automotive high-strength steel sheets (1st report). *Welding International*, *20*(8), 612–621. doi:10.1533/wint.2006.3628

Uriondo, A., Esperon-Miguez, M., & Perinpanayagam, S. (2015). The present and future of additive manufacturing in the aerospace sector: A review of important aspects. *Proceedings of the Institution of Mechanical Engineers. Part G, Journal of Aerospace Engineering*, *229*(11), 2132–2147. doi:10.1177/0954410014568797

Vaezi, M., Seitz, H., & Yang, S. (2013). A review on 3D micro-additive manufacturing technologies. *International Journal of Advanced Manufacturing Technology*, *67*(5-8), 1721–1754. doi:10.100700170-012-4605-2

Valvi, S. R., Krishnan, A., Das, S., & Narayanan, R. G. (2016). Prediction of microstructural features and forming of friction stir welded sheets using cellular automata finite element (CAFE) approach. *International Journal of Material Forming*, *9*(1), 115–129. doi:10.100712289-015-1216-0

Varis, J. P., & Lepistö, J. (2003). A simple testing-based procedure and simulation of the clinching process using finite element analysis for establishing clinching parameters. *Thin-walled Structures*, *41*(8), 691–709. doi:10.1016/S0263-8231(03)00026-0

Wang, B., Shi, M., Sadrnia, H., & Lin, F. (1995). Structural Performance of Tailor Welded Sheet Steels. *SAE Transactions*, *104*, 222–234. doi:10.4271/950376

Wang, X. S., Hu, Z. L., Yuan, S. J., & Hua, L. (2014). Influence of tube spinning on formability of friction stir welded aluminum alloy tubes for hydroforming application. *Materials Science and Engineering A*, *607*, 245–252. doi:10.1016/j.msea.2014.03.125

Weiss, M., Dingle, M. E., Rolfe, B. F., & Hodgson, P. D. (2007). The influence of temperature on the forming behavior of metal/polymer laminates in sheet metal forming. *Journal of Engineering Materials and Technology*, *129*(4), 530–537. doi:10.1115/1.2772329

Wong, K. V., & Hernandez, A. (2012). A review of additive manufacturing. *International Scholarly Research Notices*, *2012*, 1–10.

WorldAutoSteel. (2020, December 21). https://www.worldautosteel.org/

Yuan, S. J., Hu, Z. L., & Wang, X. S. (2012). Evaluation of formability and material characteristics of aluminum alloy friction stir welded tube produced by a novel process. *Materials Science and Engineering A*, *543*, 210–216. doi:10.1016/j.msea.2012.02.076

Yuqing, M., Liming, K., Chunping, H., Fencheng, L., & Qiang, L. (2016). Formation characteristic, microstructure, and mechanical performances of aluminum-based components by friction stir additive manufacturing. *International Journal of Advanced Manufacturing Technology*, *83*(9), 1637–1647. doi:10.100700170-015-7695-9

Zhang, H., & Zhang, Z. (2007). Numerical modelling of friction stir welding process by using rate-dependent constitutive model. *Journal of Materials Science and Technology*, *23*(1), 73–80.

Zhang, X., Li, L., Chen, Y., Zhu, X., & Ji, S. (2019). Numerical Simulation Analysis of Dual-Beam Laser Welding of Tailored Blanks with Different Thicknesses. *Metals*, *9*(2), 135. doi:10.3390/met9020135

Zhang, Y., Luo, R., Zhang, J., & Xiang, Q. (2011). The reinforcing mechanism of carbon fiber in composite adhesive for bonding carbon/carbon composites. *Journal of Materials Processing Technology*, *211*(2), 167–173. doi:10.1016/j.jmatprotec.2010.08.028

Zhang, Z. (2008). Comparison of two contact models in the simulation of friction stir welding process. *Journal of Materials Science*, *43*(17), 5867–5877. doi:10.100710853-008-2865-x

Zhao, K. M., Chun, B. K., & Lee, J. K. (2001). Finite element analysis of tailor-welded blanks. *Finite Elements in Analysis and Design*, *37*(2), 117–130. doi:10.1016/S0168-874X(00)00026-3

Zhu, H., Cheng, B., Ma, G., Lin, X., Zhang, Y., Wu, D., Ning, Y., & Wang, L. (2020). An eave-like model for the welding of 304 stainless-steel tailor-welded blanks with different thicknesses. *Optics and Lasers in Engineering*, *134*, 106309. doi:10.1016/j.optlaseng.2020.106309

Chapter 12
Structural Optimizations of Different Load-Carrying Members Based on Low Structural Performance Through Computational Structural Analysis:
Structural Optimizations of Sandwich Composite Through FEA Approach

Vijayanandh Raja
https://orcid.org/0000-0003-4992-3028
Kumaraguru College of Technology, India

Balaji S.
Kumaraguru College of Technology, India

Raj Kumar Gnanasekaran
Kumaraguru College of Technology, India

Naveen Kumar Kulandaiyappan
Kumaraguru College of Technology, India

Jagadeeshwaran Ponmariappan
Rajalakshmi Institute of Technology, India

Arul Prakash Raji
Kumaraguru College of Technology, India

Senthil Kumar Madasamy
Kumaraguru College of Technology, India

ABSTRACT

Load withstanding characteristics are one of the major considerations involved in structural engineering because the lifetime factor is directly proportional to load withstanding behavior. Thus, this work computationally analyzes the load withstanding behavior of various sandwich lightweight composite materials under the given flexural load. In this work, four major materials are imposed under flexural loads for two different cum prime core structures such as hexagonal cross-section and twisted cum in-

DOI: 10.4018/978-1-7998-7864-3.ch012

tegrated pentagonal cross-section. The major materials implemented for this comparative investigation are Aluminium Alloy, CFRP, GFRP, and KFRP. All the computational composite models are constructed through the advanced computational tool (i.e., ANSYS Workbench). Finally, the best structures with respect to their lightweight materials are shortlisted to withstand a high amount of flexural loads. According to this comprehensive study, the CFRP-based honeycomb sandwich composite performed better than all other lightweight materials.

INTRODUCTION

Due to their structural characteristics, man-made honeycombs are widely manufactured by using a variety of different characteristic materials, depending on the application. This chapter deals with the structural behavior of twisted pentagonal and hexagonal honeycomb and their comparison for suitable application. The honeycomb structure panel, which is columnar and hexagonal in shape, packaged arrangement, inspired from beehives. It provides a material with minimal density which has high compressive and shear properties. The strength of the sandwich panel depends on the panel size, facing material, number (density) of cells in the panel. The combination of hexagonal and pentagonal facing with pyramidal nature is jointly implemented and attained the innovative shape called twisted pentagonal structure. The building blocks of these structures are epithelial cells which are pack tightly to form the lining of blood vessels and organs in animal tissues and human skin.

LITERATURE SURVEY

In this paper, the author brought the number of degrees of freedom for FE discretization is significantly reduced by replacing the core with a homogeneous continuum. The author found that for in-plane behavior, bending and dynamic modal analysis, the agreement between predictions obtained with the continuum plate model and those with the detailed FE model is good. For the coupled behavior of the sandwich under combined bending and in-plane forces, the agreement is not good. A single-layer model-based 2-D computational approach is carried to predict the structural behaviors of cellular core sandwich panels. The 2-D approach had given acceptable predictions than the results obtained by a detailed finite element model on both static and dynamic behavior of orthotropic truss core sandwich panels such as the standard pyramidal truss core and Kagome truss core having pyramidal sub-geometry. However, the author describes that the 2-D model approach is accepted, when the size of the unit cell varies within certain ranges. Strong corrugated sandwich cores are obtained better accurate predictions than weak cores (Liu T et al., 2007).

In this paper, the author demonstrates that stiffness performance is better for a honeycomb-cored sandwich than a metal foam-cored sandwich. The author analyzed and found that the weight index of the honeycomb-cored sandwich is much smaller and the weight index of the metal foam-cored sandwich is larger. Weight reduction of beams is done in both ways, one by modifying the structural design of the beam and the other by changing the parental material. The author concluded that load-bearing capacity is more efficient for stretching-dominated 2-D lattice (Triangular, Kagome) sandwich beams than bending-dominated 2-D lattice sandwich beams for all load indices (Mirrudula P et al., 2020).

In this paper there are five important studies are described: 1. For face wrinkling and structural buckling under different loading cases, the analytical criteria are used and are optimized for sandwich cylinders using the analytical formulae. 2. The same sandwich cylinder is optimized using FEM. The FEM-based optimization results show a good agreement than analytical results. 3. The influence of the layer number and fiber orientation on the structural efficiency is studied by using the FEM model. The optimization is divided into two steps: (i) The fiber orientation is fixed, while the minimum weight is achieved by determining the optimum layer number; (ii) The thickness is fixed, and to obtain the maximum buckling load fiber orientations are optimized(Sergio D. Cardozo et al., 2011).

To improve the structural efficiency the ‘C’ frames are added to the sandwich cylinder. Different core materials are studied for weight reduction and to determine the optimum frame space and core thickness ANOVA design is used. Also, the author adds that the buckling constraint is critical for the sandwich cylinder without any stiffeners. The structural stability is improved by the addition of frames. To determine the thickness and the fiber orientation the two-step optimization is always a good choice. Sandwich structures are light, have considerable stiffness, and have a high ratio of strength to weight. In the sandwich panel, exterior surfaces transfer loads are caused by bending (flexural load and compression), while the core transfer load is caused by shearing. The properties of the laminate in the structural skin and its connection with the core of the sandwich composite give the strength of sandwich materials to the impact strength to a large extent. Higher the ratio of reinforcement weight to resin weight, the higher impact strength. The author used three methods to produce composite materials: Hand lay-up, press, and autoclave. An impact test was conducted by the author. Among these materials made from the autoclave has higher impact strength. However, materials made from this method have a total lack of structural discontinuity and visually high quality of the surface (smoothness and homogeneity). The values of distinctive strength parameters point to the efficient use of the press method as a cheaper alternative to the autoclave method. Also, the author found that separations of skin in both cases adding mechanical properties of sandwich composites produced using both methods are comparable (Li-Ming Chen et al., 2012).

In this paper, the author constructed a space frame structure and analyzed it and he compared results obtained by experimental analyses and numerical simulations. And then the structural performance of the space frame was compared to an equivalent beam of the same weight and domain height. In addition to that, a space frame structure was designed by the use of multi-objective topology optimization for minimum mass, maximum effective flexural and torsional rigidities. And under identical loads, the space frame structure shows the following than equivalent beam: 1. Lower maximum stress, 2. Greater axial buckling resistance, 3. Greater bending buckling, 4. Higher modal frequency (preventing resonance and structural damage) (Dayyani I et al., 2015).

Here to evaluate the quality of the approximation using neural networks, the parameter values are compared to the parameter values for the same design using FEM without ANN. Design variables are the characteristics that can be modified by the mathematical optimization algorithm to obtain the best structural performance. The analysis of the composite structures is carried out using the FEM. The element used is a triangular flat plate bending element with an 18 degree of freedom called DKT (Discrete Kirchhoff Triangle). Design constraints are the restrictions applied to the structure, such as a limit to avoid material failure, the maximum or minimum value of design variables, and others that depend on the problem being analyzed. The Tsai-Wu failure criterion is employed for failure prediction in a ply see. Tsai-Wu criterion derived that the constraints of the problem are the material failure (Poonam Joshikar et al., 2018).

In this paper, the author creates a corrugated material by adding two face sheets (also known as liners) as upper and lower surfaces to the corrugated sheet. (Core, medium, or fluting) new geometry is obtained called a corrugated panel. The Corrugated sandwich panel is one of the stiffness of the material of face sheets that is higher than or equal to the stiffness of the material of the corrugated core. The author concluded that the most important characteristics of a corrugated core are to keep the face sheets apart and stabilize them by resisting out-of-plane deformations. Also, the author believed that this would increase the shear strength and stiffness of the panel. Then the author explained the application of corrugated materials, they are Packaging industry, civil structures, Marine structures, Mechanical structures, and Aeronautical applications. The author concluded that the superior structural characteristics of corrugated structures are extreme anisotropic behavior and high stiffness to weight ratio, mainly arising from their geometric properties. This paper had three segments: 1. Different types of corrugated structures, their specific characteristics, and their categorized applications - Innovation and development of these materials along with combining different material properties. 2. The author presented a comprehensive set of analyses about the mechanics of these structures. FEM is used to analyze the buckling and vibration of corrugated sheets. Also, the author analyzed internal buckling, contact friction, and plastic deformation. The author found that the crack initiations and delamination were two main factors for the fracture of these structures. For that, the author took optimization problems as an important tool in the optimized design of these structures. 3. Then the author explained the use of corrugated structures in morphing applications. The applications of corrugated skin in camber morphing, winglet morphing, and span-wise morphing extension were discussed by the author. He discussed it in terms of specific boundary conditions, structural and aerodynamic loading configuration as well as the geometric and manufacturing constraints that each application poses. The corrugation geometry and Reynolds number are important in the increment of aerodynamic performance in all these applications. Then the author further coated the corrugated skin with a pre-stretched elastomeric face sheet, a segmented skin, and used foam slices filling the empty surface between the corrugated unit cells for improvement (Aneta Krzyhak et al., 2016), (Chongxin Yuan et al., 2012).

In this paper, to tackle the bending load, an Epoxy carbon wet-Honeycomb sandwich was used by the author. Even though it's a less mesh generation process, with the help of standard boundary conditions the structural simulations are having the capability to provide acceptable solutions. The change of property variations on existing material or newly invented material may chance to reflect in and around worldwide within a short duration so analysis on the material property is must be executed in a proper and standard way. The core material is normally low strength materials but its higher thickness provides the sandwich material with high bending stiffness with overall low density. The uniqueness involved in the core material is capable to provide high strength with low density, which provided the path to increase the mechanical properties of the overall composite. Carbon and Kevlar fibers were played vital roles in the fiber materials, apart from that Sand foam, PVC foam, and Honeycomb was played a vital role in the core materials (Gómez-Gálvez P et al., 2018), (Jarad Lim et al., 2020), (Komal A. Jangavali et al., 2016), .

METHODOLOGY USED – FINITE ELEMENT ANALYSIS

An advanced engineering approach was used in this optimization, in which the name of the methodology is Advanced Computational Coupled Approach. The FEM (Finite Element Method) is the base concept has been involved in this computation. Two major processes are executed in a serial manner, wherein the one way coupling plays a prime role at conjunction of these two processes. Firstly, the sandwich computational models are created with utmost care by using advanced modeling tool and then secondly the structural simulations are computed through FEA solver (Aswin Kumar V et al., 2021), (Bhagavathiyappan S et., 2020), (Indira Prasanth S et al., 2020), (Kesavan K et al., 2021).

Computational Model

The predominant contributors of this computational model are face element and core element, which fulfilled the fundamental design definition of sandwich structure. From the literature survey, the design dimensions are estimated, which are: the length of the face sheet is 215 mm, the breadth of the face sheet is 56 mm, the thickness of the face sheet is 10 mm, the thickness of the core elements of both honeycomb structure and twisted pentagon structure are 50 mm, the internal thickness of the core structures are 0.547 mm, and lengthwise six core structures [honeycomb structure and twisted pentagon structure] are assembled. With the help of the abovementioned design parameters the computational models are generated through advanced modeling tool, i.e., CATIA (Mohamed Bak K et al., 2020), (Naveen Kumar K et al., 2018, 2019), (Rajagurunathan M et al., 2018), and (Raj Kumar G et al., 2017, 2018, 2019, 2020).

Discretization

Figure 1 is revealed the complete discretized structure of sandwich composite, in which the uniform 3-D brick elements are involved. The connectivity on the conjunction regions are carefully checked and connected with ultimate care, which provided the mesh quality of 0.98 out of 1. The flexural load and its supports are clearly revealed in Figure 2, wherein the uniform distributed load of 500 Pa is given at the top surface and fixed support is provided at bottom surface of the structure. Figures 3 to 12 are revealed the structural results of honeycomb structures for various alloys and composites. The comprehensive structural outcomes of Honeycomb Structures are revealed in Figures 13 to 16. Similarly, the same computational procedures are extended for integrated pentagonal structures and thereby the results are revealed in Figures 17 to 21. The comprehensive structural outcomes of Honeycomb Structures are revealed in Figures 22 to 25. All of these computational simulations are executed with the help of advanced Finite Element Analysis solver (Ramesh M et al., 2021), (Senthil Kumar M et al., 2018, 2021), (Udhaya Prakash R et al., 2016), (Venkatesan K et al., 2020), (Vijayakumar Mathaiyan et al., 2021), and (Vijayanandh R et al., 2016, 2017, 2018, 2019, 2020).

Boundary Conditions

Details of applied boundary conditions can be seen in Figure 2.

RESULTS AND DISCUSSIONS

In the first structural computation comprises of structural analysis and their results such as total deformation, equivalent elastic strain, maximum principal stress, and equivalent stress of Honeycomb sandwich composite. As per the aforesaid boundary conditions the computations are executed and thereby the focused outcomes are captured, which are revealed in Figures 3 to 12. Figures 3 and 4 are revealed the

Figure 1. Discretized structure of honeycomb structure

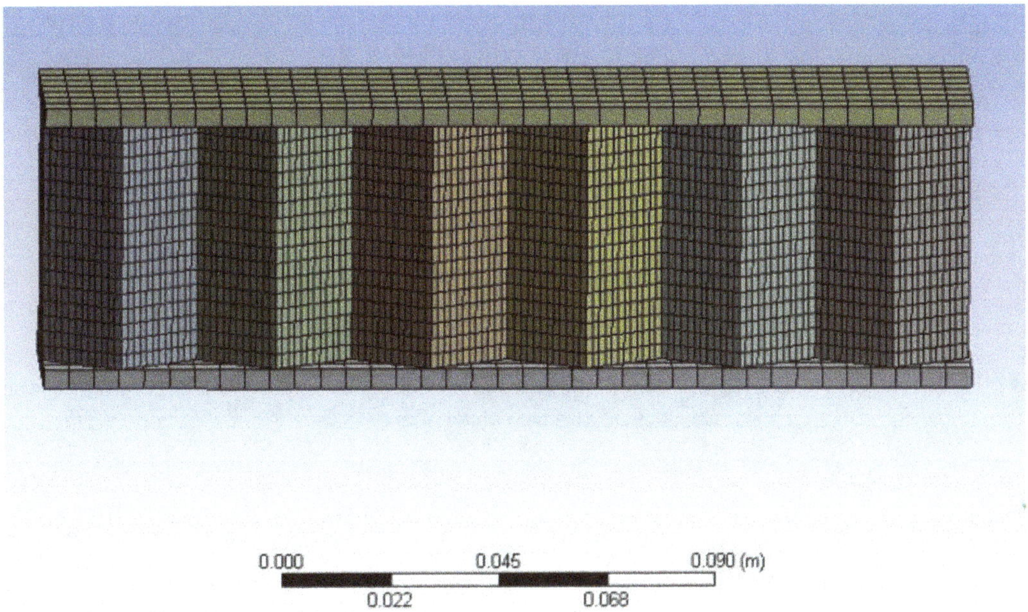

total deformation and equivalent stress results of Aluminium Alloy based sandwich composite. In the Aluminium Alloy based sandwich composite, the core and face materials are used as Aluminium Alloy.

Figure 2. Details of applied boundary conditions

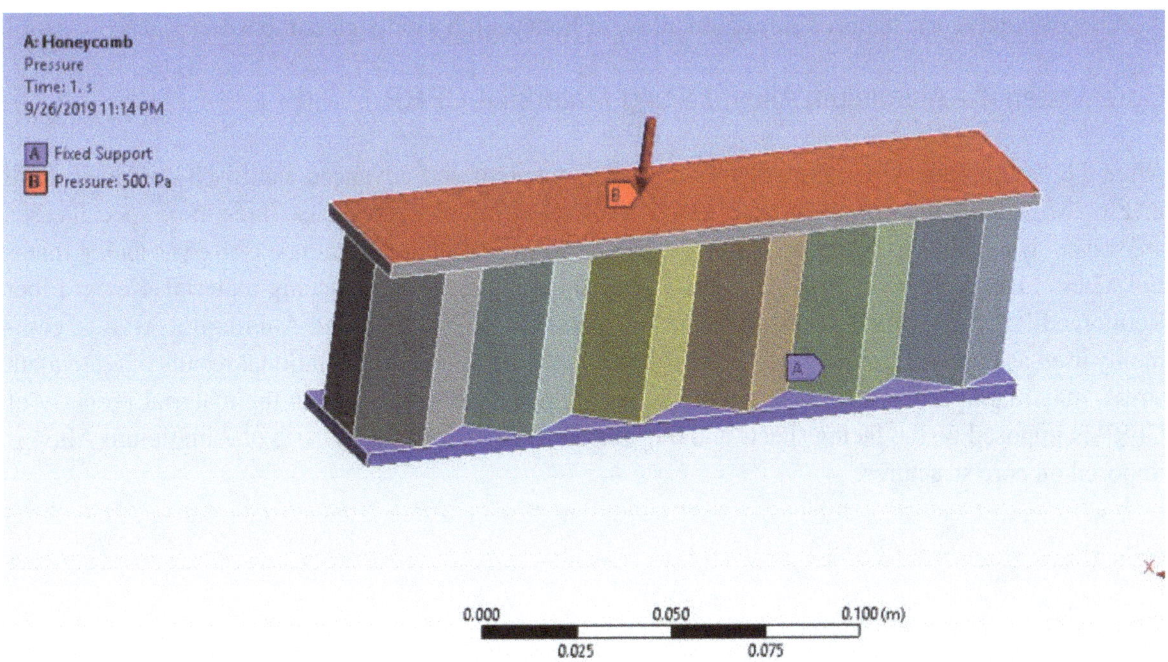

Figure 3. Deformed structure of aluminum alloy for honeycomb sandwich composite

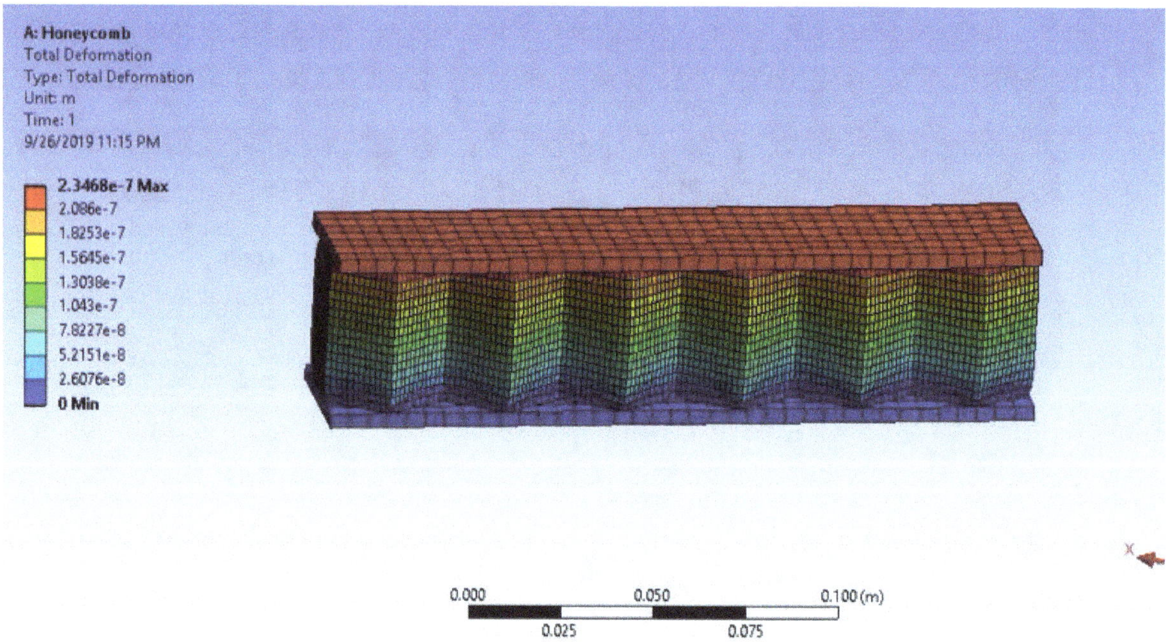

Honeycomb Structure

Core Material – Aluminium Alloy; Facing Material – Aluminium Alloy

Figure 3 shows a deformed structure of aluminum alloy for honeycomb sandwich composite and Figure 4 shows the equivalent stress of aluminum alloy for honeycomb sandwich composite.

Core Material – Aluminium Alloy; Facing Material – CFRP

After the Aluminium Alloy, the polymer matrix elements based advanced sandwich composites are investigated. Under this polymer matrix based sandwich composite categories, three more special cases are solved, which are Carbon Fiber Reinforced Polymer [CFRP] based laminate is used as facing material, Glass Fiber Reinforced Polymer [GFRP] based laminate is used as facing material, Kevlar Fiber Reinforced Polymer [KFRP] based laminate is used as facing material, and Aluminium Alloy is commonly used as core material. Figures 5 to 7 are shown the structural computational results of equivalent stress, maximum principal stress and total deformation respectively, wherein the material property of CFRP is imposed on the facing sheets and the conventional material properties of Aluminium Alloy is imposed on core structures.

Figure 4. Equivalent stress of aluminum alloy for honeycomb sandwich composite

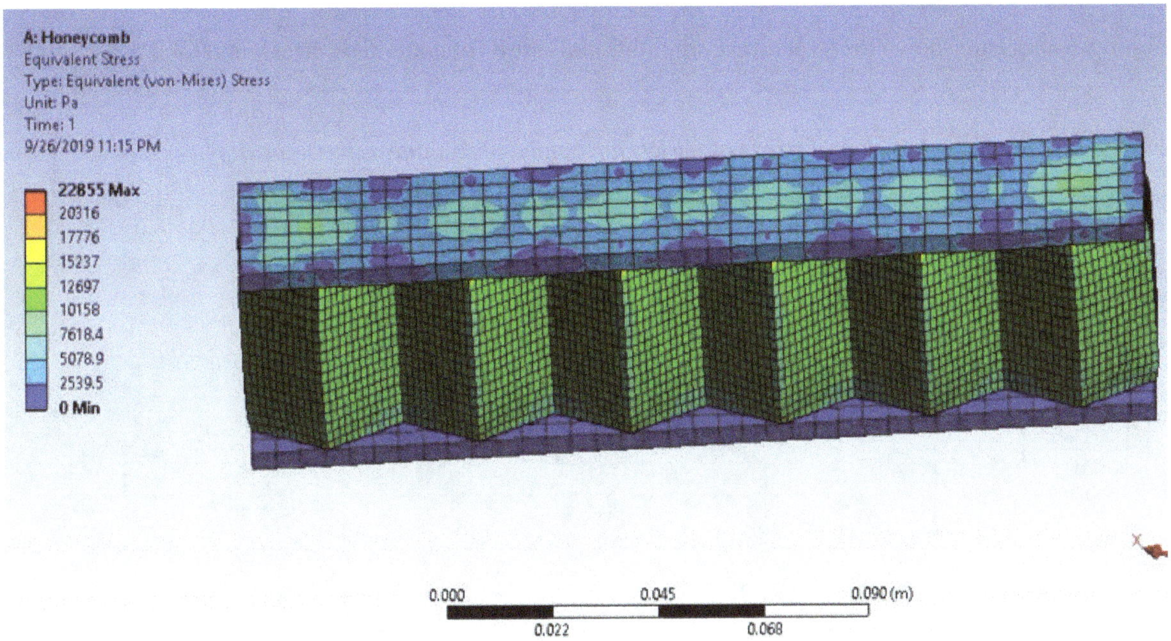

Figure 5. Equivalent stress of CFRP for honeycomb sandwich composite

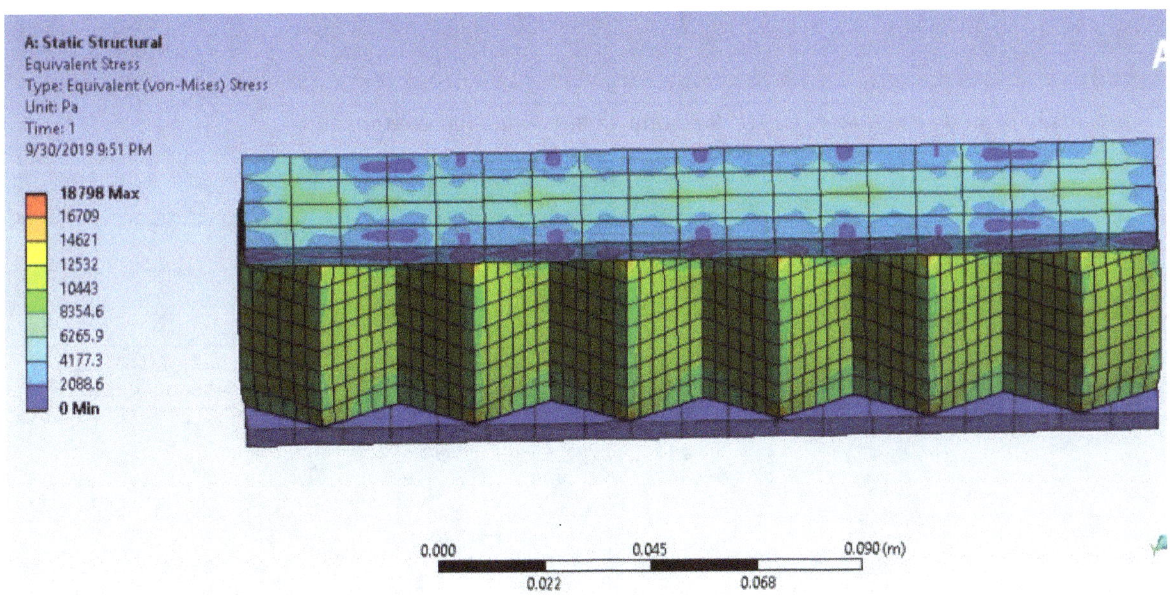

Core Material – Aluminium Alloy; Facing Material – GFRP

Figures 8 and 9 are shown the structural computational results of equivalent stress, and total deformation

Figure 6. Variation of principal stress of CFRP for honeycomb sandwich composite

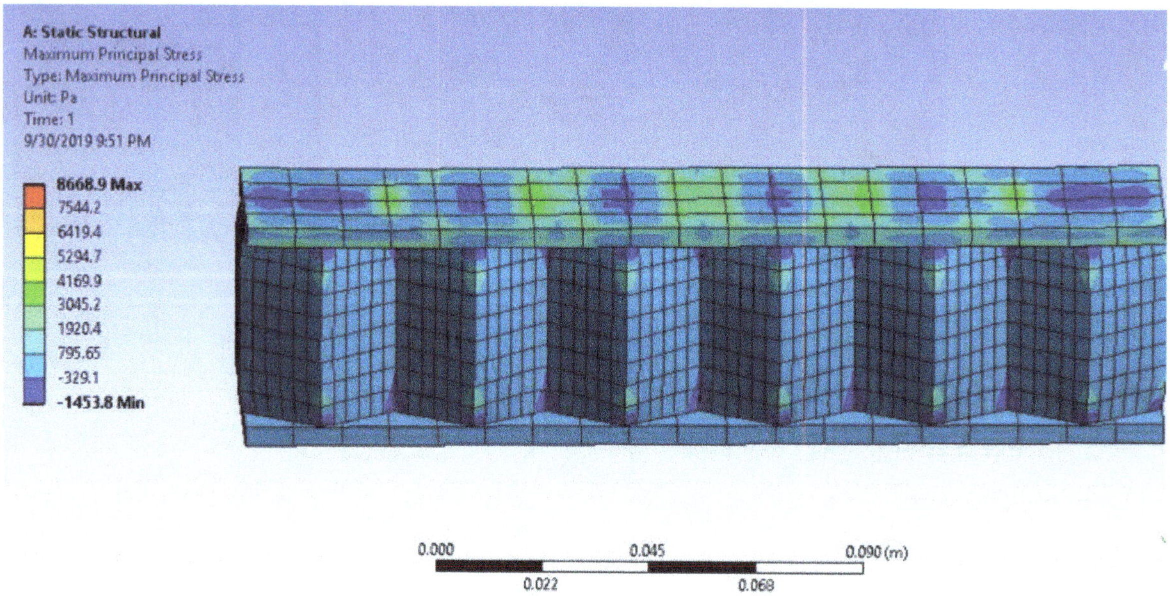

Figure 7. Deformed structure of CFRP for honeycomb sandwich composite

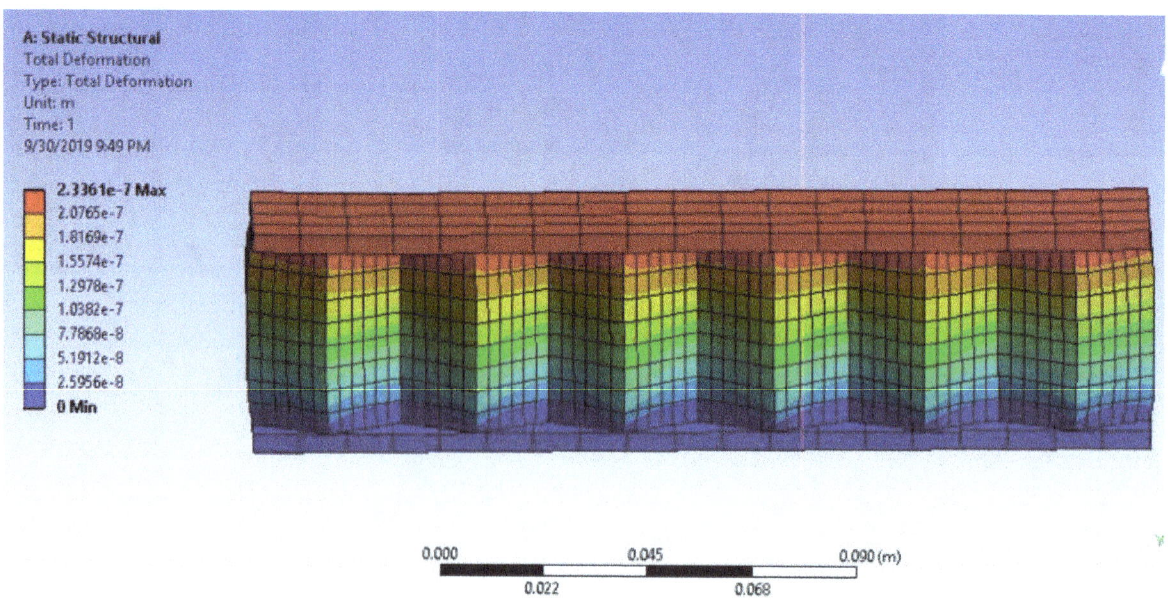

Figure 8. Distribution of equivalent stress of GFRP for honeycomb sandwich composite

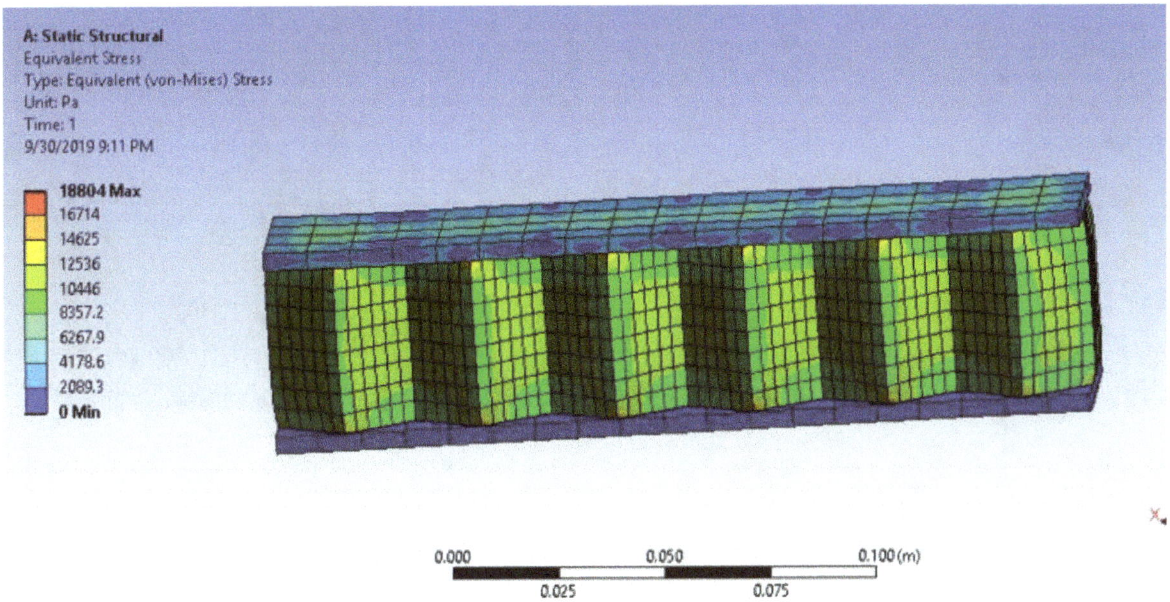

respectively, wherein the material property of GFRP is imposed on the facing sheets and the conventional material properties of Aluminium Alloy is imposed on core structures.

Figure 9. Deformed structure of GFRP for honeycomb sandwich composite

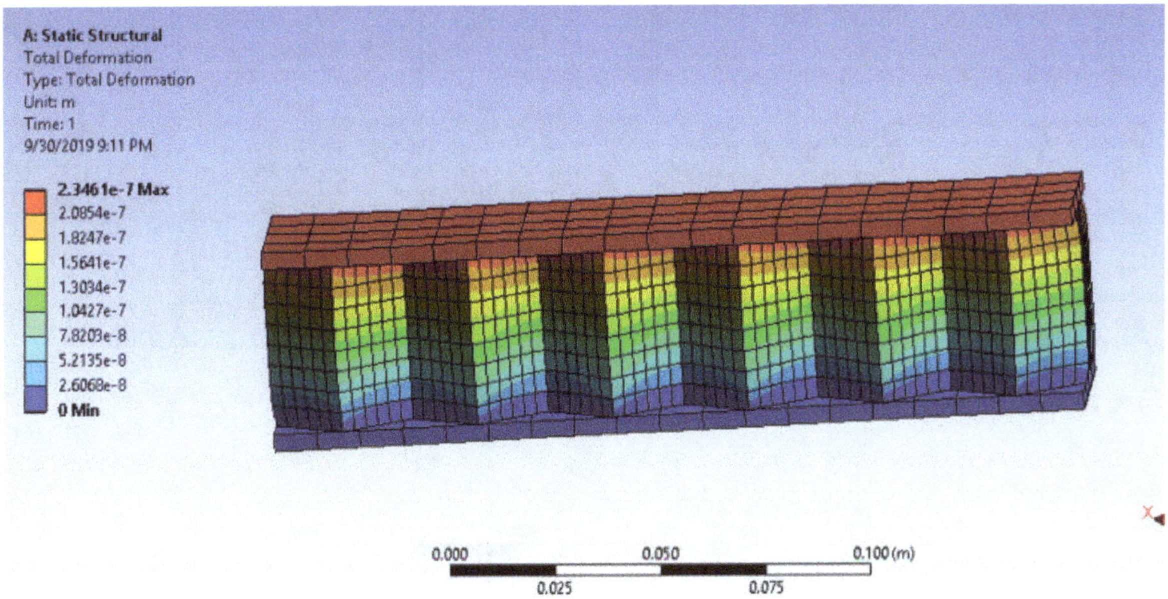

Figure 10. Variations equivalent stress of KFRP for honeycomb sandwich composite

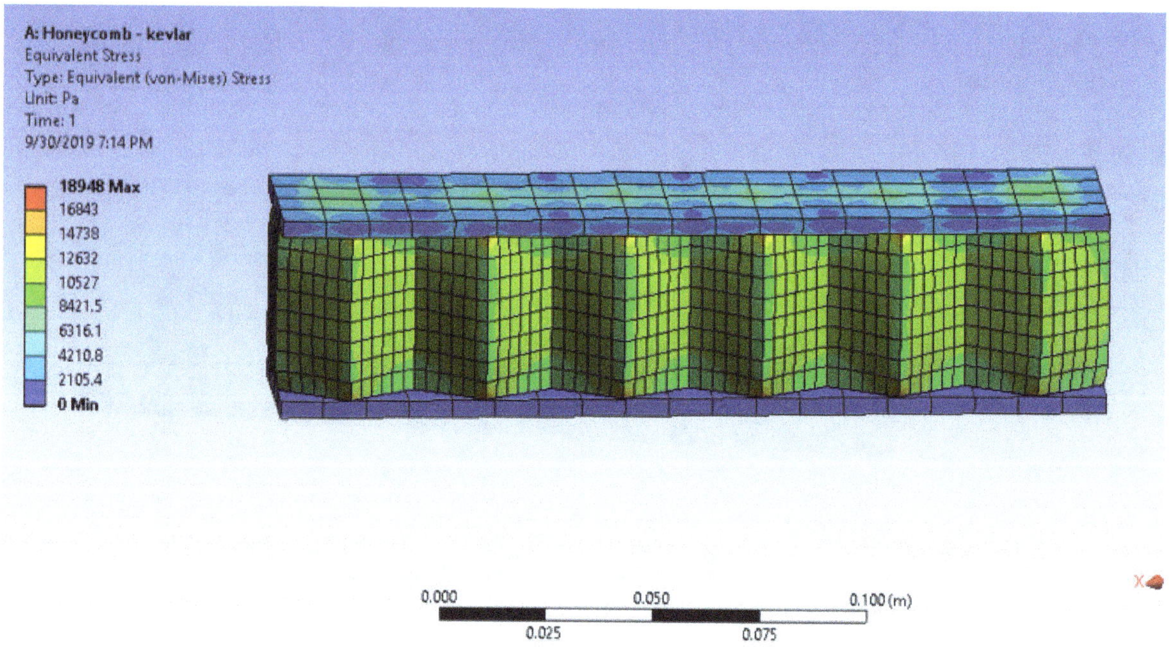

Figure 11. Variations of principal stress of KFRP for honeycomb sandwich composite

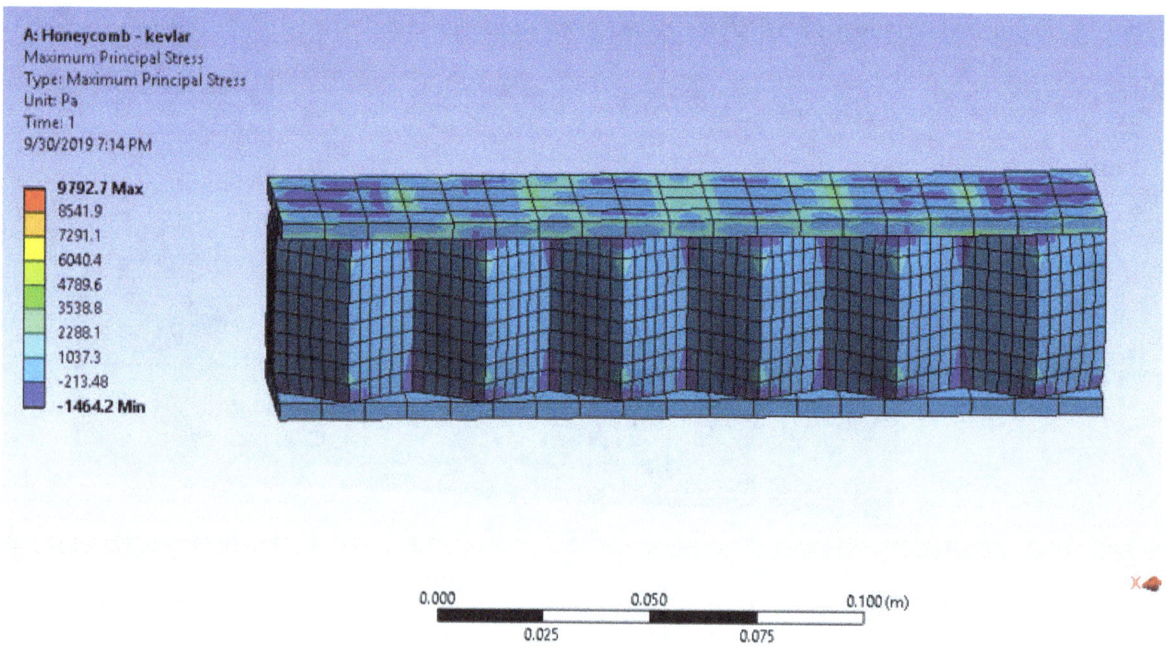

Figure 12. Deformed structure of KFRP for honeycomb sandwich composite

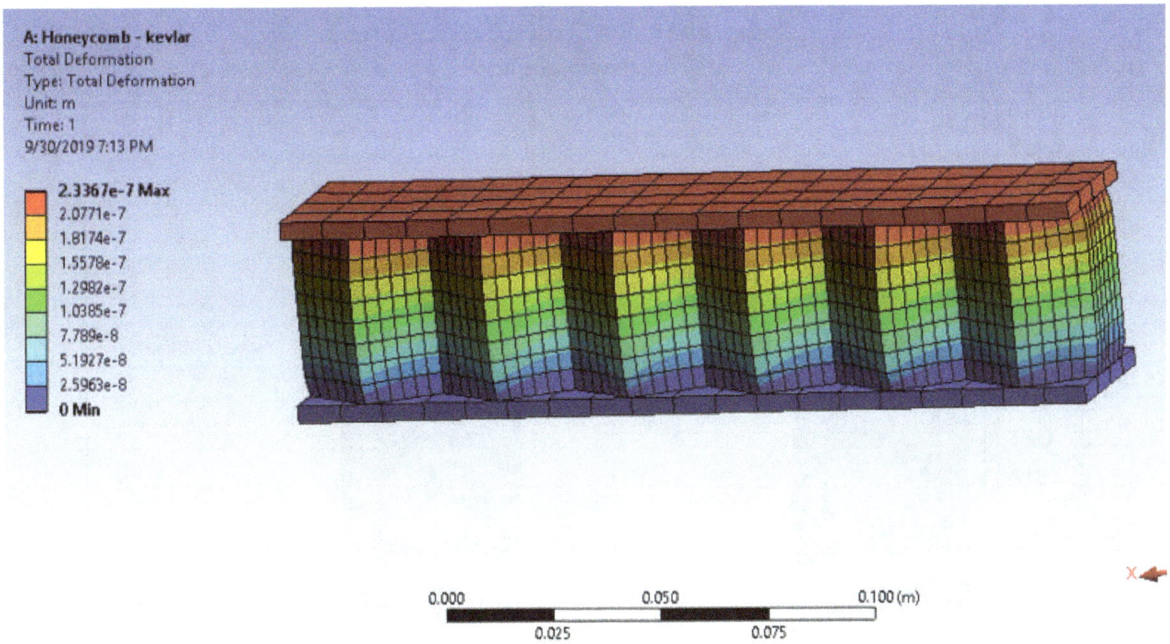

Core Material – Aluminium Alloy; Facing Material – KFRP

Figures 10 to 12 are shown the structural computational results of equivalent stress, maximum principal stress and total deformation respectively, wherein the material property of KFRP is imposed on the facing sheets and the conventional material properties of Aluminium Alloy is imposed on core structures. All the four major lightweight materials relayed sandwich composites are computed and the results are focally observed. Based on the observations, the comprehensive reports are constructed to optimize the suitable structure and lightweight material.

From the comprehensive Figures 13 to 16, the CFRP based sandwich Honeycomb structure is generated low level structural defects than other lightweight compositions thus the same material is picked as best performer. After the CFRP, the KFRP based sandwich Honeycomb structure is generated medium level structural defects than remaining two lightweight compositions thus the same material is picked as another best performer.

Integrated Pentagonal Structure

In the second structural computation comprises of structural analysis and their results such as total deformation, equivalent elastic strain, maximum principal stress, and equivalent stress of Integrated Pentagonal Structure based sandwich composite. As per the aforesaid boundary conditions the computations are executed and thereby the focused outcomes are captured, which are revealed in Figures 17 to 21. Figures 17 and 18 are revealed the equivalent stress and principal stress results of Aluminium Alloy based sandwich composite. In the Aluminium Alloy based sandwich composite, the core and face materials are used as Aluminium Alloy.

Figure 13. Comprehensive structural outcomes [total deformations] of honeycomb structures

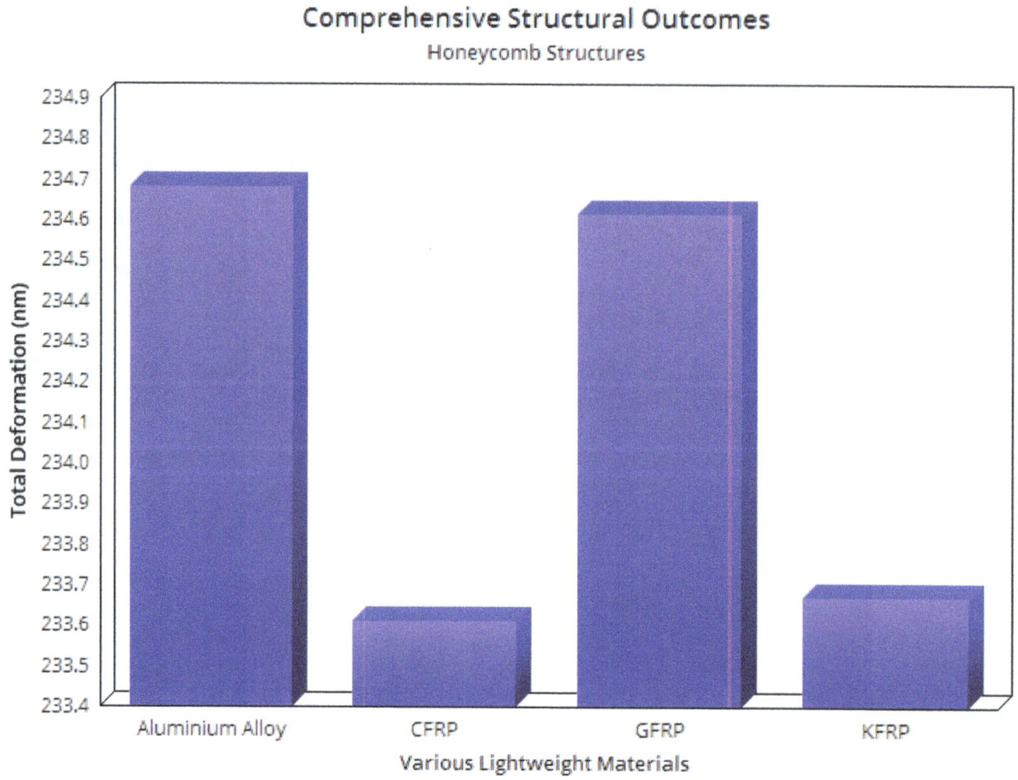

Core Material – Aluminium Alloy; Facing Material – Aluminium Alloy

Figure 17 details the maximum principal stress of aluminium alloy for integrated pentagonal structure and the equivalent stress of aluminium alloy for integrated pentagonal structure can be seen in Figure 18.

Core Material – Aluminium Alloy; Facing Material – CFRP

Figure 19 is shown the structural computational result of maximum principal stress, wherein the material property of CFRP is imposed on the facing sheets and the conventional material properties of Aluminium Alloy is imposed on core structures.

Core Material – Aluminium Alloy; Facing Material – GFRP

Figure 20 is shown the structural computational result of maximum principal stress, wherein the material property of GFRP is imposed on the facing sheets and the conventional material properties of Aluminium Alloy is imposed on core structures.

Figure 14. Comprehensive structural outcomes [equivalent stress] of honeycomb structures

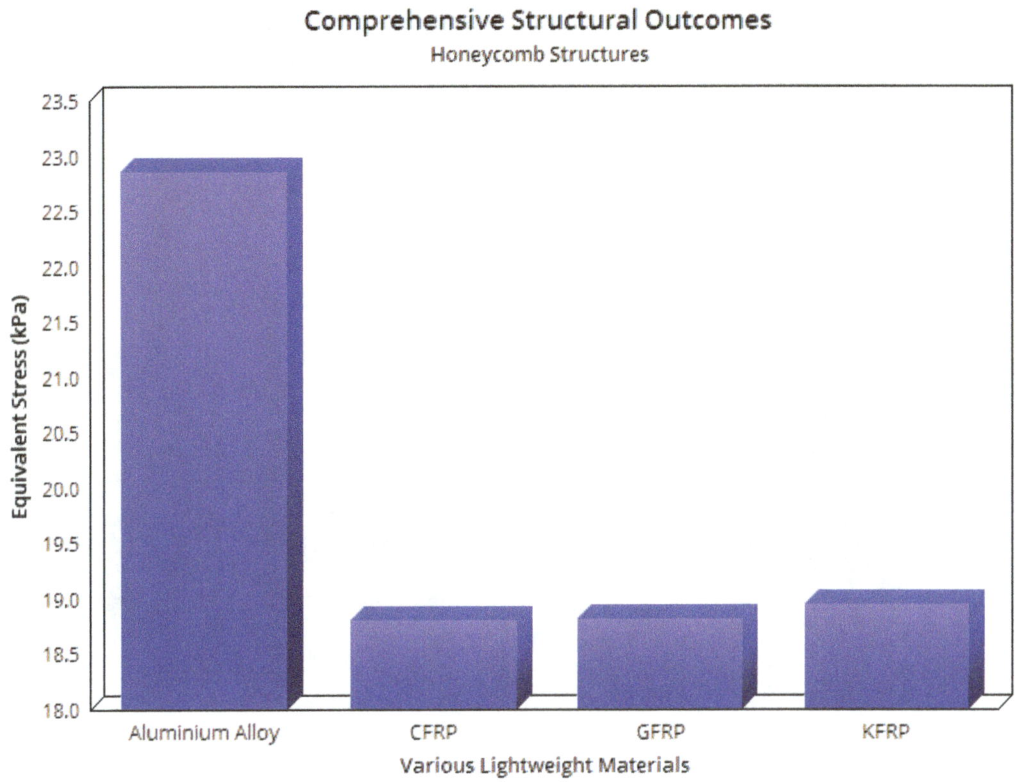

Core Material – Aluminium Alloy; Facing Material – KFRP

Figure 21 is shown the structural computational result of equivalent stress, wherein the material property of KFRP is imposed on the facing sheets and the conventional material properties of Aluminium Alloy is imposed on core structures.

From the comprehensive Figures 22 to 25, the CFRP based sandwich Integrated Pentagonal Structure is generated low level structural defects than other lightweight compositions thus the same material is picked as best performer. After the CFRP, the GFRP based sandwich Integrated Pentagonal Structure is generated medium level structural defects than remaining two lightweight compositions thus the same material is picked as another best performer.

From the comprehensive Figures 13 to 16 and 22 to 25, it is strongly observed that the structural properties of Honeycomb based Hexagonal structure have more withstanding capability than Integrated Pentagonal Structure. Through the implementation of this Honeycomb based Sandwich composite in the real-time applications, the lifetime of the imposed components are further more increased more than five times than other complicated internal structures of Sandwich composites.

Figure 15. Comprehensive structural outcomes [principal stress] of honeycomb structures

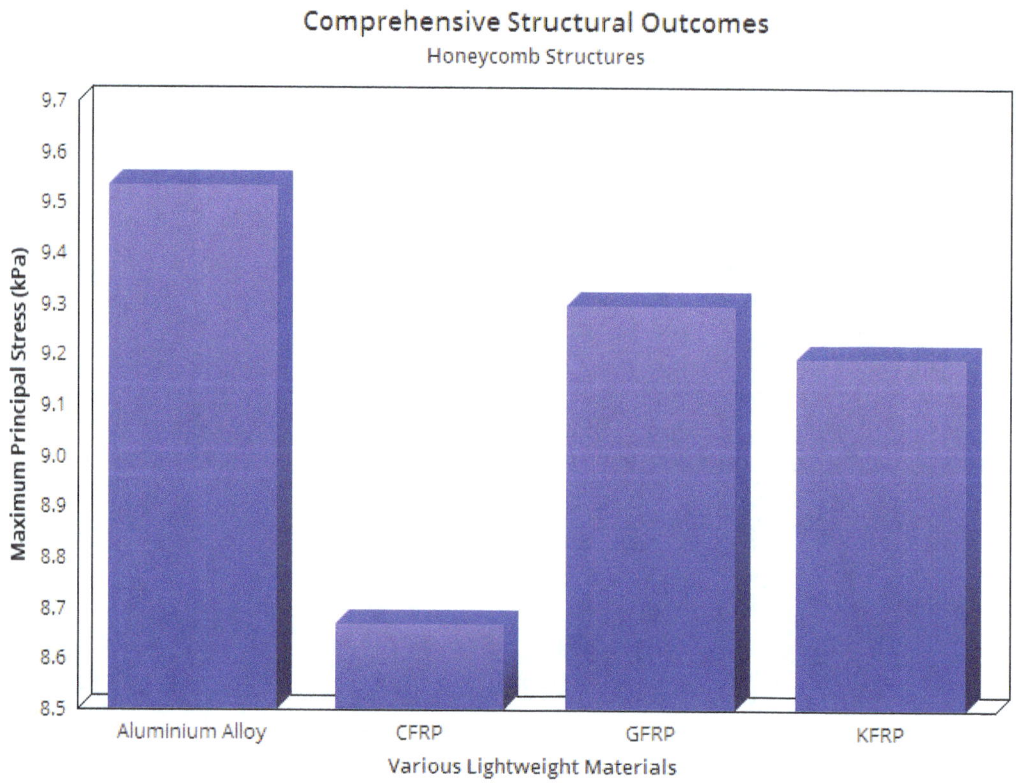

CONCLUSION

Currently, the technology inventions have been emerged greatly in all kinds of real-time problems. Especially, the structural engineering field has been upgraded hugely because of the above-mentioned emerge of inventions. To withstand such up-gradation, this work thoroughly analyzed two different perspectives, which are to pick the suitable lightweight material of internal compositions for highly resisting capacity against complicated loads and to pick the suitable internal structure for highly withstanding power against complicated loads in the sandwich composite. In this regard, the computational structural analysis based numerical approach is picked and imposed for these structural optimizations on the load carrying members. As per the literature survey and field works, the design parameters of the sandwich structure are obtained, and thereafter the conceptual design of the sandwich composite test specimens are modeled with the help of a perfect modeling tool, i.e., CATIA. Later than this modeling process, the discretization is the focused step, which converts the test model into a finite element model through an advanced discretization tool, i.e., ANSYS Mesh Tool. By using this tool, good structural and unstructural grids are formed and the obtained quality is 0.975. Next, the composite test specimens are prepared with the support of the ANSYS Composite Preprocessor, wherein the facing and core materials are assembled through the help of rosette and orientational set-ups. The structural analyses of various sandwich composites are computed in ANSYS Structural Tool. The predominant outcomes such as total deformation, equivalent elastic strain, maximum principal stress, and equivalent stress are computed for

Figure 16. Comprehensive structural outcomes [equivalent strain] of honeycomb structures

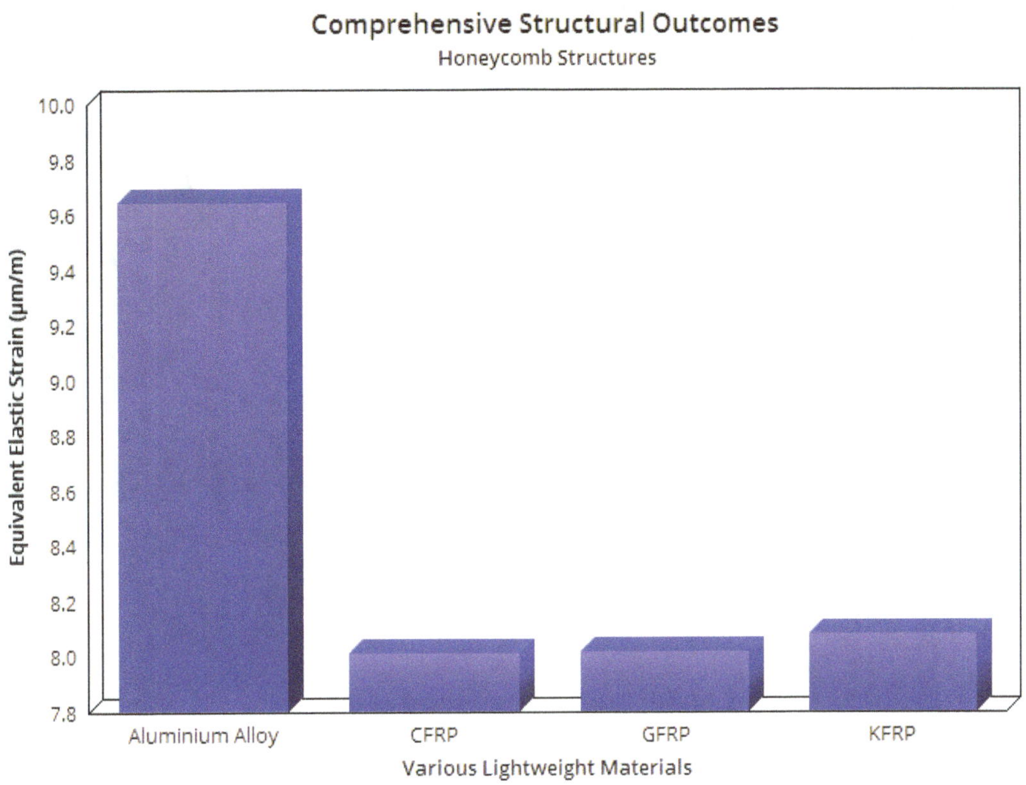

Figure 17. Maximum principal stress of aluminium alloy for integrated pentagonal structure

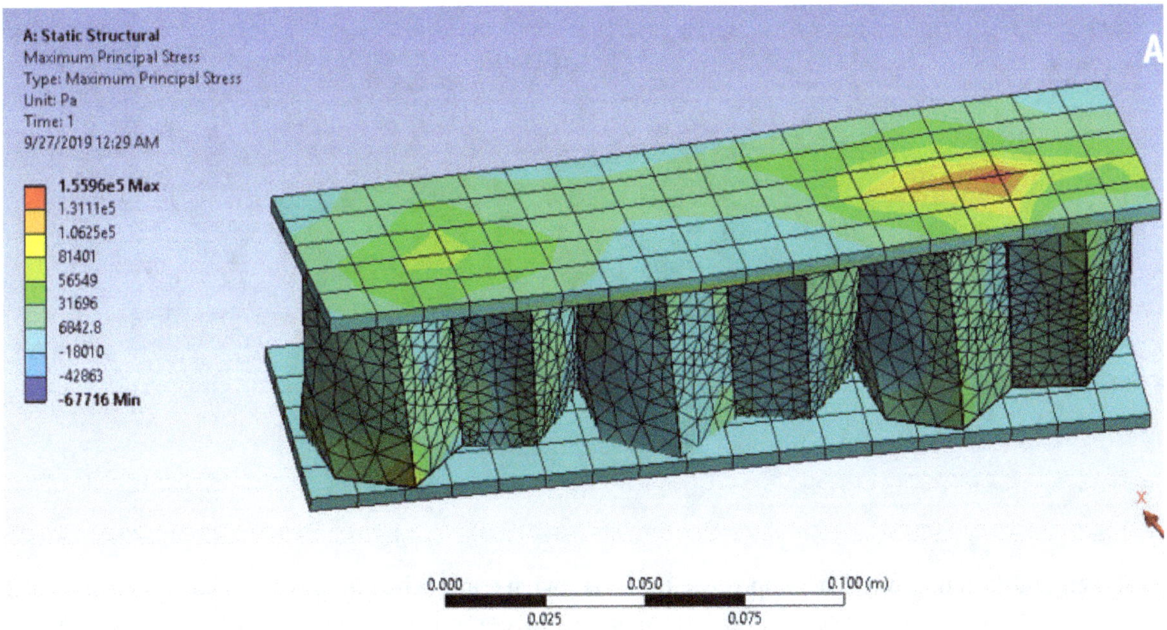

Figure 18. Equivalent stress of aluminium alloy for integrated pentagonal structure

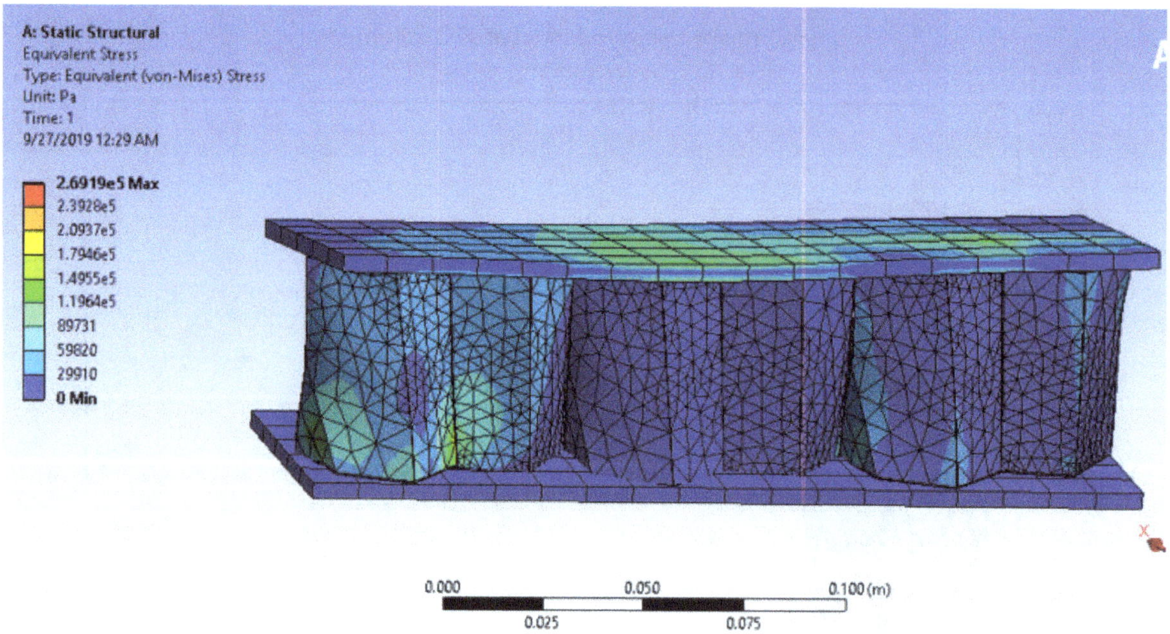

Figure 19. Maximum principal stress of CFRP for integrated pentagonal structure

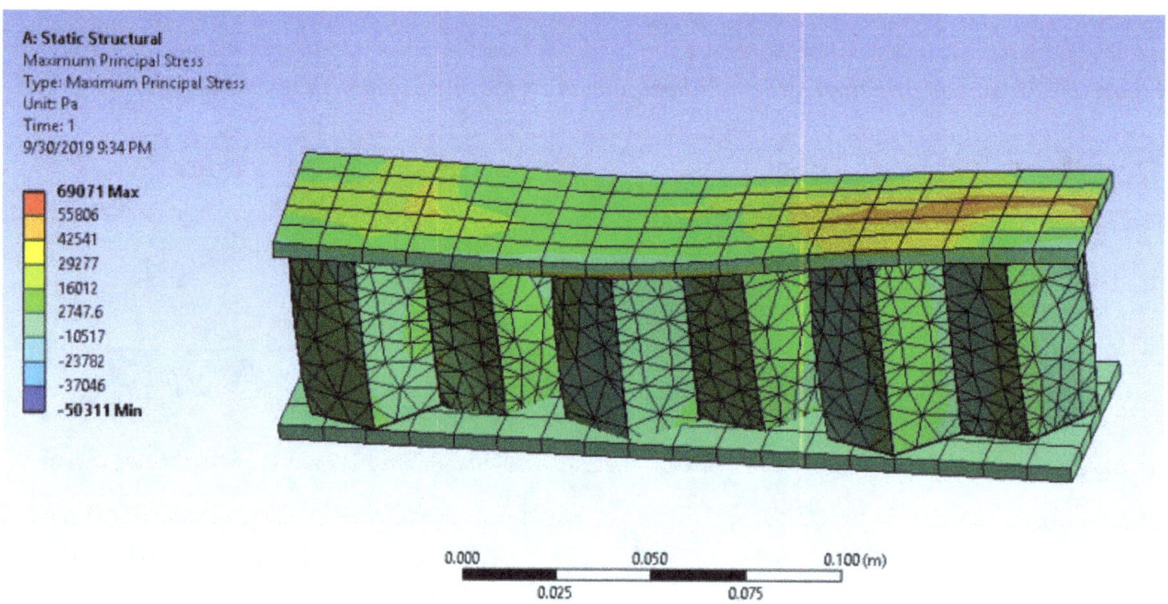

totally eight different sandwich composites. Four sandwich composite compositions have corresponded for Honeycomb based core structure and the remaining four composite sandwich compositions are belong

Figure 20. Maximum principal stress of GFRP for integrated pentagonal structure

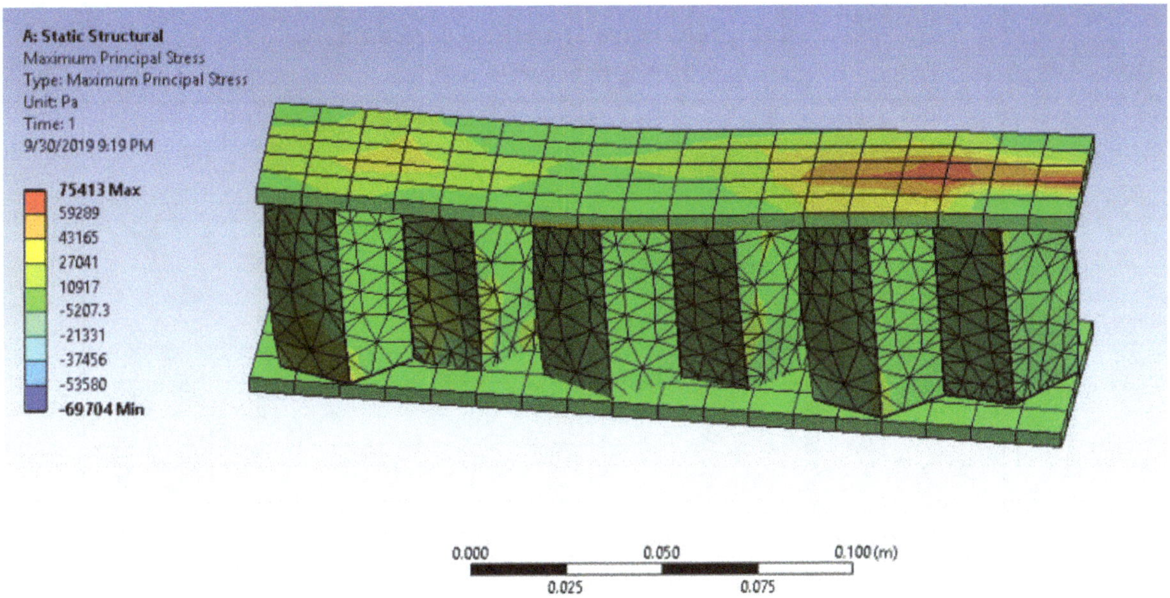

to Integrated Pentagonal Structure based core structure. Under sandwich structures, two different internal structures are imposed, which are Honeycomb based Hexagonal structure and Integrated Pentagonal Structure. In which, the advanced design is proposed in the design, which is twisted cum integrated

Figure 21. Equivalent stress of KFRP for integrated pentagonal structure

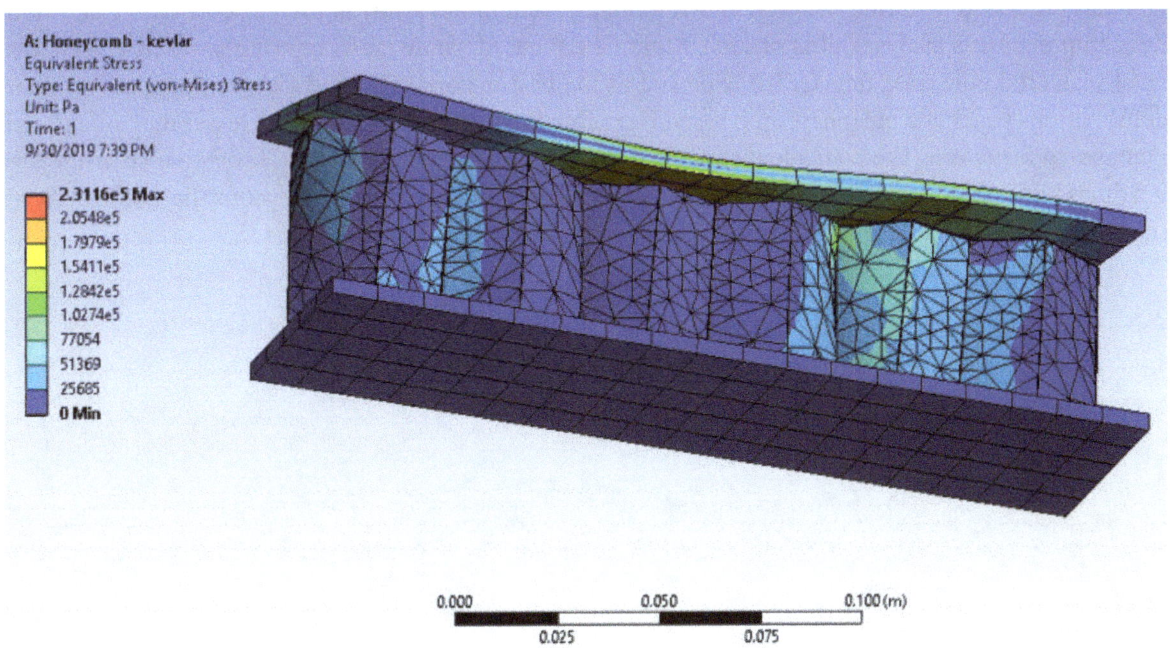

Figure 22. Comprehensive total deformation variations of integrated pentagonal structure

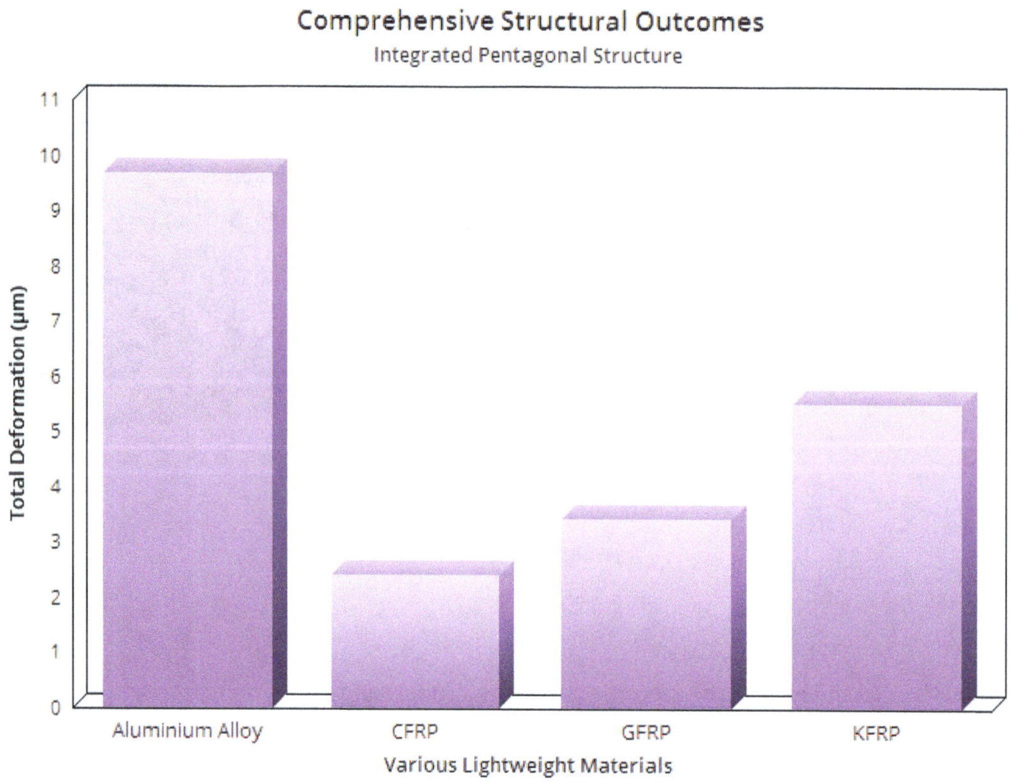

hexagonal and pentagonal shapes based core structure. The CFRP, GFRP, and KFRP based composite laminates are used as facing material and commonly Aluminium Alloy is used as core material. From the comprehensive analyses, it is observed that CFRP-based sandwich composite is reacted very low level structural outcomes thus CFRP is picked as the best material for sandwich composite. The total deformations generated in Honeycomb based Hexagonal structure are 12.84 times lesser than Integrated Pentagonal Structure. The equivalent stresses generated in Honeycomb based Hexagonal structure are 9.5 times lesser than Integrated Pentagonal Structure. Thus CFRP oriented Honeycomb based Hexagonal structure is shortlisted as the best composition for sandwich composite.

Figure 23. Comprehensive equivalent stress variations of integrated pentagonal structure

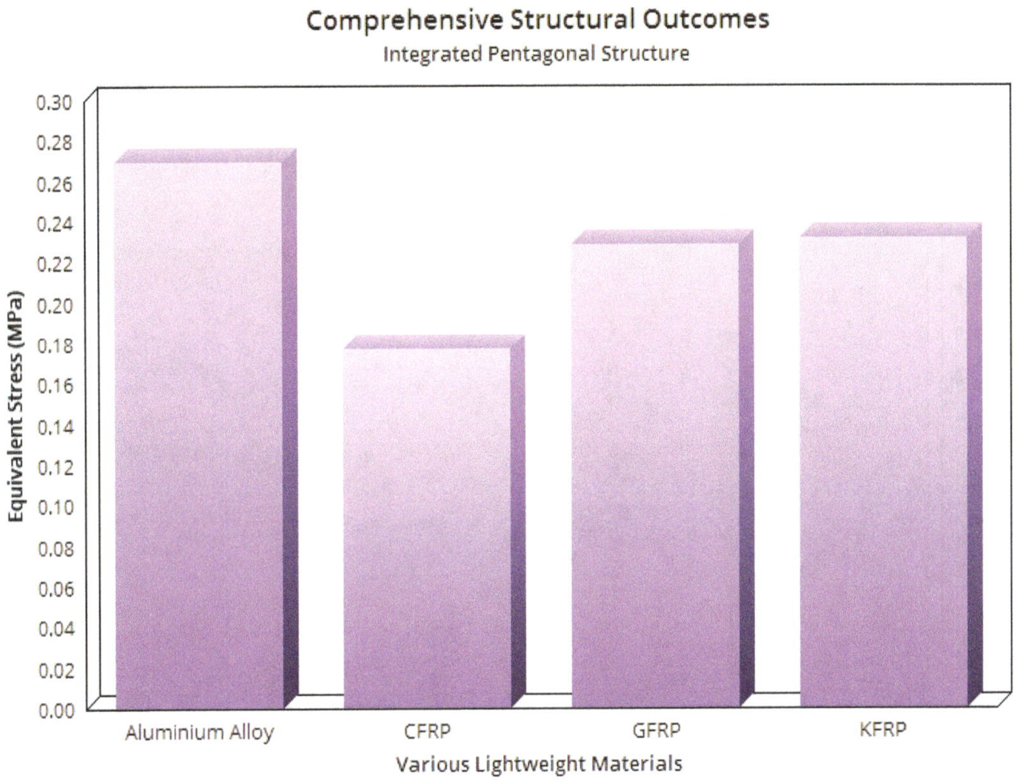

Figure 24. Comprehensive principal stress variations of integrated pentagonal structure

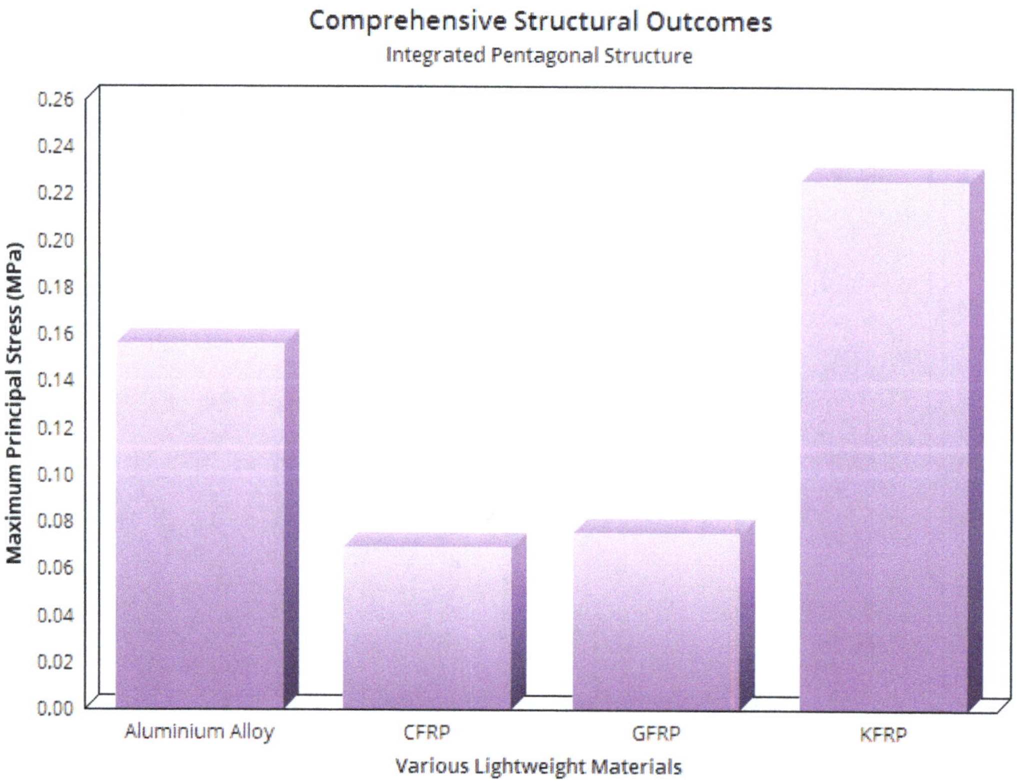

Figure 25. Comprehensive equivalent strain variations of integrated pentagonal structure

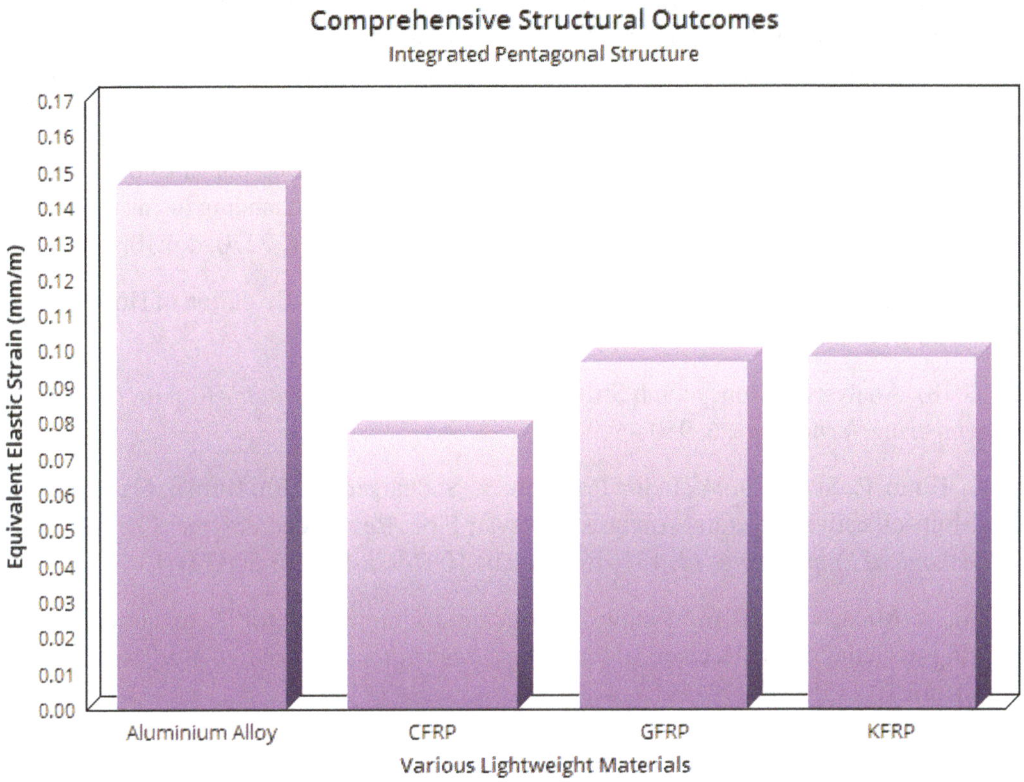

REFERENCES

Aswin Kumar, V., Sivaguru, M., Rohini Janaki, B., Sumanth Eswar, K. S., Kiran, P., & Vijayanandh, R. (2021). Structural Optimization of Frame of the Multi-Rotor Unmanned Aerial Vehicle through Computational Structural Analysis. *IOP. Journal of Physics: Conference Series*, *1849*(1), 012004. doi:10.1088/1742-6596/1849/1/012004

Bhagavathiyappan, S., Balamurugan, M., Rajamanickam, M., Vijayanandh, R., Raj Kumar, G., & Senthil Kumar, M. (2020). Comparative computational impact analysis of multi-layer composite materials. AIP Conference Proceedings, 2270. doi:10.1063/5.0019380

Chen, Chen, Pei, Zhang, & Fang. (2012). Optimal Design of Sandwich Beams With Lightweight Cores in Three-Point Bending. *International Journal of Applied Mechanics*, *4*(3). . doi:10.1142/S1758825112500330

Dayyani, I., Shaw, A. D., Saavedra Flores, E. I., & Friswell, M. I. (2015). The Mechanics of Composite Corrugated Structures: A Review with Applications in Morphing Aircraft. *Composite Structures*, *133*(1), 358–380. doi:10.1016/j.compstruct.2015.07.099

Gómez-Gálvez, P., Vicente-Munuera, P., Tagua, A., Forja, C., Castro, A. M., Letrán, M., Valencia-Expósito, A., Grima, C., Bermúdez-Gallardo, M., Serrano-Pérez-Higueras, Ó., Cavodeassi, F., Sotillos, S., Martín-Bermudo, M. D., Márquez, A., Buceta, J., & Escudero, L. M. (2018). Scutoids are a geometrical solution to three dimensional packing of epithelia. *Nature Communications*, *9*(1), 2960. doi:10.103841467-018-05376-1 PMID:30054479

Indira Prasanth, S., Kesavan, K., Kiran, P., Sivaguru, M., Sudharsan, R., & Vijayanandh, R. (2020). Advanced structural analysis on e-glass fiber reinforced with polymer for enhancing the mechanical properties by optimizing the orientation of fiber. AIP Conference Proceedings, 2270. doi:10.1063/5.0019378

Jangavali & Kamble. (2016). Finite Element Analysis and Experimental Evolution of Honeycomb Panel. *International Journal of Science and Research*, *5*(9), 1070 - 1073.

Joshikar. (2018). Analysis of Honeycomb Structure. *International Journal for Research in Applied Science & Engineering Technology*, *6*, 950 – 958.

Kesavan, K., Kiran, P., Sivaguru, M., Indira Prasanth, S., Sudharsan, R., Raj Kumar, G., & Vijayanandh, R. (2021). Multi-objective Structural Analysis of Kevlar Fiber Reinforced Polymer Composite. *Lecture Notes in Mechanical Engineering*, *11*, 137–151. doi:10.1007/978-981-16-3033-0_13

Krzyhak, A., & Micha, B. (2016). Sandwich Structured Composites for Aeronautics: Methods of Manufacturing Affecting Some Mechanical Properties. *International Journal of Aerospace Engineering*, *7816912*, 10. doi:10.1155/2016/7816912

Lim, J., You, C., & Dayyani, I. (2020). Multi-objective topology optimization and structural analysis of periodic spaceframe structures. *Materials & Design*, *190*, 108552. doi:10.1016/j.matdes.2020.108552

Liu, T., Deng, Z. C., & Lu, T. J. (2007). Structural modeling of sandwich structures with lightweight cellular cores. *Lixue Xuebao*, *23*(5), 545–559. doi:10.100710409-007-0096-z

Mathaiyan, V., Vijayanandh, R., & Jung, D. W. (2020). Determination of Strong Factor in Bird Strike Analysis using Taguchi's method for Aircraft Manufacturing guide. *Journal of Physics: Conference Series*, *1733*(012002), 1–6. doi:10.1088/1742-6596/1733/1/012002

Mirrudula, P., Kaviya Priya, P., Malavika, M., Raj Kumar, G., Vijayanandh, R., & Senthil Kumar, M. (2020). Comparative structural analysis of the sandwich composite using advanced numerical simulation. AIP Conference Proceedings, 2270. doi:10.1063/5.0019370

Mohamed Bak, K., Raj Kumar, G., Ramasamy, N., & Vijayanandh, R. (2020). Experimental and Numerical Studies on Mechanical characterization of EPDM/S-SBR with Nanoclay Composites. *IOP Conference Series. Materials Science and Engineering*, *912*(052016), 1–11. doi:10.1088/1757-899X/912/5/052016

Naveen Kumar, K., Vijayanandh, R., Bruce Ralphin Rose, J., Swathi, V., Narmatha, R., & Venkatesan, K. (2019). Research on Structural behavior of Composite Materials on different Cantilever Structures using FSI. *International Journal of Engineering and Advanced Technology*, *8*(6S3), 1075 - 1086. doi:10.35940/ijeat.F1178.0986S319

Naveen Kumar, K., Vijayanandh, R., Raj Kumar, G., Sanjeev, B., Balachander, H., & Guru Prasad, S. (2018). Comparative Approaches for Fatigue Life Estimation of Aluminium Alloy for Aerospace Applications. *International Journal of Vehicle Structures and Systems*, *10*(4), 282-286. doi:10.4273/ijvss.10.4.11

Raj Kumar, G., Balasubramaniyam, S., Senthil Kumar, M., Vijayanandh, R., Raj Kumar, R., & Varun, S. (2019). Crash Analysis on the Automotive Vehicle Bumper. *International Journal of Engineering and Advanced Technology*, *8*(6S3), 1602 - 1607. . doi:10.35940/ijeat.F1296.0986S319

Raj Kumar, G., Senthil Kumar, M., Vijayanandh, R., Raja Sekar, K., Mohamed Bak, K., & Varun, S. (2019). The Mechanical Characterization Of Carbon Fiber Reinforced Epoxy with Carbon Nanotubes. *International Journal of Mechanical and Production Engineering Research and Development*, *9*(1), 243-255.

Raj Kumar, G., & Vijayanandh, R. (2020). Comparative Investigations on the Main Elements of Carbon Fiber Based Composites Using Computational Structural Simulations. *Journal of Physics: Conference Series*, *1504*(012003), 1–11. https://iopscience.iop.org/article/10.1088/1742-6596/1504/1/012003

Raj Kumar, G., Vijayanandh, R., Kamaludeen, M. B., Balasubramanian, S., Jagadeeshwaran, P., & Ramesh, M. (2020). Comparative Structural Characterization of Fiber Reinforced Composite Rotating Disc: A Validated Investigation. *Tribology in Industry*, *42*(4), 608–620. doi:10.24874/ti.899.05.20.10

Raj Kumar, G., Vijayanandh, R., Mohammad Bak, K., Shyam Chander, R., & Arawinth, R. (2018). Experimental Testing On Mechanical Properties Effect of Aluminum Foam. *International Journal of Mechanical and Production Engineering Research and Development*, *8*(7), 1047-1059.

Raj Kumar, G., Vijayanandh, R., Senthil Kumar, M., Naveen Kumar, K., & Ahilla Bharathy, L. (2017). Conceptual design and structural analysis of integrated composite Micro Aerial Vehicle. *Journal of Advanced Research in Dynamical and Control Systems*, *9*(14), 857–881.

Raj Kumar, G., Vijayanandh, R., Senthil Kumar, M., & Sathish Kumar, S. (2017). Experimental Testing and Numerical Simulation on Natural Composite for Aerospace Applications. AIP Conference Proceedings, 1953, 090045-1–090045-5. doi:10.1063/1.5032892

Rajagurunathan, M., Raj Kumar, G., Vijayanandh, R., Vishnu, V., Rakesh Kumar, C., & Mohamed Bak, K. (2018). The Design Optimization of the Circular Piezoelectric Bimorph Actuators Using FEA. *International Journal of Mechanical and Production Engineering Research and Development*, *8*(7), 410-422.

Ramesh, M., Vijayanandh, R., & Raj Kumar, G. (2021). Comparative Structural Analysis of Various Composite Materials based Unmanned Aerial Vehicle's Propeller by using Advanced Methodologies. *IOP Conf. Series: Materials Science and Engineering*, *1017*, 1-10. 10.1088/1757-899X/1017/1/012032

Senthil Kumar, M., Krishnan, A. S., & Vijayanandh, R. (2021). Computational, Experimental And Surface Characterisation Of Glass Fibre Reinforced Plastic Composite For Wind Turbine Blade Application. *Journal of Environmental Protection and Ecology*, *22*(2), 602–616.

Senthil Kumar, M., & Vijayanandh, R. (2018). Vibrational fatigue analysis of NACA 63215 small horizontal axis wind turbine blade. *Materials Today Proceedings*, *5*(2), 6665–6674. . doi:10.1016/j.matpr.2017.11.323

Senthil Kumar, M., Vijayanandh, R., & Gopi, B. (2018). Numerical Investigation on Vibration Reduction in Helicopter Main Rotor Using Air Blown Blades. *International Journal of Mechanical and Production Engineering Research and Development*, *8*(7), 152-164.

Sergio, D. (2011). Optimization of laminated composite plates and shells using genetic algorithms, neural networks and finite elements. *Latin American Journal of Solids and Structures*, *8*(4), 413–427. doi:10.1590/S1679-78252011000400003

Udhaya Prakash, R., Raj Kumar, G., Vijayanandh, R., Senthil Kumar, M., & Ram Ganesh, T. (2016). Structural analysis of aircraft fuselage splice joint. *IOP Conference Series: Materials Science and Engineering*, *149*.

Venkatesan, K., Geetha, S., Vijayanandh, R., Raj Kumar, G., Jagadeeshwaran, P., & Raj Kumar, R. (2020). Advanced structural analysis of various composite materials with carbon nano-tubes for property enhancement. AIP Conference Proceedings, 2270. doi:10.1063/5.0019367

Venkatesan, K., Ramanathan, K., Vijayanandh, R., Selvaraj, S., Raj Kumar, G., & Senthil Kumar, M. (2020). Comparative structural analysis of advanced multi-layer composite materials. *Materials Today: Proceedings*, *27*(3), 2673–2687. doi:10.1016/j.matpr.2019.11.247

Vijayanandh, R., Naveen Kumar, K., Senthil Kumar, M., Raj Kumar, G., Naveen Kumar, R., & Ahilla Bharathy, L. (2018). Material Optimization of High Speed Micro Aerial Vehicle using FSI Simulation. *Procedia Computer Science*, *133*, 2-9. doi:10.1016/j.procs.2018.07.002

Vijayanandh, R., Raj Kumar, G., & Jagadeeshwaran, P. (2016). Comparative Numerical Analyses of Different Carbon Nanotubes Added with Carbon Fiber–Reinforced Polymer Composite. *Nanomaterials and Nanocomposites: Characterization, Processing, and Applications*, *9*, 139 – 165. doi:10.1201/9781003160946-12

Vijayanandh, R., Senthil Kumar, M., Vasantharaj, C., Raj Kumar, G., & Soundarya, S. (2016). Numerical Study on Structural Health Monitoring for Unmanned Aerial Vehicle. *Journal of Advanced Research in Dynamical and Control Systems*, *9*(6), 1937–1958.

Vijayanandh, R., Venkatesan, K., Ramesh, M., Raj Kumar, G., & Senthil Kumar, M. (2019). Optimization of Orientation Of Carbon Fiber Reinforced Polymer Based On Structural Analysis. *International Journal of Scientific & Technology Research, 8*(11), 3020 – 3029.

Vijayanandh, R., Venkatesan, K., Senthil Kumar, M., Raj Kumar, G., Jagadeeshwaran, P., & Raj Kumar, R. (2020). Comparative fatigue life estimations of Marine Propeller by using FSI. *IOP - Journal of Physics: Conference Series*, *1473*(012018), 1–8. doi:10.1088/1742-6596/1473/1/012018

Yuan, C., Bergsma, O., Koussios, S., Zu, L., & Beukers, A. (2012). Optimization of Sandwich Composites Fuselages under Flight Loads. *Applied Composite Materials*, *19*(1), 47–64. doi:10.100710443-010-9180-9

Chapter 13
The Role of Self-Assembly in Additive Manufacturing of Aerospace Applications

Rafael Vargas-Bernal
https://orcid.org/0000-0003-4865-4575
Instituto Tecnológico Superior de Irapuato, Mexico

ABSTRACT

Additive manufacturing is a strategy to produce parts with complex geometries whose process is prohibitive in cost or impossible through subtractive or formative techniques. Research groups are optimizing additive manufacturing processes to improve their performance and reduce the cost of aerospace parts. One of the emerging design techniques is self-assembly which seeks to reduce the number of parts to produce best design practices and rules. Self-assembly represents a comprehensive strategy that improves process time, product quality, cost of materials, and printability. The purpose of this chapter is to review the technological contributions that self-assembly has had in the additive manufacturing of aerospace parts. Future perspectives of the role of self-assembly in additive manufacturing are proposed. According to what was found in this research, self-assembly will facilitate the additive manufacturing of parts in various technological sectors where the manufacture of lightweight parts with high added value and restrictive regulations are required.

INTRODUCTION

In the search for techniques and technologies for the low-volume production of innovative, customized, and sustainable products, the manufacturing industry has introduced additive manufacturing in the last decades (Uriondo, 2015). Additive manufacturing is the process of joining materials to produce objects from three-dimensional model data that are additively placed layer upon layer. It is regularly used to make rapid prototypes, but thanks to the optimization of processes and material properties, it is now feasible to build aerospace parts for direct assembly purposes to systems operating in the field. Until now many techniques applied in additive manufacturing have been developed, such as stereolithography (SL), inkjet

DOI: 10.4018/978-1-7998-7864-3.ch013

printing (IJP), fused deposition modeling (FDM), selective laser sintering (SLS), selective laser melting (SLM), electron beam melting (EBM), direct metal deposition (DMD), among others, as shown in Figure 1. However, not all of them can produce metal parts. SLS, SLM, laser metal deposition (LMD), EBM, and wire and arc additive manufacturing (WAAM) processes are the most versatile processes for producing complex functional and metallic components to meet stringent requirements from the aerospace industry (Herzog, 2016; Ligon, 2017; Yusuf, 2019; Sanchez-Rexach, 2020; Alghamdi, 2021).

Figure 1. Additive manufacturing processes

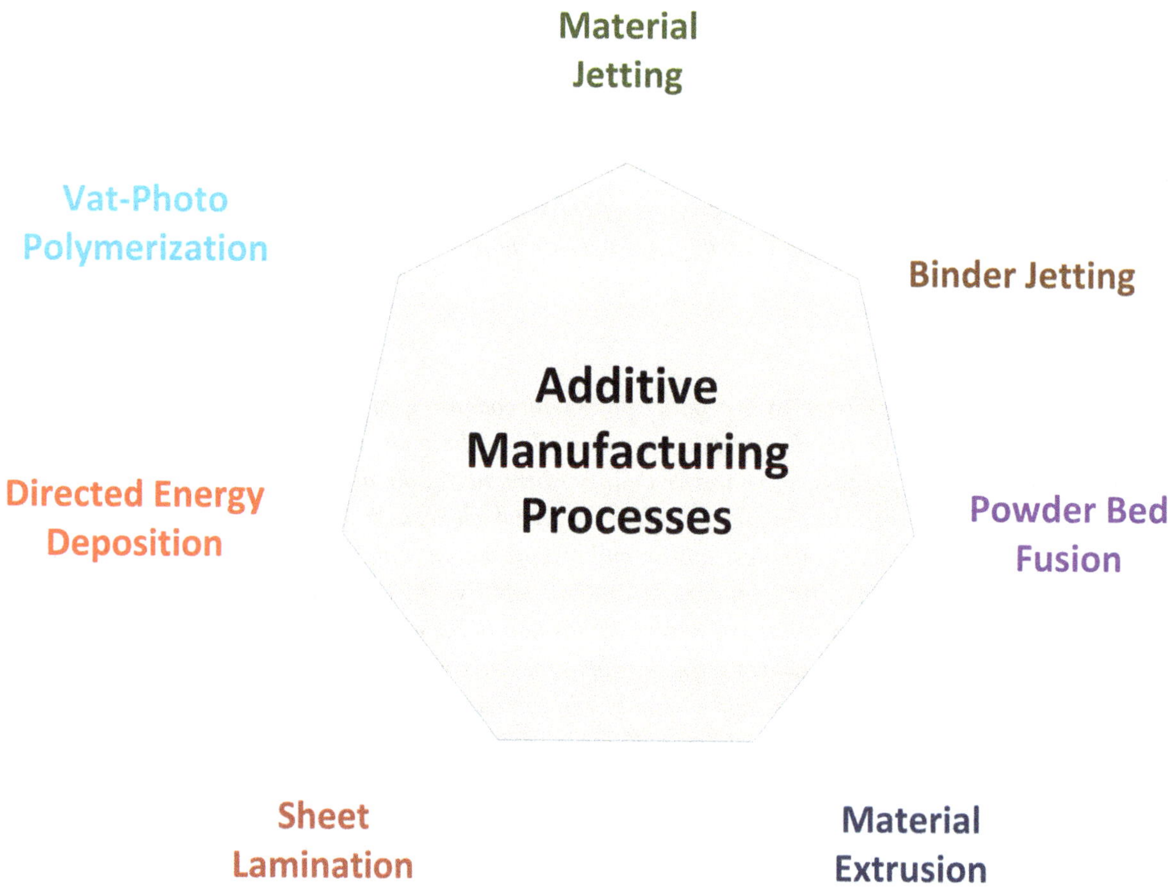

The aerospace industry consists of different players including original equipment manufacturers (OEMs), maintenance, repair, and overhaul (MRO) organizations, and commercial aerospace operators (CAOs) (Singamneni, 2019). In the manufacturing process, aircraft parts manufacturers such as Boeing, Airbus, GE Aviation, Lockheed Martin, BAE Systems, and Rolls-Royce Holdings are the main interested in bringing additive manufacturing to certification standards. Aerospace components consist of many parts, but their demand is unpredictable as they replace dispersed parts required only in scheduled or unscheduled maintenance events. This brings inventory levels to the limit of 10% for spare parts, which results in unpredictable exchange times and makes them expensive. Replacement parts are of four types: rotatable, repairable, expendable, and consumable. Each type has a different replacement policy. The policy is associated with the volume of the repair and the delivery times of the supply and needs to be updated regularly to adapt to changes in the market. Traditional manufacturing mechanisms have already reached the limits of weight reduction, but the need for better and lighter designs keeps the search for manufacturing methods to tighten these limits. Until now, significant research in terms of assembly has been focused on the design of additive manufacturing principles and their integration into design generation tools such as topology optimization, while research on part consolidation has been restricted to heuristic guidelines (Crispo, 2020).

The layer-by-layer consolidation offered by additive manufacturing leads to more complex designs, performance optimizations, lighter weight, integrated part designs, and sustainable performance attributes (Singamneni, 2019). So far, the use of additive manufacturing has been explored in the manufacture of parts such as hinges, brackets, interior components, lightweight fuselage, and airframe design, to improve fuel efficiency. Emerging applications are in engine components such as turbine blades with internal cooling channels, fuel nozzles, compressors, and integrated piping systems. The possible use of a selective variation of the material composition across the part even allows sensors or other devices to be embedded in the printed parts. Design optimization in aerospace applications has multidisciplinary criteria such as material and structural integrity, aerodynamics, weight, reliability, manufacturability, maintainability, sustainability, and cost.

Advances in computer-aided design and additive manufacturing technologies have removed the limitations described above, opening new opportunities in different areas of opportunity as discussed below (Singamneni, 2019).

- Weight reduction is a key task to improve performance and efficiency in the aerospace industry. The use of novel geometries based on cellular structures, lattices, and honeycomb or optimized structures improves the overall performance of the aerospace system. Optimized designs reduce operational costs during the assembly and maintenance stages of these systems. The topology is optimized by considering lightweight designs through finite-element analysis and iteratively removing portions of material that are operationally scrubbable. This makes additive manufacturing a unique solution for producing highly complex and optimized parts.
- Additive manufacturing can produce complex shapes which allow the optimization of parts with specific functionalities such as stress distribution, heat dissipation, or airflow patterns.
- Design freedom opens the possibility of integrating parts and manufacturing multiple components as single units. In this way, assembly effort is reduced, and component performance is improved. Reducing the number of parts and installation times leads to lower inventory pressures and operational costs. Additionally, this leads to lower logistics costs, more choice in supplier selection, and the elimination of non-recurring tooling costs, to improve the overall efficiency of the supply chain. And with it, additive manufacturing reduces the cost of component production, which gives it more flexibility related to the life cycle of products.
- Since the production volume of aerospace parts is normally limited to no more than several thousand units per part, the choice of additive manufacturing for the aerospace industry is justified.
- Additive manufacturing brings the reduction of materials and processes as well as an increase in the efficiency of energy used. The weight reduction produces considerable savings in fuel costs over the life span of an aerospace system. Material selection and geometric optimization of parts play an important role in design improvements.

- Material selection and geometric optimization of parts play an important role in design improvements. Despite advances so far, additive manufacturing technologies are not fully mature, lack repeatability, and suffer from anisotropy. Besides, the surface finish is not good as it affects the fatigue life. Building size limitations on the other hand restrict the application of the technology.

All these details continue to limit the certified use of additive manufacturing in the aerospace industry. The purpose of this chapter is to study the role that self-assembly plays in the additive manufacturing of aerospace parts. The chapter has been divided as follows: In the section entitled "*Background*", the author describes the concept of self-assembly as well as its advantages and disadvantages. Later in the section entitled "*What is the Purpose of Self-Assembly in Additive Manufacturing or 4D Printing?*", the relevance that self-assembly has for additive manufacturing is discussed. The section titled "*Solutions and Recommendations in Aerospace Applications*" describes the contribution that self-assembly-based additive manufacturing has to the aerospace industry. The section titled "*Future Research Directions*" below summarizes some of the future research trends in this field. Finally, the "*Conclusions*" section provides a breakdown of the main results of this research study.

BASIC SELF-ASSEMBLY CONCEPTS

Self-assembly is a process at which a disordered system constituted of similar or different pre-existing elements forms an organized structure or pattern through local or specific physical or chemical interactions among its elements without external stimulus. Pre-existing elements can be molecules (polymers), ions (ceramics), or atoms (metals and alloys). For example, when the pre-existing components are molecules, the process is called molecular self-assembly. Furthermore, self-assembly classifies as static (S) or dynamic (D) (Whitesides, 2002). In static self-assembly, the ordered state involves achieving systems in equilibrium by reducing their free energy or producing a solid, as shown in Figure 2. Whereas in dynamic self-assembly, the interactions among the components lead to local self-organization that does not necessarily lead to a solid. Template self-assembly forms structures by interactions between components and regular features in their environment, for example, crystallization on surfaces. Some applications of self-assembly materials are summarized in Table 1 (Whitesides, 2002).

In the classical sense, species self-assembly takes place spontaneously through non-covalent, weak interactions, and its organization is reversible. In solution, these interactions give place to covalent bonds, and a condensation reaction assembles these units in a solid. Furthermore, the introduction of tempering agents controls the morphology of the species formed. Any physical principle guides the ordering and the aggregation of species.

Entropy maximization achieves self-assembly. Through this physical principle, target structures in a controllable manner under convenient physicochemical conditions are achievable. The self-assembly of particles is possible by attracting them to the space between a pair of conductive electrodes through the application of an electric field gradient. Among the different property possibilities to produce self-assembly are magnetic fields, capillary interactions of particles trapped at interfaces, and elastic interactions for particles suspended in liquid crystals. Regardless of the mechanism chosen, material synthesis is used as the preferred self-assembly approach to avoid the problem of having to manufacture materials through one building block at a time. These approaches reduce the amount of time required to position the

Figure 2. Self-assembly process in a smart material

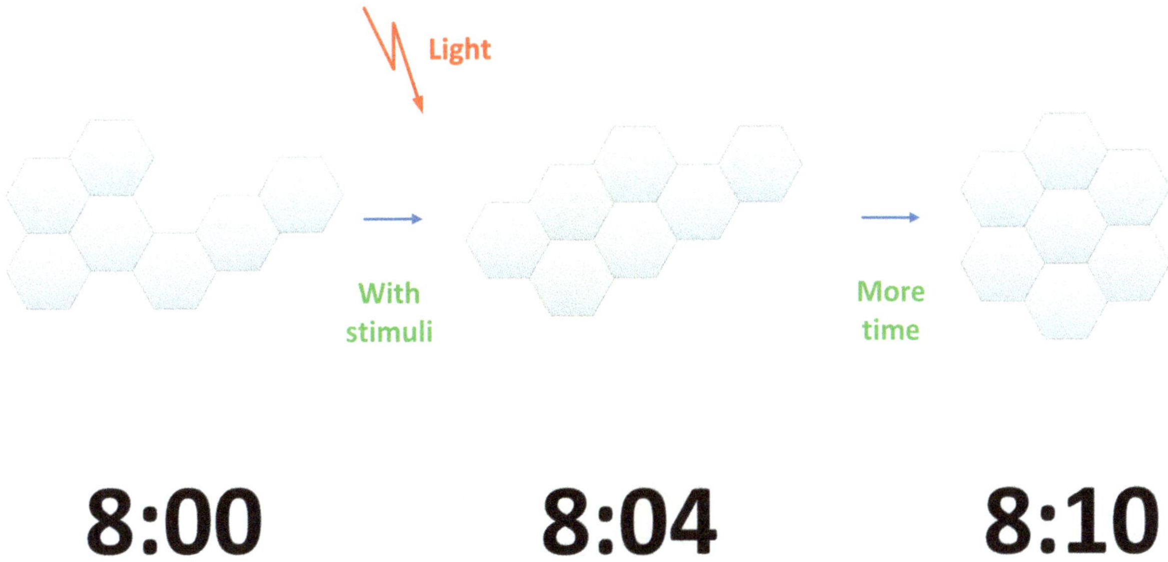

building blocks. It leads towards target structures that are macroscopic in size. When macroscopic-sized tunable materials are self-assembling, a host of high-value-added technological applications is possible.

At this point, the analysis of the process introduces the following question. Does any chemical reaction lead to the self-assembly of atoms, ions, or molecules? For example, the precipitation process leads to self-assembly. At least three distinctive features make self-assembly a unique concept: order, interactions, and building blocks, as illustrated in Figure 3.

1. The order of the components establishes the way or task in that the self-assembly uses the thermodynamic parameters present during its operation (Zheng, 2019).
2. In the self-assembly process exists two main reasons to choose the type of interaction among components. The synthesis of materials prefers the use of weak-chemical bonds over strong-chemical

Table 1. Some examples of self-assembly materials (S, static; D, dynamic; T, template)

System	Type	Applications/relevance
Atomic, ionic, and molecular crystals	S	Materials, optoelectronics
Phase-separated and ionic layered polymers	S	Catalysts, batteries, supercapacitors
Self-assembled monolayers (SAMs)	S, T	Microfabrication, sensors, nanoelectronics
Lipid bilayers and black lipid films	S	Biomembranes, emulsions
Liquid crystals	S	Displays
Colloidal crystals	S	Bandgap materials, molecular sieves
Macro- and mesoscopic structures (MESA)	S or D, T	Electronic circuits
Fluidic self-assembly	S, T	Microfabrication
Light-matter	D, T	Waveguides, quantum devices

Figure 3. Distinctive features of a self-assembly process

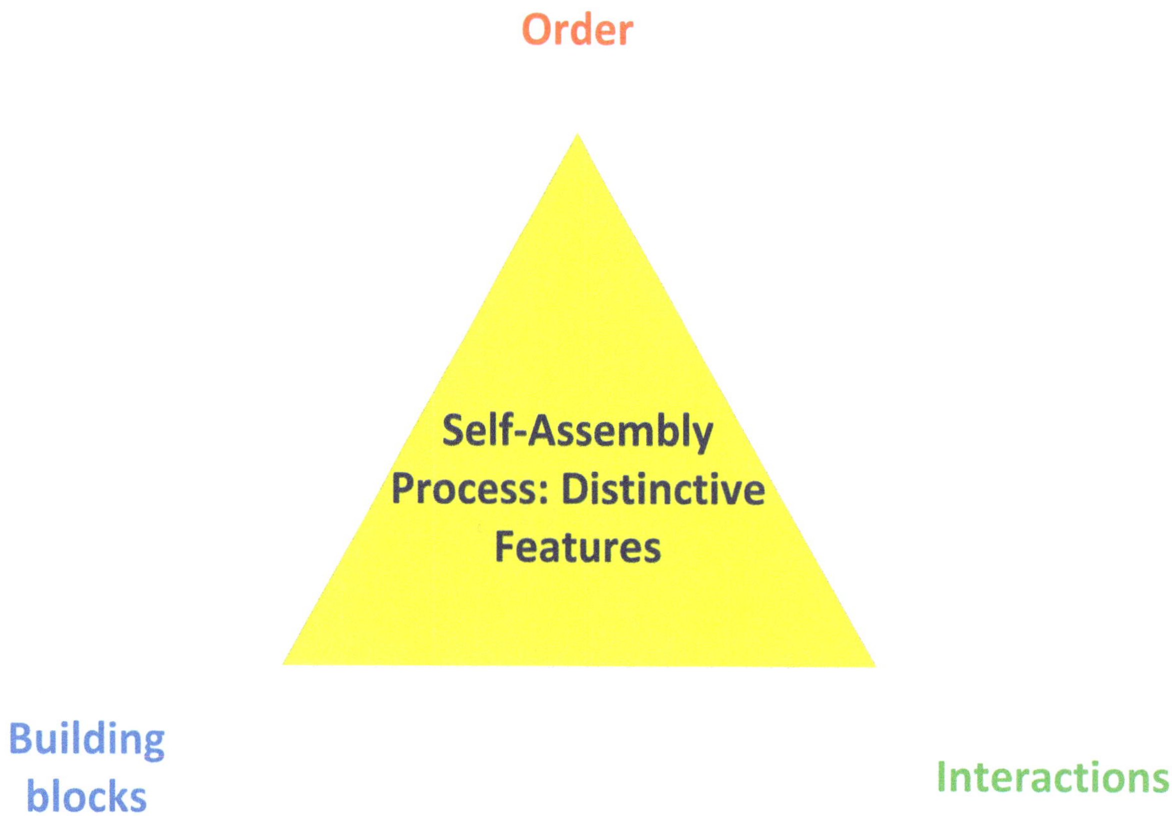

bonds. Examples of chemical-weak bonds are Van der Waals, capillaries, entropic forces, as well as π-π and hydrogen bonds (Zheng, 2019). The energy values for the different interactions are summarized in Table 2 (Swiegers, 2016). Besides, examples of chemical-strong interactions are metallic, ionic, and covalent bonds. The first reason implies that weak interactions play a prominent role in materials, especially in biological systems. These kinds of interactions determine, for example, the physical properties of liquids, the solubility of solids, and the organization of molecules in biological membranes. The second reason involves that the specificity of the interactions controls self-assembly. The highest specificity in self-assembly occurs in DNA pairing interactions, while the lowest specificity arises from the forces that produce entropy maximization (Groß, 2008).

3. Self-assembly depends on the chemical composition, functionality, and shape of the building block (Groß, 2008). Self-assembly strategies such as directional entropic forces and conventional chemical pathways synthesize the building blocks at the nanoscale. It is possible to determine the appropriate building block to produce the target self-assembly behavior through reverse engineering.

Four different types of self-assembly can be identified: (1) thermodynamic self-assembly, (2) irreversible self-assembly, (3) assisted and directed self-assembly, and (4) self-assembly with pre-, post, or intermittent modification (Swiegers, 2016), as depicted in Figure 4. Thermodynamic self-assembly involves a kinetically rapid and reversible thermodynamic equilibrium which is a widely used synthesis technique. Irreversible self-assembly uses a series of steps to generate kinetically stable products where

Table 2. The energy values for the different chemical bonds

Chemical bond	Energy value (kJ/mol)
Coordinate	~40-120
Van der Waals	~1-5
Ion-paring interactions	~12-20
Hydrophobic/hydrophilic interactions	~12-15
Hydrogen bonding	~10-20
π-π donor-acceptor interactions	~7-20

bonds must be formed correctly the first time for assembly to proceed successfully. The assisted and directed self-assemblies derived from the previous types arise from avoiding the formation of non-functional intermediates (assisted self-assembly) or stabilization of key intermediates (directed self-assembly). Finally, self-assembly with modification implies that self-assembly is followed by, preceded by, or intermixed with the formation of a covalent bond.

Figure 4. Types of self-assembly

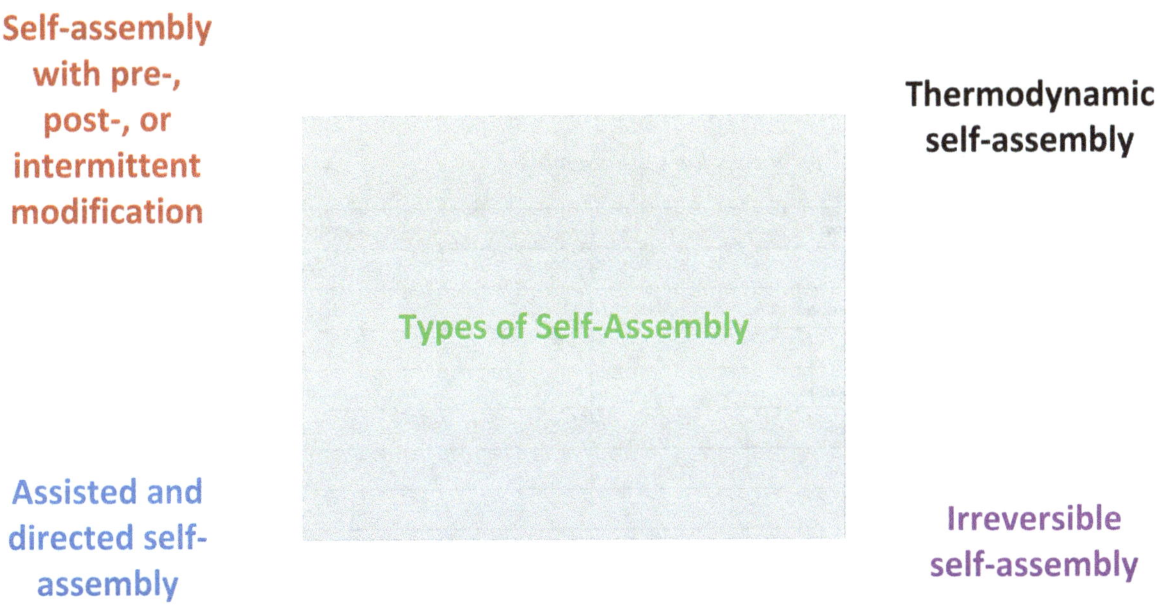

Self-assembly at the microscopic level begins with the formation of nucleation centers or seeds, followed by diffusion of the species, subsequent growth of the centers, and ends with Ostwald ripening. The free energy that drives the thermodynamic process is either enthalpic or entropic or both (Zheng, 2019). In any case, self-assembly involves the formation and breaking of bonds. A diffusion process establishes the kinetics of the self-assembly. For this process, the absorption-to-adsorption ratio follows the Langmuir adsorption model according to Fick's laws of diffusion. The desorption rate determines the bond strength of surface molecules or atoms with an activation energy barrier. The reason for growth involves the competition between these two processes.

Self-assembly seeks to optimize the use of chemistry in the synthesis process by taking advantage of the properties such as order and functionality. The order involves building blocks on all length scales. In the case of polymerization and covalent synthesis, the atoms bond with each other in any way without achieving minimal energy. The self-assembly of molecules adopts a structure that has the thermodynamic minimum. In self-assembly, design predicts the material with minimal energy regardless of the location of the atoms. One of the properties of self-assembled systems is their thermodynamic stability (Zheng, 2019). If no external force favors the self-assembly, then the system is thermodynamically stable because a lower Gibbs free energy is present. Furthermore, self-assembly leads to defect-free structures.

Any fluctuation in the thermodynamic variables modifies or compromises the material structure in self-assembly (Zheng, 2019). This event describes the sensitivity to disturbances exerted by the external environment. The weak interactions give flexibility to the arrangement and allow rearrangements of the structure according to thermodynamic events. The material will probably return to its initial condition when fluctuations in thermodynamic variables bring it to the starting point. This property of self-assembly observed in the synthesis of materials is called reversibility.

According to the previous descriptions, self-assembly involves a complex materials synthesis process due to the need to control many free parameters. The synthesis of an ordered structure by self-assembly with building blocks at the nanoscale requires the presence of attractive short-range and repulsive long-range forces, as compiled in Table 3 (Zheng, 2019). The choice of the precursors has fine control of the structures. Some physicochemical properties of the precursors can lead to more complex structures.

Table 3. Attractive and repulsive forces for self-assembly

Attractive force	Repulsive force
Bridging	Electric double-layer
Coordination bond	Hydration
Depletion	Solvation
Hydrogen bond	Steric
Hydrophobic	
π-π stacking	
Solvation	
Van der Waals	

Self-assembly involving large building units such as colloids is sensitive to external stimuli such as electric fields, magnetic fields, gravity, flow, and so on (Zheng, 2019). The spontaneous association between the components in the self-assembly covers a wide range of length scales from angstroms to centimeters, they can have different dimensions and different sources of origin, as described in Table 4. For example, in the biological case, self-assembly involves components at different scales, such as molecules, cells, and organisms. Scientists study physical models of self-assembly to understand nature and its application in advanced technology (Groß, 2008). Self-assembly processes are governed by the information encoded in the components. Coded information includes shape, surface properties, charge, polarizability, magnetic dipoles, mass, etc. (Whitesides, 2002).

Table 4. Description of the different types of self-assemblies

Type	Building units	Self-assembled systems	Characteristics	Dimension
Atomic	Metal	The epitaxial film, quantum dot	Directional, one-step, nonhierarchical	Sub-Angstroms to few nanometers
Biological	Amino acid, lipid biopolymer	DNA, RNA, protein, enzyme, membrane	Directional, stepwise, hierarchical	Nanometers to several tens of nanometers
Colloidal (mesoscopic)	Nanoparticle, nanotube, fullerene	Suspension, dispersion, sol, colloidal crystal	Random, one-step, nonhierarchical	Nanometers to a few centimeters
Interfacial	Surfactant, polymer, lipid	Surface micelle, Langmuir monolayer, Langmuir Blodgett film, Self-assembled monolayer	Directional, one-step, nonhierarchical	Sub-Angstroms to tens of nanometers
Molecular	Surfactant, polymer	Micelle, bilayer microemulsion, emulsion	Random, one-step nonhierarchical	Sub-nanometers to tens of nanometers

4D printing has generated growing interest in the development of sensitive structures for soft robotics and printable actuators used in medical devices, smart textiles, defense, and aerospace applications (Pei, 2018). Thanks to self-assembly, parts developed under additive manufacturing will no longer be restricted by the size of the print head. The parts are programmable for the post-fabrication stage, whereby the parts are self-assembled to reduce the volume of material for initial storage, paving the way to adapt smart materials to high value-added applications (Momeni, 2017). These responsive structures can be printed in a simple geometric shape before being transformed into a larger, more complex design to reduce overall printing time and manufacturing complexity. Individual parts can be printed using smaller additive manufacturing systems and then self-assembled into a larger component.

Auto-folding is a self-assembly mechanism that uses shape transformation to produce a fold, bend, or roll from a printed pattern geometry or uses thin films that are manufactured into spiral or cylindrical tube shapes (Pei, 2018). This may involve a hinged or non-hinged design. The design parameters depend on the selection of the folding mechanism, the size of the component, and the type of material. Actuation strain, actuation stress, and the ability to generate and manipulate the desired field at the chosen bend location are also considered. The active folding mechanism is obtained through an extensional fold with a variable length of the active bar or a spring connected to the two faces joined by the hinge; a torsional fold using an active torsional element in the hinge; or a flexural fold that uses an active element with a certain folded shape. These mechanisms are illustrated in Figure 5.

Conventional materials take advantage of the glass transition state of polymers for self-folding either by thermal shrinkage or the shape memory effect. Unfortunately, this makes the hinges stiff and they cannot be bent or unfolded after 4D printing. Currently, soft elastomers with metallic aggregates allow flexible and elastic hinge structures based on an origami approach, especially when subjected to numerous folding cycles (Yamamura, 2021).

Figure 5. Self-assembly mechanisms based on self-folding

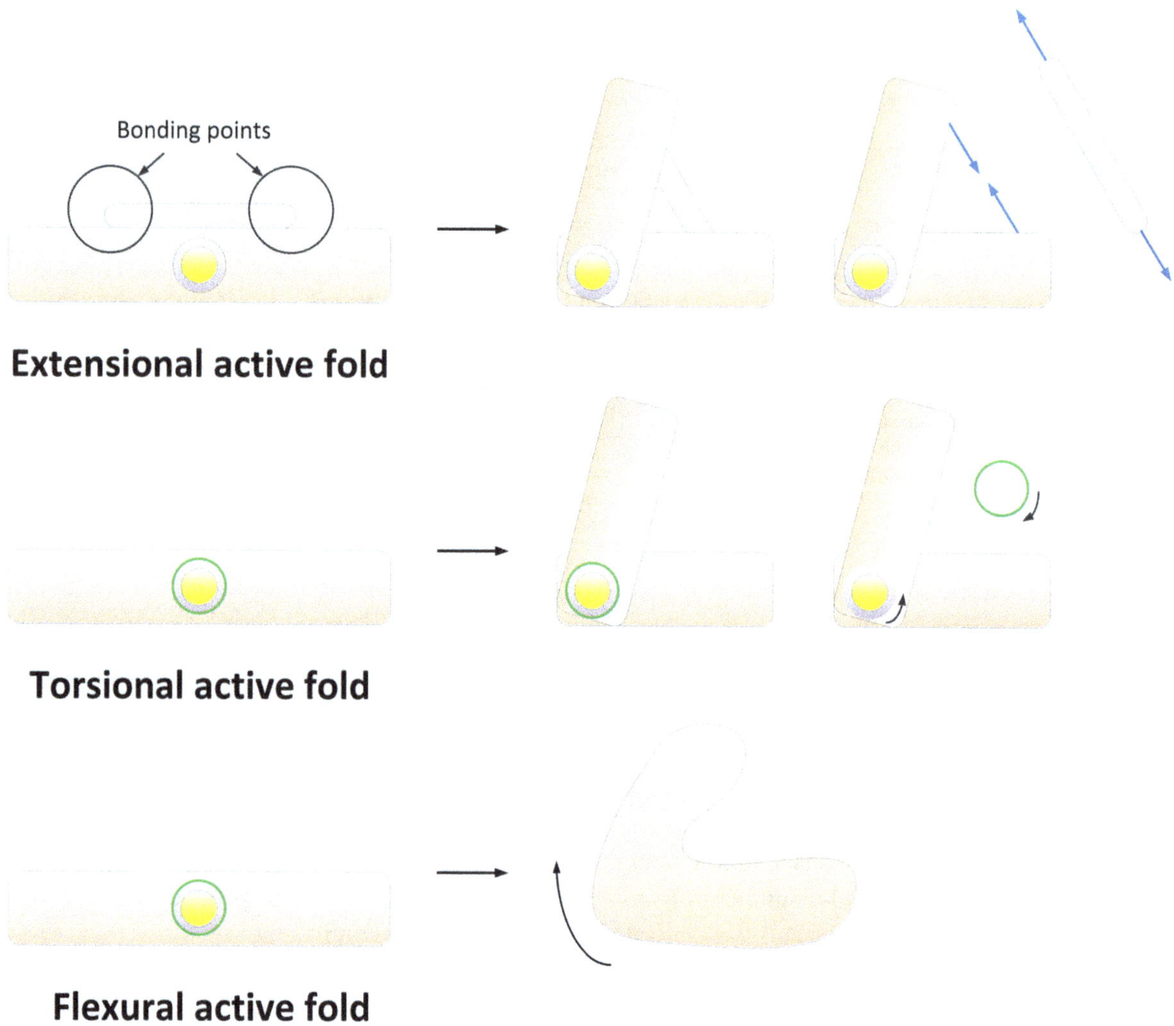

WHAT IS THE PURPOSE OF SELF-ASSEMBLY IN ADDITIVE MANUFACTURING or 4d printing?

The manufacturing of aerospace parts is often complex and time-consuming due to its stringent regulations (Singamneni, 2019). Additive manufacturing is a 3D printing used in industrial production that allows the creation of lighter and stronger systems or parts. This enables the direct production of complex parts using digital data without the need for tools or machinery. Additive manufacturing processes have 7 categories: powder bed fusion, binder jetting, direct energy deposition, extrusion, jetting, sheet lamination, and vat photo-polymerization (see Table 5, Calignano, 2017). Commonly used methods are selective laser sintering (SLS), selective laser melting (SLM), electron beam melting (EBM), stereolithography (SLA), inkjet printing (IJP), fused deposition modeling (FDM), and direct metal deposition (DMD) (Herzog, 2016; Ligon, 2017; Yusuf, 2019; Sanchez-Rexach, 2020; Alghamdi, 2021). Possible candidate materials are metals, polymers, ceramics, and composites. The main advantages of additive manufacturing over traditional manufacturing methods are lower energy consumption, better mechanical efficiency, smaller material waste, and shorter manufacturing and design lead times. Consolidation of the material at the point level or in the form of layers allows for better design flexibility, a higher level of customization, shorter lead times, and an improved supply chain solution. Due to the claims, additive manufacturing gains potential application in the aerospace industry.

Table 5. Processes, technologies, and materials used in additive manufacturing

Process	Technologies	Materials
Binder jetting	• 3D Printing • Ink-Jetting • S-Print • M-Print	• Metal • Polymer • Ceramic
Direct energy deposition	• Direct metal deposition • Laser deposition • Laser consolidation • Electron beam direct melting	• Metal: powder and wire
Material extrusion	• Fused deposition • Modeling	• Polymer
Material jetting	• Polyject • Ink-jetting • Thermojet	• Photopolymer • Wax
Powder bed fusion	• Selective laser sintering • Selective laser melting • Electron beam melting	• Metal • Polymer • Ceramic
Sheet lamination	• Ultrasonic consolidation • Laminated object manufacture	• Hybrids • Metallic • Ceramic
Vat Photopolymerization	• Stereolithography • Digital light processing	• Photopolymer • Ceramic

Self-assembly serves as an automated programmable integration method (Crane, 2011). Self-assembly processes can successfully assemble high-performance objects despite assembly errors. Despite significant errors in additive manufacturing, the self-assembly process is tolerant of significant assembly errors. The self-assembly process consists of component positioning and bonding by random interactions. This requires the formation of a spontaneous bond between the parties when they assume the desired position. Common bonding forces include chemical, electrostatic, magnetic, and surface tension. Self-assembly offers the potential for low-cost assembly at a high rate. For the self-assembly process to function as an additive manufacturing process, the programmable control of the assembly location and timing are achieved with sufficient speed and yield.

Self-assembly produces spontaneously ordered aggregates without human intervention. The structures obtained through this process normally have phases in equilibrium (Gardan, 2018). Materials regularly involved are copolymers and nanoparticles which undergo automated folding or molecular aggregations. Although its main application is in the biomedical field, its use in the aerospace sector is very attractive given the properties that can be obtained. Processes based on additive manufacturing to produce self-assembly are Fused Deposition Modeling (FDM), Multi-Jet Modeling (MJM), Digital Light Process-

Figure 6. The process of self-assembly additive manufacturing

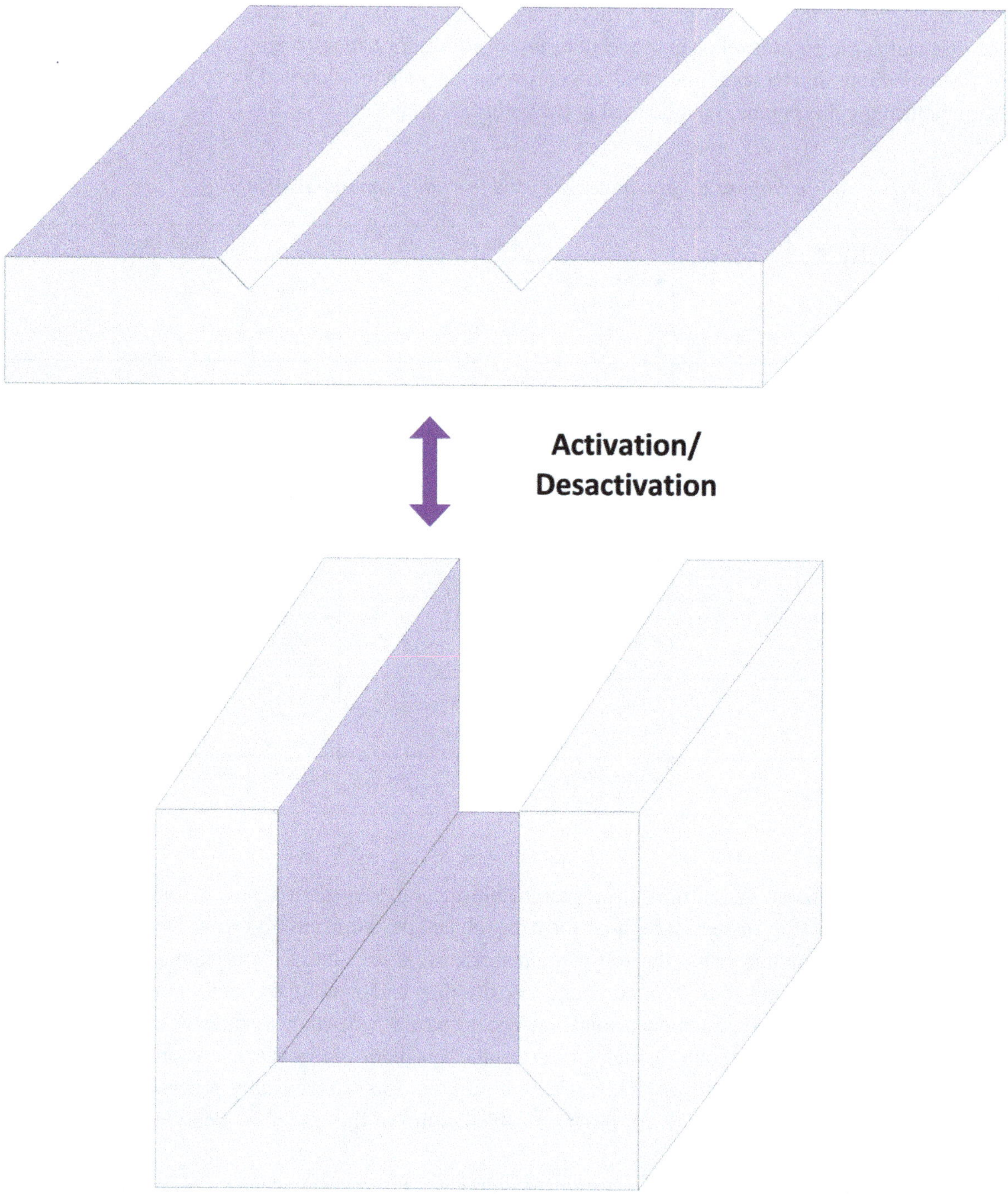

ing (DLP), Inkjet, and Stereolithography (SLA) for photo-curable methacrylate (SMP). A much more exhaustive description of the different additive manufacturing processes can be found in the following references (Herzog, 2016; Ligon, 2017; Yusuf, 2019; Sanchez-Rexach, 2020; Alghamdi, 2021).

4D printing or additive manufacturing by self-assembly combines 3D printing with variable time (Bahnini, 2018). The time variable is not related to the period used to print a part but rather to the period taken by objects to change their shapes to perform their programmed functions based simply on the properties of the materials used in the manufacture of the part. This type of self-assembly additive manufacturing uses advanced or smart materials which are programmed to reshape or include embedded functional properties that transform them when subjected to external stimuli. This strategy has been used by nature at the nanoscale as active origami, in which objects self-fold or unfold to reduce infrastructure investment to automate folding, as is depicted in Figure 6 (Gardan, 2018).

4D printing is a relatively new area of research because it has encountered numerous challenges at the level of technology, materials, and design (Bahnini, 2018). This technique represents the futuristic alternative to manufacturing processes because it provides a path to put any idea into practice. Joints and hinges for folding can be obtained by three-dimensionally printing rigid materials with smart materials (Leist, 2016). Self-assembly can be based on 1D strands to form 2D and 3D shapes, and these together with flat 2D shapes form more complex 3D objects.

4D printing or self-assembly-based 3D printing can be used to develop actuators and hinges using smart materials. The three pillars of 4D printing are morphing structures, smart materials, and 3D printing. Morphing structures are aerospace structural arrangements capable of modifying their shape without experiencing damage or permanent deformation to aerodynamic behavior (Airoldi, 2018). These allow the guidance, stabilization, and control of aerospace vehicles as well as their optimization for different missions. The most frequently used techniques for 4D printing so far are polyjet and syringe printing (Ntouanoglou, 2018). Fortunately, other 3D printing techniques discussed above are being proposed to develop 4D printing (Leist, 2016). 4D printing can be activated by water, heat-and-stress, smart memory alloys (SMAs), and by light-activated shape memory polymers. 4D printing has applications in the automotive, textile, construction, healthcare, utility, aerospace, and military industries (Leist, 2016). The defense and military sectors have the largest share, followed by the aerospace industry.

4D printing uses 3D printing and adds the fourth dimension called time. The materials used must be adapted to stimuli such as temperature, water, acidity, and light. These stimuli facilitate self-assembly, self-repair, and multiple functionalities for aerospace and medical applications (Alshahrani, 2021). During self-assembly, products use multiple parameters to change their shape and properties in a non-uniform way and return to their original shape when stimuli are removed. The common stimuli are water and temperature. However, stimuli such as pressure, humidity, light, magnetic field, gravity, heat, and light, either individually or in combination, are gaining the attention of many researchers around the world in recent years. However, the specific application dictates the appropriate choice of stimulus.

SOLUTIONS AND RECOMMENDATIONS IN AEROSPACE APPLICATIONS

Additive manufacturing technologies are suitable for many satellite applications (Calignano, 2017). These have now become capable of producing complex parts in net-form or near-net-form. Materials based on polymers, metals, alloys, ceramics, and composites are used directly in aerospace parts manufacturing. These technologies offer solutions for producing low-volume, high-value, and highly complex parts and products. When additive manufacturing integrates the aerospace industry, it solves supply chain problems and inventory barriers. The parts produced by additive manufacturing exhibit anisotropy in a particular crystalline direction due to the stacking of the layers and this can be reduced by selecting the proper orientation during the manufacture of the part.

Factors to consider about the performance of the material to be used even in the simplest aerospace component are specific resistance, fatigue resistance, resistance to plastic deformation, operating temperature, flammability, smog release, toxicity, electrical conductivity, chemical sensitivity, radiation sensitivity, appearance, sustainability of processing, and cost (Lyons, 2012). Aerospace designs must make use of products based on thin wall contours or reinforced structures, materials with low density as well as materials with a high strength-to-weight ratio, and a high stiffness-to-weight ratio to satisfy the lightweight requirement (Basheer, 2020).

Extrusion processed engineering polymers such as polyphenylsulfone (PPSU), polyetherimide (PEI), polyaryletherketone (PAEK), polyether ether ketone (PEEK), and polyarylamide (PAA) are intended for aerospace applications related to electrostatic discharge (EDS) (Calignano, 2017). Polymer inkjet printing is of considerable interest in developing prototypes for advanced applications in aerospace systems. Stereolithography (SL) enables the cost-effective production of functional components for the electronics and electromechanical sectors in the aerospace industry. In the selective laser melting process, maraging steels (aging martensitic steels) exhibit high strength and toughness properties, good weldability, and dimensional stability during aging heat treatment. This makes them attractive in the aerospace industry where these higher mechanical properties and weldability are the most important characteristics as well as their superior machinability. Aluminum alloys such as AlSi10Mg are highly desired for lightweight applications in the aerospace sector because they are age-hardenable castables and have good weldability and mechanical properties. These are used through the selective laser melting process. Similarly, the titanium alloy Ti6Al4V is used to manufacture parts with tensile strength using the SLM process for space applications (Kranz, 2015). The electron beam melting (EBM) process is used in the aerospace industry to produce γ-TiAl alloys. For example, the production of the Ti48Al1Cr2NB intermetallic alloy for low-pressure turbine blades for the gas turbine engine was designed for Boeing's 787 Dreamliner using the EBM process. The feasibility of using combined processes such as EBM and electrical discharge machining (EDM) has allowed the development of shrouded turbine blisks based on the Ti6Al4V titanium alloy to obtain a part with the shape close to the final version by EBM and its finishing process by EDM.

Different types of self-assembling arrangements can be obtained using amphiphilic block copolymers and micelles. The swelling and packing of the particles occur thanks to the compatibility and selectivity of the solvent. The physical deviations in size and shape within the particles are on the component length scale. The collective architecture is sensitive to ionic resistance, pH, thermal and redox stimuli, shear flow, ion exchange, and osmotic shock. (Basheer, 2020).

Biocompatible systems based on supramolecular polymers and silica nanoparticles have been successfully deposited in three dimensions via inkjet printing processes (Hart, 2016). These non-toxic materials self-assemble into infinite networks to deposit both simple cubic structures as well as a complex twisted pyramidal shape. This type of approach facilitates the transition between prototype production and large-scale production of advanced materials that are both processable via 3D printing and with multifunctional physical properties that are required in the final product, leading to new and novel inks.

Among the emerging materials that are being used in the aerospace industry are smart materials. The most prominent properties are its self-sensing ability, self-adaptability, memory capabilities, and multiple functions. These materials have high capacities to adapt to the presence of aerospace environmental factors such as temperature, stress, magnetic fields, chemicals, electricity, nuclear radiation, acidity, and hydrostatic pressure (Basheer, 2020). Among the smart materials that are potentially considered for aerospace applications are piezoelectric materials, magnetostrictive materials, thermo-responsive materials, shape memory alloys, electrostrictive materials, rheological materials, and photochromic materials, as shown in Figure 7.

Figure 7. Classification of smart materials

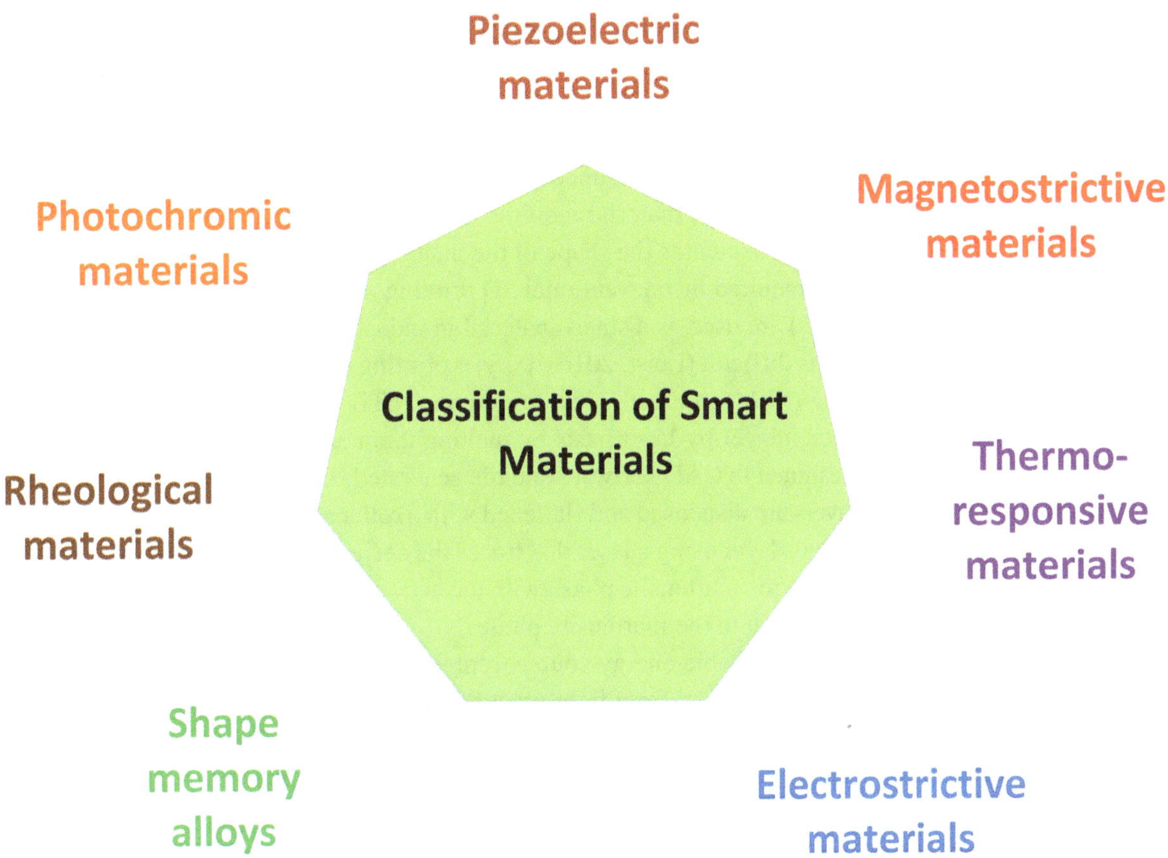

One of the prototypes based on self-assembly additive manufacturing was a 1D strand made of plastic units consisting of hinges and pins (Leist, 2016). Each hinge is designed to bend in a specific direction and connected to others to form long chains. These 1D plastic chains transform into 3D shapes when enough mechanical energy is applied to bend the material. Other versions combine embedded wires, magnets, and bending angles to form 3D objects from 2D flat shapes or assemble a 3D object from disassembled parts. This research was expanded to use multi-material 1D strands that could form different three-dimensional shapes based on rigid plastic and the hinges that cause bending are made of water-activated smart material. The smart material expands up to 150% when exposed to moisture as it is hydrophilic and hygroscopic. Rigid materials can be wood, textiles, and carbon fibers, so actuated objects can be created. 4D printing with these materials can produce automotive and aerospace parts as well as smart clothing sensitive to the user's body or environment. Some variants of this process are being tested which could even achieve some biomimetic properties found in nature.

4D printing can use high temperatures to make smart material change shape and make use of self-assembled origami (Leist, 2016). Shape memory composites can be activated by a combination of heat and stress to bend the material in different ratios and directions depending on the design of the hinges. Glassy polymer fibers show shape memory effects when heated above their glass transition temperature and can be printed three-dimensionally on an elastomeric matrix. Both the volume and the orientation of the glassy polymer fiber are specified in computer-aided design (CAD) software. Both design parameters are very important for printed active composites (PACs) like these as they influence bending angle, speed, storage modulus, deformation, and fixity. The fixity increases with the volumetric fraction of fiber due to the stiffness within the composite and the storage modulus with fiber orientation. The orientation and location of the fiber generate different geometric transformations such as bent, rolled, twisted and wavy shapes. Self-assembled origami structures such as boxes, a pyramid, a triple-hinged airplane, and a five-hinged airplane have been obtained. Until now, printing a 3D cube takes 3 hours and the post-processing to remove the support material requires numerous hours. Although heat-activated 4D printing uses a pre-deformation to change the shape of the material, still the post-processing time is reduced when compared to that required by conventional 3D printing methods.

Shape memory alloys (SMAs) are used as a smart material in industry and research, however, their application in complex designs is difficult (Leist, 2016). Polyjet printing is combined with SMAs-based wires or selective laser melting (SLM) to create nickel-titanium (NiTi) -based shape memory alloys. The SLM process prints part on a layer-by-layer basis by melting the metal powder using a laser. The desired complex shapes are designed in CAD software and the generated file is sent to the SLM process for processing. The powder layers are dispensed and flattened with a roller and the material is bonded by melting the powder to the previous layer using a laser that traces the design of the object into pieces. The material at room temperature with the martensite phase with the action of heat will pass to the austenite phase, which when cooled will return to the martensite phase.

Light, being a wireless and controllable energy source, represents an effective activation technique for self-assembly (Leist, 2016). The energy from light must be transformed into mechanical energy to be used in shape memory polymers (SMPs). Light-activated SMPs can be used in applications such as self-assembly structures, complex folding methods, transformative surface deformations, ultraviolet sensors and filters, and soft robotics. Now light-activated materials use heat from the light source to activate shape change. Regularly, pre-deformed polymer films with black ink are printed on areas of the sheet designed as hinges. The method uses an infrared light source to heat the material above the glass transition temperature of the polymer, so the black ink areas heat up faster and bend faster than the remaining material. The ink printed on the top of the sheets bends in the direction of the light source, and the ink printed on the bottom of a transparent sheet bends in the opposite direction to the light. By altering the geometry and width of the ink, as well as the number of lines printed, different bending angles, bending times, and light intensity requirements are obtained.

4D printing can be used in the production of smart textiles that have applications in the medical field, space exploration, and the military sector (Leist, 2016). Once the technique has been refined, it can be transferred to the sports area and the casual fashion industry. For these applications, the smart material must be embedded in the fibers with the desired shapes and properties. Self-assembly could allow 3D printing of space infrastructure on the International Space Station to reduce the sources required for its launch from Earth. The CubeSats can be 3D printed in a flat shape and then activated with light until obtaining the three-dimensional shape. One of the great challenges is that smart materials are selective to the activation sources since the space has different thermal and light environments.

As additive manufacturing continues to be embraced by the aerospace industry, parts consolidation is establishing itself as an emerging design technique with reduced assembly cost (Crispo, 2020). Part consolidation involves optimizing the topology for the assembly design to verify structural performance and connection layout. The methodology is based on the use of multiple domains that occupy the same space and a simple union domain to represent the assembly design. The approach proposes that in the future it is possible to calculate the costs of parts of additive manufacturing on an individual domain level. When designing an assembly, optimizing its topology requires the development of models where the geometries of multiple parts and the connections between these parts are formulated under two fundamental principles: 1) the mechanical load can only be transferred between two parts through a connection point, and 2) the stiffness of a connection point is different from the stiffness of the surrounding material.

So far, metal-based additive manufacturing is in its early stages when compared to fully mature additive plastic technology. It is considered that when fully qualified and certified it will play a critical role for applications with low part volume and where single part geometries cannot be produced by conventional manufacturing methods (Tepylo, 2019). Although it is now impossible to predict the long-term impact of additive manufacturing in the aerospace industry, aircraft structures, gas turbines, maintenance, repair, and overhaul (MRO), as well as parts and spare parts on demand, are expected to be developed. The mechanical properties of metallic materials produced by additive manufacturing are comparable to those achieved by products manufactured by conventional manufacturing methods. Post-additive manufacturing heat treatments can improve ductility even though they cause a small drop in yield strength and ultimate stress. Defects in metal parts arising from partially molten powder particles, lack of fusion between the layers, and gas entrapment can be alleviated by the hot isostatic pressing (HIP) process and must be heat treated after additive manufacturing. Additionally, post-build processes also improve the fatigue properties of additive manufacturing materials.

Hybrid manufacturing was invented as a technological alternative to additive metal manufacturing (Bahnini, 2018). The exclusive use of additive manufacturing of metals leads to unsatisfactory surface quality, poor uniformity in material properties, and loose tolerances. Therefore, post-processing is necessary to reach the required specifications, which increases the cost-to-build ratio and lead time. Hybridization or a combination of two or more processes can be used to overcome these limitations. The advantage of this approach is that by combining both material addition and subtraction it is possible to produce more complex geometries with a high buy-to-fly ratio as required for aerospace components and for manufacturing high value-added parts. Also, in environmental terms, these manufacturing systems reduce material waste and excessive tooling consumption.

The so-called matter logic approach to implementing the self-assembly of complex structures combines computer science and architectural design based on three building concepts: user input, environmental input, and material input (Tibbits, 2012). The materials are implanted with information storage which receives input, analyzes the information, and assists in the assembly of a structure. In this process, geometric constraints are placed on smaller assembly units to form larger structures when materials are exposed to mechanical energy and vibration. Realistic building materials based on metals and polymers are imbued with these intelligent properties exploiting their interactions through different activation stimuli.

Figure 8. Main challenges for the progress of additive manufacturing

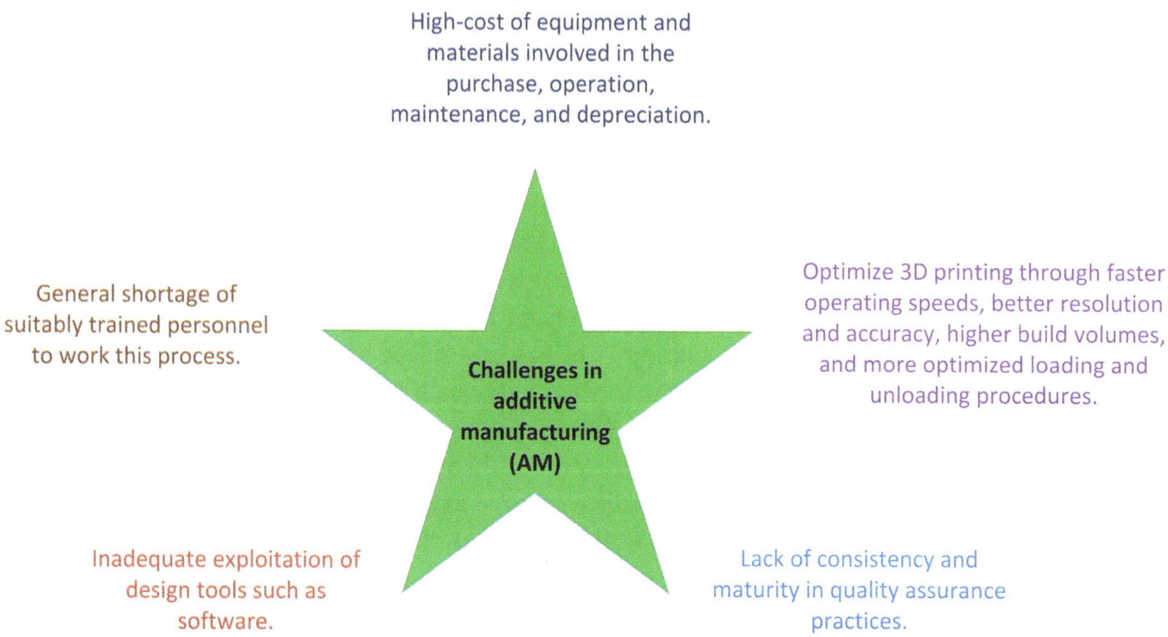

FUTURE RESEARCH DIRECTIONS

The primary goal of additive manufacturing is to produce lighter-weight parts at a lower cost in less operating time and with the same physical properties as established manufacturing techniques (Saracyakupoglu, 2019). Aerospace materials are manufactured using highly engineered techniques which makes them relatively expensive. Therefore, waste and scrap have a high significance for the associated cost. Less waste is best. To produce cost-effective parts that comply with restrictive regulations the buy-to-flight ratio must be lower. The buy-to-flight ratio is defined as the weight of the component itself over the raw material used for the component. The reduction in weight of an aerospace system reduces the material used and avoids a greater consumption of energy for transportation. Aerospace system weight reduction will continue to be the most important design parameter throughout the design and manufacturing phases. A lighter-weight system means lower fuel burn, higher payload capacity, and the ability to reach longer distances. However, even if the weight is reduced, both the strength and the performance of the parts in flight conditions must be maintained.

Given the advancement in the area of topology optimization algorithms in additive manufacturing self-assembly, it is expected that shortly, these will be able to determine the ideal part geometry, the minimum number of parts required, as well as the design of specific connections (Crispo, 2020). The use of bottom-up approximations will allow removing the bias towards the initial assembly design. The mathematical models proposed for the design of aerospace parts must combine the structural performance and the cost of the assembly for additive manufacturing. In this way, when considering the associated costs of the assembly, compromises will be established between the cost of the manufacturing process and the required performance regulations.

Some of the barriers that must be overcome for guaranteed success in the additive manufacturing process chain are improved design software, faster printing technology, increased automation, and better industry standards (Wu, 2017). Among the current research, areas are the development of new design methodologies for the optimization of parts by additive manufacturing based on multiple materials, novel printing techniques of additive manufacturing based on metals, research on the properties of materials for additive manufacturing, and the three-dimensional printing of the next generation of materials for aerospace applications. The challenges that additive manufacturing is facing for large-scale use are illustrated in Figure 8.

4D printing represents one of the techniques that will allow developing the self-assembly, the self-shape, and the self-repair of the materials used in the additive manufacturing of lightweight materials (Pei, 2018). It builds objects with smart materials, which will change additive manufacturing processes. This next-generation technology will allow companies to make products that can self-assemble, reshape themselves, or otherwise react to changing conditions, which will revolutionize the techniques for designing, manufacturing, and interacting with objects of all classes (Momeni, 2017). Smart materials cause the printed object to alter itself long after it has been made. These materials such as hydrogels or shape memory polymers are programmed to stimulate the object to change its shape, function, or color, for example, when it is subjected to water, light, heat, or electrical current. In addition, 4D printing will exponentially expand what is achievable in prototyping, design, manufacturing, and post-production adaptability and use. In this way, companies will not produce static products but products that change and grow throughout their life cycle. By combining these possibilities with advanced digital capabilities such as the Internet of Things (IoT), artificial intelligence (AI), and robotics, the potential disruption would be even more profound.

The self-assembly application provides the possibility of using products in hazardous environmental conditions such as aerospace applications (Ahmed, 2021). The materials generated can be used for transportation to space stations, as well as for the design of satellites and antennas. The strategic advantage is the ability to bend or join computer-programmed parts to increase construction speed and precision.

One of the novel self-assembly mechanisms is driven by capillary forces. It is used for the manufacture of advanced materials with wetting, adhesion, optical, mechanical, or electrical properties (Liu, 2021). Through asymmetric crosslinking densities, the printed vertical microstructures can be switched to a curved state with controlled thickness, curvature, and smooth morphology not obtainable by 3D printing. This approach is independent of distance allowing to reach a variety of self-assemblies with yields as high as 100%. Its use is aimed at emerging applications such as reversible micropatterns, switchable wetting, and dynamic actuation for micro-robots, origami, and encapsulation.

CONCLUSION

Additive manufacturing is the essential tool that will change technological development for high-tech applications in the aerospace sector that hardly one can imagine. In addition, it has the potential to offer economic, technical, and environmental advantages which reduce production costs and times, allow flexible and custom products, and reduce energy use and waste. When 4D printing is used in additive manufacturing it is possible to exploit the 3D printing process, the stimulus to activate the material to intelligent materials, and the interaction mechanisms to produce components for aerospace, medical, and intelligent textile applications with high added value. The three main capabilities that are exploited from 4D printing are self-assembly, multi-functionality, and self-repair. The use of self-assembly is directed to hazardous environments such as those found in aerospace applications. Individual parts can be printed with 3D printers to be self-assembled into larger structures such as antennas and satellites. The properties of 4D impression materials can be adapted to stimuli such as temperature, water, acidity, and light as a function of time. The use of hybrid or composite materials will allow the optimization of the three-dimensional deposition of materials that self-assemble by inkjet printing with a much more complex structure and with multifunctional properties. Self-assembly reduces the amount of material needed to build parts, optimizes product designs in terms of weight and part consolidation, as well as substantially reducing product development times. 4D printing represents the futuristic alternative to manufacturing processes because it provides a path to put any idea into practice.

ACKNOWLEDGMENT

The author is grateful for the support of the Instituto Tecnológico Superior de Irapuato (ITESI) and the Universidad de la Salle Bajío Campus Campestre for the development of this scientific research. The author wants to thank his wife and son for their support in the time to conduct this scientific research.

REFERENCES

Ahmed, A., Arya, S., Gupta, V., Furukawa, H., & Khosla, A. (2021). 4D Printing: Fundamentals, Materials, Applications and Challenges. *Polymer*, *228*, 123926. doi:10.1016/j.polymer.2021.123926

Airoldi, A., Sala, G., Di Lando, L. A., Bettini, P., & Gilardelli, A. (2018). Composite Corrugated Laminated for Morphing Applications. In A. Concilio, I. Dimino, L. Lecce, & R. Pecora (Eds.), *Morphing Wing Technologies* (pp. 247–276). Elsevier. doi:10.1016/B978-0-08-100964-2.00009-5

Alghamdi, S. S., John, S., Choudhury, N. R., & Dutta, N. K. (2021). Additive Manufacturing of Polymer Materials: Progress, Promise and Challenges. *Polymers*, *13*(5), 753. doi:10.3390/polym13050753 PMID:33670934

Alshahrani, H.A. (2021). Review of 4D printing materials and reinforced composites: Behaviors, applications and challenges. *Journal of Science: Advanced Materials and Devices*, *6*, 167-185. doi:10.1016/j.jsamd.2021.03.006

Bahnini, I., Rivette, M., Rechia, A., Siadat, A., & Elmesbahi, A. (2018). Additive Manufacturing Technology: The Status, Applications, and Prospects. *International Journal of Advanced Manufacturing Technology*, *97*(1-4), 147–161. doi:10.100700170-018-1932-y

Basheer, A. A. (2020). Advances in the Smart Materials Applications in the Aerospace Industries. *Aircraft Engineering and Aerospace Technology*, *92*(7), 1027–1035. doi:10.1108/AEAT-02-2020-0040

Calignano, F., Manfredi, D., Ambrosio, E. P., Biamino, S., Lombardi, M., Atzeni, E., Salmi, A., Minetola, P., Iuliano, L., & Fino, P. (2017). Overview on Additive Manufacturing Technologies. *Proceedings of the IEEE*, *105*(4), 593–612. doi:10.1109/JPROC.2016.2625098

Crane, N. B., Tuckerman, J., & Nielson, G. N. (2011). Self-Assembly in Additive Manufacturing: Opportunities and Obstacles. *Rapid Prototyping Journal*, *17*(3), 211–217. doi:10.1108/13552541111124798

Crispo, L., & Kim, I. Y. (2020). *Assembly Level Topology Optimization towards a Part Consolidation Algorithm for Additive Manufacturing*. AIAA SciTech 2020 Forum, Orlando, FL, United States. 10.2514/6.2020-0893

Gardan, J. (2018). Smart Materials in Additive Manufacturing: State of the Art and Trends. *Virtual and Physical Prototyping*, *14*(1), 1–18. doi:10.1080/17452759.2018.1518016

Groß, R., & Dorigo, M. (2008). Self-Assembly at the Macroscopic Scale. *Proceedings of the IEEE*, *96*(9), 1490–1508. doi:10.1109/JPROC.2008.927352

Hart, L. R., Li, S., Sturgess, C., Wildman, R., Jones, J. R., & Hayes, W. (2016). 3D Printing of Biocompatible Supramolecular Polymers and Their Composites. *ACS Applied Materials & Interfaces*, *8*(5), 3115–3122. doi:10.1021/acsami.5b10471 PMID:26766139

Herzog, D., Seyda, V., Wycisk, E., & Emmelmann, C. (2016). Additive Manufacturing of Metals. *Acta Materialia*, *117*, 371–392. doi:10.1016/j.actamat.2016.07.019

Kranz, J., Herzog, D., & Emmelmann, C. (2015). Design Guidelines for Laser Additive Manufacturing of Lightweight Structures in TiAl6V4. *Journal of Laser Applications*, *27*(S1), S14001. Advance online publication. doi:10.2351/1.4885235

Leist, S. K., & Zhou, J. (2016). Current Status of 4D Printing Technology and the Potential of Light-Reactive Smart Materials as 4D Printable Materials. *Virtual and Physical Prototyping*, *11*(4), 249–262. doi:10.1080/17452759.2016.1198630

Ligon, S. C., Liska, R., Stampfl, J., Gurr, M., & Mülhaupt, R. (2017). Polymers for 3D Printing and Customized Additive Manufacturing. *Chemical Reviews*, *117*(15), 10212–10290. doi:10.1021/acs.chemrev.7b00074 PMID:28756658

Liu, X., Wei, M., Wang, Q., Tian, Y., Han, J., Gu, H., Ding, H., Chen, Q., Zhou, K., & Gu, Z. (2021). Capillary-Force-Driven Self-Assembly of 4D-Printed Microstructures. *Advanced Materials*, *33*(22), 2100332. doi:10.1002/adma.202100332 PMID:33885192

Lyons, B. (2012). Additive Manufacturing in Aerospace: Examples and Research Outlook. *The Bridge*, *42*(1), 13–19.

Momeni, F., & Hassani,, N., Liu, X., & Ni, J. (2017). A Review of 4D Priting. *Materials & Design*, *122*, 42–79. doi:10.1016/j.matdes.2017.02.068

Ntouanoglou, K., Stavropoulos, P., & Mourtzis, D. (2018). 4D Printing Prospects for the Aerospace Industry: A Critical Review. *Procedia Manufacturing*, *18*, 120–129. doi:10.1016/j.promfg.2018.11.016

Pei, E., & Loh, G. H. (2018). Technological Considerations for 4D Printing: An Overview. *Progress in Additive Manufacturing*, *3*(1-2), 95–107. doi:10.100740964-018-0047-1

Sanchez-Rexach, E., Johnston, T. G., Jehanno, C., Sardon, H., & Nelson, A. (2020). Sustainable Materials and Chemical Processes for Additive Manufacturing. *Chemistry of Materials*, *32*(17), 7105–7119. doi:10.1021/acs.chemmater.0c02008

Saracyakupoglu, T. (2019). The Qualification of the Additively Manufactured Parts in the Aviation Industry. *American Journal of Aerospace Engineering*, *6*(1), 1–10. doi:10.11648/j.ajae.20190601.11

Singamneni, S., Lv, Y., Hewitt, A., Chalk, R., Thomas, W., & Jordison, D. (2019). Additive Manufacturing for the Aircraft Industry: A Review. *Journal of Aeronautics & Aerospace Engineering*, *8*(1), 215. doi:10.35248/2168-9792.19.8.215

Swiegers, G. F., Balakrishnan, S., & Huang, J. (2016). Assemblies and Self-Assembly. In J. Reedijk (Ed.), *Reference Module in Chemistry, Molecular Sciences and Chemical Engineering*. Elsevier. doi:10.1016/B978-0-12-409547-2.11710-X

Tepylo, N., Huang, X., & Patnaik, P. C. (2019). Laser-based Additive Manufacturing Technologies for Aerospace Applications. *Advanced Engineering Materials*, *21*(11), 1900617. doi:10.1002/adem.201900617

Tibbits, S., & Cheung, K. (2012). Programmable Materials for Architectural Assembly and Automation. *Assembly Automation*, *32*(3), 216–225. doi:10.1108/01445151211244348

Uriondo, A., Esperon-Miguez, M., & Perinpanayagam, S. (2015). The Present and Future of Additive Manufacturing in the Aerospace Sector: A Review of Important Aspects. *Proceedings of the Institution of Mechanical Engineers. Part G, Journal of Aerospace Engineering*, *229*(11), 2132–2147. doi:10.1177/0954410014568797

Whitesides, G. M., & Grzybowski, B. (2002). Self-Assembly at All Scales. *Science*, *295*(5564), 2418–2421. doi:10.1126cience.1070821 PMID:11923529

Wu, B., Myant, C., & Weider, S. (2017). *The Value of Additive Manufacturing: Future Opportunities*. Briefing paper No. 2. Institute for Molecular Science and Engineering, Imperial College London.

Yamamura, S., & Iwase, E. (2021). Hybrid Hinge Structure with Elastic Hinge on Self-Folding of 4D Printing using a Fused Deposition Modeling 3D Printer. *Materials & Design*, *203*, 109605. doi:10.1016/j.matdes.2021.109605

Yusuf, S. M., Cutler, S., & Gao, N. (2019). Review: The Impact of Metal Additive Manufacturing on the Aerospace Industry. *Metals*, *9*(12), 1286. doi:10.3390/met9121286

Zheng, Z. (2019). Fabrication on Bioinspired Surfaces. Bioinspired Design of Materials Surfaces. doi:10.1016/B978-0-12-814843-3.00003-X

ADDITIONAL READING

ASTM F2792-12a (2012). *Standard Terminology for Additive Manufacturing. ASTM International.*

Bourell, D., Kruth, J.P., Leu, M., Levy, G., Rosen, D., Beese, A.M., & Clare, A. (2017). Materials for Additive Manufacturing. *CIRP Annals – Manufacturing Technology*, 66(2): 659-681. DOI: doi:10.1016/j.cirp.2017.05.009

Di Angelo, L., Di Stefano, P., Dolatnezhadsomarin, A., Guardiani, E., & Khorram, E. (2020). A Reliable Build Orientation Optimization Method in Additive Manufacturing: The Application to DFM Technology. *International Journal of Advanced Manufacturing Technology, 108*(1-2), 263–276. doi:10.100700170-020-05359-x

Guo, N., & Leu, M. C. (2013). Additive Manufacturing: Technology, Applications and Research Needs. *Frontiers of Mechanical Engineering, 8*(3), 215–243. doi:10.100711465-013-0248-8

Kumbhar, N.N., & Mulay, A.V. (2018). Post Processing Methods used to Improve Surface Finish of Products which are Manufactured by Additive Manufacturing Technologies: A Review. *Journal of The Institution of Engineers (India): Series C*, 99(4): 481-487. doi:10.1007/s40032-016-0340-z

Narupai, B., & Nelson, A. (2020). 100th Anniversary of Macromolecular Science Viewpoint: Macromolecular Materials for Additive Manufacturing. *ACS Macro Letters, 9*(5), 627–638. doi:10.1021/acsmacrolett.0c00200

Ngo, T. D., Kashani, A., Imbalzano, G., Nguyen, K. T. Q., & Hui, D. (2018). Additive Manufacturing (3D Printing): A Review of Materials, Methods, Applications and Challenges. *Composites. Part B, Engineering, 143*, 172–196. doi:10.1016/j.compositesb.2018.02.012

Prater, T. (2017). Database Development for Additive Manufacturing. *Progress in Additive Manufacturing, 2*(1-2), 11–18. doi:10.100740964-017-0016-0

KEY TERMS AND DEFINITIONS

3D Printing: A technique of manufacturing objects through the deposition of materials through a print head, nozzle, or other printing technology.

Additive Manufacturing: A 3D printing used in industrial production that allows the creation of lighter and stronger systems or parts.

Binder Jetting: An additive manufacturing process where a liquid bonding agent is selectively deposited to bind powdered materials.

Material Extrusion: An additive manufacturing process where a flowable material is selectively dispensed through a nozzle or orifice into a die producing a constant cross-section.

Material Jetting: An additive manufacturing process that uses selectively deposited droplets as building materials.

Polyjet: 3D printing technology that produces smooth, accurate parts, prototypes, and tooling with microscopic layer resolution and accuracy down to 0.014 mm, which can produce thin walls and complex geometries using the widest range of materials available with any technology.

Powder Bed Fusion: An additive manufacturing process where heat energy selectively melts regions of a powder bed.

Self-Assembly: A process at which a disordered system constituted of similar or different pre-existing elements forms an organized structure or pattern through local or specific physical or chemical interactions among its elements without external stimulus.

Smart Memory Alloys (SMAs): Class of materials that produce a change in shape or property (rigidity, color, texture, transparency, volume) when exposed to an external stimulus.

Stereolithography: A photopolymerization process that produces parts of photo-polymeric materials in a liquid state using one or more lasers to selectively cure layer upon layer of a material to a predetermined thickness, hardness, and shape.

Chapter 14
Process Evaluation and Numerical Optimization in Friction Stir Welding of Dissimilar AMCs

Rajesh P. V.
https://orcid.org/0000-0002-4858-9032
Saranathan College of Engineering, India

Saravanan A.
https://orcid.org/0000-0003-2475-116X
Saranathan College of Engineering, India

ABSTRACT

In recent times, any engineering material is deemed worthwhile only if it satisfies functional characteristics such as weldability, formability, machinability, etc. Aluminum-based metal matrix composites have extensive usage in modern automobile parts, aircraft components, and ship structures, mainly due to their attractive properties such as low cost, high strength-to-weight ratio, excellent corrosion and wear resistance. Friction stir welding is one of the most versatile solid-state joining processes to ensure weldability between two AMC plates. In this research work, an analysis of FSW process through parameters (e.g., composition of alumina, spindle speed, feed, etc.) in joining Alumina reinforced aluminum alloy composites Al 6061 and Al 2024 together at various proportions by analyzing properties like impact strength, hardness, flatness, and ultimate tensile strength has been done. Finally, optimization is carried out to select the best possible combination using a multi-attribute decision-making technique called the complex proportional assessment of alternatives.

DOI: 10.4018/978-1-7998-7864-3.ch014

INTRODUCTION

A composite is a heterogeneous combination of different engineering materials joined together at various proportions. It has two distinctive phases: matrix and reinforcements. Matrix is the major constituent and reinforcements are minor constituents, which are added, mixed or embedded into the matrix. Composites are classified based on either the type of matrix material such as metal, ceramic, polymer, etc. or the shape and size of reinforcements like continuous fibres, whiskers, particulates, etc. (Bodunrin, M. O., Alaneme, K. K., & Chown, L. H., 2015). When more than one reinforcement is added to the matrix, the resultant material is called hybrid composite (Krishna, S. M., Shridhar, T. N., & Krishnamurthy, L., 2015). Composites are used almost everywhere in the world, rather unknowingly. They may be natural or synthetic. Their applications are immense ranging from structural components, aerospace materials, marine parts, common houseware, accessories used in elevated or cryogenic temperatures (Vengatesh, D. & Chandramohan, V., 2014). Composites are having an edge over metal alloys, owing to their improved hardness, high impact strength and superior wear resistance.

Metal Matrix Composites or simply MMCs, as they are often called are most popular and widely used among composites. They derive their name from the fact that the matrix material is usually a metal alloy. In Aluminium Matrix Composites (AMCs), the metallic matrix comprises of an aluminium alloy and reinforcements are generally non-metallic particles. In the modern era, reinforcements from diverse sources such as natural and synthetic ceramic powders like alumina and silicon carbide, agricultural wastes like Rice Husk Ash, industrial wastes like fly ash are used for strengthening the composite (Bodunrin, M.O., et al., 2015). Aluminium, the most abundant material in Earth's crust is the most preferred choice for matrix, owing to its low price, high strength-to-weight ratio, good heat withstanding capacity and excellent corrosion resistance (Pradeepkumar, J., Robinson Smart, D.S., & John Alexis, S., 2018).

Casting is one of the oldest traditional methods used for fabricating sand cast and chill cast Aluminium alloys Aluminium Metal Matrix Composites (AMMCs). It is commonly prevalent even now due to its robustness, flexibility and inexpensiveness. In this method, metal alloy to be moulded is heated till it attains semi-solid or molten state (at a temperature above its melting point) and is poured into mould cavity of required shape and dimension. In conventional casting process, a moulding box filled with green sand is used to prepare mould cavity. In recent times, mould cavity (die) made up of mild steel is used for both gravity and pressurized die casting. Stir casting is one of the categories in casting that derived its name from the process of stirring the molten metal thoroughly with propeller after the addition of reinforcement powder. This activity ensures the uniform distribution of reinforcement within the matrix (Bhandare, R.G., & Sonawane, P.M., 2013).

Welding is a metal joining process, in which two or more metal pieces are joined with the application of enormous amount of heat. The application of pressure is optional. The choice and cross section of metals may be similar or different (Wang, D., Xiao, B. L., Ni, D. R., & Ma, Z. Y., 2014). Friction Stir Welding (FSW), sometimes called as Friction Stir Processing (FSP) is the most appropriate welding technique used for Aluminium and its allied composites (Subramanya, P., Amar, M., Arun, S., Mervin, H., & Shrikantha, R., 2018). In FSW, two plates of AMCs with rectangular cross section are welded together as butt joints by means of exerting force through a rotating Stainless-Steel tool (Verma, S. & Misra, J.P., 2021). Based on the size of composite specimens, their composition, physical and mechanical properties, parameters such as welding speed, traverse feed, type and dimensional attributes of the tool material are chosen. FSW is one of the very few welding techniques that offer excellent binding strength and microstructural stability at the weldment area with relatively small Heat Affected Zone (HAZ). This versatile process enables not only the joining of similar metal alloys but also dissimilar metal alloys like Aluminium and Copper, dissimilar grades of same metal alloy and MMCs with same or different metal grades (Sadeesh, P., Venkatesh Kannan, M., Rajkumar, V., Avinash, P., Arivazhagan, N., Devendranath Ramkumar, K., & Narayanan, S., 2014).

Design of Experiments (abbreviated as DOE) is a structural method used to find out the effect or influence that a process parameter or factor exerts over the test attributes or responses (Anderson, M. J., & Whitcomb, P. J., 2010). Generally, DOE is used when multiple factors with varying levels determine the functionality of a particular manufacturing process. The factors are normally known as independent variables and the experimental design is helpful in determining the degree of closeness of relationship between independent variables and dependent variables (test attributes/properties) (Jiju Antony, 2014).

Full Factorial Design is one of the test arrays in DOE, which provides different possible combination of experimental runs with the available factors and their associated levels. A correlation among the factors is established, so that combined influence of factors on the process can be studied (Prakash, M., & Daniel Das, A., 2020).

Multi-Criteria Decision Making (MCDM) is a numerical methodology, used for the parametric optimization of a manufacturing process in which more number of contrasting criteria/attributes are used. As a result, best solution is found out from a set of various alternatives (Hwang, C.L. & Yoon, K., 1981). The quantitative MCDM is called Multi-Attribute Decision Making (MADM). This method is used repeatedly used in recent times to investigate and solve complex problems in material and tool selection, supplier selection, project management, resource management, etc (Mahmoodzadeh, S., Shahrabi, J., Pariazar, M., & Zaeri, M. S., 2007). MCDM is one of the evergreen and most evolving concepts that gave and still giving birth to numerous techniques. The common ground in all the techniques is the interaction between a set of alternatives and criteria, presented in table form as decision matrix. The optimal alternative/specimen combination is selected by ranking of alternatives (Kou, G., Lu, Y., Peng, Y., & Shi, Y., 2012).

The Complex Proportional Assessment (COPRAS) is one of the recent MCDM techniques, which was first introduced by Edmundas Kazimieras Zavadskas, a Lithuanian academician, along with his associates, Kaklauskas and Sarka in 1994. This method deals with maximizing and minimizing index values, when the effect of maximizing and minimizing indices of attributes on the output assessment is considered separately (Zavadskas, E. K., Kaklauskas, A. & Sarka, V., 1994). The COPRAS method is found to have applications in areas such as risk assessment, material selection, process selection, etc. but not limited to (Chatterjee, P., & Chakraborty, S., 2012).

BRIEF OVERVIEW ABOUT THE CHAPTER

This chapter dealt with the casting of Alumina reinforced Al 6061 and Al 2024 composite plates with varying proportion of Alumina ceramic powder followed by Friction Stir Welding of those dissimilar graded AMC plates. An experimental design consisting of 8 runs (2^3) was developed from 3 welding parameters/factors with 2 levels based on full factorial array. To evaluate the influence of FSW process parameters/factors on properties/responses, different tests were conducted. Finally, the optimized values of process parameters along with the corresponding test results were obtained through COPRAS, a numerical decision making technique.

LITERATURE REVIEW/BACKGROUND OF THE RESEARCH

Vengatesh, D., & Chandramohan, V. did an extensive survey on the available literature regarding the fabrication, processing and evaluation of aluminium alloys based MMCs (Vengatesh, D., & Chandramohan, V., 2014). Bodunrin, M. O., et al. conducted a review on Aluminium Metal Matrix Composites (AMMCs) with hybrid reinforcement. In their review article, they emphasized the potential of AMMCs, which make them suitable in various applications. They also discussed about the different combinations of reinforcement materials and their effect on mechanical properties of Aluminium composites. Various processing techniques to produce hybrid AMCs were indicated (Bodunrin, M. O., Alaneme, K. K., & Chown, L. H., 2015). Pradeepkumar, J., et al. studied the influence of non-conventional reinforcements like TaC, Ti and Si_3N_4 on mechanical properties of AMCs. The results indicated an improvement in the mechanical and tribological properties with the considerable addition of the above mentioned reinforcements (Pradeepkumar, J., Robinson Smart, D.S., & John Alexis, S., 2018). The demand of light weight materials, particularly aluminium based composites in defence, military, and aviation sector is found to be ever increasing according to Krishna, S. M., et al. They prepared silicon carbide and graphite particle reinforced hybrid Al 6061 composites by stir casting. The results indicated that the composites exhibit good thermal properties that make them eligible for cost-effective and high production aerospace structural parts (Krishna, S. M., Shridhar, T. N., & Krishnamurthy, L., 2015). Bhandare, R. G. and his associate Sonawane, P. M. suggested that stir casting is arguably the most famous and inexpensive process for creating AMCs in mass production. They also mentioned various Stir casting process parameters with their ranges (Bhandare, R. G., & Sonawane, P. M., 2013).

Verma, S., & Misra, J. P. conducted an experimental investigation on Friction Stir Welding (FSW) that employs dissimilar grades of aluminum alloy such as AA5083, AA6081, AA6082 and AA7039. The effect of material positions on performance characteristics of butt-joined aluminium alloys was discussed elaborately in their paper. The specimens were analyzed through Energy Dispersive X-ray Spectroscopy (EDAS) (Verma, S., & Misra, J. P., 2021). Rajesh, P.V. analyzed the collection of various literature available that indicate the relevance of FSW in joining AMCs. He concluded that the welding quality purely depends upon the dimensions of workpiece as well as tool configurations (Rajesh, P.V., 2020). The effect of process variables on the binding strength and microstructure of welds through FSW was described by Subramanya, P., et al. in their review article. They also briefed about the behavior of reinforcements during welding of AMCs (Subramanya, P., Amar, M., Arun, S., Mervin, H., & Shrikantha, R., 2018).

Suryanarayanan, R., & Sridhar, V.G. evaluated the influence of welding parameters in Friction Stir Welding of dissimilar aluminium alloys with varying thickness. Optimum welding parameters were determined using Response Surface Methodology (RSM) using a regression equation imported from Artificial Neural Network (ANN). Microstructure analysis was also carried out. The results suggested that improved mechanical properties are obtained at the weldment area compared to other regions (Suryanarayanan, R., & Sridhar, V.G., 2021). Cavaliere, P., et al. in their research paper carried out a detailed investigation on mechanical properties and microstructure behavior of Friction Stir Welded Al 2024- Al 7075 aluminium alloys. The grain structure of the joint is analyzed using optical microscopy. The results suggested that the FSW weld line reduces the fatigue of the joint (Cavaliere, P., Nobile, R., Panella, F. W., & Squillace, A., 2006). Dissimilar grades of aluminium alloy AA 2024 and AA 6061 were joined by FSW in the research work done by Sadeesh, P., et al. They employed five different tool designs to analyse process parameters. The results indicated that the hardness is minimum at Heat Affected Zone

(Sadeesh, P., Venkatesh Kannan, M., Rajkumar, V., Avinash, P., Arivazhagan, N., Devendranath Ramkumar, K., & Narayanan, S., 2014).

Ravi, B., et al. observed that the increase in tool rotational speed led to the enhancement in mechanical properties of welded AMCs with boron carbide reinforcement (Ravi, B., Balu naik, B., & Rajakumar, G., 2018). Junjie Zhao, et al. prepared a FSW combination by joining aluminium and magnesium alloys. They studied the effect of ultrasonic emission on tensile properties and microstructure of the welded alloys. The inference obtained from the experimental results show that the ultrasonic vibration improves the tensile strength at the middle and lower ends of the weld (Junjie Zhao, ChuanSong Wu & Hao Su, 2021). Wang, D., et al. identified that the existence of reinforcement particles led to the wear of tool material during FSW. They recommended that the reduction in the percentage composition of reinforcements as well as the usage of hard tool materials made up of tungsten carbide, cermets, etc will increase the tool life and result in better weldability (Wang, D., Xiao, B. L., Ni, D. R., & Ma, Z. Y., 2014). The flow pattern and existence of residual microstructures were visualized and analyzed in the FSW of 2024 and 6061 grades of aluminium by Li, Y., et al. in their research article (Li, Y., Murr, L. ., & McClure, J., 1999).

Sandeep Rathee, et al. optimized the process parameters in FSW of Al 6061/SiC composites using Taguchi L9 Orthogonal array. The results showed that the highest hardness value was achieved when tool rotational speed was at 1400 rpm, traverse speed was 50 mm/min and tilt angle was kept as 2.5þ (Sandeep Rathee, Sachin Maheshwari, Arshad Noor Siddiquee, Manu Srivastava & Satish Kumar Sharma, 2016). Anderson, M. J., & Whitcomb, P. J. authored a book describing the salient features of Design of Experiments (DOE). They explained the need of DOE to reduce the cost of conducting experiments separately, when more than one factor is involved in a process (Anderson, M. J., & Whitcomb, P. J., 2010). Jiju Antony, in his edited book on Design of Experiments for Engineers and Scientists elaborated the usage of latest tools and techniques in data analysis and interpretation even in non-manufacturing sector with the help of practical examples and case studies (Jiju Antony, 2014). Umanath, K., et al. fabricated composite specimens with varying percentage of silicon carbide in AA 6063 aluminium alloy by stir casting. The composite plates were further subjected to Friction Stir Processing (FSP). After undergoing several tests to confirm mechanical properties, Optimized welding parameters were identified using Taguchi technique (Umanath, K., Palanikumar, K., Veeramalai Sankaradass & Uma, K., 2021).

An experimental investigation on the impact of FSW process parameters such as tool speed, feed rate, displacement and welding time on impact strength is carried out by Prakash, M., & Daniel Das, A. A 4-factor 2-level Full Factorial Design (FFD) consisting of 16 experimental runs was considered and the quality of weld is optimized using statistical software called Minitab (Prakash, M., & Daniel Das, A., 2020). Hwang, C. L., & Yoon, K. explained various terms and terminologies used in Multiple Attribute Decision Making (MADM) in their work. They described the procedure for conducting MADM based selection of optimal specimen through working examples (Hwang, C. L., et al., 1981). Kou, G., et al. proposed a suitable approach to correlate ranking, when different classification algorithms in MCDM give contrasting ranking order (Kou, G., Lu, Y., Peng, Y., & Shi, Y., 2012). Mahmoodzadeh, S., et al. initially calculated the relative weightage given to each criterion in a decision matrix through Analytical Hierarchy Process (AHP) coupled with Fuzzy Set theory. The assessment of different alternatives and the corresponding ranking is determined using TOPSIS (Mahmoodzadeh, S., Shahrabi, J., Pariazar, M., & Zaeri, M. S., 2007).

Complex Proportional Assessment (COPRAS) was first introduced by Zavadskas, E. K., et al. in their paper titled "The new method of multi-criteria Complex Proportional Assessment of projects"

(Zavadskas, E. K., Kaklauskas, A. & Sarka, V., 1994). Organ, A. & Yalcin, E. evaluated and appraised the performance of Research assistant and teaching faculty using COPRAS. In their work, 5 different research assistants were taken into consideration based on their educational qualification and analyzed. The results inferred that the research personnel having high quality will improve the personal competency as well as university teaching efficiency (Organ, A. & Yalcin, E., 2016). A state-of-the-art literature survey to provide a deep insight about the emergence, methodology and applications of COPRAS in diverse fields was conducted and presented by Stefano, N.M., et al. (Stefano, N.M., Casarotto Filho, N., Vergara, L.G.L. & Rocha, R.U.G., 2015). Valentinas Podvezko explained the importance of ranking in MCDM methods. He also gave a detailed comparison between two such methods namely Simple Additive Weighting (SAW) and COPRAS as these two share some common traits. He sorted out the contrasting features between the two evaluating methods (Podvezko, V., 2011).

Prasenjit Chatterjee & Shankar Chakraborty released two papers dealing with material selection consecutively. In the first paper, they discussed about two MCDM methods: COPRAS and COPRAS-G used to solve complex problems for selecting materials in a cost-effective manner (Chatterjee, P., & Chakraborty, S., 2012). In their second paper, they compared the results obtained from COPRAS with that of Additive Ratio Assessment-based approach (ARAS) through some real time examples to identify the efficient one (Chatterjee, P., & Chakraborty, S., 2013). Popovic G., et al. explored the possibility of utilizing COPRAS method in management-related scenario. In their paper, they explained the ways and means of selecting projects that require huge investments. Selection of best project from a set of alternatives can be done by assessing the financial constraints in COPRAS and COPRAS-G methods, even though the system is provided with imprecise data or insufficient information (Popovic G., Stanujkic D. & Stojanovic S., 2012). Chatterjee, P., et al. employed two methods namely COPRAS and Evaluation of Mixed Data (EVAMIX) for ranking the supplied materials in ascending order from best to worst, considering various criteria (Chatterjee, P., Athawale, V. M. & Chakraborty, S., 2011). Gomathisankar, M., et al. decided that Taguchi-enabled COPRAS was best suited for predicting tensile strength and hardness values in FSW Al 6061 alloy (Gomathisankar, M., Gangatharan, M., & Pitchipoo, P., 2018). Romualdas Ginevicius, explained the importance of normalising the quantities in his paper. As the values in each criterion belong to different dimensions, it is mandatory to express them in same dimension by using different normalization techniques mentioned in his paper (Ginevicius, R., 2008). Mehdi Ajalli, et al. in their paper demonstrated the way of using hybrid AHP-COPRAS techniques for building a comprehensive decision making model to identify proper supply chain management (Mehdi Ajalli, Hossein Azimi, Abdolkarim Mohammadi Balani & Mahdi Rezaei., 2017).

RESEARCH GAP

The above literature survey left behind the following research gaps:

Various researches had been made to analyze, evaluate and optimize the FSW process. But, the experimental investigation in Friction Stir Welding of dissimilar grades of Aluminium Matrix Composites was still at its nascent stage.

The research articles dealing with the MCDM based optimization of FSW process were very meager. COPRAS enhanced selection of optimal FSW composite specimen was still an unexplored field.

In this research work, a decent attempt had been made to address the above mentioned research gaps.

Table 1. Chemical composition of Al 6061

Ingredients	Percentage
Aluminium (Al)	97.70
Magnesium (Mg)	0.97
Silicon (Si)	0.62
Copper (Cu)	0.39
Iron (Fe)	0.13
Chromium (Cr)	0.08
Manganese (Mn)	0.11

SELECTION OF MATERIALS

Material selection is the most important prerequisite for the fabrication of composites. It is mandatory to choose the apt materials for matrix and reinforcement so that the resultant component will have enhanced mechanical properties compared to individual metals (Bodunrin, M.O., et al., 2015).

Matrix

The matrix, in which the metal alloy is present at a larger proportion, provides the necessary bond and protection to the composite structure. In this research work, two different aluminum alloys: Al 6061 and Al2024, were chosen as matrix materials.

Al 6061

It is a precipitation-hardened wrought aluminum alloy with major traces of Magnesium and Silicon. It is one of the most common multi-purpose alloys extensively used in aircraft parts such as window panels, door frames, cabin interiors, etc as well as ship super-structures, utensils, household appliances, and automobile components (Krishna, S.M., et al., 2015). It is renowned for its excellent weldability, machinability and corrosion resistance.

The chemical composition of Al 6061, identified using elemental analysis, was given in Table 1 below:

Al 2024

Al 2024 is a wrought aluminum alloy with copper as its second major alloying element. It is primarily known as aircraft aluminum due to its heavy usage in aircraft parts. It is used in applications wherein high strength to weight ratio and good fatigue resistance is required.

The chemical composition of Al 2024, identified using elemental analysis was given in Table 2 below: Some common properties of Al 6061 and Al 2024 were mentioned below in Table 3.

Table 2. Chemical composition of Al 2024

Ingredients	Percentage
Aluminium (Al)	93.30
Copper (Cu)	4.30
Magnesium (Mg)	1.30
Iron (Fe)	0.40
Silicon (Si)	0.30
Chromium (Cr)	0.05
Manganese (Mn)	0.30

Reinforcement

Reinforcement is the powdered material (a non-metal in this case) that provides strength to the composite structure. It is embedded or thrust inside a metallic matrix at a comparatively smaller proportion than the matrix (Pradeepkumar, J., et al., 2018). In this research work, Alumina was chosen as reinforcement for both Al 6061 and Al 2024 alloys.

Alumina

Alumina or Aluminium Oxide (Al_2O_3) is one of the most popular ceramics repeatedly used as reinforcement in AMCs. It is a well-known abrasive, a refractory and polishing agent often used in textiles and fabric industries, manufacturing plants, material testing laboratories, etc. It provides increased strength, improved hardness, enhanced toughness, and better tribological properties to the metal alloy, when reinforced inside aluminum matrix to (Bodunrin, M.O., et al., 2015).

Table 4 indicated the general properties of Alumina.

Table 3. General properties of Al 6061 and Al 2024

Properties	Unit	Al6061	Al2024
Density	g/cm^3	2.72	2.78
Melting point	°C	582-652	500
Ultimate Tensile Strength	Mpa	130	469
Thermal conductivity	W/m-K	168	121
Brinell Hardness	-	45	120
Shear modulus	Gpa	26	28
Bulk modulus	Gpa	76	76

Table 4. General attributes of Al_2O_3

Properties	Units	Alumina (Al_2O_3)
Density	g/cm³	3.98
Vickers Hardness	-	1560
Melting Point	°C	2300
Elastic Modulus	GPa	300
Thermal conductivity	W/Mk	21
Tensile Strength	MPa	210
Fracture toughness	MPaÖm	4.9

METHODOLOGY

Fabrication Method

Following the selection of suitable materials for matrix and reinforcement, the next step was the fabrication of composite plates. Sixteen composite plates were prepared in total, with eight in Al 6061+ Al_2O_3 and Al 2024+ Al_2O_3, respectively. The reinforcement ratio was varied as follows:

- 95% Al 6061+ 5% Al_2O_3- 4 Nos.
- 90% Al 6061 + 10% Al_2O_3 - 4 Nos.
- 95% Al 2024 + 5% Al_2O_3- 4 Nos.
- 90% Al 2024 + 10% Al_2O_3 - 4 Nos.

Stir Casting

The most suitable and inexpensive method for producing AMC specimens is Stir casting. In this method, composite samples are prepared by pouring the preheated ceramic powder inside the molten metal and mixing thoroughly. When the combination, as mentioned above, starts to solidify, it is cast into a mould cavity with the help of atmospheric pressure due to gravity. Hence, this method is also called liquid metallurgy or gravity die casting.

The process started with the preheating of alumina powder in a separate furnace at about 300°C for nearly 15 minutes. The aluminum metal alloys to be melted were purchased in the form of cylindrical rods. They were cut as per required quantity, weighed, and then placed inside a graphite crucible. The crucible was then charged into a coal-fired furnace that is equipped with a blower. At about 600°C, the metal started to melt. When it reached the liquidus state at 750°C, the preheated alumina powder was poured inside the molten metal. To avoid the staying of Alumina at the bottom of the crucible as residue, the molten metal with Alumina was stirred thoroughly at a reasonable speed, either employing a manual stirrer or motorized propeller. This process enhanced the uniform mixing of ceramic powder with metal alloy, resulting in the even distribution of reinforced particles within the metallic matrix (Bhandare, R.G. &Sonawane, P.M., 2013).

Finally, the crucible was taken out and the aluminum-alumina mixture was poured into a dedicated mould cavity of dimensions 200x100x6 mm.

Figure 1. Stir casting

The process of Stir casting is shown in Figure 1 below:

Material Joining Process

Welding is a generic solid-state technique used to join metal alloys or thermoplastics by using high-intensity heat plasma supplied with the help of an electrode, thereby melting the two parts/ base materials together (Suryanarayanan, R., & Sridhar, V.G., 2021). When the two parts, which come in close contact, are melted and then allowed to cool, fusion occurs. The welding process occurs with or without heat, pressure, or filler material over the materials being combined. The resulting joint based on weld configuration (butt, fillet, tag or complete penetration) will be stronger than the base material (parent metal).

Friction Stir Welding

Friction stir welding (FSW) is a joining mechanism, which uses a non-consumable tool (preferably Stainless-Steel (SS) or High-Speed Steel (HSS) for AMCs) to join two composite workpieces without melting them (Rajesh, P.V., 2020). The two composite plates (Al 6061+ Al_2O_3 and Al 2024+ Al_2O_3) with varying proportions were placed close contact with each other and clamped in the work holding devices mounted on the work table. The entire setup and operation were done in a Vertical Milling Machine (VMM). A Stainless-Steel tool of cylindrical cross-section (Ø 15mm and length 50 mm) with nose dimensions (radius 2 mm and height 6 mm) was fixed in the tool holding spindle of VMM and made to rotate at a specific speed (Suryanarayanan, R. & Sridhar, V.G., 2021). A Uniform feed rate was given to the rotating tool. When the rotating tool moved vertically and came in contact with the stationary composite plates clamped together, friction was created and heat was generated. While the tool travelled along the traversed path, which was the joint line of composite plates, it mechanically intermixed the two pieces (providing a tailoring/stitching action) by melting the intermediate region (Heat Affected Zone or HAZ) and pressurized the hot and softened plates, thereby producing a rigid weldment. This welding action resulted in good binding strength between the two plates and a better surface finish at the weldment area.

FSW technique is primarily used in structures that require very high weld strength. In recent times, the application of FSW has exponentially increased in shipbuilding, train compartments, aerospace parts, etc (Wang, D., et al., 2014).

The combination joints of composite plates created by FSW were shown in Figure 2.

Figure 2. FSW composite joints

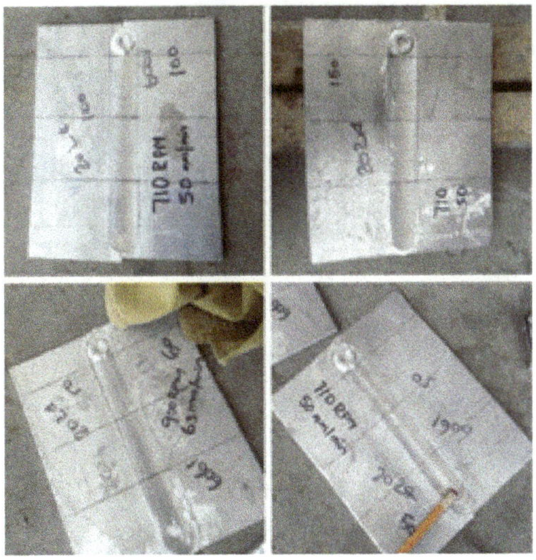

Possible combinations of weld joints to be created in FSW using different levels of available weld parameters were decided by Full Factorial Array approach in Design of Experiments.

DESIGN OF EXPERIMENTS

The methodology for Design of Experiments also called as experimental design was first postulated by R.A.Fisher in his books and research articles in the early 30's. Its main aim is to predict the output by introducing different combinations of input. Any change in the input factors or independent variables will create variation in output responses or dependent variables (Umanath, K., et al., 2021). This process is used to investigate complicated engineering problems with more than one cause/influencing factor. It is a mundane task of evaluating each factor separately. Hence, DOE helps in variating more factors at the same time. This cost-effective process thereby, saves more time (Jiju Antony, 2014).

To analyze the interaction between multiple factors affecting a particular manufacturing system, test arrays are used. One such array is Full Factorial Array that determines different combinations of parameters/factors (Gomathisankar, M., et al., 2018).

For a 3-Factor 2-Level experimental design used in this chapter, 8 experimental runs or specimen combinations (2^3) were determined and tabulated.

The different process parameters/Factors and their levels required for joining composite plates by FSW were represented in Table 5 below.

The different specimens formed as a result of combination weld joints by FSW as per Full Factorial Design (FFD) were tabulated in Table 6.

Table 5. Factors and levels for FSW

Symbols	Factors/Parameters	Experimental values	
		Level 1: low (1)	Level 2: high (2)
A	Composition of Al_2O_3 (wt. %)	5 (1)	10 (2)
B	Spindle speed (rpm)	560 (1)	710 (2)
C	Feed (mm/min)	40 (1)	50 (2)

DISCUSSION

Process Evaluation by Conducting Various Tests and Results

This section of the chapter dealt with evaluating process parameters involved in FSW by conducting various mechanical tests. The influence of process parameters, namely Wt% of Al_2O_3, Spindle speed/Tool rotation speed and Feed on mechanical properties like Hardness, Impact Strength, Flatness and Ultimate Tensile Strength, was analyzed from the test results. All the tests were performed as per ASTM Standard testing procedures, including the preparation of test specimens.

Hardness Test

The term 'hardness' is termed as the surface attribute of a material to bear heavy force, applied over its surface in the form of either indentation or abrasion (Ravi, B., et al., 2018). Hardness test is done in a machine called Hardness tester. Brinell Hardness test is normally preferred to find out the hardness index of aluminum alloys and their composites. The Friction Stir Welded composite plate was placed on the work table of the hardness tester. A static point load of 250 kgf was applied over the welded plate with the help of a 5 mm dia. steel ball indenter tip and kept constant for 10 seconds. Then the load was removed and the diameter of the impression created by the indenter tip was measured by viewing it through a microscope. The indentation/impression value in the specimen surface was calculated from Eq. (1):

Table 6. Full factorial array inspired experimental FSW composite specimens

Specimen No.	Composition of Al 6061	Composition of Al 2024	Composition of Al_2O_3 (same % in both the alloys) (A)	Spindle Speed (B)	Feed (C)
	wt %	wt %	wt %	rpm	mm/min
1	95	95	5 (1)	560 (1)	40 (1)
2	95	95	5 (1)	560 (1)	50 (2)
3	95	95	5 (1)	710 (2)	40 (1)
4	95	95	5 (1)	710 (2)	50 (2)
5	90	90	10 (2)	560 (1)	40 (1)
6	90	90	10 (2)	560 (1)	50 (2)
7	90	90	10 (2)	710 (2)	40 (1)
8	90	90	10 (2)	710 (2)	50 (2)

Table 7. Hardness test results

Specimen No.	Al 6061 portion		Weldment area		Al 2024 portion	
	d_1 (mm)	BHN_1	d_2 (mm)	BHN_2	d_3 (mm)	BHN_3
1	3.7	44.85	3.5	50.32	4.1	36.21
2	3.8	42.43	3.5	50.32	4.4	31.21
3	3.6	47.47	2.7	85.71	2.9	74.07
4	3.9	40.19	2.9	74.07	3.3	56.82
5	4.4	31.21	3.2	60.55	3.6	47.47
6	3.7	44.85	3.3	56.82	3.5	50.32
7	3.5	50.32	3.3	56.82	3.9	40.19
8	4.0	38.12	3.1	64.61	3.7	44.85

$$\text{Brinell Hardness Number} = 2P/ (\pi D\ [D - \ddot{O}(D^2 - d^2)]) \quad (1)$$

where,

D = steel ball indenter diameter (5 mm)
d = diameter of impression in mm.
P = load applied (250 kgf)

The same procedure was done repeatedly for all the remaining specimens and the results were indicated in Table 7 and Figure 3, respectively below.

In the table,

d_1 = diameter of an impression at the Al 6061 side of the FSW composite plate
d_2 = diameter of the impression at the weldment area of the FSW composite plate
d_3 = diameter of the impression at the Al 2024 side of the FSW composite plate
BHN_1 = Brinell Hardness Number at Al 6061 side of the FSW composite plate
BHN_2 = Brinell Hardness Number at the weldment area of the FSW composite plate
BHN_3 = Brinell Hardness Number at the Al 2024 side of the FSW composite plate

The above results showed that the weldment area was harder than the individual composite portions. Specimen 3, in which the percentage of Alumina was 5, spindle speed was 710 rpm and feed was 40 mm/min registered the highest BHN. The results also indicated an inverse relationship between the composition of Alumina and hardness (i.e), Hardness values decreased with the increase in the proportion of Alumina in FSW Composites. But, there was an increase in hardness values with increased tool rotation speed/spindle speed (Sandeep Rathee, et al., 2016). Hence, they were directly proportional. The effect of Feed on hardness seemed inconclusive, as no trend was traced.

Figure 3. Hardness test results

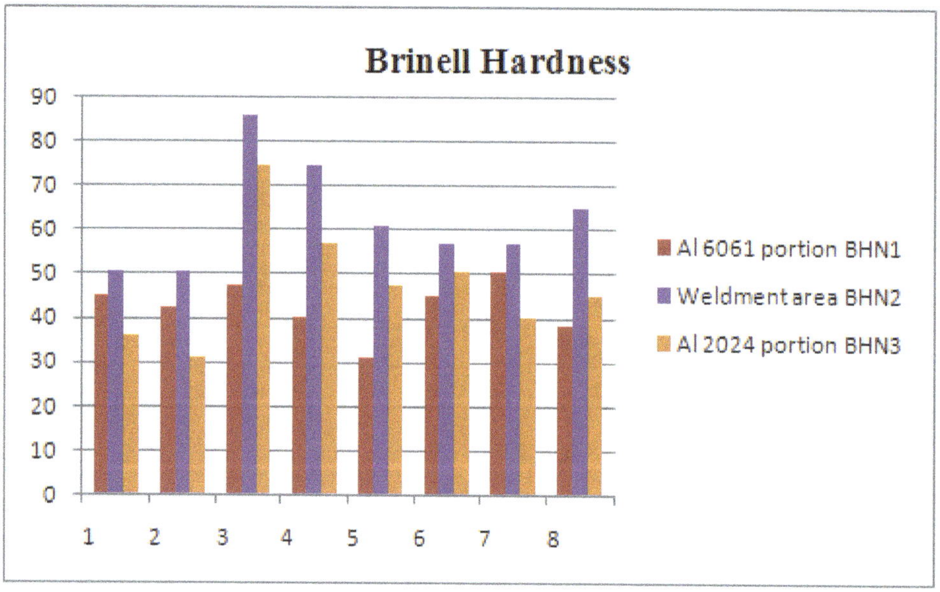

Impact Test

Impact strength of a material is a bulk property to bear suddenly applied load without fracture. It is also called a drop-weight test. This test is done with the help of the Impact tester. In this test, a notch up to 2mm is cut in the specimen with an angle of 45° and the specimen is knocked with a sudden blow from the released hammer of the testing machine. The absorbed energy, while breaking the specimen will be recorded from the graduated scale embedded in that machine. For Aluminium based specimens, Izod impact test is conducted. In Izod test, the test sample was fixed vertically in the machine like a cantilever beam and the stricker/hammer hit the specimen above the v-notch (Cavaliere, P., et al., 2006).

All the 8 FSW Composite specimens were subjected to the above-stated test procedure and the results were represented as Table 8 and Figure 4 respectively, below.

From the above results, it was certain that, specimen 8 in which the percentage of Alumina was 10, spindle speed/tool rotation speed 710 rpm and feed 50 mm/min was having the highest impact strength of 8J.

The percentage of Alumina in FSW composites was directly proportional to impact strength, that is higher the concentration of Alumina, better the impact strength. The impact strength was higher when the tool rotated faster (Ravi, B., et al., 2018). The feed was also having a positive effect on impact strength. The impact strength increased with an increase in traverse feed.

Measurement of Flatness

Perfect flatness is achieved when all the points in a surface are coplanar (lie at the same plane). Flatness is also known as Target depth. The flatness control defines the deviation of a part's original surface from its perfectly flat form (Li, Y., et al., 1999). To check the flatness of a FSW Composite specimen, its surface was measured with a spirit level and compared with the surface of an accurate Surface Table

Table 8. Impact test results

Specimen No.	Force absorbed
	J
1	4
2	6
3	2
4	4
5	4
6	6
7	6
8	8

and the deviation from the original flat reference position was identified from various points of the surface using a Dial gauge indicator.

Table 9 represented the deviations from the reference flatness value obtained at different locations of the surface in FSW Composite specimens.

Figure 5 showed the average deviations from reference flatness values of all the specimens at Al 6061 side, weldment area, and Al 2024 side, respectively.

The results obtained from flatness test indicated that the weldment region was less flat than the other two regions. Specimen 5 in which the percentage of Alumina was 10, spindle speed was 560 rpm and feed was 40 mm/min gave promising result (i.e) lowest deviation from reference flatness value. Feed was inversely proportional to flatness deviation. Larger the feed rate, flatness was more. Spindle speed

Figure 4. Impact test results

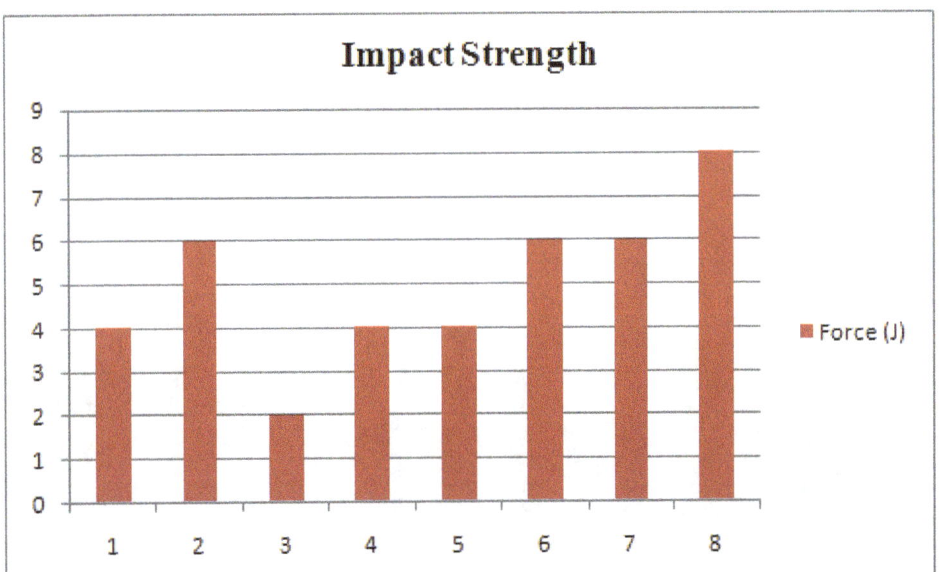

Table 9. Flatness measurement results

Specimen No.	Reference flatness Value	Average Deviation from reference value at Al 6061 side	Average Deviation from reference value at Weldment region	Average Deviation from reference value at Al 2024 side
1	2.90	0.35	0.65	0.12
2	2.62	0.30	0.60	0.28
3	2.54	0.26	0.43	0.26
4	2.34	0.20	0.44	0.19
5	2.67	0.25	0.19	0.15
6	2.70	0.17	0.40	0.24
7	2.81	0.36	0.40	0.26
8	2.75	0.09	0.26	0.20

had very less to nil influence over surface flatness. Similarly, the presence of Alumina in composite did not significantly affect the flatness condition.

Tensile Test

Tension means expansion force or pulling the two edges of material apart (in the opposite direction). Tensile strength is a function of ductility. Ductility is a property of elasticity in which a material can be drawn into long and thin wire. Therefore, tensile strength is the elongation of a material, especially metal alloy or MMC, when an external force or tensile load is applied with the reduction in cross-sectional

Figure 5. Flatness measurement results

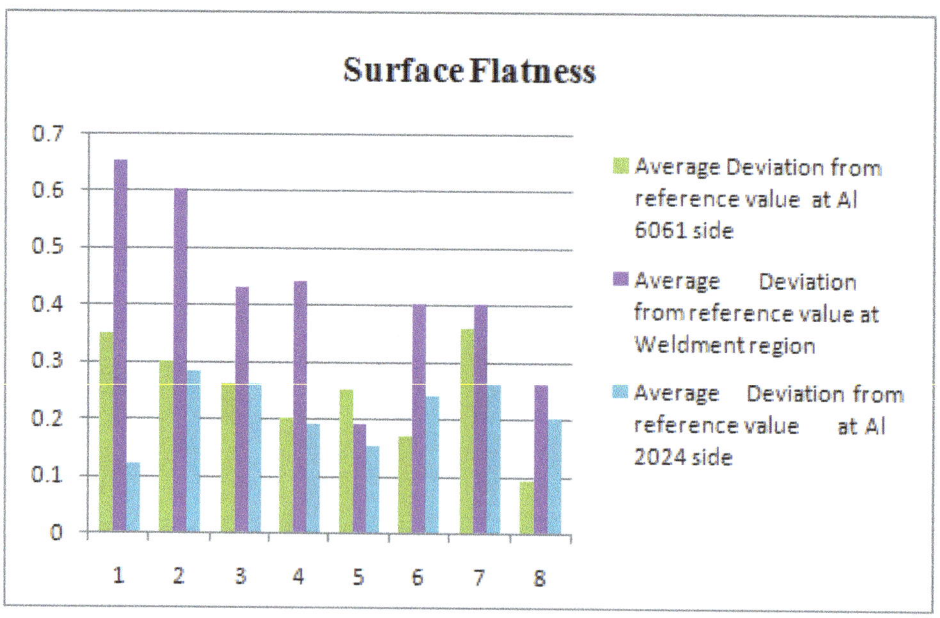

Table 10. Tensile test results

Specimen	Area of cross section at neck, A_0 (breadth x thickness)	Maximum load, P_{max}	Ultimate Tensile Strength $S=P_{max}/A_0$
	mm²	N	N/mm²
1	67.62	5139	76
2	65.82	4805	73
3	58.14	5001	86
4	58.40	5315	91
5	50.32	3472	69
6	60.30	5668	94
7	65.70	5453	83
8	64.34	6306	98

area. Tensile test is generally conducted using Universal Testing Machine (UTM). The FSW Composite specimen was fixed in between the two jaws of UTM. One jaw was fixed and the other was movable (Junjie Zhao, et al., 2021). When the jaw was moved, the specimen started to elongate, thereby entering into the plastic region from the elastic region (deformation occured even after removing external load). Maximum elongation occured at maximum load supplied to Ultimate Tensile Stress specimen. This was where the specimen obtained Ultimate Tensile Strength, after which it fractured under breaking stress.

The results of Tensile test were tabulated below in Table 10 below.

The tensile test results for all the 8 specimens were shown in Figure 6.

The above results revealed that the composition of Alumina didn't have any influence over the tensile strength, but, spindle speed and feed had a positive influence, that is, tensile strength increased with increase in spindle speed and feed.

The process parameters and the mechanical properties (Criteria) determined for the 8 specimens were collectively represented in Table 11. For further experimental work and Optimization, Brinell hardness and Average flatness deviation values at the weldment region were only considered.

NUMERICAL OPTIMIZATION USING COPRAS

Optimization is the process of identifying the best possible specimen or combination from a set of given alternatives. It gives the optimum values rather than the maximum values (Umanath, K., et al., 2021). Numerical optimization is a variety among different optimization methods, which concentrates specifically on quantitative analysis based mathematical calculations. Hence, Multi-Criteria Decision Making (MCDM) or Multi-Attribute Decision Making (MADM) comes under numerical optimization.

COPRAS

Complex Proportional Assessment is one of the modern MCDM techniques initially introduced by Zavadskas and his colleagues in their research paper. This method deals with the ranking of various alternatives by assessing the maximizing and minimizing index values (Zavadskas, E.K., et al., 1994). The procedural steps were performed as follows:

Figure 6. Tensile test results

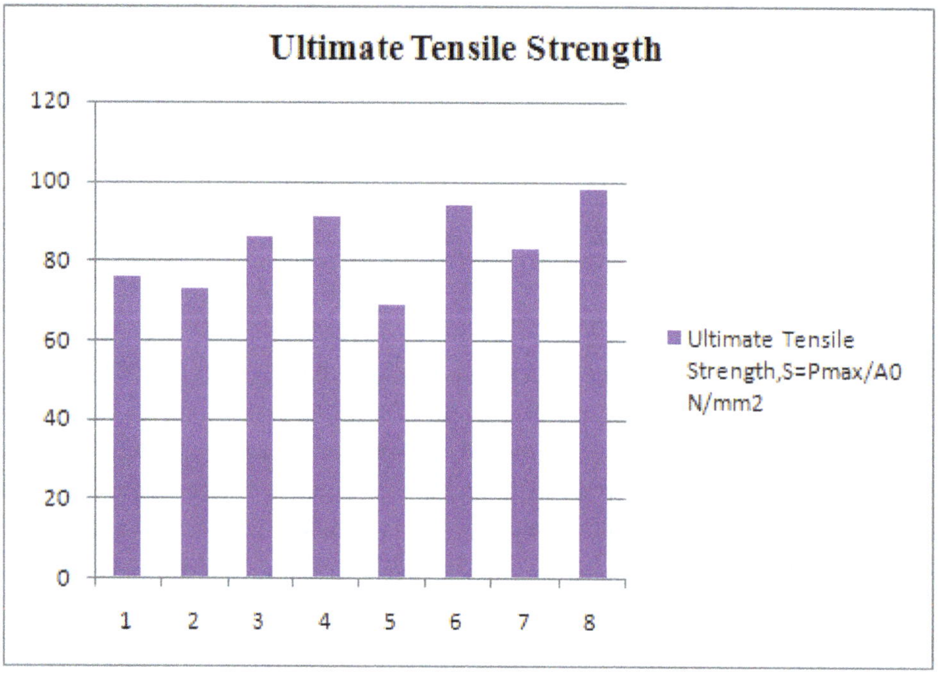

Table 11. Factors and attributes

Specimens/ Alternatives	Process parameters/Factors			Criteria/Attributes			
	Composition of Al_2O_3	Spindle Speed	Feed	Brinell Hardness	Izod Impact Strength	Average Flatness Deviation	Ultimate Tensile Strength
	wt %	rpm	mm/ min	BHN	J	-	N/mm²
1	5	560	40	50.32	4	0.65	76
2	5	560	50	50.32	6	0.60	73
3	5	710	40	85.71	2	0.43	86
4	5	710	50	74.07	4	0.44	91
5	10	560	40	60.55	4	0.19	69
6	10	560	50	56.82	6	0.40	94
7	10	710	40	56.82	6	0.40	83
8	10	710	50	64.61	8	0.26	98

Table 12. Decision matrix table

Alternatives	Brinell Hardness	Izod Impact Strength	Average Flatness Deviation	Ultimate Tensile Strength
	BHN	J	-	N/mm²
1	50.32	4	0.65	76
2	50.32	6	0.60	73
3	85.71	2	0.43	86
4	74.07	4	0.44	91
5	60.55	4	0.19	69
6	56.82	6	0.40	94
7	56.82	6	0.40	83
8	64.61	8	0.26	98

Step 1: Construction of Decision Matrix Table.

This Table showed 'm' alternatives and 'n' criteria represented in Eq. (2), in which the alternatives were placed row-wise (i) and the criteria were placed column-wise (j) (Mahmoodzadeh, S., et al., 2007).The intersection of each alternative with criteria, denoted as x_{ij} gave the criteria values. The constant 'A' indicated the alternatives.

$$D = \begin{matrix} A_1 \\ A_2 \\ \vdots \\ A_m \end{matrix} \begin{pmatrix} x_{11} & x_{12} & \cdots & x_{1n} \\ x_{21} & x_{22} & \cdots & x_{2n} \\ \vdots & \vdots & & \vdots \\ x_{m1} & x_{m2} & \cdots & x_{mn} \end{pmatrix} \tag{2}$$

Table 12 represented the decision matrix for the above-performed research work.

Step 2: Allocation of Weightage to Criteria Based on Their Relative Importance.

The allocation is presented in Table 13, which shows the weightage allotted to each criterion for the current research work.

Step 3: Normalization of the Decision Matrix

This was done for balancing the criteria as the values in each criterion will have different units of measurement and multiplication factors (Ginevicius, R., 2008). The formula for normalizing the matrix was represented using Eq. (3).

$$r_{ij} = \frac{x_{ij}}{\sum_{i=1}^{m}(x_{ij})} \tag{3}$$

Table 13. Criteria and their corresponding weights

S.No.	Criteria	Category	Objective	Weightage
1	Brinell Hardness	Beneficiary (Larger-the-better)	Maximization	0.25
2	Izod Impact Strength	Beneficiary (Larger-the-better)	Maximization	0.25
3	Flatness Deviation	Cost (Smaller-the-better)	Minimization	0.25
4	Ultimate Tensile Strength	Beneficiary (Larger-the-better)	Maximization	0.25

where,

i = 1,2,…,m
j = 1,2,…,n.

$$\begin{pmatrix} r_{11} & r_{12} & \cdots & r_{1n} \\ r_{21} & r_{22} & \cdots & r_{2n} \\ \vdots & \vdots & & \vdots \\ r_{m1} & r_{m2} & \cdots & r_{mn} \end{pmatrix} \tag{4}$$

The normalized matrix, denoted as 'R' was represented as Eq. (4):
Table 14 represented normalized values for criteria elements

Step 4: Creation of Weight Normalized Matrix

After assigning weight factor to the criteria, it is imperative to create a weight normalized decision matrix by multiplying the weightage with corresponding criteria elements/values. The sum of all weight factors should be equal to one(Mahmoodzadeh, S., et al., 2007).

A weight normalized matrix is to be constructed as per Eq. (5), in which:

$$a_{ij} = w_j * r_{ij} \tag{5}$$

The resultant matrix was represented as Eq. (6):

$$\begin{pmatrix} a_{11} & a_{12} & \cdots & a_{1n} \\ a_{21} & a_{22} & \cdots & a_{2n} \\ \vdots & \vdots & & \vdots \\ a_{m1} & a_{m2} & \cdots & a_{mn} \end{pmatrix} \tag{6}$$

Table 14. Normalized decision matrix table

Brinell Hardness	Izod Impact Strength	Average Flatness Deviation	Ultimate Tensile Strength
0.101	0.100	0.193	0.113
0.101	0.150	0.178	0.109
0.172	0.050	0.128	0.128
0.148	0.100	0.131	0.136
0.121	0.100	0.056	0.103
0.114	0.150	0.119	0.140
0.114	0.150	0.119	0.124
0.129	0.200	0.077	0.146

Table 15 represented the weight normalized matrix for current research work.

Step 5: Calculation of Maximizing and Minimizing Indices for the Eight Alternative Combinations

As we were having three beneficiary/maximization criteria (hardness, impact strength and ultimate tensile strength) and one cost/minimization criterion, each alternative should be assigned a maximizing index and a minimizing index (Stefano, N.M., et al., 2015) (Chatterjee, P., et al., 2011). The Eqs. (7) & (8) represent the formulae for calculating the same.

$$B_i = \sum_{j=1}^{k} \left(a_{ij}\right) \text{ where, j=1,2,.....,k (for calculating maximizing index)} \quad (7)$$

$$C_i = \sum_{j=k+1}^{n} \left(a_{ij}\right) \text{ where, j=k+1, k+2...,n (for calculating minimizing index)} \quad (8)$$

Table 16 represented the maximizing and minimizing indices for the present research work.

Table 15. Weight normalized decision matrix

Brinell Hardness	Izod Impact Strength	Average Flatness Deviation	Ultimate Tensile Strength
0.025	0.025	0.048	0.028
0.025	0.038	0.045	0.027
0.043	0.013	0.032	0.032
0.037	0.025	0.033	0.034
0.030	0.025	0.014	0.026
0.028	0.038	0.030	0.035
0.028	0.038	0.030	0.031
0.032	0.050	0.019	0.037

Table 16. Maximizing and minimizing indices

Maximizing index, B_i	Minimizing index, C_i
0.079	0.048
0.090	0.045
0.088	0.032
0.096	0.033
0.081	0.014
0.101	0.030
0.097	0.030
0.119	0.019

Step 6: Relative Weight Calculation for Each Alternative

Eq. (9) indicated the formula for calculating the relative weight of ith alternative, Q_i (Podvezko, V., 2011).

$$\left(\frac{\min(i)C_i * \sum_{i=1}^{m}(C_i)}{C_i * \sum_{i=1}^{m}((\min(i)C_i)/C_i)} \right) + B_i = Q_i \tag{9}$$

Table 17 indicated the relative weight for all the 8 FSW Composite plate specimens.

Step 7: Determination of Priority Order among the Alternatives

The alternative of having the highest relative weight, will be ranked first (Mehdi Ajalli, et al., 2017). Similarly, ranking will be given successively as per the descending order of the relative weight (Chatterjee, P. & Chakraborty, S., 2013).

Table 18 showed the relative weights and the corresponding priority sequence of alternatives.

Table 17. Relative weight

Q_i
0.096
0.109
0.114
0.122
0.141
0.129
0.125
0.163

Table 18. Relative weights and priority order

Alternatives	Q_i	Priority order
1	0.096	8
2	0.109	7
3	0.114	6
4	0.122	5
5	0.141	2
6	0.129	3
7	0.125	4
8	0.163	1

Step 8: Calculation of Performance Index for Each Alternative

Eq. (10) represented the formula for calculating Performance Index, P_i (Popovic, G., et al., 2012).

$$P_i = (Q_i/Q_{max})*100\% \tag{10}$$

Alternative with 100 degree meant the best one. Ranking must be performed from the best one to the worst one in the ascending order (Organ, A. & Yalcin, E., 2016).

Performance Index was calculated and ranking was done to find out the best alternative for the above-performed research work and the same was represented in Table 19.

From the above table, the optimal alternative/specimen was found out and as a result the process parameters as well as mechanical properties for the corresponding Friction Stir Welded Composite combination were tabulated as Table 20 below:

Table 19. Performance index and ranking of alternatives

P_i	Rank	Alternatives
59.03434	8	1
66.92318	7	2
70.0502	6	3
74.92331	5	4
86.6786	2	5
79.57547	3	6
77.05434	4	7
100	1	8

Table 20. Optimized welding parameters and solutions

Specimen No.	Composition of Al_2O_3	Spindle Speed	Feed	Brinell Hardness	Izod Impact Strength	Average Flatness Deviation	Ultimate Tensile Strength
	wt %	rpm	mm/ min	BHN	J	-	N/mm^2
8	10	710	50	64.61	8	0.26	98

CONCLUSION

From the above research work, it is concluded that:

- MMCs are the most sought-after engineering materials now-a-days due to their ever-increasing applications and innumerable merits (Krishna, S. M., et al., 2015).
- Aluminium Matrix Composites are scoring over individual aluminium alloys owing to their low weight, increased wear resistance, improved hardness and excellent corrosion resistance (Pradeepkumar, J., et al., 2018).
- Stir casting, also termed as liquid metallurgy is an evergreen manufacturing methodology for producing MMCs, as it is flexible, robust, less expensive and can produce large number of components at a timeAlumina is the most common synthetic ceramic used wherever resistance to scratching, abrasion, and good polishing action is required.
- Uniform and even distribution of alumina ceramic particles inside the aluminum matrix plays a pivotal role in the enhanced performance of composites.
- Friction Stir Welding is one of the most popular and precise materials joining processes, commonly used for fabricating aluminum alloys or composites joints with excellent binding strength at the molecular level and good surface finish.
- Normally, a composite specimen is worth only when it has high tensile strength, high hardness, high impact strength, and perfectly flat.
- MCDM techniques are synonymous with numerical optimization, as they involve mathematical calculations and quantitative analysis.
- COPRAS is one of the widely implemented MCDM techniques in which ranking of the best alternative is done by assessing maximizing and minimizing criteria.
- Evaluation of alternatives is carried out using COPRAS and the 8th specimen combination that has 90% composition of Al 6061 and Al 2024 each. Ten percent of alumina, a tool rotation speed of 710 rpm, and a traverse feed of 50 mm/min are found out to be the optimal parameters.

CONFLICTS OF INTEREST

The authors declare no conflict of interest.

FUTURE SCOPE

The identified scope of work may be carried out in future remains:

- Joining of aluminum alloy plate with AMC plate in FSW can be tried.
- Several other MCDM techniques can be used for optimization.
- A recursive technique that combines an algorithm-based optimization and MCDM- based optimization can also be used.

REFERENCES

Ajalli, M., Azimi, H., Balani, A. M., & Rezaei, M. (2017). Application of Fuzzy AHP and COPRAS to solve the Supplier Selection Problems. *Int. J Sup. Chain. Mgt*, *6*(3), 112–119.

Anderson, M. J., & Whitcomb, P. J. (2010). *Design of Experiments*. Kirk-Othmer Encyclopedia of Chemical Technology. doi:10.1002/0471238961.0405190908010814.a0

Antony, J. (2014). Design of experiments for engineers and scientists. Elsevier.

Bhandare, R. G., & Sonawane, P. M. (2013). Preparation of aluminium matrix composite by using stir casting method. *International Journal of Engineering and Advanced Technology*, *3*(3), 61–65.

Bodunrin, M. O., Alaneme, K. K., & Chown, L. H. (2015). Aluminium matrix hybrid composites: A review of reinforcement philosophies; mechanical, corrosion and tribological characteristics. *Journal of Materials Research and Technology*, *4*(4), 434–445. doi:10.1016/j.jmrt.2015.05.003

Cavaliere, P., Nobile, R., Panella, F. W., & Squillace, A. (2006). Mechanical and microstructural behaviour of 2024–7075 aluminium alloy sheets joined by friction stir welding. *International Journal of Machine Tools & Manufacture*, *46*(6), 588–594. doi:10.1016/j.ijmachtools.2005.07.010

Chatterjee, P., Athawale, V. M., & Chakraborty, S. (2011). Materials selection using complex propoptional assessment and evaluation of mixed data methods. *Materials & Design*, *32*(2), 851–860. https://doi.org/10.1016/j.matdes.2010.07.010

Chatterjee, P., & Chakraborty, S. (2012). Materials selection using COPRAS and COPRAS-G methods. *International Journal of Materials and Structural Integrity*, *6*(2/3/4), 111-133. doi:. doi:10.1504/ijmsi.2012.049951

Chatterjee, P., & Chakraborty, S. (2013). Gear Material Selection Using Complex Proportional Assessment and Additive Ratio Assessment-based Approaches: A Comparative Study. *International Journal of Materials Science and Engineering*, *1*(2), 104–111.

Ginevicius, R. (2008). Normalization of quantities of various dimensions. *Journal of Business Economics and Management*, *9*(1), 79–86. https://doi.org/10.3846/1611-1699.2008.9

Gomathisankar, M., Gangatharan, M., & Pitchipoo, P. (2018). A Novel Optimization of Friction Stir Welding Process Parameters on Aluminum Alloy 6061-T6. *Materials Today: Proceedings*, *5*(6), 14397–14404. doi:10.1016/j.matpr.2018.03.025

Hwang, C. L., & Yoon, K. (1981). *Multiple Attribute Decision Making–Methods and Applications. Springer-Verlag Berlin Heidelberg.*

Kou, G., Lu, Y., Peng, Y., & Shi, Y. (2012). Evaluation of classification algorithms using MCDM and rank correlation. *International Journal of Information Technology & Decision Making*, *11*(1), 197–225.

Krishna, S. M., Shridhar, T. N., & Krishnamurthy, L. (2015). Research significance, applications and fabrication of hybrid metal matrix composites. *International Journal of Innovative Science, Engineering & Technology*, *2*(5), 227–237.

Li, Y., Murr, L., & McClure, J. (1999). Flow visualization and residual microstructures associated with the friction-stir welding of 2024 aluminium to 6061 aluminium. *Materials Science and Engineering A*, *271*(1-2), 213–223. doi:10.1016/S0921-5093(99)00204-X

Mahmoodzadeh, S., Shahrabi, J., Pariazar, M., & Zaeri, M. S. (2007). Project selection by using fuzzy AHP and TOPSIS technique. *World Academy of Science, Engineering and Technology*, *30*, 333–338.

Organ, A., & Yalcin, E. (2016). Performance Evaluation of Research Assistants By Copras Method. *European Scientific Journal*, *12*(10), 102–109.

Podvezko, V. (2011). The Comparative Analysis of MCDA Methods SAW and COPRAS. *Inzinerine Ekonomika-Engineering Economics*, *22*(2), 134–146.

Popovic, G., Stanujkic, D., & Stojanovic, S. (2012). Investment Project Selection by Applying COPRAS Method and Imprecise Data. *Serbian Journal of Management*, *7*(2), 257–269.

Pradeepkumar, J., Robinson Smart, D. S., & John Alexis, S. (2018). An Aluminium Hybrid Metal Matrix Composite reinforced with TaC, Ti, Si3N4 nanoparticles - A Review. *International Journal of Mechanical and Production Engineering Research and Development*, (Special Issue), 187–194.

Prakash, M., & Daniel Das, A. (2020). Investigation on effect of FSW parameters of aluminium alloy using Full Factorial Design. *Materials Today: Proceedings*, 1–6. doi:10.1016/j.matpr.2020.05.622

Rajesh, P. V. (2020). Effects of Process Parameters in Friction Stir Welding of Aluminium Matrix Composites - A Review. *Mahasarakham International Journal of Engineering Technology*, *7*(1), 16–22.

Rathee, S., Maheshwari, S., Noor Siddiquee, A., Srivastava, M., & Kumar Sharma, S. (2016). Process parameters optimization for enhanced micro hardness of AA6061/SiC surface composites fabricated via friction stir processing. *Materials Today: Proceedings*, *3*(10), 4151–4156. doi:10.1016/j.matpr.2016.11.089

Ravi, B., Naik, B. B., & Kumar, G. R. (2018). Influence of tool rotational speed on the mechanical and microstructure properties of friction stir welded Al-B_4C MMCs. *Materials Today: Proceedings*, *5*(1), 5. doi:10.1016/j.matpr.2017.11.227

Sadeesh, P., Venkatesh Kannan, M., Rajkumar, V., Avinash, P., Arivazhagan, N., Devendranath Ramkumar, K., & Narayanan, S. (2014). Studies on Friction Stir Welding of AA 2024 and AA 6061 Dissimilar Metals. *Procedia Engineering*, *75*, 145–149. doi:10.1016/j.proeng.2013.11.031

Stefano, N. M., Casarotto Filho, N., Vergara, L. G. L., & Rocha, R. U. G. (2015). COPRAS (Complex Proportional Assessment): State of the Art Research and its Applications. *IEEE Latin America Transactions*, *13*(12), 3899–3906.

Subramanya, P., Amar, M., Arun, S., Mervin, H., & Shrikantha, R. (2018). Friction stir welding of Aluminium matrix composites – A Review. *MATEC Web of Conferences, 144*, 1-13. 10.1051/matecconf/201814403002

Suryanarayanan, R., & Sridhar, V. G. (2021). Influence of welding parameters on the weld properties in Friction stir spot welding of aluminium alloys of varying thicknesses. *Materials Today: Proceedings*, *46*, 8525–8531. Advance online publication. doi:10.1016/j.matpr.2021.03.534

Umanath, K., Palanikumar, K., Sankaradass, V., & Uma, K. (2021). Optimizations of friction stir welding process parameters of AA6063 aluminium alloy by Taguchi technique. *Materials Today: Proceedings*, *46*, 1–12. doi:10.1016/j.matpr.2021.02.539

Vengatesh, D., & Chandramohan, V. (2014). Aluminium alloy metal matrix composite: Survey paper. *International Journal of Engineering Research and General Sciences*, 2(6), 792–796.

Verma, S., & Misra, J. P. (2021). Experimental investigation on friction stir welding of dissimilar aluminium alloys. *Proceedings of the Institution of Mechanical Engineers, Part E: Journal of Process Mechanical Engineering,* 22-34. 10.1177/09544089211008694

Wang, D., Xiao, B. L., Ni, D. R., & Ma, Z. Y. (2014). Friction Stir Welding of Discontinuously Reinforced Aluminum Matrix Composites: A Review. *Acta Metallurgica Sinica. English Letters*, *27*(5), 816–824. doi:10.100740195-014-0143-2

Zavadskas, E. K., Kaklauskas, A., & Sarka, V. (1994). The new method of multi-criteria Complex Proportional Assessment of projects. *Technological and Economic Development of Economy*, *1*(3), 131–139.

Zhao, J., Wu, C. S., & Su, H. (2021). Acoustic effect on the tensile properties and metallurgical structures of dissimilar friction stir welding joints of Al/Mg alloys. *Journal of Manufacturing Processes*, *65*, 328–341. doi:10.1016/j.jmapro.2021.03.057

ADDITIONAL READING

Martin, A., Miranda Lakshmi, T., & Prasanna Venkatesan, V. (2018). A Study on Evaluation Metrics for Multi Criteria Decision Making (MCDM) Methods - TOPSIS, COPRAS & GRA. *International Journal of Computing Algorithm*, *7*(1), 29–37. doi:10.20894/IJCOA.101.007.001.006

Mousavi-Nasab, S. H., & Sotoudeh-Anvari, A. (2017). A comprehensive MCDM-based approach using TOPSIS, COPRAS and DEAas an auxiliary tool for material selection problems. *Materials & Design*, *121*, 237–253. doi:10.1016/j.matdes.2017.02.041

Pavlicic, D. (2001). Normalization affects the results of MADM methods. *Yugoslav Journal of Operations Research, 11*(2), 251–265.

Triantaphyllou, E. (2000). *Multi-Criteria Decision Making Methods: A Comparative Study*. Kluwer Academic Publishers. doi:10.1007/978-1-4757-3157-6

Zavadskas, E. K., Turskis, Z., & Kildienė, S. (2014). State of art surveys of overviews on MCDM/MADM methods. *Technological and Economic Development of Economy, 20*(1), 165–179. doi:10.3846/20294913.2014.892037

Zeleny, M. (1982). *Multiple Criteria Decision Making*. Mc-Graw-Hill.

KEY TERMS AND DEFINITIONS

Alloy: An alloy is a mixture of two or more metals homogeneously at specific proportion. The resultant material has distinctive physical and chemical properties when compared to parent metals.

Ceramic: A ceramic is a powdery material with high compressive strength that provides abrasive action and good refractoriness.

Composite: A composite is joining of two or more materials together at any proportion. Each constituent in the composite lies in its heterogeneous state. The resultant material has distinctive mechanical properties when compared to the constituents.

Criteria: Criteria are attributes or characteristics of a material, which can be determined by conducting various tests on its mechanical, physical, chemical, and tribological behaviour.

Factors: Factors are parameters used either in the fabrication or testing of an engineering material. (i.e., manufacturing or testing process).

Flatness: Flatness is a surface property in which opposite horizontal edges of a flat material lies in the same plane.

Chapter 15
Characterization and Spectroscopic Applications of Metal Foams From New Lightweight Materials

Sefa Celik
https://orcid.org/0000-0001-6216-1297
Istanbul University, Turkey

Nuray Bekoz Ullen
https://orcid.org/0000-0003-2705-2559
Istanbul University-Cerrahpasa, Turkey

Sevim Akyuz
Istanbul Kultur University, Turkey

Gizem Karabulut
https://orcid.org/0000-0003-0930-5380
Istanbul University-Cerrahpasa, Turkey

Aysen E. Ozel
Istanbul University, Turkey

ABSTRACT

Lightweight materials such as metallic foams possess good mechanical, chemical, and physical properties, which make them suitable for a wide range of functional and structural applications. Metal foams have recently gained substantial interest in both industry and academia due to their low cost, thermal conductivity, high working temperature, vibration damping, specific mechanical properties, energy absorption, and heat resistance. The use of metal foams on a large scale and successful applications depend on a detailed understanding of their characteristic properties. Metallic foams are characterized by the morphology of the porous cells (size and shape, open or closed, macro and micro), pore topology, relative density, properties of the pore wall, and the degree of anisotropy. This contribution focuses on x-ray diffraction, Fourier transform infrared (FT-IR), and Raman spectroscopic applications used for the characterization of metal foam, and also a brief of the most important applications, including a significant number of examples given.

DOI: 10.4018/978-1-7998-7864-3.ch015

GENERAL INTRODUCTION TO METAL FOAMS

Metal foams were first produced in the 1950s, and with the developments in recent years, they have attracted attention among the light-weight materials. Metallic foams have the advantages of a unique combination of mechanical properties like lower weight, high energy absorbing capacity, high stiffness, and high damping capacity. For these reasons, they are becoming the most promising class of materials in the industrial and scientific areas. Metallic foams have a role in many areas of our lives day by day. Metal foams are being used in many industrial applications such as light weight structural applications, biomedical applications, filters, electrodes, heat exchangers, home appliances etc. In addition, metallic foams are also used at high temperatures. They are used in areas such as the transpiration cooled rocket nozzles, a cooling system in the combustion chamber of gas and steam turbines, and as heat shielding for aircraft exhaust. (Ashby, 2001; Bauer et al., 2013; Qin et al., 2016; Liu and Chen, 2014; Banhart et al., 2019; Sutygina et al., 2020). Metal foams are a structure consisting of air and metal, and they can be obtained by many production methods with different pore ratio. The advancement of technology and materials science has enabled the discovery of many metallic foam production methods and the production of metallic foams with different properties. Photograph of metal foams with different porosity produced in the laboratory is given in Figure 1. This figure is taken from the PhD thesis of Nuray Beköz Üllen, one of the authors of the chapter (Beköz, 2011).

Figure 1. Steel foam specimens having different porosities (Beköz, 2011)

In fact, it is desired that the material does not contain pores in many production methods such as casting and welding, but the advantages of the pore structure are used in these materials (Degischer and Kristz, 2010). Porous structure provides thermal and sound insulation as well as impact absorption in addition to light-weight construction (Banhart, 2003). Usage areas and properties of porous metals depend on the pore type. Foam structures can be defined as several types which includes pore type: open foams (pores connected to each other so that matter can pass through them), closed foams (isolated closed pores filled with any gas), and partially open foams (Beköz, 2011; Patel et al., 2018). Metal foams with closed pores are used in light structures thanks to their high hardness and low-density properties. Due to their deformation capacity when undergoing constant stress, they come to the fore as ideal energy absorbers in packaging applications and especially in the automobile industry. They can also be used in acoustic damping applications. On the other hand, open cell foams offer a wide variety of applications, e.g. as a transmission and filtering applications, heat exchanger etc. Open cell foams have a high percentage of interconnected pores and high surface area. These features make it charming for use as heat absorbers in electronic appliances, and in transpiration cooling (Aly, 2004; Dukhan, 2013). The interconnected porosity is specific point for parts like filters, bearings, diaphragms, membranes. In addition, they also have importance in biomedical applications (Aly; 2004; Ashby et al., 2001). The properties of metallic foams vary according to the porosity, shape, size and the connection types of the pores with each other and create much different potential for usage areas (Bekoz and Oktay, 2013; Tunçer, 2006; Hakamada et al., 2007; Alhusseny and Nasser, 2018).

All aforesaid types of foams have different applications. In lightweight metals foam category Ti, Al, Ni, Fe, Zn, Cu and Mg are widely used. Metallic foams, due to having extremely low densities in comparison to solid structures, possess high specific bending strength and stiffness, they easy recycling, having non-toxic outstanding properties for sound absorption, vibration, and impact energy (Rajak et al., 2014). In this respect, characterization studies are important in terms of examining the pore properties and determining their effects on metallic foam. Therefore, this chapter includes the methods used in the characterization of metallic foams from past to present.

Characterization Methods of Metal Foams

As a result of the advancement of technology and studies, characterization methods are developed day by day. The successful application of foams depends not only upon inventing new techniques to produce foams that are cheap and of improved quality, but also upon the development of a detailed understanding of all their properties. The physical and chemical properties of metal foams can be characterized in many ways. To characterize foam structures

and to obtain useful information about their microstructure, morphology, mechanical and other properties, various techniques are used. The various methods for the characterization of metal foams will be briefly reviewed in this section.

DETERMINATION OF PORE STRUCTURE

It is known that foam properties strongly depend on the pore structure. Pore structure is a general term. What is meant by this term is pore size, shape, connections with each other and their environment, distributions of the pores and pore ratio. The structural features of foam influence its specific surface area, fluid flow resistance, and electromagnetic transmission or absorption. Various equations were suggested for the characterization and modeling of the relationship between the structure and the various properties of the foam. (Korner and Singer, 2000; Ashby et al., 2001). Metallic foams, having a wide dispersion of cell size and cell shape, are highly disordered. The terms used to characterize porous structure are given in Table 1. The characteristics of a foam material given in Table 1 are important and determine the wide variety of material properties that can be obtained from that foam. (Andure et. al., 2012).

While the pore wall cross-sectional shape and actual size are controlled by relative density, the number and nominal size of foam pore walls are controlled by pore size. The features of the base metal determine all the physical properties of the produced foam, such as melting temperature, thermal expansion coefficient, specific heat, strut hardness, chemical reactivity and oxidation limits. The metal selected to be foamed together with relative density affects the mechanical properties of the foam such

Table 1. Definition of terms used to characterize porous structure (Dukhan, 2013)

Density	**True density:** Density of a porous material except pores and inter-particle cavities, just dense part	**Apparent density:** Density of a porous material with closed and inaccessible pores	**Bulk density:** Density of a porous material including pores and inter-particle cavities
Pore Volume	Volume of the pores of (Vp)		
Pore Size	The length between two opposite walls of the pore		
Porosity	The apparent volume (V) of the particle or powder per the ratio of the total pore volume (Vp)		
Surface Area	The ratio of accessible area of solid surface to the unit mass of material		

as elastic modulus, electrical and thermal conductivity, crush strength (Aly, 2004). The mechanisms of linear elasticity depend on whether metallic foam with open (interconnected) or closed (connected) pore structure. Cell edge bending is an effective deformation mechanism in open porous metal foams. In closed-cell metal foams, deformation begins with plastic bending of the cell walls and stretching of the cell surfaces when a compression load greater than the collapsing strength of the foam is applied (Aly, 2004; Rajak and Gupta, 2020).

It is important to characterize the densities of the utilised foams, since density is the factor that significantly affects the mechanical properties of metallic foams (Aly, 2004; Degischer and Kriszt, 2010). The term relative density is used instead of density of the metallic foam. Relative density is the ratio of the densities of the metallic foam and the metal from which the metallic foam is made. Therefore, also this property has been examined in this section as it depends on the amount of pore. As mentioned before, the pore structure is the most important factor determining the usage area and properties of metallic foam. The nominal pore wall length, pore diameter and cross-sectional size are directly affected by the pore size of the metallic foam. These microstructural properties of the foam also affect different properties such as spesific surface area, electromagnetic transmission or absorption, fluid flow properties. It is required to optimize the pore size. Although small porous structures are better in terms of strength, their permeability is poor. For example, pores smaller than 200 microns and larger than 600 microns are not desired in biomaterial applications (Xue et al., 2007). The pore shape varies according to the production method and can be in various forms. It is important because it affects the mechanical properties and causes anisotropy (Harders et al., 2005; Borovinšek et al., 2016).

Density measurements of a porous metal can basically be made using Archimedes' Principle. The buoyancy in a liquid of known density is determined. If the structure does not have a closed surface, it is recommended to cover it with a polymer surface in order not to fill the pores (Banhart, 2001; Esen and Bor, 2011). In addition to this method, the Eddy current method is also used for relative density measurement. A magnetic field is created around a coil by passing alternating current, and this magnetic field changes over time. It creates a second magnetic field on the electrically conductive sample. The interaction between the two magnetic fields changes, causing a change in impedance on the coil. By measuring this impedance change, the properties of the material can be revealed (Banhart, 2001; Bor et al., 2007; Ma and Peyton, 2006).

Determination of pore size distributions or shape analysis of the pores can perform using commercial image analysis programs. Individual pores in the preparation plane can be identified with these programs. Preparation of materials should be carried out very carefully. In order to distinguish between cell membranes and inside cells, they must appear in different brightness. One of the ways to prepare is by embedding the porous material in a black resin and then polishing the surface. Manual correction may be required in order for the program to define the individual pores and calculate pore size distributions.

Many methods can be preferred for determination of other pore properties. Non-destructive methods such as porosimetry and permeametry measurement, as well as X-Ray radiography/ tomography, neutron small angle scattering and ultrasound imaging methods can be used (Banhart, 2001; Patel et al., 2018). The use of X-ray measurements explained in section 1.9. In addition, eddy current method is also used for pore size determination. Porosimetry and permeametry measurement methods are used to measure the specific surface areas, distribution of pore size and channel dimensions of metallic foams with open cell structure. Measurements are performed using methods as mercury porosimetry, gas absorption techniques, permeametry and bubble point method (Banhart, 2001).

Mercury Porosimetry: This method is based on the penetration of mercury under the effect of pressure into the pore volume and extrusion. The pressure data and penetrated volume are related to the size of the pores. It is difficult to detect pores larger than a few hundred microns with this method (Banhart, 2001; Dukhan, 2013).

Gas Absorption Techniques (BET): With this method, the total internal surface of a porous material can be determined with very high accuracy. Even very tiny pores of atomic size can be detected (Banhart, 2001; Ashby; 2001).

Permeametry: The basis of this method is to examine the flow of a gas or liquid (m^3/s) per unit time through a unit volume of a porous medium. A linear relationship is found for very slow flow by measuring the pressure drop at a foam column of height as a function of the average fluid velocity. There is a quadratic dependence for higher flow rates when the energy losses due to the inertia of the flowing fluid increase to significant degrees. If the viscosity of the fluid is known, the permeability can be derived as a coefficient of friction. Considering that the internal surface area is related to the friction of the pore volume, an estimate for the specific surface area can be obtained (Banhart, 2001; Smorygo et al., 2011).

Bubble Point Method: In this method, first the pore volume is filled with a liquid and then an increased external pressure is applied. When a certain pressure value is reached, the capillary force of the liquid is exceeded. The liquid is blown out of the sample giving rise to a resistance drop. Finally, the smaller pores are emptied until the sample is dry. Some information about the largest open pore of the specimen and the pore size distribution can be determined by comparing the flow rate as a function of pressure drop to the corresponding curve of a dry specimen (Banhart, 2001; Park, 2016).

Mechanical Testing

Mechanical testing of metallic foams to be used for any structural application is a prerequisite. Mechanical data and databases should be created to evaluate specific applications or to model porous metals or containing components in computer-aided modeling. In principle, the characterization methods are the similar as for dense materials. However, it is necessary to make adjustments to some test procedures, taking into account the different structure of porous materials. This difference is related to statistics. For example, when metallic foam made of a particular alloy is tested, the mechanical properties obtained as result of examinining several samples with the same foam density will show a much greater scatter than the test results of the dense state. The reason for this is the difference in some properties such as the mass distribution, the distribution of pores, the heterogeneous microstructure, etc. which are "hidden parameters" that make difference in the porous material samples having the same total density.For this reason, more samples may be required to obtain more accurate results in testing porous structures compared to the test standards required for non-porous materials. Also, the length scale of macroscopic heterogeneities in porous materials is sometimes in the millimeter range. This may require an enlargement of specimen dimensions in mechanical tests to current standards. Various mechanical tests can be labelled with one of the following properties (Banhart, 2001; Yang et al., 2017):

Mode of Loading: tension, compression, shear, torsion and bending

Type of Applied Stress: uniaxial, biaxial, multiaxial, quasi-static and hydrostatic

Time Dependence of Load: slowly increasing dynamic, cyclic and constant

Quasi-static and uniaxial compression tests are frequently used for characterization of porous metals. These measurements enable the mechanical stability of the porous metal to be defined. In the publication of one of the authors of this chapter (Bekoz and Oktay 2013), the effect of porosity and pore size parameters on the mechanical properties of steel foam samples was investigated. The strain-stress curve of the compression test, that applied to the steel foam samples reported in this study was given in Figure 2. Along with these tests, fracture toughness measurements on notched specimens and indentation tests with various indentor shapes are also applied porous samples. There are also tests for foam sandwich panels, such as the drum peel test (ASTM D1781), which describe the bond between the core and the face sheets. Mechanical deformation can be monitored by recording acoustic emissions with the breaking cell walls, which is a technical measurement in addition to ordinary tests (Banhart, 2001; Yang et al., 2017).

Figure 2. Compressive stress-strain curves of the steel foams having different porosities (Bekoz and Oktay, 2013)

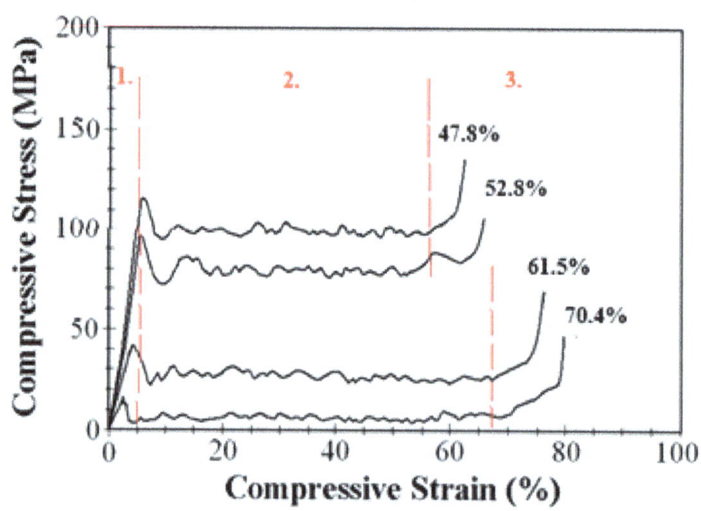

As shown in Figure 2, the stress-strain curve consists of three regions. These are the linear elastic zone (1), plateau zone (2) and densification zone (3). Linear elastic zone ends at maximum compressive strength value, after which a long plateau stress is observed where deformation occurs at almost constant stress (Aly 2004; Beköz, 2011; Dukhan, 2013). This long plateau is formed by the collapse of the cell walls and the struts in the cellular structure. Deformed pores form deformation bands in a direction perpendicular to the deformation direction (Aly, 2004). In the case of brittle materials, compression strength is defined as the first maximum stress value in the compressive stress-strain curve. However, in the case of ductile materials, the compression strength of the material is lies at the point where the slope of stress-strain curve equals zero for the first time. However, there are other ductile materials that pass directly into the plateau region without showing any maximum point (Ashby et al., 2001; Aly, 2004). Florek et al. (2010), observed these differences in brittle and ductile materials with stress-strain curves obtained as a result of the uniaxial compression test of metallic foams made of different aluminum alloys. When this happens, the intersection point of the tangent drawn to the linear elastic regime and drawn to the plateau is defined as the compressive strength of the material (Ashby et al., 2001). The stress-strain curve ends with a zone that is the stress increases quickly. This zone is known as the densification regime (Aly, 2004). The areas with large pores begin to deform at the beginning of the compression test. By increasing the compression, other pores in the deformed areas are deformed. When the compression deformation ends, it is observed that deformation bands with a thickness equal to the pore diameter are formed within the fractured areas (Aly, 2004; Haag et. al., 2003).

Porous metals have great potential for applications that require absorption of crash energy. The efficiency of the energy absorbing metals in the parts used at the points where the crash energy is desired to be absorbed is significantly affected by the stress-strain behavior under uniaxial compression. The characteristic shape of the compressive stress-strain curves applied to porous metal with an almost constant stress allows large amounts of energy to be absorbed at a relatively low stress level. A comparison of the energy absorption behavior of a fully dense elastic solid with metallic foam was given in Figure 3, taken from the PhD thesis of Nuray Beköz Üllen, one of the authors of the chapter (Beköz, 2011). Metallic foam can absorb more energy than its dense form under a specified peak stress level (Claar et al., 2000; Beköz, 2011).

Figure 3. Comparison of energy absorbtion behavior via compressive deformation for foam and dense solid materials (Beköz, 2011).

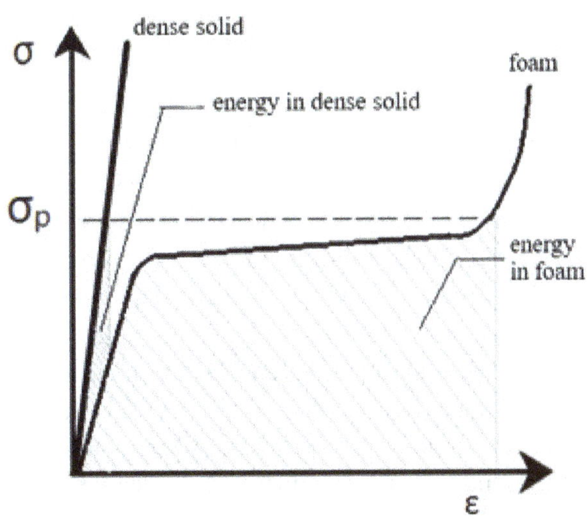

An immense range of signals can be obtained with Transmission Electron Microscope that infer diffraction patterns, images and many different types of spectra from a very small specimen area (Patel et al., 2018; Banhart, 2013). There are also different TEM types such as HRTEMs, STEMs, IVEMs, HVEMs, and AEMs. In addition, the X-ray diffraction method can be used to determine the crystal structure of all components in the specimen (Raj and Daniel; 2007). Various tests are performed on metallic foams to determine mechanical properties like compressive and tensile strength, flexural strength, hardness etc. (Bekoz and Oktay, 2012; Bekoz and Oktay, 2014; Ashby et al., 2001; Smith et al., 2012; Dukhan, 2013; Degischer and Kriszt, 2010). The tests mentioned include compression, tensile, shear, multi-axial, creep, fatigue, hardness, surface strain mapping tests. In addition to these tests, the non-destructive resonance frequency method has also been proposed in literature for tensile test and damping measurements (Patel et al., 2018).

Corrosion Testing

Metallic foams have less corrosion resistance due to the pores in their structure, as contact with air is also in the interior and have greater surface area. There is no standard corrosion test method used for metallic foams. These tests can be performed depending on the area of use (Banhart, 2001). The corrosion behavior of the material can be characterized by methods such as anodic polarisation measurements (Rossi et al., 2017), loss weight tests and electrochemical tests (Zhang et al., 2009; Ajinola, 2016; Palaniswamy et al., 2020).

Microstructural Characterization

Microstructure characterization is important for achieving the desired properties in metallic foams. As the density of metallic foams increases, their mechanical properties increase due to the increase in the pore walls. However, instead of increasing the density of a product that promises light weight, it is desirable to improve its mechanical properties by another method. Improving the mechanical properties of metallic foams is possible by optimizing the pore properties and the microstructure (Degischer and Kristz, 2010; Bekoz and Oktay, 2012; Taherishargh et al., 2017). Therefore, characterization of the microstructure is required.

Microstructural analysis of metallic foams can be determined by optical image analysis as a destructive method (Banhart, 2001; Patel et al., 2018). In the optical microscopy method, the pore microstructure of the specimen can be observed after the preparation processes. The samples are ground and polished using the normal grinding and polishing techniques. It is very important to preserve the porous structure of the material, during the preparation of the specimens for the microstructural examination. Various methods are applied for this. Resin method is one of the easiest and most frequently applied methods. The samples are infiltrated in vacuum by a low viscosity resin. The resin is pour onto the samples and on admitting the air to the chamber; it is impregnated into the small pores. Then, a sample is taken from the foam and ground and polished (Beköz, 2011). In Figure 4, which was taken from the article of Nuray Beköz Üllen, one of the chapter authors, the optical microscope image of the samples was given in the study of the steel foams produced by space holder-water leaching technique in powder metallurgy (Bekoz and Oktay 2014).

Optical microscope is not sufficient in parts such as analysis of grain structures and surface. Scanning electron microscopy (SEM) or field emission scanning electron microscopy (FESEM) may be preferred because it provides a large depth of field in such analyzes (Duarte and Banhart, 2000; Vendra et al., 2011; Jain et al., 2020; Marx and Rabiei, 2020; Oveisi and Geramipour, 2020). The differences that may occur on the surfaces can be observed with these methods. Thanks to the high depth of focus and

Figure 4. Optical microscope images of the pore walls (Bekoz and Oktay, 2014).

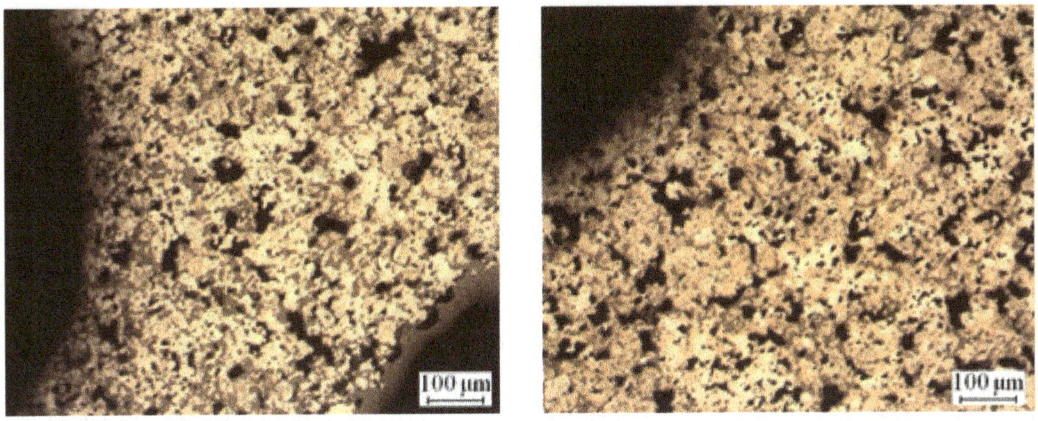

wide working distance of SEM, three-dimensional micrographs can be obtained from rough samples at all magnifications. The SEM images of the steel foams were shown in Figure 5, which was taken from another study, conducted by one of the chapter authors (Bekoz and Oktay 2012). Electron Backscatter Diffraction (EBSD) may use to determine the structure, phase, and crystal orientation of the material in SEM. No damage occurs during the analysis, but as the material requires cutting, ground or polishing in sample preparation stages, effectively a destructive technique.

Figure 5. SEM images of the steel foams (Bekoz and Oktay, 2012).

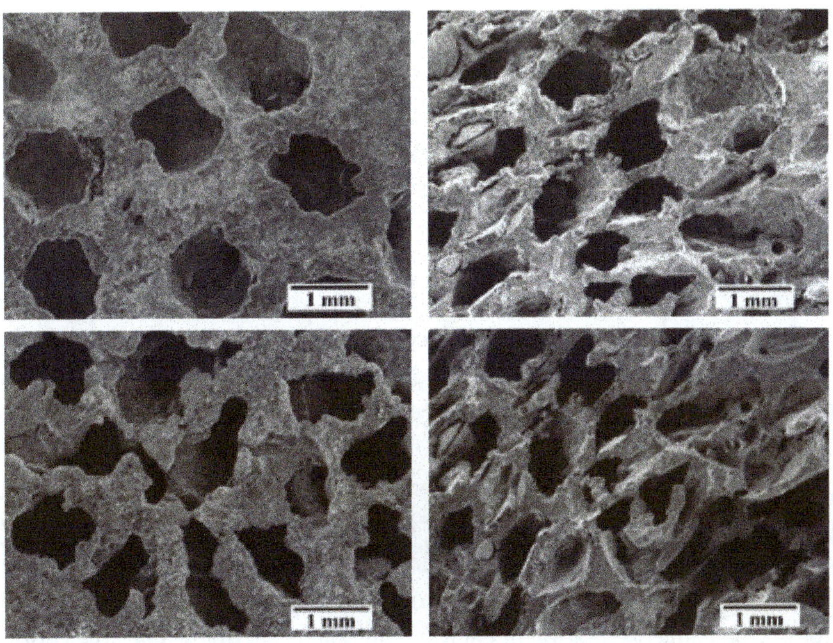

Electrical and Thermal Conductivity Measurement

The thermal conductivity of metallic foams is less than their dense form due to the pores in their structure (Dyga and Płaczek, 2015). Foam structures and pore morphology are property determining factors. Open-cell foams are ideal for use as heat exchanging materials due to their corrosion resistance combined with a large surface area and a cell wall material with high thermal conductivity properties. Metallic foams are used in heat exchangers, resistance heaters, thermal energy stores, volumetric solar receivers, and cooling machines due to their high specific surface. Mostly used closed-cell foams have a much lower thermal conductivity than open-cell foams, which makes them good for thermal shielding (Liu and Chen, 2014; Kuruneru et al., 2020; Iasiello et al., 2021). As a result of this, electrochemical activity and catalysis etc. properties of foam are better than dense form (Qin et al., 2016). For example, it is common to use lead foams in battery electrodes and nickel foams used as electrodes in electrochemical reactors. In addition, metallic foams such as Fe22Cr/Ni20Cr are used as electrode support for the fuel cell (Liu and Chen, 2014; Banhart, 2001). Methods such as low frequency eddy current measuring (Babcsán et al., 2003) and four-point technique (Goodall et al., 2006; Dharmasena and Wadley, 2002) are used for electrical conductivity measurements. Many properties of metallic foams can be measured with the eddy current method. The principles of the method mentioned in the 1.1 section. Thermal conductivity measurements can be made using general conduction tests such as the laser impulse method with diffusivity measurement (Dyga and Płaczek, 2015). This method can also be used to measure local densities and other pore parameters in the sample. Due to the low thermal resistance of metallic foam materials, a small cross-section is require. However, finite cell size affects a minimum cross-section. The contact between the the foam and the leads should be good. Senior (2017) carried out measurements by this method using samples with a cross section of 20x20 mm, following a modified standard procedure in this study on the thermal conductivity of aluminum foams.

Acoustic Measurement

Although polymer foams are generally preferred for sound insulation applications, metallic foams create great potential for these applications. In addition to sound insulation, metallic foams can insulate thermal and impact. For this reason, they are used in automobiles or machines. The unique structure of metal foams allows them to attenuate acoustic energy and sound waves by transforming them into heat energy. For this reason, open-cell metallic foams can act as acoustic absorbers. For sound absorption, cellular metals absorb sound waves once they enter the pore cavities (Sulong et al, 2014). Samples with closed pore structure show higher isolation than interconnected pores. Since the sound absorption property of metal foams can provide fast and precise measurement, it is generally measured in an impedance tube (Banhart, 2001). However, only the vertical incidence of the sound wave on a specimen can be characterized. Therefore, more tests are required for a newly developed part. In the measurement with impedance tube, the sample is prepared as disc-shaped. The tube consists of a sample near one end with changeable distance to the rear wall and a sound emitter and detector on its opposite side (Banhart, 2001). In tests on the sample of interconnected porous foam, the absorption curves are shifted to lower frequencies when the gap is expanded. However, dense specimens or porous specimens with closed pores have the same absorption because sound cannot penetrate the specimen and absorption occurs at the surface. Especially for small tube diameters, it is necessary to measure on more than one specimen. It is useful to make two measurements for one possible orientation of each side for each specimen. The absorption of closed pore foams is clearly different for the two orientations as it occurs at the surface, whereas the absorption of open pore foams spreads to the volume, so it remains unchanged when the specimen is turned (Xu et al., 2017). Lu et al., (2009) investigated the effect of different processing methods on sound absorption properties of open and closed metal cell foams. Closed-cell foams were reported poor sound absorbers due to preventing air particles to penetrate inside the material.

Vibrational Analysis

In porous structures, during the propagation the sound is attenuated due to vibration and friction losses as gas flows between cells, and thus repeated reflections within the porous structure are required for full absorption to be possible, resulting in long paths. Metal foams are used due to their sound absorption ability for the minimization of noise (Amjad, 2001). Vibration properties of materials used in automotive applications are very important. The automobile industry is move towards using more aluminum to reduce weight. However, this becomes critical because aluminum has a lower vibration damping behavior compared to cast iron and steel. However, in the tests studied collaboration with the "Karmann GmbH" and Dresden Technical University on the damping properties of aluminum sheet, steel sheet and aluminum foam sandwich (AFS), AFS demonstrated significantly better damping behavior compared to other sheets (Claar et. al.,2000).

The loss factor of metal foam and Young's modulus can be determined by performing vibrational analysis. In short, the sample (long bar of rectangular/round cross section or a thin quadratic plate) is forced to vibration. Once fixed, vibrations can be investigated by excitation (longitudinal, transverse or torsional), excited with a magnetic field or using an electromagnetic shaker without applying compression. However, applying vibration analysis to determine these properties is not entirely correct. Because foams are not homogeneous and Young modulus depends on the mass distribution (Banhart, 2001; Fenjan et al., 2019).

Biological Compatibility Tests

Metallic foams have attracted attention because many metals can be produced in porous form and the pore structure increases osseointegration in bio implant applications. Besides the porosity, the porous forms are more compatible with the mechanical properties of the bone than of the dense forms of the metals. The pore morphology and the mechanical and the corrosion properties of the metallic foam should be known for a suitable design (Wen et al., 2001; Matassi et al., 2013). Determination of pore properties is explained in Section 1.1 and measurement of mechanical properties of porous samples is explained in Section 1.2. Corrosion tests of metallic foam to be used in biological applications are different from normal corrosion tests. Corrosion behaviors are observed in-vitro and in-vivo environment (Lewis, 2013). In-vitro tests are carried out in a solution similar to body fluid prepared in a laboratory environment by examining the corrosion behavior for different times. In-vitro immersion tests are performed in different solutions such as Hank's solution (Franciska et al., 2017) and saline solution (Oriňáková et al., 2013). In addition to corrosion tests, in-vitro studies are performed in tests such as tissue/cell growth (Xue et al., 2007; Cheung et al., 2007; Ryan et al., 2008), cytotoxicity (Glorius et al., 2011) and antibacterial properties (Tong et al., 2020). In vivo tests are carried out with animal experiments (Li et al., 2007; Wegener et al., 2020) and clinical studies (Levine et al., 2006; Matsushita et al., 2017).

X-Ray Diffraction Applications on Them

After Wilhelm Conrad Röntgen discovered X-rays in 1895, there were important advances in the field of crystallography (Saravanan et al., 2011). Discovery of X-ray diffraction in crystals by Max von Laue in 1912; In 1915 W.H. Bragg and W. L. Bragg's X-ray diffraction analysis of single-crystal studies; Analysis of Polycrystalline materials proposed by Debay and P. Scherrer in 1916 and the explanation of a new chemical analysis method by Hull through X-ray diffraction (Hull, 1919; Jenkin, 2001; Borisov et al., 2012) are some of these important studies. X-ray diffraction is a highly useful method used to gain atomic scale knowledge from crystalline and amorphous materials in the field of material characterization (Suryanarayana et al., 1998). First of all, presume an isolated atom in order to illustrate the principle of this methodology. X-ray scattering is known as the process of absorption and re-emission of electromagnetic radiation (X-ray) by the atom. In the case of elastic scattering, the atom absorbs the X-ray photon, releasing another photon, with the same energy as the incident radiation (Suryanarayana et al., 1998). The constructive interference between monochromatic X-rays and crystalline samples is the basis of X-ray diffraction. When conditions obey Bragg's Law, the interaction of the incident rays with the sample generates constructive interference and as a result a diffraction pattern. In a crystalline sample, Bragg's law depends on the wavelength of electromagnetic radiation, the angle of diffraction, and the lattice spacing. The conversion of the diffraction peaks to d-spacings helps the mineral to be defined since each mineral has a specific set of d-spacings. This is usually done by comparing d-spacings to recommended samples (Dutrow et al., 2020).

The X-ray diffraction technique, which has become increasingly important in recent years, is among the techniques that are widely used in qualitative and quantitative analyzes, and in the illumination of the properties and structures of polymers.The technique provides a huge amount of details from a small amount of sample, and does not cause any damage on the sample. The X-ray diffraction technique can also be modified for quantitative purposes, in addition to being used as a qualitative analysis tool, since there is proportionality between the intensity of the diffraction peaks of a given substance in a mixture and the fraction of the material (Ryland, 1958).

In addition to ensuring deep knowledge about the crystallographic structures, chemical compositions, and physical properties of materials, XRD may also be used to classify powdered metals and mineral compounds, quantify crystal phases, quantify desired orientation, and calculate stress in metals and ceramics. (Di Gianfrancesco, 2017; Sima et al., 2016).

Since the structures of metals and alloys are interrelated with their properties, their structures must be known to improve the overall properties of materials. X-ray techniques, which include a much clearer, simpler, and more objective testing tool, have also made a major contribution to metallurgy study. If the crystal structures of metals and alloys had not been discovered by XRD, the knowledge about the shapes, defects, and properties of metals and alloys would not be available. XRD is a non-destructive tool for examining a wide variety of materials such as crystals, metals, stones, polymers, thin film coatings, and ceramics (Saravanan et al., 2011).

The following are studies of X-ray diffraction applications in this area:

In a study performed by Huang and his team in 2017, The XRD analysis of the chemical and crystalloid structure of Myristyl alcohol (MA) / metal foam composites was performed and verified that MA was physically absorbed, without chemical interaction into metal foam composites (Huang et al., 2017).

A research carried out by Michailidis et al. in 2013, the possibility of producing an intermetallic aluminum-nickel coating on Ni foam, that did not impact the original structural properties of the foam, was explored and several slurries were tested for this process. By X-ray diffraction analysis (XRD), the phases formed by the aluminized slurry phase were determined (Michailidis et al., 2013).

In another study conducted by Banciu et al. in 2020, it was reported that the graphene grown by chemical vapor deposition on metal foam substrates leads to obtaining good graphene structures suitable for different applications. In the same study graphenes grown on metal foams were examined by XRD (Banciu et al., 2020).

In the 2019 study performed by Sutygina, open-cell aluminum foams with a total porosity of ~90% were produced with a sponge copying technique. The effects of atmospheric and temperature conditions on the cell structure, porosity of the phase composition, thermal conductivity and compressive strength of these foams have been studied in this study. It has also been reported that the thermal processing of aluminum foams at temperatures up to 800°C in Ar atmosphere results in aluminum foams with lower strut porosity, lower aluminum oxide content, higher thermal conductivity, and higher compressive strength compared to foams treated thermally in the air. The lower amount of aluminum oxide in foams after thermal treatment explains these results. By performing XRD phase analysis, it was observed that aluminum oxides (α-, γ-Al_2O_3) were formed after thermal processing of Al foams (Sutygina et al., 2019).

In a 2011 study, copper reinforced aluminum foam was developed and the produced foams were characterized. The densities of open cell foams produced using mud mixtures were determined and XRD analysis was conducted; it was observed that CuO and Cu_2O structures are formed as a result of the sintering of copper-reinforced aluminum foams under normal atmospheric conditions, and these copper oxide structures are reduced by sintering in the argon atmosphere (Zaman,2011)

By vapor aluminizing, three separate Ni-Al foams were synthesized. To determine the oxidation resistance and compression properties of foams, oxidation tests and compression tests were applied below 1100 ° C. To analyze the phase composition and microstructure of Ni-Al foams before and after oxidation, an X-ray diffraction (XRD) test was conducted. The findings demonstrate that the vapor aluminization process does not affect the macro- and microstructure of the foam, and as the aluminization period increases, the phase composition changes from β / γ, phase to single β phase (Zeng et al., 2017).

In the analysis carried out by Rosip et al., in 2013, 316L stainless steel was used by the slurry process to produce metal foams. They combined the raw materials and then immersed the polyurethane foams for 24 hours in the dried mixture. The obtained samples were sintered at 1300°C in the furnace. To characterize these samples, various tests, including x-ray diffraction (XRD), were carried out, and the XRD results revealed that after sintering, the samples were oxidized (Rosip et al., 2013).

Cavuslu et al. in 2019, used the plasma electrolytic oxidation (PEO) treatment to improve the mechanical properties of the open-cell aluminum foams. The microstructure and composition of samples were revealed by scanning electron microscope (SEM) and X-ray diffraction (XRD) respectively. It was reported that the coating thickness of the struts was increased with the increasing treatment time. (Cavuslu et al., 2019).

FTIR and Raman Spectroscopic Applications on Them

In the late 1880s, when Albert A. Michelson invented the interferometer known as the Michelson interferometer, on which he and Morley conducted the well-known experiment to calculate the speed of light, the science of Fourier transform interferometry was started. Michelson was unable to take advantage of the Fourier transform spectroscopy (FTS) field because computers and electronics enstruments were not adequately usable at the time.

Practical FTS eventually began to appear in the late 1940s. Interferometers were initially used to measure light from celestial bodies, and the first Fourier transform spectrum was developed by scientists in 1949. (Jaggi et al., 2006). In the infrared part of the spectrum, molecular absorption of electromagnetic radiation accommodates the changes between the vibrational or vibrational-rotational energy levels of the ground electronic energy state of the molecule (Ismail et al., 1997). In this technique, the sample is irradiated with infrared radiation (IR) from an infrared source (containing all IR frequencies) and usually the intensity of transmitted radiation is monitored. The radiation would be absorbed at some resonant frequencies that are characteristic of the sample molecules, resulting in a sequence of peaks in the spectrum that can then be used to classify the sample. Such variations in vibrational motion contribute to bands in the vibration spectrum; their frequency and amplitude define each spectral band. (Gomez et al., 2003; . Rodriguez, 2000; Sacksteder et al., 2001; Stuart 2004; Jaggi et al., 2006). In order to classify and compare sample compositions and to evaluate the concentration of different components of a mixture, Fourier Transform Infrared (FTIR) spectroscopy can be used quantitatively. The advantage of this technique is the use of very small amounts of substance (e.g. 1 mg) and the analysis of all substances, whether crystalline or amorphous, organic or inorganic. FTIR is extremely valuable for the study of a wide variety of materials, such as determining the composition of materials such as fired clay, bone and tooth enamel, wood ash, fibers and dyes, and resins (Margaris, 2014).

The process of inelastic scattering was theoretically predicted by Brillouin (1922) and Smekal (1923), and was experimentally observed in a molecular system in 1928 by C.V.Raman and K.S.Krishnan (Jayasooriya &Jenkins, 2002). This phenomenon has since been referred to as the Raman scattering. The incident photons may be absorbed, scattered, and not interfere with the molecules of the substance. If the photon's energy correlates to the energy difference between the molecule's ground state and the excited state, the photon may interact with the molecule, can be absorbed and the molecule can be promoted to the excited state of higher energy (Smith and Dent 2019). In absorption spectroscopy, this transition is evaluated by the detection of the loss of radiation energy. However, if the energy of the photon does not match the difference between the two energy levels of the molecule, there will be no absorption, but the photon can interact with the molecule and scatter from it. In this case, the scattered photon carries information about the molecule (Smith and Dent 2019). By accumulating them at an angle to the incident light beam, the scattered photons may be detected. (Smith and Dent, 2019). Raman spectroscopy investigates the inelastic scattering of light by molecules. The technique provides information about the vibrational energy levels of the molecular systems of the environment, as a result, it offers qualitative and quantitative analysis in various media. The areas of application of Raman spectroscopy are the field of forensic science (explosives, narcotics, inorganic fillers for analysis), in the field of biology (biochemical knowledge on toxic substances, drug reactions, death or post-disease cells), as a diagnostic tool for the analysis of different biomedical materials, and it has in different fields of application, from superconductors to archaeological materials. In addition, the characterization of drug formulations, the description of the kinetic processes in drug delivery, in pharmaceutical and biopharmaceuticals, and in nanotechnology (due to the research and characterization of nanomaterials such as nanosensors, nanotubes and nanowires) are it's different application areas (Das et al., 2011).

Below are the researches of FT-IR and Raman spectroscopic applications in the subject area:

The microstructure, crystal structure and phase analysis of the composite foam of aluminium alloy (LM30), prepared by reinforcing the foam with 5 wt% of SiC particles, through melt-stir processing route, were performed by FE-SEM, EDX, XRD and FTIR analyses (Rajack et al.2018). In this study, the FTIR spectrum the foam was given and OH stretching, XH stretching, C=O stretching, -C=C- stretching, =CH bending vibration modes were determined at 3420cm^{-1}, 2910 cm^{-1}, 1720 cm^{-1}, 1605 cm^{-1} and 1100 cm^{-1}, respectively (Rajak et al., 2018).

In another study conducted by Zhang, in 2018. Microalgae were inserted into the microbial fuel cell as a cathode, for the Cd(II) extraction using nickel foam/graphene as electrodes. Graphene on nickel foam was compared with graphene oxide using FTIR spectroscopy. Since the nickel foam graphene retained the oxygen-containing functional group on the graphene oxide, the intensity of this band was significantly weakened and this was interpreted as partially decreasing the amount of graphene on the nickel foam (Zhang et al., 2018).

The synthesis of nanostructured $Ni(OH)_2$/ reduced graphene oxide (rGO) on Ni foam has been investigated using a simple and cost-effective chemical deposition method in a study conducted by Lokhande in 2019. Moreover, $Ni(OH)_2$/rGO composite's FTIR spectrum has been reported, and its vibrational bands and wavenumbers have also been determined (Lokhande et al., 2019).

Graphene was synthesized using nickel and copper foams as substrates in the chemical vapor deposition (CVD) phase with methane in a study performed by Banciu et al in 2020. The graphene grew on metal foams was analyzed using Raman spectroscopy. From the Raman intensity ratio, It was concluded that the number of graphene layers increased as time grows. The loss of the D band revealed low defect content in the case of graphene structures synthesized on Ni foam and no defects were observed in the graphene layer. (Banciu et al., 2020).

Graphene was grown on nickel foam using the chemical vapor deposition process in the study conducted by Lee et al in 2017. Raman spectroscopy technique was used to characterize the morphology and crystallization of graphene films. This research has shown that multilayer graphene grows on nickel foam (Lee et al., 2017).

In another research, Kaur et al grown carbon nanotubes on copper and nickel foams by chemical vapor deposition technique, and it was seen that the Raman spectra of all graphitic materials have G band (1580 cm^{-1} band, due to first-order Raman scattering), D band (1350 cm^{-1} band, due to one phonon second-order Raman scattering), and 2D or G' band (band at 2700 cm^{-1} due to two phonon second-order Raman scattering process) (Kaur et al., 2020).

In Mukanova's research, chemical vapor deposition technology was used and multilayer graphene (GF) thin-film synthesis was performed on Ni foam in 2018. The Raman spectroscopic investigation confirmed that the samples were GF of several layers of high consistency and purity (Mukanova et al., 2018). The MnO_2-3B graphene foam composites were developed in the 2018 by Kasap, and the Raman spectroscopy technique was also used for structural and morphological analysis. The findings showed that MnO_2 nanoparticles were effectively integrated with the foamy three-dimensional graphene structures (Kasap, 2018).

Raman spectroscopy was also used to characterize the structure of metal oxides on the surface of Field's metal foams in the 2019 study (Allioux et al., 2019).

In a study carried out by Peng et al, a series of CuO-CeO_2 / Ni metal foam composite catalysts with hierarchical structures were prepared using copper-cerium hydrosols and macroporous nickel foams. Raman spectroscopic analysis indicated that with the addition of CuO to the prepared catalysts, the band intensity observed at 465 cm^{-1} decreased significantly (Peng et al., 2014).

CONCLUSION

The characterization of metal foams from new lightweight materials is explored in this chapter. The main conclusions of this chapter can be summarized as follows:

- Metal foams are a class of materials known for their low densities and excellent strength-to-weight ratio, and good physical, chemical, and mechanical properties, which make them suitable for a wide range of industrial applications. As mentioned before; in the last years, many studies in the literature were made for metal foams, but the knowledge in the field of the characterization of metal foams is still very poor. With the development and understanding of the characterization methods of metal foams, this situation will be further clarified. As it is generally discussed in this chapter; foam metals demonstrate a wide variety of specific mechanical, microstructural, and physical properties that make them different from traditional dense metals.
- The properties of metal foams can be characterized in various ways. They are characterized by the morphology of the porous cells (cell geometry, open or closed), cell topology, relative density, pore size, pore shape, pore size distribution, pore wall defects, and degree of anisotropy. The mechanical, corrosion, and biological tests, electrical and thermal conductivity measurement, acoustic and vibrational analysis are very important in characterizing the properties of metal foams. A good understanding of all these features is necessary for a successful implementation in industrial applications.
- Foam structure and pore morphology are property determining factors. The mechanical and physical behaviors of metal foams depend on their structures. The structure of foams can be characterized in terms of geometrical and material structures.
- The details of the properties should be examined in order to use metal foams in a large scale. This chapter adds depth to the subject by considering different analysis techniques such as X-ray diffraction applications, FTIR, and Raman spectroscopic techniques and applications of metal foams.

REFERENCES

Ajinola, S. A. (2016). *Design, fabrication, and characterization of porous titanium for biological applications* (Doctoral dissertation). North Carolina Agricultural and Technical State University.

Alhusseny, A. N. M., & Nasser, A. (2018). *High-porosity metal foams: potentials, applications, and formulations. In Porosity-Process, Technologies and Applications*. IntechOpen.

Allioux, F. M., Merhebi, S., Tang, J., Idrus-Saidi, S. A., Abbasi, R., Saborio, M. G., Ghasemian, M. B., Han, J., Namivandi-Zangeneh, R., O'Mullane, A. P., Koshy, P., Daiyan, R., Amal, R., Boyer, C., & Kalantar-Zadeh, K. (2019). Catalytic Metal Foam by Chemical Melting and Sintering of Liquid Metal Nanoparticles. *Advanced Functional Materials, 30*(5), 1907879. doi:10.1002/adfm.201907879

Aly, M. S. M. A. (2004). *High temperature mechanical properties of cast as well as powder metallurgical manufactured metallic foams* (Doctoral Thesis). Von der Fakultät für Georessourcen und Materialtechnik der Rheinisch- Westfälischen Technischen Hochschule Aachen.

Amjad, S. (2001). *Thermal conductivity and noise attenuation in aluminium foams* (Master's dissertation). University of Cambridge.

Andure, M. W., Jirapure, S. C., & Dhamande, L. P. (2012). Advance automobile material for light weight future–a review. In *IJCA Proceedings on International Conference on Benchmarks in Engineering Science and Technology* (pp. 15-22). Academic Press.

Ashby, M. F., Evans, A., Fleck, N. A., Gibson, L. J., Hutchinson, J. W., Wadley, H. N. G., & Delale, F. (2001). Metal foams: A design guide. *Applied Mechanics Reviews*, *54*(6), B105–B106. doi:10.1115/1.1421119

Babcsán, N., Mészáros, I., & Hegman, N. (2003). Thermal and electrical conductivity measurements on aluminum foams. *Materialwissenschaft und Werkstofftechnik*, *34*(4), 391–394. doi:10.1002/mawe.200390081

Banciu, C., Lungulescu, M., Bara, A., Sbarcea, G., Patroi, D., & Marinescu, V. (2020, February). Graphene grown by chemical vapor deposition on metal foams. In AIP Conference Proceedings (Vol. 2206, No. 1, p. 050001). AIP Publishing LLC. doi:10.1063/5.0000296

Banhart, J. (2001). Manufacture, characterisation and application of cellular metals and metal foams. *Progress in Materials Science*, *46*(6), 559–632. doi:10.1016/S0079-6425(00)00002-5

Banhart, J. (2003). Aluminum foams: On the road to real applications. *MRS Bulletin*, *28*(4), 290–295. doi:10.1557/mrs2003.83

Banhart, J. (2013). Light-metal foams—History of innovation and technological challenges. *Advanced Engineering Materials*, *15*(3), 82–111. doi:10.1002/adem.201200217

Banhart, J., Garcia-Moreno, F., Heim, K., & Seeliger, H. W. (2019). Light-weighting in transportation and defence using aluminium foam sandwich structures. In *Light Weighting for Defense, Aerospace, and Transportation* (pp. 61–72). Springer. doi:10.1007/978-981-15-1263-6_5

Bauer, B., Kralj, S., & Bušić, M. (2013). Productıon And Application of Metal Foams In Casting Technology. *Tehnicki Vjesnik/Technical Gazette*, *20*(6).

Beköz, N. (2011). *Microstructural and Mechanical Properties of Iron Based Foamed Metals Obtained by Powder Metallurgy Method* (PhD thesis). Institute of Graduate Studies in Science and Engineering, Istanbul University.

Bekoz, N., & Oktay, E. (2012). Effects of carbamide shape and content on processing and properties of steel foams. *Journal of Materials Processing Technology*, *212*(10), 2109–2116. doi:10.1016/j.jmatprotec.2012.05.015

Bekoz, N., & Oktay, E. (2013). Mechanical properties of low alloy steel foams: Dependency on porosity and pore size. *Materials Science and Engineering A*, *576*, 82–90. doi:10.1016/j.msea.2013.04.009

Bekoz, N., & Oktay, E. (2014). High temperature mechanical properties of low alloy steel foams produced by powder metallurgy. *Materials & Design*, *53*, 482–489. doi:10.1016/j.matdes.2013.07.050

Bor, Ş., Esen, Z., Kotan, G., & Tarhan Bor, E. (2007). Toz metalurjisi yöntemiyle köpüksü saf titanyum ve Ti6AI4V alaşım üretimi ve karakterizasyonu. *TÜBİTAK*, 104-121.

Borisov, S. V., & Podberezskaya, N. V. (2012). X-ray diffraction analysis: A brief history and achievements of the first century. *Journal of Structural Chemistry*, *53*(1), 1–3. doi:10.1134/S0022476612070013

Borovinšek, M., Taherishargh, M., Vesenjak, M., Ren, Z., & Fiedler, T. (2016). Geometrical characterization of perlite-metal syntactic foam. *Materials Characterization*, *119*, 209–215. doi:10.1016/j.matchar.2016.07.024

Cavuslu, F., Korkmaz, K., & Usta, M. (2019). Mechanical properties of aluminum foams ceramic coated through plasma electrolytic oxidation. *Kovove Mater.*, *57*(02), 105–111. doi:10.4149/km_2019_1_105

Cheung, S., Gauthier, M., Lefebvre, L. P., Dunbar, M., & Filiaggi, M. (2007). Fibroblastic interactions with high-porosity Ti-6Al-4V metal foam. *Journal of Biomedical Materials Research Part B: Applied Biomaterials*, *82*(2), 440–449. doi:10.1002/jbm.b.30749 PMID:17245747

Claar, T. D., Yu, C. J., Hall, I., Banhart, J., Baumeister, J., & Seeliger, W. (2000). Ultra-lightweight aluminum foam materials for automotive applications. *SAE Transactions*, 98–106. doi:10.4271/2000-01-0335

Das, R. S., & Agrawal, Y. K. (2011). Raman spectroscopy: Recent advancements, techniques and applications. *Vibrational Spectroscopy*, *57*(2), 163–176. doi:10.1016/j.vibspec.2011.08.003

Degischer, H.-P., & Kriszt, B. (2010). *Handbook of cellular metals: Production, processing, applications*. Wiley-InterScience.

Dharmasena, K. P., & Wadley, H. N. G. (2002). Electrical conductivity of open-cell metal foams. *Journal of Materials Research*, *17*(3), 625–631. doi:10.1557/JMR.2002.0089

Di Gianfrancesco, A. (2017). Technologies for chemical analyses, microstructural and inspection investigations. In *Materials for ultra-supercritical and advanced ultra-supercritical power plants* (pp. 197–245). Woodhead Publishing. doi:10.1016/B978-0-08-100552-1.00008-7

Duarte, I., & Banhart, J. (2000). A study of aluminium foam formation—Kinetics and microstructure. *Acta Materialia*, *48*(9), 2349–2362. doi:10.1016/S1359-6454(00)00020-3

Dukhan, N. (Ed.). (2013). Metal foams: fundamentals and applications. DEStech Publications, Inc.

DutrowB. L. (2020). https://serc.carleton.edu/research_education/geochemsheets/techniques/XRD.html

Dyga, R., & Płaczek, M. (2015). Heat transfer through metal foam–fluid system. *Experimental Thermal and Fluid Science*, *65*, 1–12. doi:10.1016/j.expthermflusci.2015.02.021

Fenjan, R. M., Ahmed, R. A., Alasadi, A. A., & Faleh, N. M. (2019). Nonlocal strain gradient thermal vibration analysis of double-coupled metal foam plate system with uniform and non-uniform porosities. *Coupled Syst Mech*, *8*(3), 247–257.

Florek, R., Simancík, F., Nosko, M., & Harnuskova, J. (2010). Compression test evaluation method for aluminium foam parts of different alloys and densities. *Powder Metallurgy Progress*, *10*(4), 207–212.

Franciska, P. L., Erryani, A., Annur, D., & Kartika, I. (2017, May). Corrosion Behavior of Magnesium Based Foam Structure in Hank's Solution. In *IOP Conf. Ser. Mater. Sci. Eng.* (Vol. 202, p. 012035). 10.1088/1757-899X/202/1/012035

Glorius, S., Nies, B., Farack, J., Quadbeck, P., Hauser, R., Standke, G., Rößler, S., Scharnweber, D., & Stephani, G. (2011). Metal Foam–Bone cement Composites: Mechanical and biological properties and perspectives for bone implant design. *Advanced Engineering Materials*, *13*(11), 1019–1023. doi:10.1002/adem.201100026

Gomez, M. M., Perez, M. B., Gil, F. M., Díez, A. D., Rodríguez, J. M., Rodrıguez, P. G., ... Torres, A. R. (2003). Identification of species of Brucella using Fourier transform infrared spectroscopy. *Journal of Microbiological Methods*, *55*(1), 121–131. doi:10.1016/S0167-7012(03)00120-9 PMID:14500003

Goodall, R., Weber, L., & Mortensen, A. (2006). The electrical conductivity of microcellular metals. *Journal of Applied Physics*, 100.

Haag, M., Wanner, A., Clemens, H., Zhang, P., Kraft, O., & Arzt, E. (2003). Creep of aluminum-based closed-cell foams. *Metallurgical and Materials Transactions. A, Physical Metallurgy and Materials Science*, *34*(12), 2809–2817. doi:10.100711661-003-0182-1

Hakamada, M., Yamada, Y., Kuromura, T., Asao, Y., Chen, Y., Kusuda, H., & Mabuchi, M. (2007). Porous Metals Produced by Spacer Method as Ecomaterials. *Advanced Materials Research*, *15*, 416–421.

Harders, H., Hupfer, K., & Rösler, J. (2005). Influence of cell wall shape and density on the mechanical behaviour of 2D foam structures. *Acta Materialia*, *53*(5), 1335–1345. doi:10.1016/j.actamat.2004.11.025

Huang, X., Lin, Y., Alva, G., & Fang, G. (2017). Thermal properties and thermal conductivity enhancement of composite phase change materials using myristyl alcohol/metal foam for solar thermal storage. *Solar Energy Materials and Solar Cells*, *170*, 68–76. doi:10.1016/j.solmat.2017.05.059

Hull, A. W. (1919). A new method of chemical analysis. *Journal of the American Chemical Society*, *41*(8), 1168–1175. doi:10.1021/ja02229a003

Hüsing, N., & Schubert, U. (1998). Aerogels—airy materials: Chemistry, structure, and properties. *Angewandte Chemie International Edition*, *37*(1-2), 22–45. doi:10.1002/(SICI)1521-3773(19980202)37:1/2<22::AID-ANIE22>3.0.CO;2-I PMID:29710971

Iasiello, M., Bianco, N., Chiu, W. K., & Naso, V. (2021). The effects of variable porosity and cell size on the thermal performance of functionally-graded foams. *International Journal of Thermal Sciences*, *160*, 106696. doi:10.1016/j.ijthermalsci.2020.106696

Ismail, A. A., van de Voort, F. R., & Sedman, J. (1997). Fourier transform infrared spectroscopy: principles and applications. In *Techniques and instrumentation in analytical chemistry* (Vol. 18, pp. 93–139). Elsevier.

Jaggi, N., & Vij, D. R. (2006). Fourier transform infrared spectroscopy. In *Handbook of Applied Solid State Spectroscopy* (pp. 411–450). Springer. doi:10.1007/0-387-37590-2_9

Jain, H., Mondal, D. P., Gupta, G., Kumar, R., & Singh, S. (2020). Synthesis and characterization of 316L stainless steel foam made through two different removal process of space holder method. *Manufacturing Letters*, *26*, 33–36. doi:10.1016/j.mfglet.2020.09.005

Jayasooriya, U. A., & Jenkins, R. D. (2002). Introduction to Raman Spectroscopy. In D. L. Andrews & A. A. Demidov (Eds.), *An Introduction to Laser Spectroscopy*. Springer. doi:10.1007/978-1-4615-0727-7_3

Jenkin, J. (2001). A unique partnership: William and Lawrence Bragg and the 1915 Nobel Prize in physics. *Minerva*, *39*(4), 373–392. doi:10.1023/A:1012783802528 PMID:18335624

Kasap, S. (2018). Mangan dioksit (MnO2) katkılı üç boyutlu köpüksü yapıda grafen yapılarının üretilmesi ve karakterizasyonu. *Marmara Fen Bilimleri Dergisi*, *30*(4), 422–428. doi:10.7240/marufbd.418201

Kaur, G., Pulagara, N. V., Kumar, R., & Lahiri, I. (2020). Metal foam-carbon nanotube-reduced graphene oxide hierarchical structures for efficient field emission. *Diamond and Related Materials*, *106*, 107847. doi:10.1016/j.diamond.2020.107847

Korner, C., & Singer, R. F. (2000). Processing of metal foams—Challenges and opportunities. In Microstructural Investigation and Analysis. Wiley-VCH Verlag GmbH.

Kuruneru, S. T. W., Vafai, K., Sauret, E., & Gu, Y. (2020). Application of porous metal foam heat exchangers and the implications of particulate fouling for energy-intensive industries. *Chemical Engineering Science*, *228*, 115968. doi:10.1016/j.ces.2020.115968

Lee, Y. H., Li, S. M., Tseng, C. J., Su, C. Y., Lin, S. C., & Jhuang, J. W. (2017). Graphene as corrosion protection for metal foam flow distributor in proton exchange membrane fuel cells. *International Journal of Hydrogen Energy*, *42*(34), 22201–22207. doi:10.1016/j.ijhydene.2017.03.233

Levine, B. R., Sporer, S., Poggie, R. A., Della Valle, C. J., & Jacobs, J. J. (2006). Experimental and clinical performance of porous tantalum in orthopedic surgery. *Biomaterials*, *27*(27), 4671–4681. doi:10.1016/j.biomaterials.2006.04.041 PMID:16737737

Lewis, G. (2013). Properties of open-cell porous metals and alloys for orthopaedic applications. *Journal of Materials Science. Materials in Medicine*, *24*(10), 2293–2325. doi:10.100710856-013-4998-y PMID:23851927

Li, J. P., Habibovic, P., van den Doel, M., Wilson, C. E., de Wijn, J. R., van Blitterswijk, C. A., & de Groot, K. (2007). Bone ingrowth in porous titanium implants produced by 3D fiber deposition. *Biomaterials*, *28*(18), 2810–2820. doi:10.1016/j.biomaterials.2007.02.020 PMID:17367852

Liu, P., & Chen, G. F. (2014). *Porous materials: Processing and applications*. Elsevier.

Lokhande, P. E., & Chavan, U. S. (2019). Nanostructured Ni (OH) 2/rGO composite chemically deposited on Ni foam for high performance of supercapacitor applications. *Materials Science for Energy Technologies*, *2*(1), 52–56. doi:10.1016/j.mset.2018.10.003

Lu, M., Hopkins, C., Zhao, Y., & Seiffert, G. (2009). Sound absorption characteristics of porous steel manufactured by lost carbonate sintering. *MRS Online Proceedings Library*, *1188*(1), 184–189. doi:10.1557/PROC-1188-LL07-04

Ma, X., & Peyton, A. J. (2006). Eddy current measurement of the electrical conductivity and porosity of metal foams. *IEEE Transactions on Instrumentation and Measurement*, *55*(2), 570–576. doi:10.1109/TIM.2006.873549

Mad Rosip, N. I., Ahmad, S., Jamaludin, K. R., & Mat Noor, F. (2013). Producing of 316L stainless steel (SS316L) foam via slurry method. *Journal of Mechanical Engineering Science*, *5*, 707–712. doi:10.15282/jmes.5.2013.17.0068

Margaris, A. V. (2014). Fourier Transform Infrared Spectroscopy (FTIR): Applications in Archaeology. Encyclopedia of Global Archaeology, 2890-2893.

Marx, J., & Rabiei, A. (2020). Study on the Microstructure and Compression of Composite Metal Foam Core Sandwich Panels. *Metallurgical and Materials Transactions. A, Physical Metallurgy and Materials Science, 51*(10), 5187–5197. doi:10.100711661-020-05964-1

Matassi, F., Botti, A., Sirleo, L., Carulli, C., & Innocenti, M. (2013). Porous metal for orthopedics implants. *Clinical Cases in Mineral and Bone Metabolism, 10*(2), 111. PMID:24133527

Matsushita, T., Fujibayashi, S., & Kokubo, T. (2017). Titanium foam for bone tissue engineering. In *Metallic Foam Bone* (pp. 111–130). Woodhead Publishing. doi:10.1016/B978-0-08-101289-5.00004-4

Michailidis, N., Stergioudi, F., Omar, H., Missirlis, D., Vlahostergios, Z., Tsipas, S., Albanakis, C., & Granier, B. (2013). Flow, thermal and structural application of Ni-foam as volumetric solar receiver. *Solar Energy Materials and Solar Cells, 109*, 185–191. doi:10.1016/j.solmat.2012.10.021

Mukanova, A., Zharbossyn, A., Nurpeissova, A., Kim, S. S., Myronov, M., & Bakenov, Z. (2018). Electrochemical Study of Graphene Coated Nickel Foam as an Anode for Lithium-Ion Battery. *Eurasian Chemico-Technological Journal, 20*(2), 91–97. doi:10.18321/ectj694

Oveisi, H., & Geramipour, T. (2020). High mechanical performance alumina-reinforced aluminum nanocomposite metal foam produced by powder metallurgy: Fabrication, microstructure characterization, and mechanical properties. *Materials Research Express, 6*(12), 1250c2.

Palaniswamy, S., Arunagiri, K., & Prakash, S. (2020, December). Corrosion behaviour of closed cell magnesium foam with rare earth elements by powder metallurgy process. In. AIP Conference Proceedings: Vol. 2311. *No. 1* (p. 040010). AIP Publishing LLC.

Park, S. (2016). *Gas Diffusion Layer. Fuel Cells: Data*. Facts, and Figures.

Patel, P., Bhingole, P. P., & Makwana, D. (2018). Manufacturing, characterization and applications of lightweight metallic foams for structural applications. *Materials Today: Proceedings, 5*(9), 20391–20402. doi:10.1016/j.matpr.2018.06.414

Peng, P. Y., Jin, I., Yang, T. C. K., & Huang, C. M. (2014). Facile preparation of hierarchical CuO–CeO2/Ni metal foam composite for preferential oxidation of CO in hydrogen-rich gas. *Chemical Engineering Journal, 251*, 228–235. doi:10.1016/j.cej.2014.04.077

Qin, J., Chen, Q., Yang, C., & Huang, Y. (2016). Research process on property and application of metal porous materials. *Journal of Alloys and Compounds, 654*, 39–44. doi:10.1016/j.jallcom.2015.09.148

Raj, R. E., & Daniel, B. S. S. (2007). Aluminum melt foam processing for light-weight structures. *Materials and Manufacturing Processes, 22*(4), 525–530. doi:10.1080/10426910701236072

Rajak, D. K., & Gupta, M. (2020). Manufacturing Methods of Metal Foams. In *An Insight Into Metal Based Foams* (pp. 39–52). Springer. doi:10.1007/978-981-15-9069-6_3

Rajak, D. K., Kumaraswamidhas, L. A., & Das, S. (2014). An energy absorption behaviour of foam filled structures. *Procedia Materials Science, 5*, 164–172. doi:10.1016/j.mspro.2014.07.254

Rajak, D. K., Kumaraswamidhas, L. A., & Das, S. (2018). Experimental fabrication and compression analysis characterization of LM30 Al alloy foam with 5wt% SiCp at room temperature. *Materials Research Express*, *5*(6), 066526. doi:10.1088/2053-1591/aac9a0

Rodriguez, M. P. Q. (2000). Fourier transform infrared (FTIR) technology for the identification of organisms. *Clinical Microbiology Newsletter*, *22*(8), 57–61. doi:10.1016/S0196-4399(00)88850-9

Rossi, S., Fedel, M., Da Col, L., Deflorian, F., & Petrolli, S. (2017). Coatings to increase the corrosion behaviour of aluminium foam. *Surface Engineering*, *33*(6), 405–409. doi:10.1080/02670844.2016.1276700

Ryan, G. E., Pandit, A. S., & Apatsidis, D. P. (2008). Porous titanium scaffolds fabricated using a rapid prototyping and powder metallurgy technique. *Biomaterials*, *29*(27), 3625–3635.

Ryland, A. L. (1958). X-ray diffraction. *Journal of Chemical Education*, *35*(2), 80–83.

Sacksteder, C., & Barry, B. A. (2001). Fourier transform infrared spectroscopy: A molecular approach to an organismal question. *Journal of Phycology*, *37*(2), 197–199. doi:10.1046/j.1529-8817.2001.037002197.x

Saravanan, R., & Rani, M. P. (2011). *Metal and alloy bonding-an experimental analysis: Charge density in metals and alloys*. Springer Science & Business Media.

Senior, F. (2017). *The Development of Aluminium Foams for Enhanced Heat Transfer* (Doctoral dissertation). University of Sheffield.

Sima, F., Ristoscu, C., Duta, L., Gallet, O., Anselme, K., & Mihailescu, I. N. (2016). Laser thin films deposition and characterization for biomedical applications. In *Laser Surface Modification of Biomaterials* (pp. 77–125). Woodhead Publishing. doi:10.1016/B978-0-08-100883-6.00003-4

Smith, B. H., Szyniszewski, S., Hajjar, J. F., Schafer, B. W., & Arwade, S. R. (2012). Steel foam for structures: A review of applications, manufacturing and material properties. *Journal of Constructional Steel Research*, *71*, 1–10. doi:10.1016/j.jcsr.2011.10.028

Smith, E., & Dent, G. (2019). *Modern Raman spectroscopy: A practical approach*. Wiley. doi:10.1002/9781119440598

Smorygo, O., Mikutski, V., Marukovich, A., Ilyushchanka, A., Sadykov, V., & Smirnova, A. (2011). An inverted spherical model of an open-cell foam structure. *Acta Materialia*, *59*(7), 2669–2678. doi:10.1016/j.actamat.2011.01.005

Stuart, B. H. (2004). Infrared Spectroscopy: Fundamentals and Applications. John Wiley & Sons, Ltd. doi:10.1002/0470011149

Sulong, M. A., Vesenjak, M., Belova, I. V., Murch, G. E., & Fiedler, T. (2014). Compressive properties of Advanced Pore Morphology (APM) foam elements. *Materials Science and Engineering A*, *607*, 498–504. doi:10.1016/j.msea.2014.04.037

Suryanarayana, C., & Norton, M. G. (1998). *X-ray diffraction: A practical approach*. Springer. doi:10.1007/978-1-4899-0148-4

Sutygina, A., Betke, U., Hasemann, G., & Scheffler, M. (2020, July). Manufacturing of Open-Cell Metal Foams by the Sponge Replication Technique. *IOP Conference Series. Materials Science and Engineering, 882*(1), 012022. doi:10.1088/1757-899X/882/1/012022

Sutygina, A., Betke, U., & Scheffler, M. (2019). Open-cell aluminum foams by the sponge replication technique. *Materials (Basel), 12*(23), 3840. doi:10.3390/ma12233840 PMID:31766482

Tong, X., Shi, Z., Xu, L., Lin, J., Zhang, D., Wang, K., Li, Y., & Wen, C. (2020). Degradation behavior, cytotoxicity, hemolysis, and antibacterial properties of electro-deposited Zn–Cu metal foams as potential biodegradable bone implants. *Acta Biomaterialia, 102*, 481–492. doi:10.1016/j.actbio.2019.11.031 PMID:31740321

Vendra, L. J., Brown, J. A., & Rabiei, A. (2011). Effect of processing parameters on the microstructure and mechanical properties of Al–steel composite foam. *Journal of Materials Science, 46*(13), 4574–4581. doi:10.100710853-011-5356-4

Wegener, B., Sichler, A., Milz, S., Sprecher, C., Pieper, K., Hermanns, W., ... Wegener, V. (2020). Development of a novel biodegradable porous iron-based implant for bone replacement. *Scientific Reports, 10*(1), 1–10. doi:10.103841598-020-66289-y PMID:32499489

Wen, C. E., Mabuchi, M., Yamada, Y., Shimojima, K., Chino, Y., & Asahina, T. (2001). Processing of biocompatible porous Ti and Mg. *Scripta Materialia, 45*(10), 1147–1153. doi:10.1016/S1359-6462(01)01132-0

Xu, C., Mao, Y., & Hu, Z. (2017). Tonal and broadband noise control of an axial-flow fan with metal foams: Design and experimental validation. *Applied Acoustics, 127*, 346–353. doi:10.1016/j.apacoust.2017.06.018

Xue, W., Krishna, B. V., Bandyopadhyay, A., & Bose, S. (2007). Processing and biocompatibility evaluation of laser processed porous titanium. *Acta Biomaterialia, 3*(6), 1007–1018. doi:10.1016/j.actbio.2007.05.009 PMID:17627910

Yang, F., Ma, Y., Tao, N., & He, X. (2017). Experimental study on nonlinear vibrating of aluminum foam using electronic speckle pattern interferometry. In *Fifth International Conference on Optical and Photonics Engineering* (Vol. 10449). International Society for Optics and Photonics. 10.1117/12.2270397

Zaman, E. (2011). *Açık Hücreli Alüminyum Köpük Üretimi Ve Karakterizasyonu* (MSc Thesis). Institute of Science and Technology, İstanbul Technical University.

Zdravkov, B., Čermák, J., Šefara, M., & Janků, J. (2007). Pore classification in the characterization of porous materials: A perspective. *Open Chemistry, 5*(2), 385–395. doi:10.247811532-007-0017-9

Zeng, W. W., Hou, S. H., Ding, X. J., Duan, D. L., Li, S., & Zhang, S. H. (2017). Synthesis and Compression Property of Oxidation-Resistant Ni–Al Foams. *Acta Metallurgica Sinica. English Letters, 30*(10), 965–972. doi:10.100740195-017-0569-4

Zhang, Y., He, Q., Xia, L., Li, Y., & Song, S. (2018). Algae cathode microbial fuel cells for cadmium removal with simultaneous electricity production using nickel foam/graphene electrode. *Biochemical Engineering Journal, 138*, 179–187. doi:10.1016/j.bej.2018.07.021

Zhang, Y. M., Chu, X. M., Wang, H., He, S. Y., & He, D. P. (2009). Fabrication of Al–Mg–Re foams and their corrosion resistance properties. *Corrosion Science*, *51*(6), 1436–1440. doi:10.1016/j.corsci.2009.03.032

Chapter 16
Machining of Poly Methyl Methacrylate (PMMA) and Other Olymeric Materials: A Review

Fredrick Mwema
https://orcid.org/0000-0001-6116-5587
University of Johannesburg, South Africa

Job Maveke Wambua
Dedan Kimathi University of Technology, Kenya

ABSTRACT

Polymers have been adopted industrially in the manufacture of lenses for optical applications due to their attractive properties such as high hardness, high strength, high ductility, high fracture toughness, and also their low thermal and electrical conductivities. However, they have limited machinability and are therefore classified as hard-to-machine materials. This study conducts a critical review on the machining of various polymers and polymeric materials, with particular focus on poly (methyl methacrylate) (PMMA). From the review it was concluded that various machining parameters affect the output qualities of polymers and which include the spindle speed, the feed rate, vibrations, the depth of cut, and the machining environment. These parameters tend to affect the surface roughness, the cutting forces, delamination, cutting temperatures, tool wear, precision, vibrations, material removal rate, and the mechanical properties such as hardness, among others. A multi-objective optimization of these machining parameters is therefore required, especially in the machining of PMMA.

INTRODUCTION

Hard to machine materials have been categorized as materials exhibiting high hardness of above 45 HRC (Astakhov, 2011). These materials, also termed as difficult to machine materials are mainly super-alloys

DOI: 10.4018/978-1-7998-7864-3.ch016

and other refractory metals including titanium, tungsten, nickel, niobium, steel, rhenium, tantalum, cobalt and chromium alloys. In addition to these, structural ceramics, composites and polymers are also considered to be hard to machine materials. The categorization of these materials is based on the production of extreme tool wear, high temperatures at the cutting points and elevated cutting forces during their machining (Shokrani, Dhokia & Newman, 2012). Polymers are chemically mixed compounds with large macromolecules made up of repetitive units, and which are created through polymerization. Such polymers are poly ethylene, poly propylene, polyvinyl chloride, polyethylene terephthalate, poly imide, polystyrene, polycarbonate, polyurethane, polytetrafluoroethylene and poly (methyl methacrylate). These polymers are often characterized as elastomers and have poor conductivity of heat, resulting in high cutting zone generation of heat, and which lead to high adhesion. This adhesive property prevents the movement of tools on the workpiece surfaces, hence limiting their machinability (Shokrani et al., 2012).

These polymers discussed above have been employed in such fields as in the making of wire and cable insulation (poly ethylene), making of plastic chairs (poly propylene), manufacture of magnetic tapes (polyethylene terephthalate), making of mechanical stress buffers (poly imide), making of license plate frames (polystyrene), manufacture of eye protection items (polycarbonate), manufacture of lenses and optical media (PMMA), and manufacture of plain bearings and gears (polytetrafluoroethylene), among others (Sohail, 2012). The materials have also found applications in the automotive, military, and aerospace industries due to their high strength-to-weight ratio. These applications owe to the chemical and mechanical properties of these materials, which include high hardness, strength, ductility, fracture toughness and also the low thermal and electrical conductivities. These properties affect the machinability of these materials, with a very reduced useful life of the tool, high surface coarseness, low accuracy in terms of dimensions and poor generation and movement of chips, which makes them hard to machine materials (Shokrani et al., 2012).

In this chapter, the theory of machining which will encompass the general machining aspects, terminology and processes will be examined. Moreover, a critical review on published materials on PMMA, other polymers, as well as other polymeric related materials and different machining and optimization methods will be discussed. This will be concluded with a summary of the review findings. Several applications of these materials in making of lightweight components will also be evaluated and presented.

BACKGROUND: THEORY OF MACHINING

Machining involves material removal in the form of chips, and which is achieved by shear deformation on the workpiece under machining. This process is characterized by the inter-relation of five machining elements, that is, the machining tool, tool holder and guiding element, the material holder, the workpiece and the machine. The cutting tools are categorized as either single or multi-edged, and may either perform the machining process in a linear or rotary direction, and these tools are designed depending on the cutting operation intended (Nee, Dufraine, Evans & Hill, 2010). The sharp edges on the machining tools allow for the chip formation as the tool interacts with the workpiece, and acts as a major determinant of tool life, the integrity of the surface finish, the shearing force required and the temperatures generated in the process. The cutting ability of a tool is dependent on the tool geometry, which is a product of the face angle, the flank, the rake angle and the clearance angle (Nee et al., 2010).

In milling, the material removal is achieved using a rotating tool which is equipped with multiple cutting edges. In the machining of composites and polymers, peripheral milling (edge trimming), which

Figure 1. Polymer milling operations (Patel, 2008)

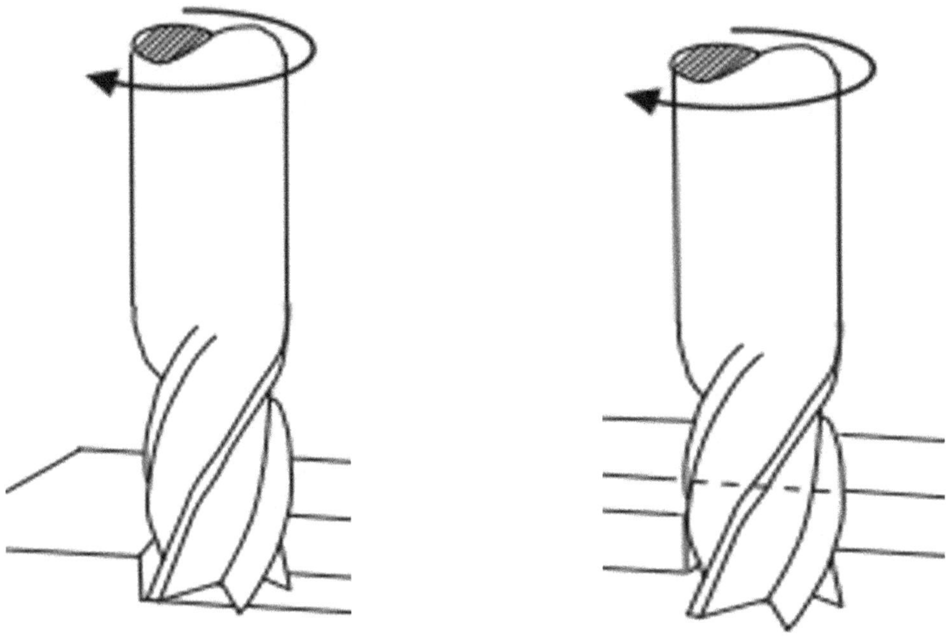

involves cutting using the edges on the tool periphery, and end milling (Patel, 2008), as shown in figures 1 (a) and (b) are majorly adopted. These milling operations are also grouped into either up milling or down milling. In up milling, the direction of rotation of the cutting edge is generally opposite to that of the workpiece and the federate. On the contrary, down milling involves a parallel motion between the direction of the edge of cut and the and the workpiece feed. The formation of chips in these practices resemble a comma. These up and down milling operations are displayed in figure 2 (a) and (b) respectfully.

The milling tools are characterized by several parameters and configurations which include the clearance angles, the pitch, the rake angles, and the helix angles, among others. The pitch of a milling tool is the distance in terms of an angle of the adjacent teeth. The face of the tooth is the front side of the tooth which normally becomes the cutting edge. The cutting edge is the tooth angle which carries out the cutting operation. The surface behind the cutting edge of every tooth is the land. Moreover, the angle formed between the tooth face and the cutter centerline is the rake angle, which determines the edge of the cutting point, and also gives way to the chips formed (DeGeare, 2015). This geometry of up and down cutting tools is shown in figure 3.

The composition of the workpiece material and its shape determines the selection of the holding device. This material also influences the method of machining and the choice of the direction of movement of the tool. The workpiece holding device also depends on the anticipated generation of forces, as it acts as the support and workpiece location. These devices may be incorporated with guidance devices such as jigs and fixtures (Nee et al., 2010). The machining elements exhibit different impacts on the machinability of various materials, as well as the effectiveness of machining processes.

The tool geometry is an important determinant of the output surface finish, stress development, tool wear, chip formation, heat generation, cutting forces and surface damage (Dogra, Sharma & Dureja,

Figure 2. Up and down milling (Patel, 2008)

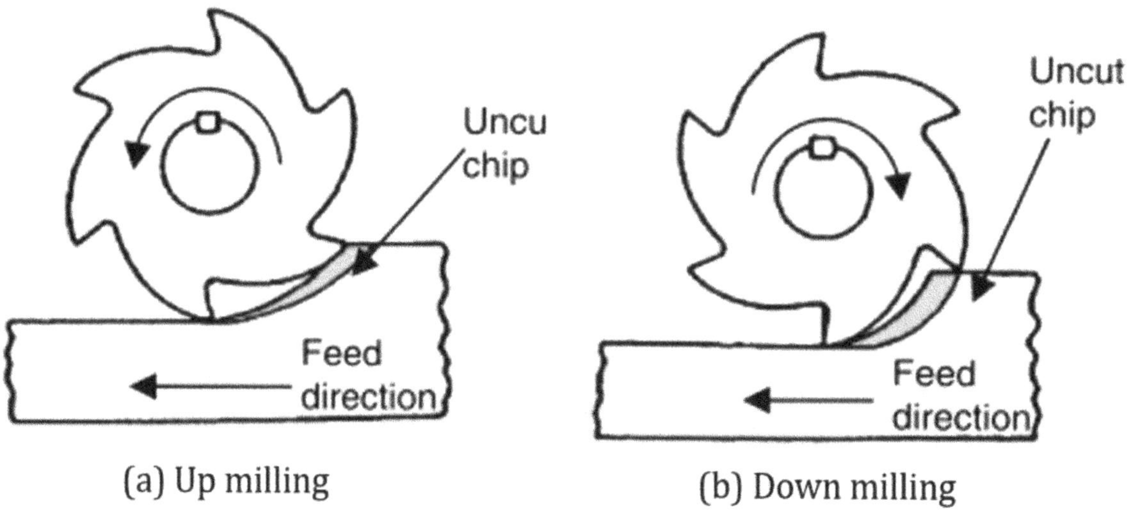

2011). The cutting forces on a tool are a key determinant of the power consumption, tool holders and fixtures and vibrations produced during machining (Sharma, Fromentin, Poulachon & Brendlen, 2014). The cutting edge also affects the formation of chips, which is closely related to the shear angle and the shear plane. Increasing the elevation angle of the shear plane for the tool reduces the shear plane which means a reduced force due to shear. This also reduces the cutting forces, the power consumed during the machining process, and temperatures generated (Conradie, Oosthuizen, Dimitrov & Saxer, 2015). The negative rake angle on the other end increases friction, and hence the generation of cutting forces

Figure 3. Milling cutter geometry (DeGeare, 2015)

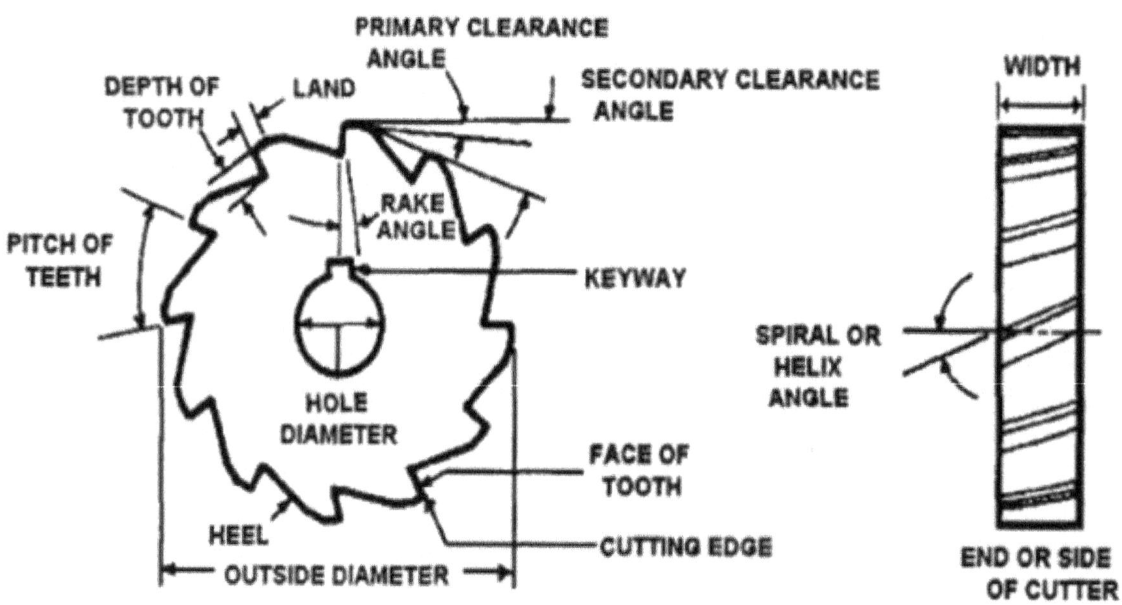

through the increase of the contact area of the tool chips. The rake face also affects the flow of chips, which determines the level of the cutting resistance, the formation of chips, their removal, the machining temperatures and also the tool life (Durkovic, Mladenovic, Tanovic & Danon, 2018).

Moreover, increasing the positive rake angle of a tool creates a sharp cutting edge which reduces the power consumed during machining, but also exposes the cutting edge to fracture. An increase of 1° on the rake angle has been observed to reduce the power consumption of the machine by 1% (Conradie et al., 2015). The impact of rake angle is however depended on the material properties. Generally, an increase in the rake angle has been identified to decrease the cutting force significantly (Durkovic et al., 2018). The lead angle influences the productivity and the integrity of the surface finish. It also affects the burring properties of the tool through the imposition of free forces. Increasing the lead angle decreases the radial force with a significant increase of the axial force. Moreover, high lead angles create room for an increased clearance and hence a reduced tool life. In addition, a small lead angle provides for more contact area as well as an increase in the tool life (Daniyan, Tlhabadira, Daramola & Mpofu, 2019).

In addition, the helix angle affects the surface integrity of the workpiece, may reduce or increase the vibrations, and may lead to either increased or reduced machining efficiency. An increase in the helix angle of a milling tool increases the axial force while decreasing the radial force and the tool vibrations. This allows for maximum chip removal. These components of force tend to balance at an angle of 45°, and at this point, the tool is exposed to breakages and fracture (Daniyan et al., 2019). Reducing the helix angle increases the tool strength but reduces the surface quality produced. Lastly, the end relief angle reduces the cutting tool-workpiece interference during machining. The smaller the angle, the thinner are the chips formed, and hence a reduced impact, and the converse applies (Daniyan et al., 2019).

A REVIEW OF MACHINING OF POLYMERIC AND RELATED MATERIALS

The low cost, low thermal conductivity, optical and biocompatibility characteristics of polymers has attracted their industrial consideration and application in making of microfluidic items, electronic chips, drug delivery elements, in energy devices, and optical applications. The materials have also been of great interest in the areas requiring light but strong materials, such as the military, automotive, and aerospace industry. Several methods have been employed in the manufacture of these items for industrial purposes, and which include rapid prototyping, plasma etching, ultrasonic welding, iron beam milling, and injection molding, among others. Various studies have been presented on the machining of these materials through different methods (Korkmaz et al., 2017).

The machining of polycarbonate for the use in optical applications was conducted by Bolat (2013), which depicted that an increase in vibrations led to a poor surface finish of the machined component. Regarding the machining parameters, the study found out that a feeding rate of 2 μm/rev and below, a cutting depth of lower than 5μm as well as a spindle speed of lower than 2000 rev per minute were best for high quality polycarbonate surfaces. Further, a study evaluating the impact of cutting factors to the integrity of the surface of PMMA in the diamond single point turning process was conducted by Jagtap & Pawade (2014). The spindle speed was identified to be the major contributing factor towards the surface quality. Castricum, Yang, Bakker & Deursen (1997) evaluated the milling process of uncontaminated selected polymers and the process of transformation of polyethylene. The high molecular weight polymers were found to undergo a total phase transformation, that is, from the orthorhombic structure to the monoclinic structure during the planetary and vibratory milling. This was attributed to shear and temperatures produced by the milling process. Singh, Ahuja & Kapoor (2017) conducted a mathematical modelling study for the material removal rate of Polycarbonate and Acrylic Glass, and the optimization of the surface roughness during ultrasonic drilling. From the study, optimum drilling of polycarbonate bullet proof glass required a 40% abrasive concentration, with 600 grit abrasive added to 1.5% of the hydrofluoric acid, while the optimum drilling of acrylic glass required an abrasive material grit of 600 and a hydrofluoric acid percentage of one. A high grit amount could probably lead to an increased material removal rate but on the other hand affect the surface quality of the machined components.

Jagtap, Pawade & Balasubramaniam (2012) investigated the impact of machining factors on the surface integrity and flatness during the turning process of Nylon and Polypropylene in CNC. The design of experiments and experimentation followed an L9 Taguchi design with the input parameters being the rate of feed, rotational speed, the insert angle of clearance and the cutting depth. It was identified that for both polymers, the rate of feeding was the major determinant of the surface quality of the workpieces. The study also revealed that an increase in the clearance angle increased the surface integrity of both polymers. The minimum depth of cut also produced a better surface finish, with Polypropylene having a better machinability than Nylon (Jagtap et al., 2012). Lower depths of cut reduce the machining forces, and hence reduce cracking and breakages of surfaces during machining. This reduces the roughness. High clearance angles also tend to reduce the surface roughness as chips are able to flow smoothly, hence reducing built up edges (BUE) on the tools.

A laser machining study conducted by Elsheikh, Deng & Showaib (2019) in the investigation of the cutting quality of PMMA through optimization displayed that increasing the cutting parameters increased the top and bottom kerf widths. The kerf taper also increased as the cutting speed increased. From the study, the minimum kerf taper was obtained from the maximum sheet thickness, while the maximum kerf taper was obtained from the cutting of the minimum sheet thickness (Elsheikh et al., 2019). Another optimization study conducted by Sanci, Halis & Kaplan (2017) on the optimization of the machining parameters during the turning process of carbon and fiber filled polytetrafluoroethylene samples. The 0.25 and 0.25 concentrations of the carbon and glass fiber filled in the PTFE singled out the rate of feed as the major contributing parameter towards the integrity of the surface, in addition to the speed of cut. The machining rate of feed and cutting speeds tend to create higher cutting forces and increase the shear factor, which in turn affects the surface integrity. The cutting depth depicted an insignificant impact on the surface integrity of the blend.

Lazarevic, Madic, Jankovic & Lazarevic (2012) conducted an investigation on the optimum machining factors for the best surface quality in the turning of Polyethylene. From the results, the researchers concluded that the rate of material feed was the highly significant determinant of the integrity of the workpiece surface, with the tool nose radius coming second and finally speed of cut. The impact of cutting depth was identified to be negligible. The optimum parameters, according to the researchers, were a cutting speed of 213.88 meter per min, a rate of feed of 0.049 mm per rev, a cutting depth of 2 mm and lastly a nose radius of 0.8 mm (Lazarević et al., 2012).

Moreover, during the machining of tough polyethylene material in the optimization of surface coarseness and machining temperature, the rate of feed was identified to be the major contributing parameter towards surface integrity, as well as the cutting speed, cutting depth and their interactions affected the cutting temperatures (Hamlaoui, Azzouz, Chaoui, Azari & Yallese, 2017). Higher cutting depths and speeds tend to increase the tool contact and therefore creating zones of high temperatures. A single response optimization study conducted by Raj, Vijayakumar, Kannan, Kumar & Ragavan (2016) on the turning parameters of Poly Tetra Fluoro Ethylene (PTFE) rods depicted that the maximum rate of material removal was achieved at 3000 rpm rotational speed, a rate of feed of 0.09 mm and a cutting depth of 0.6 mm. Further, the authors identified the cutting depth to be the dominant factor in the micro-drilling process. This could be related to the large contact area in large depth of cut, which further increases the material removal rate. Sanjeev, Kaviarasan & Venkatesan (2012) also evaluated the optimization of the surface roughness of PTFE. Using the Genetic Algorithm optimization technique, the optimum surface roughness was obtained using the speed of cut of 158 m per min, a feeding rate of 0.16 mm per rev and a cutting depth of 1.72 mm.

Chen, Pan, Lee & Li (2014) also optimized the roughness of polycarbonate substrate during a micro-milling operation, through the investigation of the impacts of the machining parameters to the surface integrity. The study found the minimum surface roughness at a rotational spindle speed of 20,000 rpm, a rate of feed of 300 mm per min and a cutting depth of 10 micrometer. The spindle speed was found out to be the largest contributor, while the cutting depth had the least contribution to the surface integrity (Chen et al., 2014). Another micro-milling optimization study was conducted by Zhao, Jia, Gao & Ding (2016) on copper clad polyimide, and three machining parameters, that is, the rotational speed, the rate of feed per tooth and the diameter of the tool were evaluated against the workpiece quality and efficiency of machining through minimization of the burr height. The outputs depicted that the optimum micro-milling parameters were 26,110 r/min for the spindle speed, 0.7 μm/z and 200 μm for the feed per tooth and the diameter of tool respectively. The burr height was significantly reduced from 100 μm to 13.9 μm. Increasing the feed per tooth above the optimum value led to an increased burr height, while increasing the spindle speed above 45,000 r/min increased the burr height as well. This can be attributed to the increase in machining forces and stresses on the machining points and hence poor quality.

Quadrini, Squeo & Tagliaferri (2007) also investigated the machining parameters of 30 wt% glass fiber reinforced polyamide through the drilling operation. The thrust was compared through drilling at different speeds, feed rates and the drill diameters. The study showed that the chip formation process affected the relationship between the machining parameters and the thrust produced. For discontinuous chips, the thrust was identified to increase with the increase in the speeds of drilling and reduce with the feed rate. The machining mode was also identified to be a significant element, as the force was identified to increase with as much as a factor of 8 as the drill moved from the center to the outer circumference (Quadrini et al., 2007). A precision study involving the turning of PA66 polyamide without and with the incorporation of glass fiber reinforcement was undertaken by Davim, Silva, Festas & Abrao (2009), and which aimed at investigating the effect of the machining parameters on the two materials using two distinct tools from different materials. It was identified that the radial force had the highest value, with the cutting and feed forces following. It was also identified that the PCD tool gave out the best surface finish and the lowest generated force, which was followed by the ISO grade K15 carbide tool, and which was uncoated.

Another experimental investigation and optimization study on the glass fiber-reinforced plastic composite material turning was conducted by Kumar, Meenu & Satsangi (2012) using the regression analysis. The study evaluated the effects of the machining factors on the integrity of the surface was evaluated using an L16 Taguchi methodology approach. It was observed that the roughness of the surface of the composite material was largely dependent on the rate of feed (63.32%) as well as the cutting environment (14.09%), that is, dry or wet (Kumar et al., 2012). This could be attributed to the generation of high tool forces and friction in high feed rates. The extent of these forces and friction is also affected by the presence or absence of a cooling medium.

Ali, Mohamed, Khan, Asfana, Lutfi & Fahmi (2013) evaluated the impacts of end milling parameters on the vibration levels during the machining of PMMA. The study used the rotational spindle speed, the rate of machine feeding and the cutting depth as the factors, with vibrations being the response. From the results, it was identified that for minimum vibrations, the optimum parameters were a cutting speed (rotational) of 2500 rev per min, a 2mm per min rate of feed and a 1.5mm cutting depth. The study also showed that the rate of feed and the cutting depths were the largest contributors towards the vibration levels. This was as result of high forces of cutting and friction as the rate of cutting increases as the cutting depth and rate of feed increase.

Korkmaz, Onler & Ozdoganler (2017) evaluated the micro-milling process of PMMA using a full factorial experimental design and a single crystal diamond cutting tool. The study involved varying spindle speeds, rates of feed and also depths of cut evaluated axially. The impacts of these parameters were evaluated on the forces in the machining process, the roughness of the surface, the formation of burr and the shape retention. From the results, it was conclusive that a single crystal diamond micro-end machining tool reduced the roughness of the outputs, the formation of burr and the amounts of machining forces, which improved the overall quality of outputs (Korkmaz et al., 2017). The surface roughness increased with the feed rate (81.2% contribution) and decreased as the spindle speed increased (17.4% contribution). The depth of cut (50.1% contribution) and the feed rate (43.4% contribution) were the major contributors towards the cutting forces, which could be related to the production of cutting forces and machining temperatures.

Consequently, Guan, Cherrill, Pai & Priest (2019) evaluated the impact of the mold roughness on injection molded PMMA products. The surface roughness of a micro-milled mold inserts of PMMA and that from an injection molding process were compared, and the wettability evaluated. Increasing the factor of roughness led to an increase in the hydrophilicity of the molds (Guan et al., 2019). Ranganathan, Rajkumar, Bensingh, Lawrence, Kader & Nayak (2020) conducted a study to investigate the difference in tribological characteristics between a laser ablated PMMA and a micro-milled PMMA, basing the study on the grooved surface roughness, the fracture toughness and fatigue endurance. The results showed that an increase in wear time led to a decreased surface roughness of both materials, with an increase in the coefficient of friction after the minimum possible roughness. High fracture toughness was obtained on the laser ablated PMMA, as well as the endurance to fatigue.

Agrawal, Patidar, Dixit & Sharma (2010) evaluated the impacts of changes in temperature to the thermo-mechanical properties of PMMA. This study displayed that an increase in temperature reduced the storage modulus of PMMA, the young modulus, the tensile strength and the toughness. The break elongation however increased with an increase in the temperature up to the glass transition temperature (Agarwal et al., 2010). A PMMA facing optimization study conducted by Dhakad, Mahajan & Mitra (2017) on the CNC lathe machine showed that the optimum values of facing the material were 274.89 meter per min for the cutting speed, 0.05 mm per rev for the rate of feed and 0.1mm for cutting depth. The study confirmed that the feed rate (74.01%) was the largest influencer factor on the quality of the surface, with the cutting depth (2.03%) having a negligible impact on the surface roughness (Dhakad et al., 2017).

Another study investigating the impacts of vibrations on the roughness of the surface of PMMA using ANN model was conducted by Mahajan, Pawade, Mandale, Chikorde, Nawale & Birari (2019). The ANN outputs of the optimum parameters in the design were paralleled with the actual outputs from the experiments and then identifying a match between the two. The model had a surface roughness error of approximately 5%, which displayed that it could be used for modelling. The study depicted that an increase in vibration increased the roughness of the surfaces. This was as a result of the increase in acceleration towards the Z-direction.

Various multi-objective optimization studies have been conducted using different machining methods as well as different optimization techniques. Zhou, Gupta & Ray, (2000) conducted a non-dominated sorting genetic algorithm of continuous casting of PMMA. The process employed two objective functions which aimed at maximizing the cross-section average value of the monomer conversion of the casting and minimizing the length of the film reactor. Yaser & Shunmugesh (2019) conducted a multi-objective optimization of a polymer, which was reinforced with glass fiber, and using the grey relational analysis coupled with the desirability function. This study found out that a lower cutting speed resulted to high surface roughness. It also depicted that the delamination factor was inversely related to the speed of cut and directly related to the rate of feed and the cutting depth.

In a multi-objective optimization study presented by Jenarthanan, Gujjalapudi & Venkatraman (2017) on the glass fiber end-milling, under a reinforcement with polymer composites, the authors employed the desirability function and the grey relational analysis (GRA) in coming up with the optimal milling factors. The machining factors, that is, the rotational spindle speed, the orientation of the fiber angle, the tool helix angle and the rate of feed were considered, and their impact on the force of machining, the surface integrity and the delamination factor was evaluated. The results depicted that fiber orientation was the largest contributor, followed by the rate of feed, the helix angle, and finally the rotational spindle speed, with 66.75%, 15.05%, 7.76% and 0.30% (Jenarthanan et al., 2017). Additionally, Azmi (2012) conducted a multi-objective study on the optimization of the end milling of glass fiber-reinforced polymer through the Taguchi array and GRA. The study identified the rate of feed to be the major contributing parameter impacting on all the responses (84.39%). This was followed by the cutting depth (13.12%), and which are all closely related to cutting forces which affect the surface integrity of components.

Izamshah, Zuraidah, Kasim, Hadzley & Amran (2015) evaluated the impact of machining parameters, that is, the spindle speed, the feed rate and the depth of cut on the roughness of the surfaces and the delamination of cellulose-based hybrid composite in a multi-objective optimization study. The study revealed rate of feed to be the major parameter impacting on the responses, and the depth of cut least affected the responses. Combining the maximum spindle speed and the least rate of feed reduced the responses. The optimum parameters for the composite machining were therefore identified to be 200 mm per min, 2 mm and 1924 rpm for the rate of feed, the cutting depth and the spindle speed respectively.

Machining of polymers for optical applications has further been conducted by various researchers. Kobayashi & Hirakawa (2006) evaluated the ultraprecision machining of PMMA to obtain the optical grade surface quality. The study showed that increasing the feed rate led to an increase in the surface roughness. It also found out that the maximum surface roughness obtained was about 5 nm, which was acceptable in terms of optical applicability. The best surface quality was obtained using a speed of cut of 730 m/min, a depth of cut of 10 μm and a feed rate of 3 μm/rev. In a single point diamond turning study presented by Khatri, Mishra & Sarepaka (2012) on the Nano-quality of PMMA, the machining parameters were identified to be key determinants of the surface quality. The tool feed was the major factor affecting the surface integrity, with the depth of cut having the minimum impact. For a good optical quality of the surface, lower depth of cut was desired. The optimal parameters for a precision machining of PMMA were identified as a spindle speed of 2000 rpm, a depth of cut of 2 μm and a feed rate of 1 μm/rev.

Grey relational analysis has been employed in conducting multi-objective studies in collaboration with the Taguchi approach in several studies. Prayogo & Lusi (2016) conducted a study on the optimization

Table 1. Summary of review of literature

STUDY (S)	AUTHOR (S) (YEAR)	FACTOR (S)	RESPONSE (S)	CONCLUSIONS
Polymeric materials				
Polycarbonate machining for optical applications	Bolat (2013)	Spindle speed, cutting depth and rate of feed	Vibrations and surface finish	-An increase in vibrations result in poor surface finish -Optimum parameters are 2 μm/rev feed rate, a cutting depth of lower than 5μm and a spindle speed of less than 2000 rpm
Mathematical modelling of material removal of Polycarbonate and Acrylic Glass and optimization of surface milling in Ultrasonic drilling.	Singh, Ahuja & Kapoor (2017)	Abrasive concentration, grit abrasive size and hydrofluoric acid.	Surface roughness and MRR	-Optimum drilling parameters are 40% abrasive concentration, 600 grit size and a hydrofluoric acid percentage of 1.5% for Polycarbonate. -Optimum drilling parameters are a grit size of 600 and a hydrofluoric acid of 1%.
Impact of machining parameters on precision turning of Nylon and Polypropylene in CNC	Jagtap, Pawade & Balasubramaniam (2012)	Spindle speed, cutting depth, rate of feed and insert clearance angle	Surface roughness and surface flatness	-The rate of feed has the major impact on the surface roughness of both polymers. -An increase in clearance angle increased the surface integrity of both polymers. -The minimum depth of cut produced the best surface finish.
Parameter optimization during turning of carbon filled and fiber filled polytetrafluoroethylene	Sanci, Halis & Kaplan (2017)	Spindle speed, cutting depth and rate of feed	Surface roughness	- The 0.25 and 0.25 concentrations of the carbon and glass fiber filled in the PTFE depicted the surface roughness to be most contributing factor towards the surface integrity followed by the speed of cut. - The cutting depth had insignificant impact on the surface integrity of the blend
Optimum machining factors for the best surface roughness in Polyethylene turning.	Madic, Jankovic & Lazarevic (2012)	Spindle speed, cutting depth, rate of feed and tool nose radius	Surface roughness	- The rate of feed was the dominant factor towards surface integrity (90%), then the nose radius and the *ss*
Machining of tough polyethylene material for the optimization of the roughness of the surfaces and the generation of heat	Hamlaoui, Azzouz, Chaoui, Azari & Yallese (2017)	Machining (cutting) speed, cutting depth and rate of feeding	Surface roughness and cutting temperature	-The feed rate was the main contributing factor towards surface roughness (92.22%). -The interactions of the factors of machining had high impact towards the generation of heat.
Single response optimization in turning of PTFE.	Raj, Vijayakumar, Kannan, Kumar & Ragavan (2016)	Rotational speed, cutting depth and rate of feeding	MRR	- The maximum material removal rate was obtained at 3000 rpm rotational speed, a rate of feed of 0.09 mm and a cutting depth of 0.6 mm
Optimization of the surface roughness of PTFE using GA technique.	Sanjeev, Kaviarasan & Venkatesan (2012)	Speed of cut, cutting depth and rate of feeding	Surface roughness	-The optimum quality of surface was obtained under a speed of cut of 158 m/min, a rate of feeding of 0.16 mm/rev and a cutting depth of 1.72 mm.
Optimization of the roughness of surface in the micro-milling of polycarbonate substrate	Chen, Pan, Lee & Li (2014)	Spindle speed, cutting depth and rate of feed.	Surface roughness	-The lowest value of the surface roughness was obtainable at a rotational speed of 20,000 rpm, a rate of feed of 300 mm per min and a cutting depth of 10 μm. -The spindle speed was found out to be the largest contributor, while the cutting depth had the least contribution to surface finish.
Optimization of workpiece quality and machining efficiency in the Copper clad polyimide micro-milling process.	Zhao, Jia, Gao & Ding (2014)	Spindle speed, rate of feed per tooth and the tool diameter.	Burr height	-The optimum micro-milling parameters were 26,110 r/min for the spindle speed, 0.7 μm/z and 200 μm for the rate of feed per tooth and the tool diameter respectively. - The burr height was significantly reduced from 100 μm to 13.9 μm.
Investigating the machining parameters of 30 wt% glass fiber reinforced polyamide through drilling.	Quadrini, Squeo & Tagliaferri (2007)	Drilling speed, feed rate, drill diameter and machining mode.	Chip formation and thrust.	-For discontinuous chips, the thrust was identified to increase with the increase in the speeds of drilling and reduce with the feed rate. - The machining force was identified to increase with as much as a factor of 8 as the drilling moved from the center to the outer circumference
Turning of PA66 Polyamide with and without glass fiber reinforcement	Davim, Silva, Festas & Abrao (2009)	Radial force, cutting force and feed force	Surface finish and generated force	- The radial force had the highest value, with the cutting and feed forces following. -The PCD tool gave out the best surface finish and the lowest generated force.
Optimization of glass fibre-reinforced plastic composite material turning process.	Kumar, Meenu & Satsangi (2012)	Speed of cut, cutting depth, feed rate and machining environment.	Surface roughness	-The speed of cut, the environment and the feed rate had the highest impacts on the surface roughness variation, while the cutting depth had the least impact.
Poly (methyl methacrylate), PMMA.				
Impact of single point diamond turning parameters on surface integrity of PMMA	Jagtap & Pawade (2014)	Spindle speed, cutting depth, rate of feed and cutting environment	Surface roughness and surface flatness	-The spindle speed is the largest contributor towards the quality of surface. -The rate of feed is the most impactful on the flatness (32.99%).
Optimization of laser machining of PMMA	Elsheikh, Deng & Showaib (2019)	Speed of cutting, assisted gas pressure, power of the laser beam and thickness of sheet	Top and bottom kerf widths and the kerf taper	-An increase in all the machining factors increased the top and bottom kerf widths. -Maximum sheet thickness leads to minimum kerf taper.
Impact of PMMA end-milling parameters on the vibration levels.	Ali, Mohamed, Khan, Asfana, Lutfi & Fahmi (2013)	Spindle speed, cutting depth and rate of feed.	Vibrations	-For the minimum vibrations, the optimum factors were a spindle speed of 2500 rpm, a 2mm/min rate of feed and a 1.5mm cutting depth. -The rate of feed and the cutting depth were the largest contributors towards the vibration levels.

continued on following page

Table 1. Continued

STUDY (S)	AUTHOR (S) (YEAR)	FACTOR (S)	RESPONSE (S)	CONCLUSIONS
Micro-milling of PMMA using single crystal diamond tool.	Korkmaz, Onler & Ozdoganler (2017)	Spindle speed, axial cutting depth and rate of feed.	Machining forces, surface roughness, bur formation and shape retention	-Single crystal diamond micro-end tool reduced the roughness of the outputs, the formation of burr and the amounts of machining forces. -The surface roughness increased as the rate of feed increased (81.2% contribution). -The surface roughness decreased with an increase in the spindle speed (17.4% contribution). -The impacts of the rate of feed and the cutting depths on the machining forces were significant (43.4% and 50.1% respectively). The impacts of spindle speed on the forces were negligible.
PMMA facing optimization on CNC lathe machine.	Dhakad, Mahajan & Mitra (2017)	Speed of cutting, cutting depth and rate of feed.	Surface roughness	-Optimum values of facing the material were 274.89 m/min speed of cutting, 0.05mm/rev for the rate of feed and 0.1mm for the cutting deoth. -The feed rate (74.01%) was the largest influencer factor on surface roughness, with the cutting depth (2.03%) having a negligible impact on the surface roughness.
Modeling the impacts of vibrations on the surface roughness of PMMA.	Mahajan, Pawade, Mandale, Chikorde, Nawale & Birari (2019)	Cutting velocity, cutting depths and rates of feed.	Surface roughness	-An increase in vibration increased the surface roughness, which was attributed to the increase in acceleration in the Z-direction. -The model had a surface roughness error of approximately 5%, which displayed that it could be used for modelling.
Ultraprecision machining of PMMA for optical grade surface roughness.	Kobayashi & Hirakawa (2006)	Speed of cutting, cutting depth and rate of feed.	Surface roughness	-The surface roughness was high for an increasing rate of feed. -The highest value of surface roughness was 5 nm, which was acceptable in terms of optical applicability. -The best surface quality was obtained using a speed of cutting of 730 m/min, a cutting depth of 10 micro-meter and a feed rate of 3 micro-meter/rev.
Single point diamond precision turning of PMMA	Khatri, Mishra & Sarepaka (2012)	Spindle speed, cutting depth and rate of feed.	Surface roughness	- The tool feed was the major factor affecting the surface integrity, with the cutting depth having the minimum impact. -The optimal parameters for a precision machining of PMMA were a spindle speed of 2000 rpm, a cutting depth of 2 μm and a rate of feed of 1 μm/rev.
Multi-Objective Studies				
Multi-objective optimization on end milling of glass fibre-reinforced polymer using the Taguchi array and GRA	Azmi (2012)	Speed of cutting, cutting depth and rate of feed.	Cutting forces, surface roughness and tool wear.	- The rate of feed was identified as the major contributing parameter impacting on all the responses with a contribution of 84.39% followed by the depth of cut with a contribution of 13.12%.
Multi-objective optimization of a glass fiber reinforced polymer using grey relational analysis and the desirability function	Yaser & Shunmugesh (2019)	Spindle speed, cutting speed and rate of feed	Surface roughness, machining force and delamination factor	-Lower cutting speed resulted to high surface roughness. -Delamination factor was inversely related to the speed of cutting and directly related to the rate of feed and the cutting depth.
Multi-objective optimization of glass fiber end-milling with polymer composite reinforcement.	Jenarthanan, Gujjalapudi & Venkatraman (2017)	Spindle speed, fiber orientation angle, tool helix angle and feed rate.	Surface integrity, machining force and delamination factor	-The results depicted that fiber orientation was the largest contributor (66.75%), followed by the feed rate (15.05%), the helix angle (7.76%), and finally the spindle speed (0.30%).
Effects of machining parameters on the surface roughness and delamination of cellulose-based hybrid composite (MOO).	Izamshah, Zuraidah, Kasim, Hadzley & Amran (2015)	Spindle speed, cutting depth and rate of feed.	Surface roughness and delamination.	-The rate of feed was the major factor affecting the responses, and the cutting depth was the least contributor. - Combining the maximum spindle speed and the least feed rate reduced the responses. - The optimum parameters were identified to be a rate of feed of 200 mm/min, a cutting depth of 2 mm and the spindle speed of 1924 rpm.
Taguchi based GRA				
Optimization of the multi-performance of the EDM machining parameters on steel	Prayogo & Lusi (2016)	Pulse Current, On-time, Off-time and the time to discharge.	MRR and overcut	- The parameter combination in the study was identified to improve the performance of the EDM process, with a 27% increase in the MRR and a 9% decrease in overcut.
Multiple optimizations of the turning factors using GRA and Taguchi approach to reduce the surface roughness and increase the MRR	Puh, Jurkovic, Perinic, Brezocnic & Buljan (2016)	Cutting speed, depth of cut and feed rate.	Surface roughness and MRR	-The multi-objective optimization process led to a significant reduction in the roughness with an improvement in the MRR.
CNC optimization turning of steel 316	Ramu, Srinivas & Vekatesh, (2018)	Speed of cutting, cutting depth and rate of feed.	Surface roughness and MRR.	- In maximizing the MRR and minimizing the roughness, the factors 240 rpm, 15mm/min and 0.3mm for the speed of cutting, rate of feed and cutting depth respectively were identified.
Multi-objective optimization study of the severe plastic deformation of Al 6061 alloy using the Taguchi based GRA.	Girish, Siddesh & Satish (2019)	Displacement, thickness and number of passes.	Micro-hardness, tensile strength and grain size.	-The number of passes and the rate of displacement has the largest contribution to the responses. -The interaction of parameters and the thickness contributed the least. -The optimum factors were identified to be 1.5 mm/min, 4 mm and 5 for the displacement, thickness and number of passes respectively.
Multi-optimization of wear process in Co-Continuous composite using the Taguchi GRA	Sylajakumari, Ramakrishnasamy & Palaniappan (2018)	Load applied, speed of sliding and the distance of sliding.	Friction coefficient and specific wear rate.	-The best wear performance was obtainable under a load application of 60 N, a speed of sliding of 1 m/s and a distance of sliding of 1000 m.

of the multi-performance of the EDM machining parameters on steel. The study coupled the Taguchi methodology and the Grey Relational Analysis (GRA) in coming up with the optimal parameters, and a combination of these parameters was identified to improve the performance of the EDM process, with a 27% increase in the MRR and a 9% decrease in overcut. Another study conducted by Puh, Jurkovic, Perinic, Brezocnic & Buljan (2016) evaluated the multiple optimal parameters during turning using the GRA and Taguchi approach in order to reduce the surface coarseness and increase the material removal rate. The study depicted that the speed of cut and the depth of cut were the most significant factors during machining. The surface roughness was significantly reduced, while the material removal rate improved after the multi-objective optimization (Puh et al., 2016). In CNC turning of steel 316, Ramu, Srinivas & Vekatesh, (2018) used the Taguchi based GRA to come up with the optimal the factors. The optimal machining factors for the lowest surface roughness and highest rate of material removal were 240 rpm, 15mm/min and 0.3mm for the speed of cut, the rate of feed and cutting depth respectively.

Girish, Siddesh & Satish (2019) also conducted a multi-objective optimization study of the severe plastic deformation process using the Taguchi based GRA, and the optimum factors were identified to be 1.5 mm/min, 4 mm and 5 for the displacement, thickness and number of passes respectively. Lastly but not least, Sylajakumari, Ramakrishnasamy & Palaniappan (2018) studied the multi-objective optimization of the wear process in Co-Continuous composite using the Taguchi GRA. The optimal parameters were obtained to be 60 N, 1 m/s and 1000 m for the load applied, the speed of sliding and the distance of sliding respectively. The confirmatory study depicted a 35% improvement on the grey relational code. This means that the Taguchi method coupled with the GRA could effectively be employed in conducting a multi-objective analysis.

LIGHTWEIGHT APPLICATIONS OF POLYMERS AND OTHER POLYMERIC MATERIALS

Polymers, polymeric materials, ceramics, and composites have attracted a wide range of applications in the aerospace and automotive industries. These materials exhibit high tensile strengths and have low densities which increase their strength to weight ratio (Campbell, 2012). This high ratio is highly desirable in the manufacture of aerospace and automotive parts and machinery. Currently, more than 1000 car parts are made from these lightweight materials, with more than 50% of the modern car interior being made of plastics (Begum, Rane & Kanny, 2019). This has significantly led to a reduction of fuel consumption since it is directly related to the weight of the automobile. According to Begum, Rane & Kanny (2019), a 10% reduction in the weight of a car translates to an approximate reduction of fuel use of between 5% and 7%.

In the automotive industry, polycarbonates have been adopted in the manufacture of light components including lights and lenses. This is due to their desirable properties of heat resistance, transparency, good toughness, and excellent electric insulation properties (Strumberger, Gospocic & Bartulic, 2005). Car manufacturers especially in Europe, the United States, and Asia have adopted lightweight lenses made from polycarbonates. When compared to the previous used glasses in making of these lenses and lighting devices, polycarbonates provide weight savings of up to 1.5 kg in every vehicle (Gestermann et al., 2005). These materials also offer added advantages over glass since they can easily be molded and folded to attain desirable lighting and lens' finishes in automotive. Polycarbonates have also been used in blends, aimed at maximizing their potential and boosting their mechanical and thermal properties.

These blends include glass fiber and mineral reinforced polycarbonates which can be coated easily and possess excellent surface qualities as explained by Gestermann et al. (2005). Optical grade polycarbonates have also been used in the automotive industry to make glazing panels, which are fixed on the roof of vehicles for better light penetration (Krishnamoorthy et al., 2009).

In addition to the applications of polycarbonates, polypropylene (PP), which is generated from a combination of several propylene monomers, is the mostly used of all the polymers, taking more than 55% of the plastics used in the automotive industry (Strumberger et al., 2005). This material has replaced fabric which are heavier and which are less durable. In interior applications, PP has been used in making car seats, linings of roofs and side-plates. Other polymeric materials employed in the making of interior parts of vehicles include polyurethane and resin reinforced polycarbonates foam which has found applications in making of vehicle interiors roofs (Strumberger et al., 2005). In the manufacture of anterior parts, thermoplastics and polyurethanes have found a wide range of applications. These include the manufacture of intake manifolds, valve lids, noise dampers, fan housing, fuel ducts, air ducts, driving belts, seals, and membranes. Polymeric materials have also been adopted in the making of fuel tanks which further reduce the weight of vehicles (Strumberger et al., 2005).

Polymer composites have also been adopted in the aviation industry in making of windows, canopies, and interior parts. From a comparison of two plane models made with different volumes of polymer composites conducted by Atique, Probha & Nafi (2014), Boeing 787 was made with about 50% composite material as compared to Boeing 777 which adopted about 12% of composites. This led to a reduction of Boeing 787's weight by more than 15,875 kg in weight leading to a 20% savings on fuel consumption. The use of polymer composites in the planes also increased their operational life from about six years to approximately 12 years (Kesarwani, 2017).

PMMA has also attracted applications in the aviation, marine, and automotive industry owing to its low density, high toughness, durability, and low moisture and water absorbing capacity, as well as its resistance to wear in saline conditions (Begum et al., 2019). The material has been used in the manufacture of ship, vehicle, and plane windows. Its formability has also increased its beauty in these applications. Recently, researchers have been attracted to the material. An investigation conducted by Raghuwanshi & Verma (2014) on the possibility of applying the aero grade PMMA in the making of aircraft canopies depicted that the material can withstand all the stresses and environmental conditions experienced in aerospace applications without any degradation for a long period. The material was found to resist the extensive inside cabin pressures and also remain stable at extreme UV radiations.

CONCLUSION

Polymers and polymeric materials have found a wide range of applications especially in the weight sensitive fields such as aviation, marine, and automotive. This is due to their high strength-to-weight ratio, which allows their use to reduce weight in components while maintaining the desired properties such as strength, stability, durability, toughness, and impact resistance. PMMA and other polymeric materials have been used in the automotive industry to manufacture lighting components, body parts, interior elements, engine parts, and other elements such as car linings. In the aviation industry, they have been used in the manufacture of windows, canopies, and other interior elements. They have been adopted in the marine industry due to their low density and resistance to wear in saline and wet conditions.

Different studies exist regarding the machining of polymers and polymeric materials. Generally, it can be observed from literature that these studies could be categorized into investigation of machinability, single-objective optimization machining and multi-objective optimization of machining of polymers and other related materials as summarized in table 1. From the summary of these studies, various machining parameters (factors) affect the outputs (responses) differently. The major considered input parameters in the machining of these polymers are the spindle speed, the feed rate, vibrations, the depth of cut, and the environment of machining. The impacts of these input parameters to the various responses, that is, surface roughness, cutting forces, delamination, cutting temperatures, tool wear, precision, vibrations, mechanical properties and material removal rate, among others, have been investigated and presented. In the machining of PMMA, it has been observed that the major machining factors affecting the quality of output are the spindle or cutting speed, the rate of feed, and the cutting depth. Therefore, multi-objective methods should be adopted in the machining of these materials to improve the quality of products while maintaining their desirable qualities.

REFERENCES

Agarwal, S., Patidar, D., Dixit, M., & Saxena, N. S. (2010). *Investigation of Thermomechanical Properties of PMMA*. doi:10.1063/1.3466571

Ali, M. Y., Mohamed, A. R., Khan, A. A., Asfana, B., Lutfi, M., & Fahmi, M. I. (2013). Empirical modelling of vibration in micro end milling of PMMA. *World Applied Sciences Journal, 21*(S1), 73–78. doi:10.5829/idosi.wasj.2013.21.mae.99925

Astakhov, V. P. (2011). Machining of hard materials - Definitions and industrial applications. *Machining of Hard Materials*, 1–32. doi:10.1007/978-1-84996-450-0_1

Atique, M. S. A., Probha, N. N., & Nafi, A. S. (2014). Polymer composites: a blessing to modern aerospace engineering. *International Conference on Mechanical, Industrial and Energy Engineering*, 1–6. https://www.researchgate.net/publication/336411271

Azmi, A. I. (2012). Multi-objective Optimisation of Machining Fibre Reinforced Composites. *Journal of Applied Sciences, 12*(23), 2360–2367. doi:10.3923/jas.2012.2360.2367

Begum, S. A., Rane, A. V., & Kanny, K. (2019). Applications of Compatibilized Polymer Blends in Automobile Industry. In *Compatibilization of Polymer Blends: Micro and Nano Scale Phase Morphologies, Interphase Characterization, and Properties*. Elsevier Inc. doi:10.1016/B978-0-12-816006-0.00020-7

Bolat, M. (2013). *Machining of Polycarbonate for Optical Applications.* Academic Press.

Campbell, F. C. (2012). Introduction and Uses of Lightweight Materials. In Lightweight Materials: Understanding the Basics (p. 4). ASM International. doi:10.31399/asm.tb.lmub.t53550001

Castricum, H. L., Yang, H., Bakker, H., & Van Deursen, J. H. (1997). A study of milling of pure polymers and a structural transformation of polyethylene. *Materials Science Forum, 235–238*(1), 211–216. doi:10.4028/www.scientific.net/msf.235-238.211

Chen, P. C., Pan, C. W., Lee, W. C., & Li, K. M. (2014). Optimization of micromilling microchannels on a polycarbonate substrate. *International Journal of Precision Engineering and Manufacturing*, *15*(1), 149–154. doi:10.100712541-013-0318-1

Conradie, P. J., Oosthuizen, G., Dimitrov, D., & Saxer, M. (2015). Effect of Milling Strategy and Tool Geometry on Machining Cost when Cutting Titanium Alloys. *South African Journal of Industrial Engineering*, *26*(3), 137–151. doi:10.7166/26-3-1172

Daniyan, I. A., Tlhabadira, I., Daramola, O. O., & Mpofu, K. (2019). Design and optimization of machining parameters for effective AISI P20 removal rate during milling operation. *Procedia CIRP*, *84*, 861–867. doi:10.1016/j.procir.2019.04.301

DeGeare, J. (2015). Milling Operations. *The Guide to Oilwell Fishing Operations*, 91–101. doi:10.1016/B978-0-12-420004-3.00012-5

Dhakad, M. R., Mahajan, K. A., & Mitra, A. C. (2017). Experimental Analysis and Optimization of Cutting Parameters for the Surface Roughness in the Facing Operation of PMMA Material. *IOSR Journal of Mechanical and Civil Engineering*, *17*(01), 52–60. doi:10.9790/1684-17010015260

Dogra, M., Sharma, V. S., & Dureja, J. (2011). Effect of tool geometry variation on finish turning –. *RE:view*, *4*(1), 1–13.

Durkovic, M., Mladenovic, G., Tanovic, L., & Danon, G. (2018). Impact of feed rate, milling depth and tool rake angle in peripheral milling of oak wood on the cutting force. *Maderas. Ciencia y Tecnología*, *20*(1), 25–34. doi:10.4067/S0718-221X2018005001301

Elsheikh, A. H., Deng, W., & Showaib, E. A. (2019). Improving laser cutting quality of polymethylmethacrylate sheet: Experimental investigation and optimization. *Journal of Materials Research and Technology*, *x*, 1–15. doi:10.1016/j.jmrt.2019.11.059

Gestermann, S., Koppchen, W., Krause, V., Mothrath, M., Pophusen, D. W., Sandquist, A., & Zollner, O. (2005). Polycarbonate and its Blends for Car Body Parts Authors. *Materials (Basel)*, *107*, 22–24. https://link.springer.com/article/10.1007%2FBF03224786

Girish, B. M., Siddesh, H. S., & Satish, B. M. (2019). Taguchi grey relational analysis for parametric optimization of severe plastic deformation process. *SN Applied Sciences*, *1*(8), 937. Advance online publication. doi:10.100742452-019-0982-6

Guan, B., Cherrill, M., Pai, J. H., & Priest, C. (2019). Effect of mould roughness on injection moulded poly (methyl methacrylate) surfaces: Roughness and wettability. *Journal of Manufacturing Processes*, *48*(October), 313–319. doi:10.1016/j.jmapro.2019.10.024

Hamlaoui, N., Azzouz, S., Chaoui, K., Azari, Z., & Yallese, M. A. (2017). Machining of tough polyethylene pipe material: Surface roughness and cutting temperature optimization. *International Journal of Advanced Manufacturing Technology*, *92*(5–8), 2231–2245. doi:10.100700170-017-0275-4

Izamshah, R., Zuraidah, Z., Kasim, M. S., Hadzley, M., & Amran, M. (2015). Multi Objective Optimization of Cutting Parameters in Machining Cellulose Based Hybrid Composites. *Applied Mechanics and Materials*, *761*, 287–292. . doi:10.4028/www.scientific.net/AMM.761.287

Jagtap, K., & Pawade, R. (2014). Experimental Investigation on the Influence of Cutting Parameters on Surface Quality obtained in SPDT of PMMA. *Advanced Design and Manufacturing Technology, 7*(2). www.SID.ir

Jagtap, K. A., Pawade, R. S., & Balasubramaniam, R. (2012). Some Investigations on Surface characteristics in Precision Turning of Nylon and Polypropylene. *Special Issue of International Journal of Electronics, Communication & Soft. Computing in Science & Engineering, 1*, 240. http://www.ijecscse.org/papers/SpecialIssue/mech/04.pdf

Jenarthanan, M. P., Gujjalapudi, V. S. S., & Venkataraman, V. (2017). Multi-objective optimization in end-milling of glass fiber reinforced polymer composites using desirability functional analysis and grey relational analysis. *Multidiscipline Modeling in Materials and Structures, 13*(3), 391–408. doi:10.1108/MMMS-11-2016-0059

Kesarwani, S. (2017). Polymer Composites in Aviation Sector. *International Journal of Engineering Research & Technology (Ahmedabad), 6*(06), 518–525. doi:10.17577/IJERTV6IS060291

Korkmaz, E., Onler, R., & Ozdoganlar, O. B. (2017). Micromilling of Poly(methyl methacrylate, PMMA) Using Single-Crystal Diamond Tools. *Procedia Manufacturing, 10*, 683–693. doi:10.1016/j.promfg.2017.07.017

Krishnamoorthy, N., Nayak, S., Surisetty, G. K., Chaturvedi, M., & Shuler, S. (2009). Polycarbonate Glazing Body Panels for Automotive Applications. *SAE Technical Papers*. Advance online publication. doi:10.4271/2009-26-0085

Kumar, S. (2015, December). Experimental Investigation and Optimization in Turning of UD-GFRP Composite Material by Regression Analysis Using PCD Tool Address for Correspondence. *International Journal of Advances in Engineering and Technology*.

Lazarević, D., Madić, M., Janković, P., & Lazarević, A. (2012). Cutting parameters optimization for surface roughness in turning operation of polyethylene (PE) using Taguchi method. *Tribology in Industry, 34*(2), 68–73.

Mahajan, K. A., Pawade, R. S., Mandale, A. M., Chikorde, P. P., Nawale, M. A., & Birari, S. N. (2019). Prediction of vibration effect on Surface Roughness of Poly methyl methacrylate (PMMA) by using ANN. *IOSR Journal of Engineering*, 55–60. www.iosrjen.org

Nee, J. G., Dufraine, W., Evans, J. W., & Hill, M. (2010). *Fundamentals of Tool Design* (6th ed.). Society of Manufacturing Engineers. https://books.google.com.my/books/about/Fundamentals_of_Tool_Design_Fifth_Editio.html?id=4Y3x5Fq9TJkC&pgis=1

Patel, Y. (2008). *The machining of polymers* (PhD thesis). Department of Mechanical Engineering, Imperial College London.

Prayogo, G. S., & Lusi, N. (2016). Application of Taguchi technique coupled with grey relational analysis for multiple performance characteristics optimization of EDM parameters on ST 42 steel. *AIP Conference Proceedings, 1725*, 020061. Advance online publication. doi:10.1063/1.4945515

Puh, F., Jurkovic, Z., Perinic, M., Brezocnik, M., & Buljan, S. (2016). Optimizacija parametara obrade tokarenja s više kriterija kvalitete uporabom Grey relacijske analize. *Tehnicki Vjesnik (Strojarski Fakultet)*, *23*(2), 377–382. doi:10.17559/TV-20150526131717

Quadrini, F., Squeo, E. A., & Tagliaferri, V. (2007). Machining of glass fiber reinforced polyamide. *Express Polymer Letters*, *1*(12), 810–816. doi:10.3144/expresspolymlett.2007.112

Raghuwanshi, R. K., & Verma, V. K. (2014). Mechanical and Thermal Characterization of Aero Grade Polymethyl Metha Acrylate Polymer used in Aircraft Canopy. *International Journal of Engineering and Advanced Technology*, *3*, 2249–8958.

Raj, I. J. A., Vijayakumar, P., Kannan, T., Kumar, P., & Ragavan, R. V. (2016). Design optimization of turning parameters of PTFE (Teflon) cylindrical rods using ANOVA Methodology. *International Journal of Applied Engineering Research: IJAER*, *11*(3). http/www.ripublication.com/ijaer.htm%0ADesign

Ramu, I., Srinivas, P., & Vekatesh, K. (2018). Taguchi based grey relational analysis for optimization of machining parameters of CNC turning steel 316. *IOP Conference Series. Materials Science and Engineering*, *377*(1), 012078. Advance online publication. doi:10.1088/1757-899X/377/1/012078

Sanjeev, M., Kaviarasan, V., & Venkatesan, R. (2012). Machining Parameter Optimization of Poly Tetra Fluoro Ethylene (PTFE) Using Genetic Algorithm. *International Journal of Modern Engineering Research*, *2*(1), 143–149.

Sharma, V. S., Fromentin, G., Poulachon, G., & Brendlen, R. (2014). Investigation of tool geometry effect and penetration strategies on cutting forces during thread milling. *International Journal of Advanced Manufacturing Technology*, *74*(5-8), 963–971. doi:10.100700170-014-6040-z

Shokrani, A., Dhokia, V., & Newman, S. T. (2012). Environmentally conscious machining of difficult-to-machine materials with regard to cutting fluids. *International Journal of Machine Tools & Manufacture*, *57*, 83–101. doi:10.1016/j.ijmachtools.2012.02.002

Sohail, M. (2012). *Applications of Polymers, Elastomers, Fibers and Composites*. https://books.google.com/books?id=o1raBwAAQBAJ&pgis=1

Strumberger, N., Se, D., & Zagreb, F. (2005). Polymeric Metarials in Automobiles. *Promet-Traffic-Traffico*, *17*(3), 149–160.

Sylajakumari, P. A., Ramakrishnasamy, R., & Palaniappan, G. (2018). Taguchi grey relational analysis for multi-response optimization of wear in co-continuous composite. *Materials (Basel)*, *11*(9), 1–17. doi:10.3390/ma11091743 PMID:30223617

Compilation of References

Ilangovan, M., Rajendra Boopathy, S., & Balasubramanian, V. (2015). Effect of tool pin profile on microstructure and tensile properties of friction stir welded dissimilar AA 6061eAA 5086 aluminum alloy joints. Elsevier. *Defence Technology, 11*(2), 174–184. doi:10.1016/j.dt.2015.01.004

Martinez, N., Kumar, N., Mishra, R. S., & Doherty, K. J. (2017). Microstructural variation due to heat gradient of a thick friction stir welded aluminum 7449 alloy. *Journal of Alloys and Compounds, 713*, 51–63. doi:10.1016/j.jallcom.2017.04.147

Tecotzky, M., Blasse, G., & Meijerink, A. (1992). *US Patent 5,140,163*. US Patent Office.

Bo, Y. H., & Yoo, J. S. (2009). Korean. *Chemical Engineering Research & Design, 47*, 150–156.

Khanna, Sharma, Bharati, & Badheka. (2020). Friction Stir Welding of Dissimilar Aluminum Alloys AA6061-T6 and AA8011-h14: A Novel Study. *Journal of the Brazilian Society of Mechanical Sciences and Engineering*.

Pantelis, D. I., Karakizis, P. N., Daniolos, N. M., Charitidis, C. A., Koumoulos, E. P., & Dragatogiannis, D. A. (2016). Microstructural study and mechanical properties of dissimilar friction stir welded AA5083-H111 and AA6082-T6 reinforced with SiC nanoparticles. *Materials and Manufacturing Processes, 31*(3), 264–274. doi:10.1080/10426914.2015.1019095

Meijerink, A., & Blasse, G. (1990). Luminescence and temperature dependent decay behaviour of divalent europium in Ba5SiO4X6 (X = Cl, Br). *Journal of Luminescence, 47*(1-2), 471–475. doi:10.1016/0022-2313(90)90052-D

Patel Satyam. (2017). A Review on an Investigation on the Effect of Immersed. Medium in Friction Stir Welding of Aluminium Alloys (AA 6061-T6). *Journal of Material & Metallurgical Engineering, Volume, 07*(1).

Pourali, M., Abdollah-Zadeh, A., Saeid, T., & Kargar, F. (2017). Influence of welding parameters on intermetallic compounds formation in dissimilar steel/aluminum friction stir welds. *Journal of Alloys and Compounds, 715*, 1–8. doi:10.1016/j.jallcom.2017.04.272

Abed, M., & Mtiller-Buschbaum, H. (1992). Zur Kristallstruktur von Ba5SiO4Cl6. *Journal of Alloys and Compounds, 190*(1), 61–66. doi:10.1016/0925-8388(92)90174-8

Cavaliere, P. (2013). Friction stir welding of Al alloys: Analysis of processing parameters affecting mechanical behavior. *Procedia CIRP, 11*, 139–144. doi:10.1016/j.procir.2013.07.039

Yu, C., Hua, D., Li, J. Z., Zhao, J. W., Fu, M. J., & Li, X. H. (2015). Effect of welding heat input and post-welded heat treatment on hardness of stir zone for friction stir-welded 2024-T3 aluminum alloy. *Transactions of Nonferrous Metals Society of China, 25*(8), 2524–2532. doi:10.1016/S1003-6326(15)63871-7

Kumbhar, N. T., Sahoo, S. K., Samajdar, I., Dey, G. K., & Bhanumurthy, K. (2011). Microstructure and microtextural studies of friction stir welded aluminium alloy 5052. *Materials & Design, 32*(3), 1657–1666. doi:10.1016/j.matdes.2010.10.010

Mishra, R. S., & Ma, Z. Y. (2005). Friction stir welding and processing. Elsevier. *Materials Science and Engineering R Reports*, *50*(1-2), 1–78. doi:10.1016/j.mser.2005.07.001

Yang & Zhang. (2003). The relationship between chemical bond properties and Stokes shift of Eu2+ in some silicate host lattices. *Journal of Physics and Chemistry Solids, 64*, 213-217.

Shanavas, S., & Edwin Raja Dhas, J. (2017). Parametric Optimization of Friction Stir Welding Parameters of Marine Grade Aluminium Alloy using Response Surface Methodology. *Transactions of Nonferrous Metals Society of China, Volume*, *27*(11), 2334–2344. doi:10.1016/S1003-6326(17)60259-0

Tozaki, Y., Uematsu, Y., & Tokaji, K. (2007). Effect of tool geometry on microstructure and static strength in friction stir spot welded aluminium alloys. *International Journal of Machine Tools & Manufacture*, *47*(15), 2230–2236. doi:10.1016/j.ijmachtools.2007.07.005

Zeng, Q., Tanno, H., Egoshi, K., Tanamachi, N., & Zhang, S. (2006). Ba5SiO4Cl6:Eu2+ Ba5SiO4Cl6:Eu2+: An intense blue emission phosphor under vacuum ultraviolet and near-ultraviolet excitation. *Applied Physics Letters*, *88*, 051906.

Elangovan, K., & Balasubramanian, V. (2007). Influences of pin profile and rotational speed of the tool on the formation of friction stir processing zone in AA2219 aluminium alloy. *Materials Science and Engineering A*, *459*(1-2), 7–18. doi:10.1016/j.msea.2006.12.124

Lai, S., Yang, Z., Wang, R., Wu, H., Liao, J., Qiu, J., Song, Z., Yang, Y., & Zhou, D. (2013). Preparation and blue–white luminescence properties of Bi^{3+}-doped $Ba_5SiO_4Cl_6$. *Journal of Materials Science*, *48*, 8566–8571.

Ugender. (2018). Influence of tool pin profile and rotational speed on the formation of friction stir welding zone in AZ31 magnesium alloy. *Journal of Magnesium and Alloys, 6*, 205–213.

Gharavi, F., Matori, K. A., Yunus, R., Othman, N. K., & Fadaeifard, F. (2015). Corrosion behavior of Al6061 alloy weldment produced by friction stir welding process. *Journal of Materials Research and Technology*, *4*(3), 314–322. doi:10.1016/j.jmrt.2015.01.007

Raj Kumar, Venkatesh Kannan, Sadeesh, Arivazhagan, & Devendranath Ramkumar. (2014). Studies on the effect of tool design and welding parameters on the friction stir welding of dissimilar aluminum alloys AA 5052 – AA 6061. *Procedia Engineering, 75*, 93-97.

Yang, J.-Z., Qiu, J.-B., Yang, Z.-W., Song, Z.-G., Yong, Y., & Zhou, D.-C. (2015). Preparation and up conversion luminescence properties of Ba5SiO4Cl6: Yb (3+), Er (3+), Li+ phosphors. *Wuli Xuebao*, *64*, 138101.

Boumerzoug, Z., & Helal, Y. (2017). *Friction stir welding of dissimilar materials Aluminum Al 6061-T6 to Ultra Low Carbon Steel*. MDPI, Metals.

Park, J. H., Min, D. J., & Song, H. S. (2004). Amphoteric behavior of alumina in viscous flow and structure of CaO-SiO_2 (-MgO)-Al_2O_3 slags. *Metallurgical and Materials Transactions. B, Process Metallurgy and Materials Processing Science*, *35B*, 269.

Soto-Díaz, R., Sandoval-Amador, A., & Unfried-Silgado, J. (2021). Experimental evaluation of rotational and traverse speeds effects on corrosion behavior of friction stir welded joints of aluminum alloy AA5052-H32. *International Journal of Advanced Manufacturing Technology*, 1–11.

Abbas, W. S., Alali, M., & Salim, J. A. (2021, March). Effect of rotational speed and tool pin profile on the corrosion rate of friction stir welded AA6061-T3. *IOP Conference Series. Materials Science and Engineering*, *1090*(1), 012124. doi:10.1088/1757-899X/1090/1/012124

Mishra, D., & Das, A. K. (2021). Linear model analysis of fused deposition modeling process parameters for obtaining the maximum tensile strength in acrylonitrile butadiene styrene (ABS) and carbon fiber polylactic acid (PLA) materials. Multidiscipline Modeling in Materials and Structures. doi:10.1108/MMMS-09-2020-0239

Sohn, I., & Min, D. J. (2012). A review of the relationship between viscosity and the structure of calcium-silicate-based slags in ironmaking. *Steel Research International*, *83*, 611–617.

Sharma, N., Khan, Z. A., & Siddiquee, A. N. (2017). Friction stir welding of aluminum to copper—An overview. *Transactions of Nonferrous Metals Society of China*, *27*(10), 2113–2136. doi:10.1016/S1003-6326(17)60238-3

Venkata Rao, Ch., Madhusudhan Reddy, G., & Srinivasa Rao, K. (2015). Influence of Tool Pin Profile on Microstructure and Corrosion Behaviour. Elsevier. *Defence Technology*, *11*, 197–208. doi:10.1016/j.dt.2015.04.004

Zhu, G., Wang, Y., Ci, Z., Liu, B., Shi, Y., & Xin, S. (2012). Ca8Mg(SiO4)4Cl2:Ce3+, Tb3+: A potential single-phased phosphor for white-light-emitting diodes. *Journal of Luminescence*, *132*(2), 531–536. doi:10.1016/j.jlumin.2011.09.029

Raturi, M., & Bhattacharya, A. (2021). Mechanical strength and corrosion behavior of dissimilar friction stir welded AA7075-AA2014 joints. *Materials Chemistry and Physics*, *262*, 124338. doi:10.1016/j.matchemphys.2021.124338

Zhang, X., Wang, X., Huang, J., Shi, J., & Gong, M. (2009). Near UV-based LED fabricated with Ba5SiO4(F, Cl)6: Eu2+ as blue- and green-emitting phosphor. *Optical Materials*, *32*, 75–78.

Lawson, J. K., & Payne, S. A. (1993). Excited-state absorption of Eu2+ doped materials. *Physical Review. B*, *47*, 14003.

Dorenbos, P. (2003). Energy of the first 4f7-4f65d transition of Eu2+ in inorganic compounds. *Journal of Luminescence*, *104*, 239–244.

Blasse. (1978). Luminescence of Inorganic Solids. *Luminescence of inorganic solids: From isolated centres to concentrated systems*, 457-453.

Hoffman, M. V. (1971). Alkaline Earth Aluminum Fluoride Compounds with Eu+ 2 Activation. *Journal of the Electrochemical Society*, *118*, 933–939.

Hoffman, M. V. (1972). Eu+ 2 emission in ternary alkaline earth aluminum fluorides. *Journal of the Electrochemical Society*, *119*, 905–911.

Du, Sun, Xia, & Sun. (2009). High numerical aperture microlens arrays of close packing. *Applied Physics Letter B, 96*, 459–463.

Mishra, D., & Tulasi, T. (2020). *Experimental Investigation on Stir Casting Processing and Properties of Al 6082/SiC Metal Matrix Composites. In Lecture Notes in Mechanical Engineering*. Springer.

Sivashanmugam, M., Kumar, T., Ravikumar, S., Rao, V. S., & Muruganandam, D. (2010, November). A review on friction stir welding for aluminium alloys. *Frontiers in Automobile and Mechanical Engineering*, *2010*, 216–221. doi:10.1109/FAME.2010.5714839

Kuang, J., Liu, Y., & Lei, B. (2006). Effect of RE3+ as a co-dopant in long-lasting phosphorescence CdSiO3:Mn2+ (RE =Y, La, Gd, Lu). *Journal of Luminescence*, *118*(1), 33–38. doi:10.1016/j.jlumin.2005.06.005

Maneiah, D., & Debashis Mishra, K. (2020). Process parameters optimization of friction stir welding for optimum tensile strength in Al 6061-T6 alloy butt welded joints. *Materials Today: Proceedings*, *27*, 904–908. doi:10.1016/j.matpr.2020.01.215

Salem, H. G. (2003). Friction stir weld evolution of dynamically recrystallized AA 2095 weldments. *Scripta Materialia*, *49*(11), 1103–1110. doi:10.1016/j.scriptamat.2003.08.010

Chen, Y., Liu, B., Kirm, M., Qib, Z., Shi, C., True, M., Vielhauer, S., & Zimmerer, G. (2006). Luminescent properties of blue-emitting long afterglow phosphors Sr2xCax MgSi2O7:Eu2+, Dy3+ (x =0, 1). *Journal of Luminescence*, *118*(1), 70–78. doi:10.1016/j.jlumin.2005.05.011

Desai, P.S., & Kapopara, S.M. (2009). *Inhibiting effect of anisidines on corrosion of aluminium in hydrochloric acid.* Academic Press.

Ugrasen, G., Bharath, G., Kishor Kumar, G., Sagar, R., Shivu, P. R., & Keshavamurthy, R. (2018). Optimization of Process Parameters for Al6061-Al7075 alloys in Friction Stir Welding using Taguchi's Technique. *Materials Today: Proceedings*, *5*(1), 3027–3035. doi:10.1016/j.matpr.2018.01.103

Ating, E. I., Umoren, S. A., Udousoro, I. I., Ebenso, E. E., & Udoh, A. P. (2010). Leaves extract of Ananassativum as green corrosion inhibitor for aluminium in hydrochloric acid solutions. *Green Chemistry Letters and Reviews*, *3*(2), 61–68. doi:10.1080/17518250903505253

Elatharasan & Kumar Senthil. (2013). An experimental analysis and optimization of process parameters on friction stir welding of AA6061-T6 aluminum alloys using RSM. *Procedia Engineering, 64*, 1227-1234.

Murayama, Y. (1996). Nikkei Science. *Scientific American in Japanese*, *26*, 20.

Jiang, L., Chang, C., & Mao, D. (2003). Luminescent properties of CaMgSi2O6 and Ca2MgSi2O7 phosphors activated by Eu2+, Dy3+ and Nd3+. *Journal of Alloys and Compounds*, *360*(1-2), 19–24. doi:10.1016/S0925-8388(03)00361-X

Obi-Egbedi, N. O., Obot, I. B., & Umoren, S. A. (2012). Spondias mombin L. as a green corrosion inhibitor for aluminium in sulphuric acid: Correlation between inhibitive effect and electronic properties of extracts major constituents using density functional theory. *Arabian Journal of Chemistry*, *5*(3), 361–373. doi:10.1016/j.arabjc.2010.09.002

Padhy, Wu, & Gao. (2018). Friction Stir Based Welding and Processing Technologies -Processes, Parameters, Microstructures and Applications: A review. *Journal of Materials Science and Technology.*

Garcia, Latourrette, & Fouassier. (1979). Ba5SiO4Cl6: Eu, A New Blue-Emitting Photoluminescent Material with High Quenching Temperature. *Journal of Electrochemical Society, 126*, 1734-1741.

Liu, H. J., Hou, J. C., & Guo, H. (2013). Effect of Welding Speed on Microstructure and Mechanical Properties of self-Reacting Friction Stir Welded 6061-T6 Aluminum Alloy. *Materials and Design, Elsevier*, *50*, 872–878. doi:10.1016/j.matdes.2013.03.105

Lueth, R. C., & Hale, T. E. (1970). Compressive Strength of Cemented Carbides—Failure Mechanics and Testing Methods. *Materials Research and Standards*, *10*(2), 23–28.

Ramanjaneyulu, Madhusudhan Reddy, & Gokhale. (2015). Optimization of process parameters of aluminum alloy AA2014 - T6 friction stir welds by response surface methodology. *Defence Technology, 11*, 209 - 212.

Tang, J., & Shen, Y. (2017). Effects of preheating treatment on temperature distribution and material flow of aluminum alloy and steel friction stir welds. *Journal of Manufacturing Processes*, *29*, 29–40. doi:10.1016/j.jmapro.2017.07.005

Von, Winkler, Ziemer, & Wieker. (1983). Article. *Journal of Inorganic and General Chemistry, 604,* 89-92.

Garcia, C. F., & Hagenmuller, P. (1980). Charge transfer excitation of the Nd^{3+}, Sm^{3+}, Dy^{3+}, Ho^{3+}, Er^{3+} and Tm^{3+} emission in $CaGa_2S_4$ C. R. *Academy Science*, *290*, 33–36.

Ghosh, M., Kumar, K., Kailas, S. V., & Ray, A. K. (2010). Optimization of friction stir welding parameters for dissimilar aluminum alloys. Elsevier. *Materials & Design*, *31*(6), 3033–3037. doi:10.1016/j.matdes.2010.01.028

Guo, N., Fu, Y., Wang, Y., Meng, Q., & Zhu, Y. (2017). Microstructure and mechanical properties in friction stir welded 5A06 aluminum alloy thick plate. *Materials & Design*, *113*, 273–283. doi:10.1016/j.matdes.2016.10.030

Aal, A. A., & Aly, M. S. (2009). Electroless Ni–Cu–P plating onto open cell stainless steel foam. *Applied Surface Science*, *255*(13-14), 6652–6655. doi:10.1016/j.apsusc.2009.02.073

Abbas, T., Ya, H. H., & Abdullah, M. Z. (2016). Energy absorption capability of aluminium tube partially wrapped with glass/epoxy subjected to quasi-static loading. *Journal of Engineering and Applied Sciences (Asian Research Publishing Network)*, *11*(24), 14313–14318.

Abbas, T., Ya, H. H., & Abdullah, M. Z. (2017). Comparison of Energy Absorption of Aluminiumcomposite Tubes Subjected to Axial Loading. *Materials Science and Engineering*, •••, 205.

Abbas, T., Ya, H. H., & Abdullah, M. Z. (2017). Experimental Investigation of Energy Absorption of Partially Wrapped Thin-Walled Aluminium-Glass/Epoxy Tube Subjected to Quasi-Static Loading. *Materials Science Forum*, *904*, 61–67. doi:10.4028/www.scientific.net/MSF.904.61

Abd El Salam Shakal, M., & Pfeiffer, P. (1998). Effect of different surface preparations on the tensile strength of adhesively bonded metal joints. *Journal of Adhesion Science and Technology*, *12*(3), 349–361. doi:10.1163/156856198X00920

Abdullah, K., Wild, P. M., Jeswiet, J. J., & Ghasempoor, A. (2001). Tensile testing for weld deformation properties in similar gage tailor welded blanks using the rule of mixtures. *Journal of Materials Processing Technology*, *112*(1), 91–97. doi:10.1016/S0924-0136(01)00555-6

Abdullah, M. R., & Cantwell, W. J. (2006). The Impact Resistance of Fiber–metal Laminates Based on Glass Fiber Reinforced Polypropylene. *Polymer Composites*, *27*(6), 700–708. doi:10.1002/pc.20240

Abdulrahman, K. O., Akinlabi, E. T., & Mahamood, R. M. (2018). Additive Manufacturing- Laser Metal Deposition and effect of preheating on properties of deposited Ti-4822-4 alloy. In *Additive Manufacturing Technologies from an Optimization Perspective*. Advances in Logistics, Operations, and Management Science. IGI Global.

Abdulrahman, K. O., Akinlabi, E. T., & Mahamood, R. M. (2019). Characteristics of laser metal deposited titanium aluminide. *Materials Research Express*, *6*(4), 046504. Advance online publication. doi:10.1088/2053-1591/aaf9c8

Abdulrahman, K. O., Akinlabi, E. T., Mahamood, R. M., Pityana, S., & Tlotleng, M. (2018). Laser Metal Deposition of Titanium Aluminide Composites: A Review. *Materials Today: Proceedings*, *5*(9), 19738–19746. doi:10.1016/j.matpr.2018.06.336

Abdulrahman, K. O., Mahamood, R. M., Akinlabi, E. T., & Adediran, A. A. (2021). Effect of laser power on the microstructure and mechanical properties of laser deposited titanium aluminide composite. *Advances in Materials and Processing Technologies*, *00*(00), 1–12. doi:10.1080/2374068X.2021.1945268

Abe, H., Chung, J. C., Mori, T., Hosoi, A., Jespersen, K. M., & Kawada, H. (2019). The effect of nanospike structures on direct bonding strength properties between aluminum and carbon fiber reinforced thermoplastics. *Composites. Part B, Engineering*, *17*, 226–232. doi:10.1016/j.compositesb.2019.05.025

Abhilash, M., Prakash, G. N., & Kasthurirengan, S. (2018). Cryogenic deflashing for rubber products. *MATEC Web of Conferences, 144, 03004*. 10.1051/matecconf/201814403004

Adam, H. (1997). Carbon fibre in automotive applications. *Materials & Design*, *4*(6), 349–355. doi:10.1016/S0261-3069(97)00076-9

Adel, M., Abdelaal, O., Gad, A., Nasr, A. B., & Khalil, A. (2018). Polishing of fused deposition modeling products by hot air jet: Evaluation of surface roughness. *Journal of Materials Processing Technology*, *251*, 73–82. doi:10.1016/j.jmatprotec.2017.07.019

Adeniji, D., Schoop, J., Gunawardena, S., Hanson, C., & Jahan, M. (2020). Characterization and modeling of surface roughness and burr formation in slot milling of polycarbonate. *Journal of Manufacturing and Materials Processing*, *4*(2), 59. doi:10.3390/jmmp4020059

Agarwal, S., Patidar, D., Dixit, M., & Saxena, N. S. (2010). *Investigation of Thermomechanical Properties of PMMA.* doi:10.1063/1.3466571

Aghchai, A. J., Shakeri, M., & Mollaei-Dariani, B. (2008). Theoretical and experimental formability study of two-layer metallic sheet (Al1100/St12). *Proceedings of the Institution of Mechanical Engineers. Part B, Journal of Engineering Manufacture*, *222*(9), 1131–1138. doi:10.1243/09544054JEM1140

Agius, D., Kourousis, K. I., Wallbrink, C., Hu, W., Wang, C. H., & Dafalias, Y. F. (2017). Aluminum Alloy 7075 Ratcheting and Plastic Shakedown Evaluation with the Multiplicative Armstrong–Frederick Model. *AIAA Journal*, *55*(7), 2461–2470. doi:10.2514/1.J055833

Agrawal, A. K., & Narayanan, R. G. (2017). Joining of a tube to a sheet through end curling. *Journal of Materials Processing Technology*, *246*, 291–304. doi:10.1016/j.jmatprotec.2017.03.035

Agrawal, A. K., & Narayanan, R. G. (2018). Pull-out tests on tube to sheet joints fabricated by endforming. *Journal of Constructional Steel Research*, *144*, 186–197. doi:10.1016/j.jcsr.2018.01.027

Agrawal, A. K., & Narayanan, R. G. (2020a). Friction stir processing of AA6063-T6 tubes and end forming characterization at varying tool pin profiles. *Materials Performance and Characterization*, *9*(1), 497–517. doi:10.1520/MPC20190248

Agrawal, A. K., & Narayanan, R. G. (2020b). Experimental and numerical studies on joining steel tubes by end forming. *Journal of Constructional Steel Research*, *167*, 105792. doi:10.1016/j.jcsr.2019.105792

Agrawal, A. K., Narayanan, R. G., & Kailas, S. V. (2017). End forming behaviour of friction stir processed Al6063-T6 tubes at different tool rotational speeds. *Journal of Strain Analysis for Engineering Design*, *52*(7), 434–449. doi:10.1177/0309324717724662

Agrawal, A. K., Narayanan, R. G., & Kailas, S. V. (2019). Experimental evaluation and prediction of end forming behaviour of friction stir processed Al6063T6 tubes at different tool traverse speeds. *International Journal of Advanced Manufacturing Technology*, *104*(9-12), 3607–3627. doi:10.100700170-019-04068-4

Agrawal, A. K., Narayanan, R. G., & Kailas, S. V. (2021). Formability and instability evaluation of friction stir processed AA6063-T6 tubes during end forming and numerical prediction. *Journal of Materials Engineering and Performance*, *30*(2), 973–993. doi:10.100711665-020-05400-w

Ahmed, A., Arya, S., Gupta, V., Furukawa, H., & Khosla, A. (2021). 4D Printing: Fundamentals, Materials, Applications and Challenges. *Polymer*, *228*, 123926. doi:10.1016/j.polymer.2021.123926

Ahmetoglu, M. A., Brouwers, D., Shulkin, L., Taupin, L., Kinzel, G. L., & Altan, T. (1995). Deep drawing of round cups from tailor-welded blanks. *Journal of Materials Processing Technology*, *53*(3-4), 684–696. doi:10.1016/0924-0136(94)01767-U

Aida, S. F., Zuhailawati, H., & Anasyida, A. S. (2017). The effect of space holder content and sintering temperature of magnesium foam on microstructural and properties prepared by sintering dissolution process (SDP) using carbamide space holder. *Procedia Engineering*, *184*, 290–297. doi:10.1016/j.proeng.2017.04.097

Airoldi, A., Sala, G., Di Lando, L. A., Bettini, P., & Gilardelli, A. (2018). Composite Corrugated Laminated for Morphing Applications. In A. Concilio, I. Dimino, L. Lecce, & R. Pecora (Eds.), *Morphing Wing Technologies* (pp. 247–276). Elsevier. doi:10.1016/B978-0-08-100964-2.00009-5

Ajalli, M., Azimi, H., Balani, A. M., & Rezaei, M. (2017). Application of Fuzzy AHP and COPRAS to solve the Supplier Selection Problems. *Int. J Sup. Chain. Mgt*, *6*(3), 112–119.

Ajinola, S. A. (2016). *Design, fabrication, and characterization of porous titanium for biological applications* (Doctoral dissertation). North Carolina Agricultural and Technical State University.

Akano, T. T., & Fakinlede, O. A. (2013). Fatigue Failure Model for Polymeric Compliant Systems. *ISRN Polymer Science*, *2013*, 1–11. doi:10.1155/2013/321489

Akinlabi, E. T., & Akinlabi, S. A. (2016). *Powder Flow Rate Influence on Laser Metal Deposited TiC on Ti-6Al-4V.* Lmd.

Akinlabi, E. T., Mahamood, R. M., Shukla, M., & Pityana, S. (2012b). Effect of Scanning Speed on Material Efficiency of Laser Metal Deposited Ti6Al4V. World Academy of Science, Engineering and Technology, 1531–1535.

Akinlabi, E. T., Mahamood, R. M., Shukla, M., & Pityana, S. (2012a). Effect of Scanning Speed on Material Efficiency of Laser Metal Deposited Ti6Al4V. *International Journal of Mechanical and Mechatronics Engineering*, *6*(11), 2509–2513.

Akinlabi, E. T., Mahamood, R. M., Akinlabi, S. A., & Ogunmuyiwa, E. (2014). Processing parameters influence on wear resistance behaviour of friction stir processed Al-TiC composites. *Advances in Materials Science and Engineering*, *2014*, 1–12. Advance online publication. doi:10.1155/2014/724590

Aksonov, Y., Kombarov, V., Fojtů, O., Sorokin, V., & Kryzhyvets, Y. (2019). Investigation of processes in high-speed equipment using CNC capabilities. *MM Science Journal*, *2019*(04), 3271–3276. doi:10.17973/MMSJ.2019_11_2019081

Aktarer, S. M., Sekban, D. M., Kucukomeroglu, T., & Purcek, G. (2019). Microstructure, mechanical properties and formability of friction stir welded dissimilar materials of IF-steel and 6061 Al alloy. *International Journal of Minerals Metallurgy and Materials*, *26*(6), 722–731. doi:10.100712613-019-1783-z

Alantali, A., Alia, R., Umer, R., & Cantwell, W. (2017). Energy absorption in aluminium honeycomb cores reinforced with carbon fibre reinforced plastic tubes. *The Journal of Sandwich Structures & Materials*. Advance online publication. doi:10.1177/1099636217727145

Albiez, M., Vallée, T., Fricke, H., & Ummenhofer, T. (2019). Adhesively bonded steel tubes—Part I: Experimental investigations. *International Journal of Adhesion and Adhesives*, *90*, 199–210. doi:10.1016/j.ijadhadh.2018.02.005

Alghamdi, S. S., John, S., Choudhury, N. R., & Dutta, N. K. (2021). Additive Manufacturing of Polymer Materials: Progress, Promise and Challenges. *Polymers*, *13*(5), 753. doi:10.3390/polym13050753 PMID:33670934

Alhashmy, H. (2012). Fabrication of Aluminium Matrix Composites (AMCs) by Squeeze Casting Technique Using Carbon Fiber as Reinforcement. *24th Materials Science Conference.*

Alhashmy, H. A., & Nganbe, M. (2015). Laminate squeeze casting of carbon fiber reinforced aluminum matrix composites. *Materials & Design*, *67*, 154–158. doi:10.1016/j.matdes.2014.11.034

Alhusseny, A. N. M., & Nasser, A. (2018). *High-porosity metal foams: potentials, applications, and formulations. In Porosity-Process, Technologies and Applications.* IntechOpen.

Ali, M. Y., Mohamed, A. R., Khan, A. A., Asfana, B., Lutfi, M., & Fahmi, M. I. (2013). Empirical modelling of vibration in micro end milling of PMMA. *World Applied Sciences Journal*, *21*(S1), 73–78. doi:10.5829/idosi.wasj.2013.21.mae.99925

Allavikutty, R., Gupta, P., Santra, T. S., & Rengaswamy, J. (2021). Additive Manufacturing of Mg alloys for Biomedical Applications: Current Status and Challenges. *Current Opinion in Biomedical Engineering, 18*, 100276. doi:10.1016/j.cobme.2021.100276

Allioux, F. M., Merhebi, S., Tang, J., Idrus-Saidi, S. A., Abbasi, R., Saborio, M. G., Ghasemian, M. B., Han, J., Namivandi-Zangeneh, R., O'Mullane, A. P., Koshy, P., Daiyan, R., Amal, R., Boyer, C., & Kalantar-Zadeh, K. (2019). Catalytic Metal Foam by Chemical Melting and Sintering of Liquid Metal Nanoparticles. *Advanced Functional Materials, 30*(5), 1907879. doi:10.1002/adfm.201907879

Almonti, D., & Ucciardello, N. (2019). Design and thermal comparison of random structures realized by indirect additive manufacturing. *Materials (Basel), 12*(14), 2261. doi:10.3390/ma12142261 PMID:31337088

Alshahrani, H.A. (2021). Review of 4D printing materials and reinforced composites: Behaviors, applications and challenges. *Journal of Science: Advanced Materials and Devices, 6*, 167-185. doi:10.1016/j.jsamd.2021.03.006

Altan, M., & Altan, E. (2014). Investigation of burr formation and surface roughness in drilling engineering plastics. *Journal of the Brazilian Society of Mechanical Sciences and Engineering, 36*(2), 347–354. doi:10.100740430-013-0089-8

Alves, L. M., Dias, E. J., & Martins, P. A. (2011). Joining sheet panels to thin-walled tubular profiles by tube end forming. *Journal of Cleaner Production, 19*(6-7), 712–719. doi:10.1016/j.jclepro.2010.12.014

Aly, M. S. M. A. (2004). *High temperature mechanical properties of cast as well as powder metallurgical manufactured metallic foams* (Doctoral Thesis). Von der Fakultät für Georessourcen und Materialtechnik der Rheinisch- Westfälischen Technischen Hochschule Aachen.

Amancio, S.T. (2009). *Friction Spot Joining of Lightweight Metals and Fiber-Reinforced Polymer Hybrid Structures.* Academic Press.

Amancio-Filho, S., & dos Santos, J. (2009). Joining of Polymers and Polymer-Metal Hybrid Structures: Recent Developments and Trends. *Polymer Engineering and Science, 49*(8), 1461–1476. doi:10.1002/pen.21424

Amigo, N., Palza, H., Canales, D., Sepúlveda, F., Vasco, D. A., Sepúlveda, F., & Zapata, P. A. (2019). Effect of starch nanoparticles on the crystallization kinetics and photodegradation of high density polyethylene. *Composites. Part B, Engineering, 174*, 106979. doi:10.1016/j.compositesb.2019.106979

Aminzadeh, A., Parvizi, A., & Moradi, M. (2020). Multi-objective topology optimization of deep drawing dissimilar tailor laser welded blanks; experimental and finite element investigation. *Optics & Laser Technology, 125*, 106029. doi:10.1016/j.optlastec.2019.106029

Amjad, S. (2001). *Thermal conductivity and noise attenuation in aluminium foams* (Master's dissertation). University of Cambridge.

AnandaRao, M., Reddy, K.V.K., & Seshaiah, T. (2014). Finite Element Analysis of Epoxy (Ly-556) Glass Fiber Filled Aluminium Circular Section Hollow Column. *International Organization of Scientific Research, 4*(7), 7-12.

Anderson, M. J., & Whitcomb, P. J. (2010). *Design of Experiments.* Kirk-Othmer Encyclopedia of Chemical Technology. doi:10.1002/0471238961.0405190908010814.a0

Andre, N. M., Goushegir, S. M., dos Santos, J. F., Canto, L. B., & Amancio-Filho, S. T. (2016). Friction Spot Joining of aluminum alloy 2024-T3 and carbon-fiber-reinforced poly (phenylene sulfide) laminate with additional PPS film interlayer: Microstructure, mechanical strength and failure mechanisms. *Composites. Part B, Engineering, 94*, 197–208. doi:10.1016/j.compositesb.2016.03.011

Andure, M. W., Jirapure, S. C., & Dhamande, L. P. (2012). Advance automobile material for light weight future–a review. In *IJCA Proceedings on International Conference on Benchmarks in Engineering Science and Technology* (pp. 15-22). Academic Press.

Andure, M. W., Jirapure, S. C., & Dhamande, L. P. (2012). Advance automobile material for light weight future—a review. In *IJCA Proceedings on International Conference on Benchmarks in Engineering Science and Technology* (pp. 15-22). Academic Press.

Angelo & Ravisankar. (2018). *Non Ferrous Alloys: Structures, Properties, and Engineering Applications*. Cengage.

Anik, S. (1991). *Gedik Eğitim Vakfı Kaynak Teknolojisi Eğitim Araştırma ve Muayene Enstitüsü*. Gedik Education Foundation.

Annoni, M., & Carboni, M. (n.d.). *Science and Technology of Welding and Joining Ultrasonic metal welding of AA 6022-T4 lap joints: Part I-Technological characterisation and static mechanical behaviour*. doi:10.1179/136217181 0Y.0000000014

Antony, J. (2014). Design of experiments for engineers and scientists. Elsevier.

Anzalone, G. C., Zhang, C., Wijnen, B. A. S., Sanders, P. G., & Pearce, J. M. (2013). A Low-Cost Open-Source Metal 3-D Printer. *IEEE Access: Practical Innovations, Open Solutions*, *1*, 803–810. doi:10.1109/ACCESS.2013.2293018

Apalak, M. K., Davies, R., & Apalak, Z. G. (1995). Analysis and design of adhesively bonded double-containment corner joints. *Journal of Adhesion Science and Technology*, *9*(2), 267–293. doi:10.1163/156856195X01166

Ashby, M. F., Evans, A., Fleck, N. A., Gibson, L. J., Hutchinson, J. W., Wadley, H. N. G., & Delale, F. (2001). Metal foams: A design guide. *Applied Mechanics Reviews*, *54*(6), B105–B106. doi:10.1115/1.1421119

Ashby, M. F., Evans, T., Fleck, N. A., Hutchinson, J. W., Wadley, H. N. G., & Gibson, L. J. (2000). *Metal foams: a design guide*. Elsevier.

Aşık, E. E., & Bor, Ş. (2015). Fatigue behavior of Ti–6Al–4V foams processed by magnesium space holder technique. *Materials Science and Engineering A*, *621*, 157–165. doi:10.1016/j.msea.2014.10.068

Askin, M. Y., & Turen, Y. (2019). The effect of GNP addition on mechanical and residual stress properties of 2024-T3 aluminum and carbon fiber reinforced FML. *Materials Research Express*, *6*(12), 126546. Advance online publication. doi:10.1088/2053-1591/ab575b

Asnafi, N., Langstedt, G., Andersson, C. H., Ostergren, N., & Hakansson, T. (2000). A new lightweight metal-composite-metal panel for applications in the automotive and other industries. *Thin-walled Structures*, *36*(4), 289–310. doi:10.1016/S0263-8231(00)00004-5

Assovskii, I. G., & Merzhanov, A. G. (2013). Validity of experimental and theoretical modeling of combustion of high-energy materials. *Combustion, Explosion, and Shock Waves*, *49*(3), 264–272. doi:10.1134/S0010508213030027

Assunção, E., Quintino, L., & Miranda, R. (2010). Comparative study of laser welding in tailor blanks for the automotive industry. *International Journal of Advanced Manufacturing Technology*, *49*(1-4), 123–131. doi:10.100700170-009-2385-0

Astakhov, V. P. (2011). Machining of hard materials - Definitions and industrial applications. *Machining of Hard Materials*, 1–32. doi:10.1007/978-1-84996-450-0_1

Aswin Kumar, V., Sivaguru, M., Rohini Janaki, B., Sumanth Eswar, K. S., Kiran, P., & Vijayanandh, R. (2021). Structural Optimization of Frame of the Multi-Rotor Unmanned Aerial Vehicle through Computational Structural Analysis. *IOP. Journal of Physics: Conference Series*, *1849*(1), 012004. doi:10.1088/1742-6596/1849/1/012004

Atique, M. S. A., Probha, N. N., & Nafi, A. S. (2014). Polymer composites: a blessing to modern aerospace engineering. *International Conference on Mechanical, Industrial and Energy Engineering*, 1–6. https://www.researchgate.net/publication/336411271

ATL Anlagentechnik Luhden GmbH. (2021). *TEM applications*. https://atl-luhden.de/en/tem-machinery-construction/tem-deburring/tem-applications/

Atwater, M. A., Guevara, L. N., Darling, K. A., & Tschopp, M. A. (2018). Solid state porous metal production: A review of the capabilities, characteristics, and challenges. *Advanced Engineering Materials*, *20*(7), 1700766. doi:10.1002/adem.201700766

Aydin, K., Kaya, Y., & Kahraman, N. (2012). Experimental study of diffusion welding/bonding of titanium to copper. *Materials & Design*, *37*, 356–368. doi:10.1016/j.matdes.2012.01.026

Azmi, A. I. (2012). Multi-objective Optimisation of Machining Fibre Reinforced Composites. *Journal of Applied Sciences*, *12*(23), 2360–2367. doi:10.3923/jas.2012.2360.2367

Babcsán, N., & Banhart, J. (2006). Metal foams: towards high-temperature colloid chemistry. *Colloidal Particles at Liquid Interfaces*, 445-499.

Babcsán, N., Banhart, J., & Leitlmeier, D. (2003). Metal foams–manufacture and physics of foaming. In *Proceedings of the International Conference Advanced Metallic Materials* (pp. 5-15). Academic Press.

Babcsán, N., Mészáros, I., & Hegman, N. (2003). Thermal and electrical conductivity measurements on aluminum foams. *Materialwissenschaft und Werkstofftechnik*, *34*(4), 391–394. doi:10.1002/mawe.200390081

Baburaja, K., VenkataSubbaiah, K., & Kalluri, R. (2016). Hybrid materials of aluminium. *Materials Today: Proceedings*, *3*, 4140–4145. doi:10.1016/j.matpr.2016.11.087

Bagheri, B., Abbasi, M., & Hamzeloo, R. (2021). Comparison of different welding methods on mechanical properties and formability behaviors of tailor welded blanks (TWB) made from AA6061 alloys. Proc. of IMechE, Part C. *Journal of Mechanical Engineering Science*, *235*(12), 2225–2237. doi:10.1177/0954406220952504

Bag, S., & Akinlabi, E. T. (2020). Eco Friendly Aspects in Hybridization of Friction Stir Welding Technology for Dissimilar Metallic Materials. In *Encyclopedia of Renewable and Sustainable Materials* (pp. 225–236). Elsevier. doi:10.1016/B978-0-12-803581-8.11153-1

Bahl, S., Hu, X., Sisco, K., Haynes, J. A., & Shyam, A. (2020). Influence of copper content on the high temperature tensile and low cycle fatigue behavior of cast Al-Cu-Mn-Zr alloys. *International Journal of Fatigue*, *140*, 105836. doi:10.1016/j.ijfatigue.2020.105836

Bahl, S., Singh, T., Kumar, V., Sehgal, S., & Bagha, A. K. (2021). A systematic review on recent progress in advanced joining techniques of the lightweight materials. *AIMS Materials Science*, *8*(1), 62–81. doi:10.3934/matersci.2021005

Bahnini, I., Rivette, M., Rechia, A., Siadat, A., & Elmesbahi, A. (2018). Additive Manufacturing Technology: The Status, Applications, and Prospects. *International Journal of Advanced Manufacturing Technology*, *97*(1-4), 147–161. doi:10.100700170-018-1932-y

Baird, R. (2017). *Thermal deburring of plastic components with infra-red – a new technology*. https://www.linkedin.com/pulse/thermal-deburring-plastic-components-infra-red-new-technology-baird

Balaji, G., & Annamalai, K. (2018). Crushing response of square aluminium column filled with carbon fibre tubes and aluminium honeycomb. *Thin-walled Structures*, *132*, 667–681. doi:10.1016/j.tws.2018.07.037

Balle, F., & Eifler, D. (2012). Statistical test planning for ultrasonic welding of dissimilar materials using the example of aluminum-carbon fiber reinforced polymers (CFRP) joints. *Materialwissenschaft und Werkstofftechnik*, *43*(4), 286–292. doi:10.1002/mawe.201200943

Balle, F., Emrich, S., Wagner, G., Eifler, D., Brodyanski, A., & Kopnarski, M. (2013). Improvement of Ultrasonically Welded Aluminum/Carbon Fiber Reinforced Polymer-Joints by Surface Technology and High Resolution Analysis. *Advanced Engineering Materials*, *15*(9), 814–820. doi:10.1002/adem.201200282

Balle, F., Huxhold, S., Wagner, G., & Eifler, D. (2011). Damage monitoring of ultrasonically welded aluminum/CFRP-joints by electrical resistance measurements. *Procedia Engineering*, *10*, 433–438. doi:10.1016/j.proeng.2011.04.074

Balle, F., Wagner, G., & Eifler, D. (2007). Ultrasonic spot welding of aluminum sheet/carbon fiber reinforced polymer – joints. *Materialwissenschaft und Werkstofftechnik*, *38*(11), 934–938. doi:10.1002/mawe.200700212

Balle, F., Wagner, G., & Eifler, D. (2009). Ultrasonic Metal Welding of Aluminium Sheets to Carbon Fibre Reinforced Thermoplastic Composites. *Advanced Engineering Materials*, *11*(1-2), 35–39. doi:10.1002/adem.200800271

Balzarotti, R., Beretta, A., Groppi, G., & Tronconi, E. (2019). A comparison between washcoated and packed copper foams for the intensification of methane steam reforming. *Reaction Chemistry & Engineering*, *4*(8), 1387–1392. doi:10.1039/C9RE00125E

Banciu, C., Lungulescu, M., Bara, A., Sbarcea, G., Patroi, D., & Marinescu, V. (2020, February). Graphene grown by chemical vapor deposition on metal foams. In AIP Conference Proceedings (Vol. 2206, No. 1, p. 050001). AIP Publishing LLC. doi:10.1063/5.0000296

Bandyopadhyay, K., Panda, S. K., & Saha, P. (2014). Investigations Into the Influence of Weld Zone on Formability of Fiber Laser-Welded Advanced High Strength Steel. *Journal of Materials Engineering and Performance*, *23*(4), 1465–1479. doi:10.100711665-014-0881-3

Banerjee, D., & Williams, J. C. (2013). Perspectives on Titanium Science and Technology. *Acta Materialia*, *61*(3), 844–879. doi:10.1016/j.actamat.2012.10.043

Banhart, J. (2004). Industrialisation of aluminium foam technology. In *Proceedings of the ninth International Conference on aluminium alloys* (pp. 764-770). Academic Press.

Banhart, J. (2000). Manufacturing routes for metallic foams. *JOM*, *52*(12), 22–27. doi:10.100711837-000-0062-8

Banhart, J. (2001). Manufacture, characterisation and application of cellular metals and metal foams. *Progress in Materials Science*, *46*(6), 559–632. doi:10.1016/S0079-6425(00)00002-5

Banhart, J. (2003). Aluminum foams: On the road to real applications. *MRS Bulletin*, *28*(4), 290–295. doi:10.1557/mrs2003.83

Banhart, J. (2013). Light-metal foams—History of innovation and technological challenges. *Advanced Engineering Materials*, *15*(3), 82–111. doi:10.1002/adem.201200217

Banhart, J. (2018). Production of Metal Foams. *Comprehensive Composite Materials*, *II*, 347–363. doi:10.1016/B978-0-12-803581-8.09976-8

Banhart, J., Garcia-Moreno, F., Heim, K., & Seeliger, H. W. (2019). Light-weighting in transportation and defence using aluminium foam sandwich structures. In *Light Weighting for Defense, Aerospace, and Transportation* (pp. 61–72). Springer. doi:10.1007/978-981-15-1263-6_5

Banhart, J., & Weaire, D. (2002). On the road again: Metal foams find favor. *Physics Today*, *55*(7), 37–42. doi:10.1063/1.1506749

Barnes, T. A., & Pashby, I. R. (2000). Joining techniques for aluminum spaceframes used in automobiles. Part II - adhesive bonding and mechanical fasteners. *Journal of Materials Processing Technology*, *99*(1), 72–79. doi:10.1016/S0924-0136(99)00361-1

Basheer, A. A. (2020). Advances in the Smart Materials Applications in the Aerospace Industries. *Aircraft Engineering and Aerospace Technology*, *92*(7), 1027–1035. doi:10.1108/AEAT-02-2020-0040

Baudana, G., Biamino, S., Ugues, D., Lombardi, M., Fino, P., Pavese, M., & Badini, C. (2016). Titanium aluminides for aerospace and automotive applications processed by Electron Beam Melting: Contribution of Politecnico di Torino. *Metal Powder Report*, *71*(3), 193–199. Advance online publication. doi:10.1016/j.mprp.2016.02.058

Bauer, B., Kralj, S., & Bušić, M. (2013). Production and application of metal foams in casting technology. *Tehnicki Vjesnik/Technical Gazette*, *20*(6).

Bauer, B., Kralj, S., & Bušić, M. (2013). Productıon And Application of Metal Foams In Casting Technology. *Tehnicki Vjesnik/Technical Gazette*, *20*(6).

Bayode, A., Akinlabi, E. T., & Pityana, S. (2018). Fabrication of stainless steel-based FGM by laser metal deposition. In K. Kumar & P. Davim (Eds.), *Hierarchical composite materials: Materials, Manufacturing and Engineering*. Walter de Gruyter. doi:10.1515/9783110545104-004

Baysore, J. K., Williamson, M. S., Adonyi, Y., & Milian, J. L. (1995). Laser beam welding and formability of tailored blanks. *Welding Research Supplement*, 345-351.

Begum, S. A., Rane, A. V., & Kanny, K. (2019). Applications of Compatibilized Polymer Blends in Automobile Industry. In *Compatibilization of Polymer Blends: Micro and Nano Scale Phase Morphologies, Interphase Characterization, and Properties*. Elsevier Inc. doi:10.1016/B978-0-12-816006-0.00020-7

Begum, S., Chen, D. L., Xu, S., & Luo, A. A. (2008). Strain-Controlled Low-Cycle Fatigue Properties of a Newly Developed Extruded Magnesium Alloy. *Metallurgical and Materials Transactions. A, Physical Metallurgy and Materials Science*, *39*(12), 3014–3026. doi:10.100711661-008-9677-0

Beköz, N. (2011). *Microstructural and Mechanical Properties of Iron Based Foamed Metals Obtained by Powder Metallurgy Method* (PhD thesis). Institute of Graduate Studies in Science and Engineering, Istanbul University.

Bekoz, N., & Oktay, E. (2012). Effects of carbamide shape and content on processing and properties of steel foams. *Journal of Materials Processing Technology*, *212*(10), 2109–2116. doi:10.1016/j.jmatprotec.2012.05.015

Bekoz, N., & Oktay, E. (2013). Effect of heat treatment on mechanical properties of low alloy steel foams. *Materials & Design*, *51*, 212–218. doi:10.1016/j.matdes.2013.03.098

Bekoz, N., & Oktay, E. (2013). Mechanical properties of low alloy steel foams: Dependency on porosity and pore size. *Materials Science and Engineering A*, *576*, 82–90. doi:10.1016/j.msea.2013.04.009

Bekoz, N., & Oktay, E. (2014). High temperature mechanical properties of low alloy steel foams produced by powder metallurgy. *Materials & Design*, *53*, 482–489. doi:10.1016/j.matdes.2013.07.050

Bekoz, N., & Oktay, E. (2014). The role of pore wall microstructure and micropores on the mechanical properties of Cu–Ni–Mo based steel foams. *Materials Science and Engineering A*, *612*, 387–397. doi:10.1016/j.msea.2014.06.064

Belov, N. A., Eskin, D. G., & Aksenov, A. A. (2005). Introduction. In *Multicomponent Phase Diagrams* (pp. v–vi). Elsevier. doi:10.1016/B978-008044537-3/50000-7

Benedyk, J. C. (2010). Aluminum alloys for lightweight automotive structures. *Materials. Design and Manufacturing for Lightweight Vehicles*, *3*, 79–113. doi:10.1533/9781845697822.1.79

Benouali, A. H., Froyen, L., Delerue, J. F., & Wevers, M. (2002). Mechanical analysis and microstructural characterisation of metal foams. *Materials Science and Technology*, *18*(5), 489–494. doi:10.1179/026708302225002056

Bhagavathiyappan, S., Balamurugan, M., Rajamanickam, M., Vijayanandh, R., Raj Kumar, G., & Senthil Kumar, M. (2020). Comparative computational impact analysis of multi-layer composite materials. AIP Conference Proceedings, 2270. doi:10.1063/5.0019380

Bhagwan, A. V., Kridli, G. T., & Friedman, P. A. (2002). *Influence of weld characteristics on numerically predicted deformation behavior of aluminum tailor welded blanks* (No. 2002-01-0386). SAE Technical Paper.

Bhandare, R. G., & Sonawane, P. M. (2013). Preparation of aluminium matrix composite by using stir casting method. *International Journal of Engineering and Advanced Technology*, *3*(3), 61–65.

Bhardwaj, N., Narayanan, R. G., Dixit, U. S., & Hashmi, M. S. J. (2019). Recent developments in friction stir welding and resulting industrial practices. *Advances in Materials and Processing Technologies*, *5*(3), 461–496. doi:10.1080/2374068X.2019.1631065

Bhaskar, V. V., Narayanan, R. G., & Narasimhan, K. (2004, June). Effect of thickness ratio on formability of tailor welded blanks (TWB). In. AIP Conference Proceedings: Vol. 712. *No. 1* (pp. 863–868). American Institute of Physics. doi:10.1063/1.1766635

Bhatt, A., Khanna, M., & Pimoli, B. S. (2015). Metal foaming of aluminium alloys. *IOSR Journal of Mechanical and Civil Engineering*, *12*(1), 40–44.

Bhatt, K., Eisenmenger, M., & Shi, M. (1995). Formability of Mash Seam Welded Blanks: Effects of Welding Set-Up Conditions. *SAE Transactions*, *104*, 889–895. doi:10.4271/950923

Bikas, H., Stavropoulos, P., & Chryssolouris, G. (2016). Additive manufacturing methods and modelling approaches: A critical review. *International Journal of Advanced Manufacturing Technology*, *83*(1-4), 389–405. doi:10.100700170-015-7576-2

Bilym, P., Rogozin, A., Firsov, P., & Zolotov, S. (2021). Carbon fiber plastics for vehicles manufactured by resource-saving formation technology. *IOP Conference Series: Materials Science and Engineering*, *1021*(1), 012034. doi:10.1088/1757-899X/1021/1/012034

Biron, M. (2018). *Thermoplastics and thermoplastic composites*. William Andrew.

Bisht, A., Patel, V. K., & Gangil, B. (2019). Future of metal foam materials in automotive industry. In *Automotive Tribology* (pp. 51–63). Springer. doi:10.1007/978-981-15-0434-1_4

Blaga, L., dos Santos, J. F., Bancila, R., & Amancio-Filho, S. T. (2015). Friction Riveting (FricRiveting) as a new joining technique in GFRP lightweight bridge construction. *Construction & Building Materials*, *80*, 167–179. doi:10.1016/j.conbuildmat.2015.01.001

Bodunrin, M. O., Alaneme, K. K., & Chown, L. H. (2015). Aluminium matrix hybrid composites: A review of reinforcement philosophies; mechanical, corrosion and tribological characteristics. *Journal of Materials Research and Technology*, *4*(4), 434–445. doi:10.1016/j.jmrt.2015.05.003

Bolat, M. (2013). *Machining of Polycarbonate for Optical Applications.* Academic Press.

Bolt, S. (2014). *Ultrasonic Plastic Welding of Carbon Fiber Reinforced Polyamide 6 to Aluminium and Steel An Experimental Study.* Master of Science Thesis.

Bor, Ş., Esen, Z., Kotan, G., & Tarhan Bor, E. (2007). Toz metalurjisi yöntemiyle köpüksü saf titanyum ve Ti6AI4V alaşım üretimi ve karakterizasyonu. *TÜBİTAK,* 104-121.

Borba, N.Z., Blaga, L., Santos, J.F.d., Canto, L. B., & Filho S. T.A. (2014). *Friction Riveting of Pultruded Thermoset Glass Fiber Reinforced Polyester Composite and Ti6al4v Hybrid Joints.* Academic Press.

Borisov, S. V., & Podberezskaya, N. V. (2012). X-ray diffraction analysis: A brief history and achievements of the first century. *Journal of Structural Chemistry, 53*(1), 1–3. doi:10.1134/S0022476612070013

Borovinšek, M., Taherishargh, M., Vesenjak, M., Ren, Z., & Fiedler, T. (2016). Geometrical characterization of perlite-metal syntactic foam. *Materials Characterization, 119*, 209–215. doi:10.1016/j.matchar.2016.07.024

Borysova, O. S. (2011). *Improving methods dosage energy at the finish thermopulse cleaning of precision parts of flying vehicle* [Ph.D. dissertation]. National Aerospace University "Kharkiv Aviation Institute."

Boschetto, A., & Bottini, L. (2015). Surface improvement of fused deposition modeling parts by barrel finishing. *Rapid Prototyping Journal, 21*(6), 686–696. doi:10.1108/RPJ-10-2013-0105

Boschetto, A., Bottini, L., & Veniali, F. (2016). Finishing of fused deposition modeling parts by CNC machining. *Robotics and Computer-integrated Manufacturing, 41*, 92–101. doi:10.1016/j.rcim.2016.03.004

Botelho, E. C., Silva, R. A., Pardini, L. C., & Rezende, M. C. (2006). A Review on the Development and Properties of Continuous Fiber/epoxy/aluminum Hybrid Composites for Aircraft Structures. *Materials Research, 9*(3), 248–256. doi:10.1590/S1516-14392006000300002

Bozhko, V. P., Losev, A. V., Takoriants, M. P., Pleshkov, V. I., Strizhenko, V. E., Levityansky, I. A., Filippov, B. L., & Kononenko, T. I. (1989). *Thermopulse apparatus for deburring parts* (U.S. Patent No. 4 802 654). U.S. Patent and Trademark Office.

Braiek, S., Khalifa, A. B., Zitoune, R., & Zidi, M. (2018). Experimental and numerical investigation of adhesively bonded±55° filament wound tubular specimens under internal pressure. *Engineering Fracture Mechanics, 199*, 461–475. doi:10.1016/j.engfracmech.2018.06.009

Bremen, S., Meiners, W., & Diatlov, A. (2012, April). Selective Laser Melting A manufacturing technology for the future? *Laser-Technik-Journal, 9*(2), 33–38. Advance online publication. doi:10.1002/latj.201290018

Brumm, S., & Bürkner, G. (2015). Gas Metal arc Pulse Welding with Alternating Current for Lightweight Materials. *Materials Today: Proceedings, 2*, S179–S187. doi:10.1016/j.matpr.2015.05.008

Buste, A., Lalbin, X., Worswick, M. J., Clarke, J. A., Altshuller, B., Finn, M., & Jain, M. (1999). Prediction of strain distribution in aluminum tailor welded. *Proceedings of the 4th International conference and workshop on Numerical simulation of 3D sheet forming process, NUMISHEET'99.*

Caiazzo, F., Campanelli, S. L., Cardaropoli, F., Contuzzi, N., Sergi, V., & Ludovico, A. D. (2017). Manufacturing and characterization of similar to foam steel components processed through selective laser melting. *International Journal of Advanced Manufacturing Technology, 92*(5), 2121–2130. doi:10.100700170-017-0311-4

Calignano, F., Manfredi, D., Ambrosio, E. P., Biamino, S., Lombardi, M., Atzeni, E., Salmi, A., Minetola, P., Iuliano, L., & Fino, P. (2017). Overview on Additive Manufacturing Technologies. *Proceedings of the IEEE, 105*(4), 593–612. doi:10.1109/JPROC.2016.2625098

Campbell, F. C. (2012). Introduction and Uses of Lightweight Materials. In Lightweight Materials: Understanding the Basics (p. 4). ASM International. doi:10.31399/asm.tb.lmub.t53550001

Campean, R., & Hancu, L. (2019). Proposal for cryogenic deburring equipment's design and thermal insulation. *MATEC Web of Conferences, 299, 01013.* 10.1051/matecconf/201929901013

Cao, X., Shi, Q., Liu, D., Feng, Z., Liu, Q., & Chen, G. (2018). Fabrication of in situ carbon fiber/aluminum composites via friction stir processing: Evaluation of microstructural, mechanical and tribological behaviors. *Composites. Part B, Engineering, 139*, 97–105. doi:10.1016/j.compositesb.2017.12.001

Čapek, J., Vojtěch, D., & Oborná, A. (2015). Microstructural and mechanical properties of biodegradable iron foam prepared by powder metallurgy. *Materials & Design, 83*, 468–482. doi:10.1016/j.matdes.2015.06.022

Carcel, B., Carcel, A. C., Perez, I., Fernandez, E., Barreda, A., Sampedro, J., & Ramos, J. A. (2009, April). Manufacture of metal foam layers by laser metal deposition. In *XVII International Symposium on Gas Flow, Chemical Lasers, and High-Power Lasers* (Vol. 7131, p. 713123). International Society for Optics and Photonics. 10.1117/12.816702

Carley, L. (2015). *Valve Selection: Hot Valve Materials for Hot Engines*. https://www.enginebuildermag.com/2005/12/valve-selection-hot-valve-materials-for-hot-engines/

Casalino, G., El Mehtedi, M., Forcellese, A., & Simoncini, M. (2018). Effect of Cold Rolling on the Mechanical Properties and Formability of FSWed Sheets in AA5754-H114. *Metals, 8*(4), 223. doi:10.3390/met8040223

Castricum, H. L., Yang, H., Bakker, H., & Van Deursen, J. H. (1997). A study of milling of pure polymers and a structural transformation of polyethylene. *Materials Science Forum, 235–238*(1), 211–216. doi:10.4028/www.scientific.net/msf.235-238.211

Cavaliere, P., Nobile, R., Panella, F. W., & Squillace, A. (2006). Mechanical and microstructural behaviour of 2024–7075 aluminium alloy sheets joined by friction stir welding. *International Journal of Machine Tools & Manufacture, 46*(6), 588–594. doi:10.1016/j.ijmachtools.2005.07.010

Çavdar, K., & Durmuş, A. (2018). Effects of pressure and time on radial riveting process. *Journal of the Faculty of Engineering and Architecture of Gazi University, 33*(1), 312–322. doi:10.17341/gazimmfd.406803

Cavuslu, F., Korkmaz, K., & Usta, M. (2019). Mechanical properties of aluminum foams ceramic coated through plasma electrolytic oxidation. *Kovove Mater., 57*(02), 105–111. doi:10.4149/km_2019_1_105

Chabot, K. A., & Brescia, J. A. (1993). Evaluation of primers for adhesively-bonded aircraft repair. *Journal of Adhesion Science and Technology, 7*(11), 1183–1194. doi:10.1163/156856193X00042

Chai, Y., Li, R. W., Perriman, D. M., Chen, S., Qin, Q. H., & Smith, P. N. (2018). Laser polishing of thermoplastics fabricated using fused deposition modelling. *International Journal of Advanced Manufacturing Technology, 96*(9), 4295–4302. doi:10.100700170-018-1901-5

Chandran, K. S. R. (2005). Duality of fatigue failures of materials caused by Poisson defect statistics of competing failure modes. *Nature Materials, 4*(4), 303–308. doi:10.1038/nmat1351

Changdar, A., & Chakraborty, S. S. (2021). Laser processing of metal foam-A review. *Journal of Manufacturing Processes, 61*, 208–225. doi:10.1016/j.jmapro.2020.10.012

Chang, S. Y., Lee, K. S., Lee, S. H., Hong, S. K., Park, K. T., & Shin, D. H. (2003). Effect of Al Content and Pressing Temperature on ECAP of Cast Mg Alloys. *Materials Science Forum, 419–422*, 491–496. . doi:10.4028/www.scientific.net/MSF.419-422.491

Chan, L. C., Cheng, C. H., Chan, S. M., Lee, T. C., & Chow, C. L. (2005). Formability analysis of tailor-welded blanks of different thickness ratios. *Journal of Manufacturing Science and Engineering*, *127*(4), 743–751. doi:10.1115/1.2034518

Chan, S. M., Chan, L. C., & Lee, T. C. (2003). Tailor-welded blanks of different thickness ratios effects on forming limit diagrams. *Journal of Materials Processing Technology*, *132*(1-3), 95–101. doi:10.1016/S0924-0136(02)00407-7

Chatterjee, P., & Chakraborty, S. (2012). Materials selection using COPRAS and COPRAS-G methods. *International Journal of Materials and Structural Integrity*, *6*(2/3/4), 111-133. doi:. doi:10.1504/ijmsi.2012.049951

Chatterjee, P., Athawale, V. M., & Chakraborty, S. (2011). Materials selection using complex propoptional assessment and evaluation of mixed data methods. *Materials & Design*, *32*(2), 851–860. https://doi.org/10.1016/j.matdes.2010.07.010

Chatterjee, P., & Chakraborty, S. (2013). Gear Material Selection Using Complex Proportional Assessment and Additive Ratio Assessment-based Approaches: A Comparative Study. *International Journal of Materials Science and Engineering*, *1*(2), 104–111.

Chen, Chen, Pei, Zhang, & Fang. (2012). Optimal Design of Sandwich Beams With Lightweight Cores in Three-Point Bending. *International Journal of Applied Mechanics*, *4*(3). . doi:10.1142/S1758825112500330

Chen, M.-A., Li, H.-Z., & Zhang, X.-M. (2007). *Improvement of shear strength of aluminium polypropylene lap joints by grafting maleic anhydride onto polypropylene*. Academic Press.

Cheng, C. H., Chan, L. C., Chow, C. L., & Lee, T. C. (2005). Experimental investigation on the weldability and forming behavior of aluminum alloy tailor-welded blanks. *Journal of Laser Applications*, *17*(2), 81–88. doi:10.2351/1.1848521

Cheng, C. H., Chan, L. C., Tang, C. Y., & Chow, C. L. (2005). Determination of true stress-strain curve for the weldment of aluminum laser-welded blanks. *Journal of Laser Applications*, *17*(3), 159–170. doi:10.2351/1.1961652

Cheng, D., & Li, L. (2020). An overview of laser-based multiple metallic material additive manufacturing: From macro- to micro-scales. *International Journal of Extreme Manufacturing*.

Chen, H. B., Yan, K., Lin, T., Chen, S. B., Jiang, C. Y., & Zhao, Y. (2006). The investigation of typical welding defects for 5456 aluminum alloy friction stir welds. *Materials Science and Engineering A*, *433*(1-2), 64–69. doi:10.1016/j.msea.2006.06.056

Chen, H., Han, Q., Wang, C., Liu, Y., Chen, B., & Wang, J. (2020). Porous Scaffold Design for Additive Manufacturing in Orthopedics: A Review. *Frontiers in Bioengineering and Biotechnology*, *8*, 609. doi:10.3389/fbioe.2020.00609 PMID:32626698

Chen, M.-A., Li, H.-Z., & Zhang, X.-M. (2007, April). Improvement of shear strength of aluminium-polypropylene lap joints by grafting maleic anhydride onto polypropylene. *International Journal of Adhesion and Adhesives*, *27*(3), 175–187. doi:10.1016/j.ijadhadh.2006.01.008

Chen, P. C., Pan, C. W., Lee, W. C., & Li, K. M. (2014). Optimization of micromilling microchannels on a polycarbonate substrate. *International Journal of Precision Engineering and Manufacturing*, *15*(1), 149–154. doi:10.100712541-013-0318-1

Chen, Y., Jin, K., Li, H., Lin, Y., Lu, Y., Hua, X., Yuan, Q., Wang, H., & Tao, J. (2018). Effect of peer ply and cooling rate on the tensile properties of Al/Gf/PP laminate prepared by hot pressing. *The Journal of Sandwich Structures & Materials*, *0*(0), 1–15. doi:10.1177/1099636218802473

Chen, Y., Yang, X., Li, M., & Mei, M. (2019). Influence of working temperatures on mechanical behavior of hybrid joints with carbon fiber reinforced plastic/aluminum lightweight materials for automotive structure. *Journal of Manufacturing Processes*, *45*, 392–407. doi:10.1016/j.jmapro.2019.07.022

Cherevko, S., & Chung, C. H. (2011). Direct electrodeposition of nanoporous gold with controlled multimodal pore size distribution. *Electrochemistry Communications*, *13*(1), 16–19. doi:10.1016/j.elecom.2010.11.001

Cheung, S., Gauthier, M., Lefebvre, L. P., Dunbar, M., & Filiaggi, M. (2007). Fibroblastic interactions with high-porosity Ti-6Al-4V metal foam. *Journal of Biomedical Materials Research Part B: Applied Biomaterials*, *82*(2), 440–449. doi:10.1002/jbm.b.30749 PMID:17245747

Cho, H.-Y., & Kim, D. (2015). Finite element analysis for joining glass fiber reinforced plastic and aluminium alloy sheets. *Journal of Welding and Joining*, *33*(2), 78–84. doi:10.5781/JWJ.2015.33.2.78

Choi, D. H., Lee, C. Y., Ahn, B. W., Choi, J. H., Yeon, Y. M., Song, K., Hong, S. G., Lee, W. B., Kang, K. B., & Jung, S. B. (2011). Hybrid Friction Stir Welding of High-carbon Steel. *Journal of Materials Science and Technology*, *27*(2), 127–130. doi:10.1016/S1005-0302(11)60037-6

Choi, Y., Heo, Y., Kim, H. Y., & Seo, D. (2000). Investigations of weld-line movements for the deep drawing process of tailor welded blanks. *Journal of Materials Processing Technology*, *108*(1), 1–7. doi:10.1016/S0924-0136(00)00536-7

Cho, J. H., Boyce, D. E., & Dawson, P. R. (2005). Modeling strain hardening and texture evolution in friction stir welding of stainless steel. *Materials Science and Engineering A*, *398*(1-2), 146–163. doi:10.1016/j.msea.2005.03.002

Cho, J. H., Kang, S. H., Han, H. N., & Oh, K. H. (2008). Modeling friction stir welding process of aluminum alloys. *Metals and Materials International*, *14*(2), 247–258. doi:10.3365/met.mat.2008.04.247

Chozha Rajan, B. M., Kumar, A. S., & Sornakumar, T. (2018). Impact Response and Damage Characteristics of Carbon Fibre Reinforced Aluminium Laminates (CARAL) under Low Velocity Impact Tests. *Materials Today: Proceedings*, *5*(9), 20070–20077. doi:10.1016/j.matpr.2018.06.373

Christke, S., Gibson, A. G., Grigoriou, K., & Mouritz, A. P. (2016). Multi-layer polymer metal laminates for the fire protection of lightweight Structures. *Materials & Design*, *97*, 349–356. doi:10.1016/j.matdes.2016.02.105

Civjan, S. A., Guihan, T., & Peterman, K. (2020). Testing of oxyacetylene weld strength. *Journal of Constructional Steel Research*, *168*, 105921. Advance online publication. doi:10.1016/j.jcsr.2019.105921

Claar, T. D., Yu, C. J., Hall, I., Banhart, J., Baumeister, J., & Seeliger, W. (2000). Ultra-lightweight aluminum foam materials for automotive applications. *SAE Transactions*, 98–106. doi:10.4271/2000-01-0335

Clausen, J., Kelch, M., Wöstmann, F. J., & Busse, M. (2018). Mechanical characterization of integral aluminum-FRP-structures produced by high pressure die-casting. *Production Engineering*, *12*(2), 269–278. doi:10.100711740-018-0811-6

Cole, G. S., & Sherman, A. M. (1995). Light weight materials for automotive applications. *Materials Characterization*, *35*(1), 3–9. doi:10.1016/1044-5803(95)00063-1

Compston, P., Cantwell, W. J., Jones, C., & Jones, N. (2001). Impact Perforation Resistance and Fracture Mechanisms of A Thermoplastic Based Fiber-Metal Laminate. *Journal of Materials Science Letters*, *20*(7), 597–599. doi:10.1023/A:1010904930497

Conradie, P. J., Oosthuizen, G., Dimitrov, D., & Saxer, M. (2015). Effect of Milling Strategy and Tool Geometry on Machining Cost when Cutting Titanium Alloys. *South African Journal of Industrial Engineering*, *26*(3), 137–151. doi:10.7166/26-3-1172

Contorno, D., Filice, L., Fratini, L., & Micari, F. (2006). Forming of aluminum foam sandwich panels: Numerical simulations and experimental tests. *Journal of Materials Processing Technology*, *177*(1–3), 364–367. doi:10.1016/j.jmatprotec.2006.04.028

Cox, M. E., & Dunand, D. C. (2011). Bulk gold with hierarchical macro-, micro-and nano-porosity. *Materials Science and Engineering A*, *528*(6), 2401–2406. doi:10.1016/j.msea.2010.11.072

Crane, N. B., Tuckerman, J., & Nielson, G. N. (2011). Self-Assembly in Additive Manufacturing: Opportunities and Obstacles. *Rapid Prototyping Journal*, *17*(3), 211–217. doi:10.1108/13552541111124798

Crispo, L., & Kim, I. Y. (2020). *Assembly Level Topology Optimization towards a Part Consolidation Algorithm for Additive Manufacturing*. AIAA SciTech 2020 Forum, Orlando, FL, United States. 10.2514/6.2020-0893

Crupi, V., Kara, E., Epasto, G., Guglielmino, E., & Aykul, H. (2014). Prediction model for the impact response of glass fiber reinforced aluminium foam sandwiches. *International Journal of Impact Engineering*.

Cui, Z., Liu, Q., Sun, Y., & Li, Q. (2020). On crushing responses of filament winding CFRP/aluminum and GFRP/CFRP/aluminum hybrid structures. *Composites. Part B, Engineering*, *200*, 108341. doi:10.1016/j.compositesb.2020.108341

D'Urso, G., & Maccarini, G. (2012). The formability of aluminum foam sandwich panels. *International Journal of Material Forming*, *5*(3), 243–257. doi:10.100712289-011-1036-9

Dahil, L., Katirci, R., & Sümbül, H. İ. (2020). Effect of Artificial Aging Process on Aluminum Foam Made of Etial 160. *Transactions of the Indian Institute of Metals*, *73*(11), 2739–2745. doi:10.100712666-020-02081-w

Dani, I., Drossel, W. G., Milaev, N., Korn, H., Hannemann, C., Hohlfeld, J., & Wertheim, R. (2020). Sustainability of industrial components using additive manufacturing and foam materials. *Procedia Manufacturing*, *43*, 10–17. doi:10.1016/j.promfg.2020.02.102

Daniyan, I. A., Tlhabadira, I., Daramola, O. O., & Mpofu, K. (2019). Design and optimization of machining parameters for effective AISI P20 removal rate during milling operation. *Procedia CIRP*, *84*, 861–867. doi:10.1016/j.procir.2019.04.301

Das, B., Pal, S., & Bag, S. (2017). Torque based defect detection and weld quality modelling in friction stir welding process. *Journal of Manufacturing Processes*, *27*, 8–17. doi:10.1016/j.jmapro.2017.03.012

Das, R. S., & Agrawal, Y. K. (2011). Raman spectroscopy: Recent advancements, techniques and applications. *Vibrational Spectroscopy*, *57*(2), 163–176. doi:10.1016/j.vibspec.2011.08.003

Das, S. (2000). Aluminium tailor welded blanks. *Advanced Materials & Processes*, 41–49.

Das, S. (2010). Modeling mixed microstructures using a multi-level cellular automata finite element framework. *Computational Materials Science*, *47*(3), 705–711. doi:10.1016/j.commatsci.2009.10.012

Das, S., Palmiere, E. J., & Howard, I. C. (2003). CAFE: a potential tool for modeling thermomechanical processes. *Proceedings of International Conference on Thermomechanical Processing: Mechanics, Microstructure and Control.*

Davies, C. H. J. (1995). The effect of neighbourhood on the kinetics of a cellular automaton recrystallisation model. *Scripta Metallurgica et Materialia*, *33*(7), 1139–1143. doi:10.1016/0956-716X(95)00335-S

Dayyani, I., Shaw, A. D., Saavedra Flores, E. I., & Friswell, M. I. (2015). The Mechanics of Composite Corrugated Structures: A Review with Applications in Morphing Aircraft. *Composite Structures*, *133*(1), 358–380. doi:10.1016/j.compstruct.2015.07.099

de Assis, S. L., Wolynec, S., & Costa, I. (2006). Corrosion characterization of titanium alloys by electrochemical techniques. *Electrochimica Acta*, *51*(8–9), 1815–1819. doi:10.1016/j.electacta.2005.02.121

De Freitas, W. J., Takeya, G. S., Casteletti, L. C., Neto, A. L., & Totten, G. E. (2015). *Influence of sintering atmosphere in the hardness and corrosion resistance of 17-4 PH stainless steel shaped from powder injection molding process.* Academic Press.

DeGeare, J. (2015). Milling Operations. *The Guide to Oilwell Fishing Operations*, 91–101. doi:10.1016/B978-0-12-420004-3.00012-5

Degischer, H. P., Körner, C., Singer, R. F., Banhart, J., Baumgärtner, F., Rausch, G., Arnold, M., Thies, M., San Marchi, C., Mortensen, A., Andersen, O., & Stephani, G. (2002). Material definitions, processing, and recycling. Handbook of Cellular Metals: Production, Processing, Applications, 5-70.

Degischer, H.-P., & Kriszt, B. (2010). *Handbook of cellular metals: Production, processing, applications.* Wiley-InterScience.

Devi, G. R., & Palanikumar, K. (2015). Tensile Property Evaluation of Woven Glass Fiber Reinforced Plastic and Aluminium Stack. *Applied Mechanics and Materials, 766-767*, 44–49. doi:10.4028/www.scientific.net/AMM.766-767.44

Dhakad, M. R., Mahajan, K. A., & Mitra, A. C. (2017). Experimental Analysis and Optimization of Cutting Parameters for the Surface Roughness in the Facing Operation of PMMA Material. *IOSR Journal of Mechanical and Civil Engineering, 17*(01), 52–60. doi:10.9790/1684-17010015260

Dhaliwal, G. S., & Newaz, G. M. (2016). Experimental and numerical investigation of flexural behavior of carbon fiber reinforced aluminum laminates. *Journal of Reinforced Plastics and Composites, 35*(12), 945–956. doi:10.1177/0731684416632606

Dhaliwal, G. S., & Newaz, G. M. (2016). Modeling Low Velocity Impact Response of Carbon Fiber Reinforced Aluminum Laminates (CARALL). *Journal of Dynamic Behavior of Materials, 2*(2), 181–193. doi:10.100740870-016-0057-3

Dhaliwal, G. S., & Newaz, G. M. (2020). Experimental and numerical investigation of flexural behavior of hat sectioned aluminum/carbon fiber reinforced mixed material composite beam. *Composites. Part B, Engineering, 182*, 107642. doi:10.1016/j.compositesb.2019.107642

Dharmasena, K. P., & Wadley, H. N. G. (2002). Electrical conductivity of open-cell metal foams. *Journal of Materials Research, 17*(3), 625–631. doi:10.1557/JMR.2002.0089

Di Franco, G., Fratini, L., Pasta, A., & Ruisi, V. F. (2013). On the self-piercing riveting of aluminium blanks and carbon fibre composite panels. *International Journal of Material Forming, 6*(1), 137–144. doi:10.100712289-011-1067-2

Di Gianfrancesco, A. (2017). Technologies for chemical analyses, microstructural and inspection investigations. In *Materials for ultra-supercritical and advanced ultra-supercritical power plants* (pp. 197–245). Woodhead Publishing. doi:10.1016/B978-0-08-100552-1.00008-7

Dialami, N., Chiumenti, M., Cervera, M., Segatori, A., & Osikowicz, W. (2017). Enhanced friction model for Friction Stir Welding (FSW) analysis: Simulation and experimental validation. *International Journal of Mechanical Sciences, 133*, 555–567. doi:10.1016/j.ijmecsci.2017.09.022

Dieter, G. E. (1986). *Mechanical metallurgy.* McGraw-Hill.

Dilberoglu, U. M., Gharehpapagh, B., Yaman, U., & Dolen, M. (2017). The role of additive manufacturing in the era of Industry 4. 0. *Procedia Manufacturing, 11*(June), 545–554. doi:10.1016/j.promfg.2017.07.148

Dilip, J. J. S., Janaki Ram, G. D., & Stucker, B. E. (2012). Additive manufacturing with friction welding and friction deposition processes. *International Journal of Rapid Manufacturing, 3*(1), 56–69. doi:10.1504/IJRAPIDM.2012.046574

Dinda, G. P., Dasgupta, A. K., & Mazumder, J. (2009). Laser aided direct metal deposition of iconel 625 superalloy: Microstructural evolution and thermal stability. *Materials Science and Engineering A*, *509*(1), 98–104. doi:10.1016/j.msea.2009.01.009

Dogra, M., Sharma, V. S., & Dureja, J. (2011). Effect of tool geometry variation on finish turning –. *RE:view*, *4*(1), 1–13.

Dong, Y., Shang, W., Yang, J., Zhang, L., Zhang, W., Li, Z., Guo, L., Zhan, X., Du, H., Deng, B., & Pu, Y. (2013). The impact of low-Z impurities on x-ray conversion efficiency from laser-produced plasmas of low-density gold foam targets. *Physics of Plasmas*, *20*(12), 123305. doi:10.1063/1.4859215

Dorward, R. C., & Pritchett, T. R. (1988). Advanced Aluminium Alloys for Aircraft and Aerospace Applications. *Materials & Design*, *9*(2), 63–69. doi:10.1016/0261-3069(88)90076-3

Drégelyi-Kiss, A., & Horváth, A. (2018). Investigations on the accuracy of additive and conventional manufacturing. *IOP Conference Series: Materials Science and Engineering*, *393*(1), 012098. doi:10.1088/1757-899X/393/1/012098

Dry, D., Hughes, D., & Owen, R. (2001). Methods of assessing influence of weld properties on formability of laser welded tailored blanks. *Ironmaking & Steelmaking*, *28*(2), 89–95. doi:10.1179/030192301677984

du Plessis, A., Broeckhoven, C., Yadroitsava, I., Yadroitsev, I., Hands, C. H., Kunju, R., & Bhate, D. (2019). Beautiful and functional: A review of biomimetic design in additive manufacturing. *Additive Manufacturing*, *27*, 408–427. doi:10.1016/j.addma.2019.03.033

Duarte, I., & Banhart, J. (2000). A study of aluminium foam formation—Kinetics and microstructure. *Acta Materialia*, *48*(9), 2349–2362. doi:10.1016/S1359-6454(00)00020-3

Dukhan, N. (Ed.). (2013). Metal foams: fundamentals and applications. DEStech Publications, Inc.

Durkovic, M., Mladenovic, G., Tanovic, L., & Danon, G. (2018). Impact of feed rate, milling depth and tool rake angle in peripheral milling of oak wood on the cutting force. *Maderas. Ciencia y Tecnología*, *20*(1), 25–34. doi:10.4067/S0718-221X2018005001301

Dursun, T., & Soutis, C. (2014). Recent developments in advanced aircraft aluminium alloys. *Materials & Design*, *56*, 862–871. doi:10.1016/j.matdes.2013.12.002

DutrowB. L. (2020). https://serc.carleton.edu/research_education/geochemsheets/techniques/XRD.html

Dwibedi, S., Jain, N. K., & Pathak, S. (2018). Investigations on joining of stainless steel tailored blanks by μ-PTA process. *Materials and Manufacturing Processes*, *33*(16), 1851–1863. doi:10.1080/10426914.2018.1476766

Dyga, R., & Płaczek, M. (2015). Heat transfer through metal foam–fluid system. *Experimental Thermal and Fluid Science*, *65*, 1–12. doi:10.1016/j.expthermflusci.2015.02.021

Eisenmenger, M., Bhatt, K. K., & Shi, M. F. (1995). Influence of laser welding parameters on formability and robustness of blank manufacturing: An application to a body side frame. *SAE Transactions*, 877–888. doi:10.4271/950922

Ekşi, S., & Genel, K. (2015). Three Point Bending Behavior of Woven Glass, Aramid and Carbon Fiber Reinforced Hybrid Composite Tube. *Acta Physica Polonica A*, *128*(2), 59–60. doi:10.12693/APhysPolA.128.B-59

Elsheikh, A. H., Deng, W., & Showaib, E. A. (2019). Improving laser cutting quality of polymethylmethacrylate sheet: Experimental investigation and optimization. *Journal of Materials Research and Technology*, *x*, 1–15. doi:10.1016/j.jmrt.2019.11.059

Esen, Z., & Bor, Ş. (2007). Processing of titanium foams using magnesium spacer particles. *Scripta Materialia*, *56*(5), 341–344. doi:10.1016/j.scriptamat.2006.11.010

Esen, Z., & Bor, Ş. (2011). Characterization of Ti–6Al–4V alloy foams synthesized by space holder technique. *Materials Science and Engineering A*, *528*(7-8), 3200–3209. doi:10.1016/j.msea.2011.01.008

Espach, A., & Gupta, K. (2020). Mechanical performance in fused deposition modeling manufactured parts – an additive manufacturing review. In *Proceedings of the International Conference on Industrial Engineering and Operations Management* (pp. 3201–3209). IEOM Society International. http://www.ieomsociety.org/detroit2020/papers/640.pdf

Esteves, J. V., Goushegir, S. M., dos Santos, J. F., Canto, L. B., Hage, E. Jr, & Amancio-Filho, S. T. (2015). Friction spot joining of aluminum AA6181-T4 and carbon fiber-reinforced poly(phenylene sulfide): Effects of process parameters on the microstructure and mechanical strength. *Materials & Design*, *66*, 437–445. doi:10.1016/j.matdes.2014.06.070

Ezugwu, E. O., & Wang, Z. M. (1997). Titanium alloys and their machinability—A review. *Journal of Materials Processing Technology*, *68*(3), 262–274. doi:10.1016/S0924-0136(96)00030-1

Fadzil, M., Abdullah, A. B., Samad, Z., Yusof, F., & Manurung, Y. H. P. (2021). Application of lightweight materials toward design for sustainability in automotive component development. In *Design for Sustainability* (pp. 435–463). Elsevier. doi:10.1016/B978-0-12-819482-9.00006-X

Farhadi, S. (2018). *Dynamic Characterization and Modelling of Metallic Foam Material* (Doctoral dissertation). Politecnico di Torino.

Feng, M., Li, Y., Zhao, C., & Luo, Z. (2016). Mechanical properties and interface morphology of Mg/Al ultrasonic spot weld bonding welds. *Science and Technology of Welding and Joining*, *21*(8), 688–699. doi:10.1080/13621718.2016.1144262

Fenjan, R. M., Ahmed, R. A., Alasadi, A. A., & Faleh, N. M. (2019). Nonlocal strain gradient thermal vibration analysis of double-coupled metal foam plate system with uniform and non-uniform porosities. *Coupled Syst Mech*, *8*(3), 247–257.

Fergus, J. W., Mishra, B., Anderson, D., Sarver, E. A., & Neelameggham, N. R. (Eds.). (2015). *Engineering solutions for sustainability : Materials and resources II : John Wiley & Sons*. Retrieved August 25, 2021, from https://books.google.com/books/about/Engineering_Solutions_for_Sustainability.html?hl=tr&id=XV0avgAACAAJ

Florek, R., Simancík, F., Nosko, M., & Harnuskova, J. (2010). Compression test evaluation method for aluminium foam parts of different alloys and densities. *Powder Metallurgy Progress*, *10*(4), 207–212.

Franciska, P. L., Erryani, A., Annur, D., & Kartika, I. (2017, May). Corrosion Behavior of Magnesium Based Foam Structure in Hank's Solution. In *IOP Conf. Ser. Mater. Sci. Eng.* (Vol. 202, p. 012035). 10.1088/1757-899X/202/1/012035

Franco, G. D., Fratinib, L., Pasta, A., & Ruisi, V. F. (2010). On the self-piercing riveting of aluminium blanks and carbon fibre composite panels. *Journal Materials Form*, *3*(1), 1035–1038. doi:10.100712289-010-0947-1

Franco, G.-D., Fratini, L., & Pasta, A. (2012). Influence of the distance between rivets in self-piercing riveting bonded joints made of carbon fiber panels and AA2024 blanks. *Materials & Design*, *35*, 342–349. doi:10.1016/j.matdes.2011.09.036

Frazier, C., Lamnaouer, M., Divo, E., Kassab, A., & Petersen, E. (2011). Effect of wall heat transfer on shock-tube test temperature at long times. *Shock Waves*, *21*(1), 1–17. doi:10.100700193-010-0282-y

Frazier, W. E. (2014). Metal additive manufacturing: A review. *Journal of Materials Engineering and Performance*, *23*(6), 1917–1928. doi:10.100711665-014-0958-z

Friedman, P. A., & Kridli, G. T. (2000). Microstructural and mechanical investigation of aluminum tailor-welded blanks. *Journal of Materials Engineering and Performance*, *9*(5), 541–551. doi:10.1361/105994900770345674

Fritz, A., Sekol, L., Koroskenyi, J., Walch, B., Minear, J., Fernandez, V., & Liu, L. (2012). Experimental analysis of thermal energy deburring process by design of experiment. In *Proceedings of the ASME 2012 International Mechanical Engineering Congress and Exposition* (Vol. 3, pp. 2035–2041). ASME. 10.1115/IMECE2012-88411

Froes, F. H., & Dutta, B. (2014). The Additive Manufacturing (AM) of Titanium Alloys. *Advanced Materials Research, 1019*, 19–25. . doi:10.4028/www.scientific.net/AMR.1019.19

Furumoto, T., Koizumi, A., Alkahari, M. R., Anayama, R., Hosokawa, A., Tanaka, R., & Ueda, T. (2015). Permeability and strength of a porous metal structure fabricated by additive manufacturing. *Journal of Materials Processing Technology, 219*, 10–16. doi:10.1016/j.jmatprotec.2014.11.043

Gábora, A., & Mankovits, T. (2019). Quality control of closed-cell metal foam produced by direct foaming. *IOP Conference Series. Materials Science and Engineering, 659*(1), 012037. doi:10.1088/1757-899X/659/1/012037

Galliková, J., Grenčík, J., Barta, D., & Barlok, M. (2020). FMECA analysis of thermal deburring machine EXTRUDE HONE TEM P-350. *Scientific Journals of the Maritime University of Szczecin, 63*, 9–16. doi:10.17402/434

Galyshev, S., Gomzin, A., Gallyamova, R., Khodos, I., & Musin, F. (2019). On the liquid-phase technology of carbon fiber/aluminum matrix composites. *International Journal of Minerals Metallurgy and Materials, 26*(12), 1578–1584. doi:10.100712613-019-1877-7

Gandin, C. A., & Rappaz, M. (1994). A coupled finite element-cellular automaton model for the prediction of dendritic grain structures in solidification processes. *Acta Metallurgica et Materialia, 42*(7), 2233–2246. doi:10.1016/0956-7151(94)90302-6

Gao, X., Zhou, J., Du, R., Xie, Z., Deng, S., Liu, R., Liu, Z., & Zhang, J. (2016). Robust superhydrophobic foam: A graphdiyne-based hierarchical architecture for oil/water separation. *Advanced Materials, 28*(1), 168–173. doi:10.1002/adma.201504407 PMID:26551876

García-Moreno, F. (2016). Commercial applications of metal foams: Their properties and production. *Materials (Basel), 9*(2), 85. doi:10.3390/ma9020085 PMID:28787887

Gardan, J. (2018). Smart Materials in Additive Manufacturing: State of the Art and Trends. *Virtual and Physical Prototyping, 14*(1), 1–18. doi:10.1080/17452759.2018.1518016

Garg, P., Jamwal, A., Kumar, D., Sadasivuni, K. K., Hussain, C. M., & Gupta, P. (2019). Advance research progresses in aluminium matrix composites: Manufacturing & applications. *Journal of Materials Research and Technology, 8*(5), 4924–4939. doi:10.1016/j.jmrt.2019.06.028

Garre, P., & Rohan, G. (2018). Design and analysis of Al6063 T6 and glass fiber reinforced composite material. *Materials Science and Engineering, 455*, 012011–012015.

Gasper, A. N. D., Catchpole-Smith, S., & Clare, A. T. (2017). In-situ synthesis of titanium aluminides by direct metal deposition. *Journal of Materials Processing Technology, 239*, 230–239. doi:10.1016/j.jmatprotec.2016.08.031

Gaya, A., Lefebvrec, F., Bergamoa, S., Valiorgueb, F., Chalandond, P., Michela, P., & Bertrand, P. (2015). Fatigue of aluminum/glass fiber reinforced polymer composite assembly joined by self-piercing riveting. *Procedia Engineering, 133*, 501–507. doi:10.1016/j.proeng.2015.12.620

Geen, H. C., & Rice, E. E. (1969). *Process for treating articles of manufacture to eliminate superfluous projections* (U.S. Patent No. 3 475 229). U.S. Patent and Trademark Office.

Gesing, A. (2004). Assuring the continued recycling of light metals in end-of-life vehicles: A global perspective. *JOM, 56*(8), 18–27. doi:10.100711837-004-0176-5

Gestermann, S., Koppchen, W., Krause, V., Mothrath, M., Pophusen, D. W., Sandquist, A., & Zollner, O. (2005). Polycarbonate and its Blends for Car Body Parts Authors. *Materials (Basel), 107*, 22–24. https://link.springer.com/article/10.1007%2FBF03224786

Ghoo, B. Y., Back, S. J., Keum, Y. T., & Kang, S. Y. (1998). Finite element analysis of tailored sheet forming processes considering laser welding zone. *Metals and Materials, 4*(4), 862–870. doi:10.1007/BF03026414

Ghoo, B. Y., Keum, Y. T., & Kim, Y. S. (2001). Evaluation of the mechanical properties of welded metal in tailored steel sheet welded by CO_2 laser. *Journal of Materials Processing Technology, 113*(1-3), 692–698. doi:10.1016/S0924-0136(01)00674-4

Ghoo, B. Y., Park, S. W., & Keum, Y. T. (2001). New forming limit diagram of laser tailored blank. *Journal of Strain Analysis for Engineering Design, 36*(2), 143–152. doi:10.1243/0309324011512694

Gibson, I., Rosen, D. W., & Stucker, B. (2010a). *Additive Manufacturing Technologies*. Springer., doi:10.1007/978-1-4419-1120-9

Gibson, I., Rosen, D. W., & Stucker, B. (2010b). Printing Processes. In *Additive Manufacturing Technologies*. Springer. doi:10.1007/978-1-4419-1120-9_7

Gillespie, L. K. (1999). *Deburring and edge finishing handbook*. Society of Manufacturing Engineers.

Ginevicius, R. (2008). Normalization of quantities of various dimensions. *Journal of Business Economics and Management, 9*(1), 79–86. https://doi.org/10.3846/1611-1699.2008.9

Girish, B. M., Siddesh, H. S., & Satish, B. M. (2019). Taguchi grey relational analysis for parametric optimization of severe plastic deformation process. *SN Applied Sciences, 1*(8), 937. Advance online publication. doi:10.100742452-019-0982-6

Glorius, S., Nies, B., Farack, J., Quadbeck, P., Hauser, R., Standke, G., Rößler, S., Scharnweber, D., & Stephani, G. (2011). Metal Foam–Bone cement Composites: Mechanical and biological properties and perspectives for bone implant design. *Advanced Engineering Materials, 13*(11), 1019–1023. doi:10.1002/adem.201100026

Godovsky, Y. K. (2012). *Thermophysical Properties of Polymers*. Springer. doi:10.1007/978-3-642-51670-2

Golański, G., & Mroziński, S. (2013). Low cycle fatigue and cyclic softening behaviour of martensitic cast steel. *Engineering Failure Analysis, 35*, 692–702. doi:10.1016/j.engfailanal.2013.06.019

Gomathisankar, M., Gangatharan, M., & Pitchipoo, P. (2018). A Novel Optimization of Friction Stir Welding Process Parameters on Aluminum Alloy 6061-T6. *Materials Today: Proceedings, 5*(6), 14397–14404. doi:10.1016/j.matpr.2018.03.025

Gómez-Gálvez, P., Vicente-Munuera, P., Tagua, A., Forja, C., Castro, A. M., Letrán, M., Valencia-Expósito, A., Grima, C., Bermúdez-Gallardo, M., Serrano-Pérez-Higueras, Ó., Cavodeassi, F., Sotillos, S., Martín-Bermudo, M. D., Márquez, A., Buceta, J., & Escudero, L. M. (2018). Scutoids are a geometrical solution to three dimensional packing of epithelia. *Nature Communications, 9*(1), 2960. doi:10.103841467-018-05376-1 PMID:30054479

Gomez, M. M., Perez, M. B., Gil, F. M., Díez, A. D., Rodríguez, J. M., Rodrıguez, P. G., ... Torres, A. R. (2003). Identification of species of Brucella using Fourier transform infrared spectroscopy. *Journal of Microbiological Methods, 55*(1), 121–131. doi:10.1016/S0167-7012(03)00120-9 PMID:14500003

Goodall, R., Weber, L., & Mortensen, A. (2006). The electrical conductivity of microcellular metals. *Journal of Applied Physics*, 100.

Gornet, T., & Wohlers, T. (2014). *History of additive manufacturing*. http://wohlersassociates.com/history2014.pdf

Goushegir, S. M. (2016). Friction spot joining (FSpJ) of aluminum-CFRP hybrid structures. *Welding in the World, 60*(6), 1073–1093. doi:10.100740194-016-0368-y

Goushegir, S. M., dos Santos, J. F., & Amancio-Filho, S. T. (2014). Friction spot joining of aluminum AA2024/carbon-fiber reinforced poly(phenylene sulfide) composite single lap joints: Microstructure and mechanical performance. *Materials & Design, 50*, 196–206. doi:10.1016/j.matdes.2013.08.034

Goushegir, S. M., dos Santos, J. F., & Amancio-Filho, S. T. (2015). Influence of process parameters on mechanical performance and bonding area of AA2024/carbon-fiber reinforced poly(phenylene sulfide) friction spot single lap joints. *Materials & Design, 83*, 431–442. doi:10.1016/j.matdes.2015.06.044

Graf, B., Marko, A., Petrat, T., Gumenyuk, A., & Rethmeier, M. (2018). 3D laser metal deposition: Process steps for additive manufacturing. *Welding in the World, 62*(4), 877–883. doi:10.100740194-018-0590-x

Grigore, D. R., & Popp, O. T. (1991). Erratum: The complete classification of generalised homogeneous symplectic manifolds. *Journal of Mathematical Physics, 32*(2), 564–564. doi:10.1063/1.529395

Groß, R., & Dorigo, M. (2008). Self-Assembly at the Macroscopic Scale. *Proceedings of the IEEE, 96*(9), 1490–1508. doi:10.1109/JPROC.2008.927352

Guan, B., Cherrill, M., Pai, J. H., & Priest, C. (2019). Effect of mould roughness on injection moulded poly (methyl methacrylate) surfaces: Roughness and wettability. *Journal of Manufacturing Processes, 48*(October), 313–319. doi:10.1016/j.jmapro.2019.10.024

Guddati, S., Kiran, A. S. K., Leavy, M., & Ramakrishna, S. (2019). Recent advancements in additive manufacturing technologies for porous material applications. *International Journal of Advanced Manufacturing Technology, 105*(1), 193–215. doi:10.100700170-019-04116-z

Gülsoy, H. Ö., & German, R. M. (2008). Sintered foams from precipitation hardened stainless steel powder. *Powder Metallurgy, 51*(4), 350–353. doi:10.1179/174329008X286703

Gunnarsson, L., & Schedin, E. (2001). Improving the properties of exterior body panels in automobiles using variable blank holder force. *Journal of Materials Processing Technology, 114*(2), 168–173. doi:10.1016/S0924-0136(01)00727-0

Guo, C. Q., Sun, Y. D., Zhou, Y., Xie, B., Wang, T. Y., & Zuo, X. Q. (2018). Fabrication, structure and property of copper foam. *Materials Science Forum, 933*, 41–48. doi:10.4028/www.scientific.net/MSF.933.41

Guo, X., Liu, X., Shan, S., Zhao, W., Su, H., & Jia, Q. (2019). Green approach toward sustainable adhesive: Synthesis and characterization of poly (myrcene sulfone). *Journal of the Taiwan Institute of Chemical Engineers, 95*, 208–216. doi:10.1016/j.jtice.2018.07.004

Gupta, K. (2020). A review on green machining techniques. *Procedia Manufacturing, 51*, 1730–1736. doi:10.1016/j.promfg.2020.10.241

Gupta, M., & Meenashisundaram, G. K. (2015). Synthesis of Magnesium-Based Biomaterials. In *Insight into Designing Biocompatible Magnesium Alloys and Composites* (pp. 17–34). Springer.

Guruşçu, A. (2009). *Joining and Interfacial Properties of Aluminum/Glass Fiber Reinforced Polypropylene Sandwich Composites*. Academic Press.

Gussone, J., Hagedorn, Y. C., Gherekhloo, H., Kasperovich, G., Merzouk, T., & Hausmann, J. (2015). Microstructure of γ-titanium aluminide processed by selected laser melting at elevated temperatures. *Intermetallics, 66*, 133–140. Advance online publication. doi:10.1016/j.intermet.2015.07.005

Haag, M., Wanner, A., Clemens, H., Zhang, P., Kraft, O., & Arzt, E. (2003). Creep of aluminum-based closed-cell foams. *Metallurgical and Materials Transactions. A, Physical Metallurgy and Materials Science, 34*(12), 2809–2817. doi:10.100711661-003-0182-1

Habibi, M., Hashemi, R., Fallah Tafti, M., & Assempour, A. (2018). Experimental investigation of mechanical properties, formability and forming limit diagrams for tailor-welded blanks produced by friction stir welding. *Journal of Manufacturing Processes, 31*, 310–323. doi:10.1016/j.jmapro.2017.11.009

Hackert, A., Müller, S., & Kroll, L. (2017). Lightweight Wheel Disc with Carbon Aluminium Foam Sandwich. *Lightweight Design Worldwide, 10*(1), 6–11. doi:10.100741777-017-0016-2

Haghshenas, M., & Gerlich, A. P. (2018). Joining of automotive sheet materials by friction-based welding methods: A review. *Engineering Science and Technology, an International Journal, 21*(1), 130–148. doi:10.1016/j.jestch.2018.02.008

Hakamada, M., Yamada, Y., Kuromura, T., Asao, Y., Chen, Y., Kusuda, H., & Mabuchi, M. (2007). Porous Metals Produced by Spacer Method as Ecomaterials. *Advanced Materials Research, 15*, 416–421.

Hamlaoui, N., Azzouz, S., Chaoui, K., Azari, Z., & Yallese, M. A. (2017). Machining of tough polyethylene pipe material: Surface roughness and cutting temperature optimization. *International Journal of Advanced Manufacturing Technology, 92*(5–8), 2231–2245. doi:10.100700170-017-0275-4

Hang, Z. Y., Jones, M. E., Brady, G. W., Griffiths, R. J., Garcia, D., Rauch, H. A., Cox, C. D., & Hardwick, N. (2018). Non-beam-based metal additive manufacturing enabled by additive friction stir deposition. *Scripta Materialia, 153*, 122–130. doi:10.1016/j.scriptamat.2018.03.025

Han, L., Thornton, M., Li, D., & Shergold, M. (2011). Effect of governing metal thickness and stack orientation on weld quality and mechanical behaviour of resistance spot welding of AA5754 aluminium. *Materials & Design, 32*(4), 2107–2114. doi:10.1016/j.matdes.2010.11.047

Han, Z., Li, H., Xu, X., Wang, H., Li, H., & Tao, J. (2020). Crushing characteristics of aluminum/CFRP/aluminum hybrid tubes prepared by spinning forming. *Composite Structures, 112551*. Advance online publication. doi:10.1016/j.compstruct.2020.112551

Hao, G. L., Han, F. S., & Li, W. D. (2009). Processing and mechanical properties of magnesium foams. *Journal of Porous Materials, 16*(3), 251–256. doi:10.100710934-008-9194-y

Harders, H., Hupfer, K., & Rösler, J. (2005). Influence of cell wall shape and density on the mechanical behaviour of 2D foam structures. *Acta Materialia, 53*(5), 1335–1345. doi:10.1016/j.actamat.2004.11.025

Hardouin Duparc, O. B. M. (2010). The Preston of the Guinier-Preston Zones. Guinier. *Metallurgical and Materials Transactions. B, Process Metallurgy and Materials Processing Science, 41*(5), 925–934. doi:10.100711663-010-9387-z

Harper, C. A., & Petrie, E. M. (2003). *Plastics materials and processes*. John Wiley & Sons. doi:10.1002/0471459216

Hartl, R., Praehofer, B., & Zaeh, M. F. (2020). Prediction of the surface quality of friction stir welds by the analysis of process data using Artificial Neural Networks. *Proceedings of the Institution of Mechanical Engineers, Part L: Journal of Materials: Design and Applications, 234*(5), 732-751. 10.1177/1464420719899685

Hart, L. R., Li, S., Sturgess, C., Wildman, R., Jones, J. R., & Hayes, W. (2016). 3D Printing of Biocompatible Supramolecular Polymers and Their Composites. *ACS Applied Materials & Interfaces, 8*(5), 3115–3122. doi:10.1021/acsami.5b10471 PMID:26766139

Harun, W. S. W., Kamariah, M. S. I. N., Muhamad, N., Ghani, S. A. C., Ahmad, F., & Mohamed, Z. (2018). A review of powder additive manufacturing processes for metallic biomaterials. *Powder Technology*, *327*, 128–151. doi:10.1016/j.powtec.2017.12.058

Hasan, T. U., & Ansari, R. (2013). Study of the Impact Response of Glass Fiber Reinforced Aluminium Laminates for Conical and Blunt Nose Projectiles. *International Journal of Engineering Research and Applications*, *3*(4), 580–587.

Hashim, U. R., Jumahat, A., Ismail, M. H., & Razali, R. N. M. (2014). Fabrication and characterization of carbon fibre reinforced polymer rods with aluminium foam core. *Materials Research Innovations*, *18*(6), 204–208.

Hasirci, V., & Hasirci, N. (2018). Metals as Biomaterials. In *Fundamentals of Biomaterials* (pp. 35–49). Springer. doi:10.1007/978-1-4939-8856-3_3

Hassan, M. A., Suenaga, R., Takakura, N., & Yamaguchi, K. (2005). A novel process on friction aided deep drawing using tapered blank holder divided into four segments. *Journal of Materials Processing Technology*, *159*(3), 418–425. doi:10.1016/j.jmatprotec.2004.06.006

Hassan, M. A., Takakura, N., & Yamaguchi, K. (2003). A novel technique of friction aided deep drawing using a blank-holder divided into four segments. *Journal of Materials Processing Technology*, *139*(1-3), 408–413. doi:10.1016/S0924-0136(03)00514-4

Hayashi, H., Pak, S. W., & Usuda, M. (1998). Metal flow of a tailored blank in square cup deep drawing. *SAE Transactions*, 273–279. doi:10.4271/980447

Heckert, A., & Zaeh, M. F. (2015). Laser surface pre-treatment of aluminum for hybrid joints with glass fiber reinforced thermoplastics. *Journal of Laser Applications*, *27*(2), 29005-1-29005-5.

He, J., Xian, G., & Zhang, Y. X. (2020). Effect of moderately elevated temperatures on bond behaviour of CFRP-to-steel bonded joints using different adhesives. *Construction & Building Materials*, *241*, 118057. doi:10.1016/j.conbuildmat.2020.118057

Heo, Y., Choi, Y., Kim, H. Y., & Seo, D. (2001). Characteristics of weld line movements for the deep drawing with drawbeads of tailor-welded blanks. *Journal of Materials Processing Technology*, *111*(1-3), 164–169. doi:10.1016/S0924-0136(01)00503-9

Herderick, E. (2011). Additive Manufacturing of Metals : A Review. *Materials Science and Technology*, *176252*, 1413–1425.

Hernández-Nava, E., Smith, C. J., Derguti, F., Tammas-Williams, S., Léonard, F., Withers, P. J., Todd, I., & Goodall, R. (2015). The effect of density and feature size on mechanical properties of isostructural metallic foams produced by additive manufacturing. *Acta Materialia*, *85*, 387–395. doi:10.1016/j.actamat.2014.10.058

Herzog, D., Seyda, V., Wycisk, E., & Emmelmann, C. (2016). Additive Manufacturing of Metals. *Acta Materialia*, *117*, 371–392. doi:10.1016/j.actamat.2016.07.019

He, S., Wu, X., & Hu, S. J. (2003). Formability enhancement for tailor-welded blanks using blank holding force control. *Journal of Manufacturing Science and Engineering*, *125*(3), 461–467. doi:10.1115/1.1580853

Hevner, A. R. (2007). A Three Cycle View of Design Science Research. *Scandinavian Journal of Information Systems*, *19*(2).

He, W., Yao, L., Meng, X., Sun, G., Xie, D., & Liu, J. (2019). Effect of structural parameters on low-velocity impact behavior of aluminum honeycomb sandwich structures with CFRP face sheets. *Thin-walled Structures*, *137*, 411–432. doi:10.1016/j.tws.2019.01.022

Hirsch, J. (2010). Aluminium sheet fabrication and processing. In Fundamentals of Aluminium Metallurgy: Production, Processing and Applications (pp. 719–746). Elsevier Ltd. doi:10.1533/9780857090256.3.719

Hodge, A. M., Hayes, J. R., Caro, J. A., Biener, J., & Hamza, A. V. (2006). Characterization and mechanical behavior of nanoporous gold. *Advanced Engineering Materials*, *8*(9), 853–857. doi:10.1002/adem.200600079

Honarpardaz, O., Trangärd, A., Derkx, J., & Feng, X. (2015). Benchmark of Advanced Structural Materials for Light-weight Design of Industrial Robots. IEEE, 3178-3183.

Hsu, C. M., Tzeng, Y. C., Chen, S. F., Chen, Y. L., & Lee, H. L. (2021). Fabrication of 17-4PH Stainless Steel Foam by a Pressureless Powder Space Holder Technique. *Advanced Engineering Materials*, *23*(6), 2001202. doi:10.1002/adem.202001202

Huang, H., Sun, T., Zhang, G., Sun, L., & Zong, Z. (2018). Modeling and computation of turbulent slot jet impingement heat transfer using RANS method with special emphasis on the developed SST turbulence model. *International Journal of Heat and Mass Transfer*, *126*, 589–602. doi:10.1016/j.ijheatmasstransfer.2018.05.121

Huang, R., Ma, S., Zhang, M., Xu, J., & Wang, Z. (2019). Dynamic deformation and failure process of quasi-closed-cell aluminum foam manufactured by direct foaming technique. *Materials Science and Engineering A*, *756*, 302–311. doi:10.1016/j.msea.2019.04.050

Huang, X., Lin, Y., Alva, G., & Fang, G. (2017). Thermal properties and thermal conductivity enhancement of composite phase change materials using myristyl alcohol/metal foam for solar thermal storage. *Solar Energy Materials and Solar Cells*, *170*, 68–76. doi:10.1016/j.solmat.2017.05.059

Hufenbach, W., Gude, M., Czulak, A., Śleziona, J., Dolata-Grosz, A., & Dyzia, M. (2009). Development of textile-reinforced carbon fibre aluminium composites manufactured with gas pressure infiltration methods. *Journal of Achievements in Materials and Manufacturing Engineering*, *35*(2), 177–183.

Hull, A. W. (1919). A new method of chemical analysis. *Journal of the American Chemical Society*, *41*(8), 1168–1175. doi:10.1021/ja02229a003

Huo, P. C., Hao, Z. Y., Bai, P. K., Zhang, L. Z., Wu, L. Y., Du, W. B., & Han, B. (2020). The Compressive Behavior of Porous AlSi10Mg Prepared by Selective Laser Melting (SLM). Lasers in Engineering, 47.

Hurina, L., Vyshnepolskyi, Y., Pavlenko, D., & Stepanov, D. (2020). Investigation of the printing parameters influence on the bond lines length in fused filament fabrication. In *2020 IEEE 10th International Conference Nanomaterials: Applications & Properties (NAP) (pp. 02SAMA07-1)*. IEEE. 10.1109/NAP51477.2020.9309668

Hüsing, N., & Schubert, U. (1998). Aerogels—airy materials: Chemistry, structure, and properties. *Angewandte Chemie International Edition*, *37*(1-2), 22–45. doi:10.1002/(SICI)1521-3773(19980202)37:1/2<22::AID-ANIE22>3.0.CO;2-I PMID:29710971

Hu, Y., Wang, H., Ning, F., & Cong, W. (2016). Laser Engineered Net Shaping of Commercially Pure Titanium : Effects of Fabricating Variables. *Proceedings of the ASME 2016 International Manufacturing Science and Engineering Conference*. 10.1115/MSEC2016-8812

Hu, Y., Yuan, B., Cheng, F., & Hu, X. (2019). NaOH etching and resin pre-coating treatments for stronger adhesive bonding between CFRP and aluminium alloy. *Composites. Part B, Engineering*, *178*, 107478. doi:10.1016/j.compositesb.2019.107478

Hwang, C. L., & Yoon, K. (1981). *Multiple Attribute Decision Making–Methods and Applications. Springer-Verlag Berlin Heidelberg.*

Iasiello, M., Bianco, N., Chiu, W. K., & Naso, V. (2021). The effects of variable porosity and cell size on the thermal performance of functionally-graded foams. *International Journal of Thermal Sciences*, *160*, 106696. doi:10.1016/j.ijthermalsci.2020.106696

Immarigeon, J. P., Beddoes, J. C., Holt, R. T., Koul, A. K., Zhao, L., & Wallace, W. (1995). Lightweight Materials for Aircraft Applications. *Materials Characterization*, *35*(1), 41–67. doi:10.1016/1044-5803(95)00066-6

Indira Prasanth, S., Kesavan, K., Kiran, P., Sivaguru, M., Sudharsan, R., & Vijayanandh, R. (2020). Advanced structural analysis on e-glass fiber reinforced with polymer for enhancing the mechanical properties by optimizing the orientation of fiber. AIP Conference Proceedings, 2270. doi:10.1063/5.0019378

Irving, B. (1995). Welding tailored blanks is hot issue for automakers. *Welding Journal*, *74*(8), 49–52.

Ismail, A. A., van de Voort, F. R., & Sedman, J. (1997). Fourier transform infrared spectroscopy: principles and applications. In *Techniques and instrumentation in analytical chemistry* (Vol. 18, pp. 93–139). Elsevier.

Ismail, M. F., Jumahat, A., Ahmad, N., & Ismail, M. H. (2015). Low Velocity Impact of Aluminium Foam - Glass Fibre Reinforced Plastic Sandwich Panels. *Advanced Materials Research*, *1113*, 74–79. doi:10.4028/www.scientific.net/AMR.1113.74

Izamshah, R., Azam, M. A., Hadzley, M., Ali, M. M., Kasim, M. S., & Aziz, M. A. (2013). Study of surface roughness on milling unfilled-polyetheretherketones engineering plastics. *Procedia Engineering*, *68*, 654–660. doi:10.1016/j.proeng.2013.12.235

Izamshah, R., Zuraidah, Z., Kasim, M. S., Hadzley, M., & Amran, M. (2015). Multi Objective Optimization of Cutting Parameters in Machining Cellulose Based Hybrid Composites. *Applied Mechanics and Materials*, *761*, 287–292. . doi:10.4028/www.scientific.net/AMM.761.287

Jaggi, N., & Vij, D. R. (2006). Fourier transform infrared spectroscopy. In *Handbook of Applied Solid State Spectroscopy* (pp. 411–450). Springer. doi:10.1007/0-387-37590-2_9

Jagtap, K., & Pawade, R. (2014). Experimental Investigation on the Influence of Cutting Parameters on Surface Quality obtained in SPDT of PMMA. *Advanced Design and Manufacturing Technology*, *7*(2). www.SID.ir

Jagtap, K. A., Pawade, R. S., & Balasubramaniam, R. (2012). Some Investigations on Surface characteristics in Precision Turning of Nylon and Polypropylene. *Special Issue of International Journal of Electronics, Communication & Soft. Computing in Science & Engineering*, *1*, 240. http://www.ijecscse.org/papers/SpecialIssue/mech/04.pdf

Jahn, R., Cooper, R., & Wilkosz, D. (2007). The effect of anvil geometry and welding energy on microstructures in ultrasonic spot welds of AA6111-T4. *Metallurgical and Materials Transactions. A, Physical Metallurgy and Materials Science*, *38*(3), 570–583. doi:10.100711661-006-9087-0

Jain, H., Mondal, D. P., Gupta, G., & Kumar, R. (2021). Effect of compressive strain rate on the deformation behaviour of austenitic stainless steel foam produced by space holder technique. *Materials Chemistry and Physics*, *259*, 124010. doi:10.1016/j.matchemphys.2020.124010

Jain, H., Mondal, D. P., Gupta, G., Kumar, R., & Singh, S. (2020). Synthesis and characterization of 316L stainless steel foam made through two different removal process of space holder method. *Manufacturing Letters*, *26*, 33–36. doi:10.1016/j.mfglet.2020.09.005

Jain, N. K., & Jain, V. K. (2003). Process selection methodology for advanced machining processes. *Journal of Advanced Manufacturing Systems*, *2*(1), 5–45. doi:10.1142/S0219686703000204

Jakubowicz, J., Adamek, G., & Dewidar, M. (2013). Titanium foam made with saccharose as a space holder. *Journal of Porous Materials*, *20*(5), 1137–1141. doi:10.100710934-013-9696-0

Jangavali & Kamble. (2016). Finite Element Analysis and Experimental Evolution of Honeycomb Panel. *International Journal of Science and Research*, *5*(9), 1070 - 1073.

Jayasooriya, U. A., & Jenkins, R. D. (2002). Introduction to Raman Spectroscopy. In D. L. Andrews & A. A. Demidov (Eds.), *An Introduction to Laser Spectroscopy*. Springer. doi:10.1007/978-1-4615-0727-7_3

Jenarthanan, M. P., Gujjalapudi, V. S. S., & Venkataraman, V. (2017). Multi-objective optimization in end-milling of glass fiber reinforced polymer composites using desirability functional analysis and grey relational analysis. *Multidiscipline Modeling in Materials and Structures*, *13*(3), 391–408. doi:10.1108/MMMS-11-2016-0059

Jenkin, J. (2001). A unique partnership: William and Lawrence Bragg and the 1915 Nobel Prize in physics. *Minerva*, *39*(4), 373–392. doi:10.1023/A:1012783802528 PMID:18335624

Jha, N., Mondal, D. P., Majumdar, J. D., Badkul, A., Jha, A. K., & Khare, A. K. (2013). Highly porous open cell Ti-foam using NaCl as temporary space holder through powder metallurgy route. *Materials & Design*, *47*, 810–819. doi:10.1016/j.matdes.2013.01.005

Jiang, B., Chen, Q., & Yang, J. (2020). Advances in joining technology of carbon fiber-reinforced thermoplastic composite materials and aluminum alloys. *International Journal of Advanced Manufacturing Technology*, *110*(9-10), 2631–2649. Advance online publication. doi:10.100700170-020-06021-2

Jiang, H., Luo, T., Li, G., Zhang, X., & Cui, J. (2017). Fatigue life assessment of electromagnetic riveted carbon fiber reinforces plastic/aluminum alloy lap joints using Weibull distribution. *International Journal of Fatigue*, *105*, 180–189. doi:10.1016/j.ijfatigue.2017.08.026

Jiang, J., Xu, X., & Stringer, J. (2018). Support structures for additive manufacturing: A review. *Journal of Manufacturing and Materials Processing*, *2*(4), 64. doi:10.3390/jmmp2040064

Jia, Q., Guo, W., Li, W., Zhu, Y., Peng, P., & Zou, G. (2016). Microstructure and tensile behavior of fiber laser-welded blanks of DP600 and DP980 steels. *Journal of Materials Processing Technology*, *236*, 73–83. doi:10.1016/j.jmatprotec.2016.05.011

Jin, Y., Wan, Y., Zhang, B., & Liu, Z. (2017). Modeling of the chemical finishing process for polylactic acid parts in fused deposition modeling and investigation of its tensile properties. *Journal of Materials Processing Technology*, *240*, 233–239. doi:10.1016/j.jmatprotec.2016.10.003

Joost, W. J. (2012). Reducing Vehicle Weight and Improving U.S. Energy Efficiency Using Integrated Computational Materials Engineering. *JOM*, *64*(9), 1032–1038. Advance online publication. doi:10.100711837-012-0424-z

Joshikar. (2018). Analysis of Honeycomb Structure. *International Journal for Research in Applied Science & Engineering Technology*, *6*, 950 – 958.

Jung, K-W., Kawahito, Y., Takahashi, M., & Katayama, S. (2013). Laser direct joining of carbon fiber reinforced plastic to aluminum alloy. *Journal of Laser Applications*, *25*(3), 032003-(1) - 032003-(6).

Kader, B. A. (1981). Temperature and concentration profiles in fully turbulent boundary layers. *International Journal of Heat and Mass Transfer*, *24*(9), 1541–1544. doi:10.1016/0017-9310(81)90220-9

Kagan, L. S., Gordon, P. V., & Sivashinsky, G. I. (2012). A minimal model for end-gas autoignition. *Combustion Theory and Modelling*, *16*(1), 1–12. doi:10.1080/13647830.2011.610947

Kalaiselvan, K., Dinaharan, I., & Murugan, N. (2021). Routes for the Joining of Metal Matrix Composite Materials. In *Reference Module in Materials Science and Materials Engineering*. Elsevier. doi:10.1016/B978-0-12-803581-8.11899-5

Kampuš, Z., & Balič, J. (2003). Deep drawing of tailored blanks without a blankholder. *Journal of Materials Processing Technology*, *133*(1-2), 128–133. doi:10.1016/S0924-0136(02)00215-7

Kanani, A. Y., Hou, X., & Ye, J. (2020). The influence of notching and mixed-adhesives at the bonding area on the strength and stress distribution of dissimilar single-lap joints. *Composite Structures*, *241*, 112136. doi:10.1016/j.compstruct.2020.112136

Kang, J., Rao, H., Zhang, R., Avery, K., & Su, X. (2016). Tensile and fatigue behaviour of self-piercing rivets of CFRP to aluminium for automotive application. *Materials Science and Engineering, 137,* 1-9.

Kang, B., Cai, W., & Tan, C. A. (2013). Dynamic response of battery tabs under ultrasonic welding. *Journal of Manufacturing Science and Engineering*, *135*(5), 051013. Advance online publication. doi:10.1115/1.4024535

Kantor, B. Y., Smetankina, N. V., & Shupikov, A. N. (2001). Analysis of non-stationary temperature fields in laminated strips and plates. *International Journal of Solids and Structures*, *38*(48-49), 8673–8684. doi:10.1016/S0020-7683(01)00099-3

Kara, E., Crupi, V., Epasto, G., Guglielmino, E., & Aykul, H. (n.d.). Mechanical Behaviour of Glass Fiber Reinforced aluminium Honeycomb Sandwiches. *The 19th International Conference on Composite Materials.*

Kara, E., Crupi, V., Epasto, G., Guglielmino, E., & Aykul, H. (2012). Low Velocity Impact Response of Glass Fiber Reinforced Aluminium Foam Sandwich. *European Conference on Composite Materials*, 24-28.

Karami, V., Dariani, B. M., & Hashemi, R. (2021). Investigation of forming limit curves and mechanical properties of 316 stainless steel/St37 steel tailor-welded blanks produced by tungsten inert gas and friction stir welding method. *CIRP Journal of Manufacturing Science and Technology*, *32*, 437–446. doi:10.1016/j.cirpj.2021.02.002

Karthikeyan, P., Sabarianand, D. V., & Suganthan, S. (2019). Investigation on Adaptability of Carbon Fiber Tube for Serial Manipulator. *FME Transactions*, *47*(3), 412–417. doi:10.5937/fmet1903412K

Karunakaran, K. P., Suryakumar, S., Pushpa, V., & Akula, S. (2010). Low cost integration of additive and subtractive processes for hybrid layered manufacturing. *Robotics and Computer-integrated Manufacturing*, *26*(5), 490–499. doi:10.1016/j.rcim.2010.03.008

Kasap, S. (2018). Mangan dioksit (MnO2) katkılı üç boyutlu köpüksü yapıda grafen yapılarının üretilmesi ve karakterizasyonu. *Marmara Fen Bilimleri Dergisi*, *30*(4), 422–428. doi:10.7240/marufbd.418201

Kaur, G., Pulagara, N. V., Kumar, R., & Lahiri, I. (2020). Metal foam-carbon nanotube-reduced graphene oxide hierarchical structures for efficient field emission. *Diamond and Related Materials*, *106*, 107847. doi:10.1016/j.diamond.2020.107847

Kaur, G., Pulagara, N. V., & Lahiri, I. (2020). Three-Dimensional Graphene Materials for Supercapacitors. *Graphene as Energy Storage Material for Supercapacitors*, *64*, 77–128. doi:10.21741/9781644900550-4

Kee, K. J., Rhee, M. H., Choi, B. I., Kim, C. W., Sung, C. W., Han, C. P., Kang, K. W., & Won, S. T. (2009). Development of application technique of aluminum sandwich sheets for automotive hood. *International Journal of Precision Engineering and Manufacturing*, *10*(4), 71–75. doi:10.100712541-009-0073-5

Kellens, K., Mertens, R., Paraskevas, D., Dewulf, W., & Duflou, J. R. (2017). Environmental Impact of Additive Manufacturing Processes: Does AM contribute to a more sustainable way of part manufacturing ? *Procedia CIRP*, *61*(3), 582–587. doi:10.1016/j.procir.2016.11.153

Kelley, D. G., & Schwarz, K. H. (1991). Thermal Energy Deburring. Technical Paper MR91-136. Society of Manufacturing Engineers

Kennedy, A. (2012). Porous metals and metal foams made from powders. *Powder Metallurgy*, 124.

Kesarwani, S. (2017). Polymer Composites in Aviation Sector. *International Journal of Engineering Research & Technology (Ahmedabad)*, *6*(06), 518–525. doi:10.17577/IJERTV6IS060291

Kesavan, K., Kiran, P., Sivaguru, M., Indira Prasanth, S., Sudharsan, R., Raj Kumar, G., & Vijayanandh, R. (2021). Multi-objective Structural Analysis of Kevlar Fiber Reinforced Polymer Composite. *Lecture Notes in Mechanical Engineering*, *11*, 137–151. doi:10.1007/978-981-16-3033-0_13

Kesharwani, R. K., Panda, S. K., & Pal, S. K. (2015). Experimental investigations on formability of aluminum tailor friction stir welded blanks in deep drawing process. *Journal of Materials Engineering and Performance*, *24*(2), 1038–1049. doi:10.100711665-014-1361-5

Khalili, S. M. R., Mittal, R. K., & Gharibi Kalibar, S. (2005). A study of the mechanical properties of steel/aluminium/GRP laminates. *Materials Science and Engineering A*, *412*(1-2), 137–140. doi:10.1016/j.msea.2005.08.016

Khalili, S. M. R., Shokuhfar, A., Hoseini, S. D., Bidkhori, M., Khalili, S., & Mittal, R. K. (2008). Experimental study of the influence of adhesive reinforcement in lap joints for composite structures subjected to mechanical loads. *International Journal of Adhesion and Adhesives*, *28*(8), 436–444. doi:10.1016/j.ijadhadh.2008.04.009

Khan, N. Z., Siddiquee, A. N., Khan, Z. A., & Shihab, S. K. (2015). Investigations on tunneling and kissing bond defects in FSW joints for dissimilar aluminum alloys. *Journal of Alloys and Compounds*, *648*, 360–367. doi:10.1016/j.jallcom.2015.06.246

Khan, S., Sarang, S., & Hiratsuka, I. (2016). Study of Bending Strength for Aluminum Reinforced with Epoxy Composite. *SAE International Journal of Materials and Manufacturing*, *9*(3), 781–787. doi:10.4271/2016-01-0516

Kim, B. G., Dong, S. L., & Park, S. D. (2001). Effects of thermal processing on thermal expansion coefficient of a 50 vol.% SiCp/Al composite. *Materials Chemistry and Physics*, *72*(1), 42–47. doi:10.1016/S0254-0584(01)00306-6

Kim, D., Lee, W., Kim, J., Kim, C., & Chung, K. (2010). Formability evaluation of friction stir welded 6111-T4 sheet with respect to joining material direction. *International Journal of Mechanical Sciences*, *52*(4), 612–625. doi:10.1016/j.ijmecsci.2010.01.001

Kim, H. C., Shin, D. K., & Lee, J. J. (2013). Characteristics of aluminum/CFRP short square hollow section beam under transverse quasi-static loading. *Composites. Part B, Engineering*, *51*, 345–358. doi:10.1016/j.compositesb.2013.03.020

Kim, H. C., Shin, D. K., Lee, J. J., & Kwon, J. B. (2014). Crashworthiness of aluminum/CFRP square hollow section beam under axial impact loading for crash box application. *Composite Structures*, *112*, 1–10. doi:10.1016/j.compstruct.2014.01.042

Kim, Y., Hossain, A., & Nakamura, Y. (2013). Numerical study of melting of a phase change material (PCM) enhanced by deformation of a liquid–gas interface. *International Journal of Heat and Mass Transfer*, *63*, 101–112. doi:10.1016/j.ijheatmasstransfer.2013.03.052

Kinsey, B. L., & Cao, J. (2003). An analytical model for tailor welded blank forming. *Journal of Manufacturing Science and Engineering*, *125*(2), 344–351. doi:10.1115/1.1537261

Kinsey, B., Liu, Z., & Cao, J. (2000). A novel forming technology for tailor-welded blanks. *Journal of Materials Processing Technology*, *99*(1-3), 145–153. doi:10.1016/S0924-0136(99)00412-4

Kiratisaevee, H., & Cantwell, W. J. (2004). The Impact Response of Aluminum Foam Sandwich Structures Based on a Glass Fiber-Reinforced Polypropylene Fiber-Metal Laminate. *Polymer Composites*, *25*(5), 499–509. doi:10.1002/pc.20043

Kleiner, M., Geiger, M., & Klaus, A. (2003). Manufacturing of Lightweight Components by Metal Forming. *CIRP Annals*, *52*(2), 521–542. doi:10.1016/S0007-8506(07)60202-9

Klocke, F., & König, W. (2007). Chemisches abtragen. In Fertigungsverfahren 3. VDI-Buch (pp. 127–131). Springer. doi:10.1007/978-3-540-48954-2_3

Kobryn, P. A., Ontko, N. R., Perkins, L. P., & Tiley, J. S. (2006). Additive Manufacturing of Aerospace Alloys for Aircraft Structures. *Meeting Proceedings RTO-AVT-139*, 3-1-3–14. http://www.rto.nato.int/abstracts.asp

Koike, R., Matsumoto, T., Aoyama, T., & Kondo, M. (2020). Fabrication method for stainless steel foam block in directed energy deposition. *CIRP Annals*, *69*(1), 173–176. doi:10.1016/j.cirp.2020.04.060

Koike, R., Matsumoto, T., Kakinuma, Y., Aoyama, T., Oda, Y., Kuriya, T., & Kondo, M. (2018). A basic study on metal foam fabrication with titanium hydride in direct energy deposition. *Procedia Manufacturing*, *18*, 68–73. doi:10.1016/j.promfg.2018.11.009

Kombarov, V., Sorokin, V., Tsegelnyk, Y., Plankovskyy, S., Aksonov, Y., & Fojtů, O. (2021). Numerical control of machining parts from aluminum alloys with sticking minimization. *International Journal of Mechatronics and Applied Mechanics*, *I*(9), 209–216. doi:10.17683/ijomam/issue9.30

Komissarchuk, O., Hao, H., Zhang, X. L., & Karpov, V. (2018). Fabrication of Al–Si Gasar by mold casting technique. *International Journal of Materials Research*, *109*(4), 332–340. doi:10.3139/146.111612

Kondratiev, A., Melnikov, S., Nabokina, T., & Tsaritsynskyi, A. (2021). Effect of parameters of adhesive application by intaglio printing on honeycomb core bonding strength. In M. Nechyporuk, V. Pavlikov, & D. Kritskiy (Eds.), Integrated Computer Technologies in Mechanical Engineering – 2020 (pp. 221–233). Springer. doi:10.1007/978-3-030-66717-7_18

Kopp, A., Derra, T., Müther, M., Jauer, L., Schleifenbaum, J. H., Voshage, M., & Kröger, N. (2019). Influence of design and postprocessing parameters on the degradation behavior and mechanical properties of additively manufactured magnesium scaffolds. *Acta Biomaterialia*, *98*, 23–35. doi:10.1016/j.actbio.2019.04.012 PMID:30959185

Korkmaz, E., Onler, R., & Ozdoganlar, O. B. (2017). Micromilling of Poly(methyl methacrylate, PMMA) Using Single-Crystal Diamond Tools. *Procedia Manufacturing*, *10*, 683–693. doi:10.1016/j.promfg.2017.07.017

Korner, C., & Singer, R. F. (2000). Processing of metal foams—Challenges and opportunities. In Microstructural Investigation and Analysis. Wiley-VCH Verlag GmbH.

Koruk, A. N., Cretteur, L., & Corostola, J. R. (2001). Laser welded tailored blanks made of stainless steel/carbon steel. Proceedings of Laser Assisted Net Shape Engineering 01, LANE 2001.

Kosturek, R., Śnieżek, L., Torzewski, J., & Wachowski, M. (2020). Low Cycle Fatigue Properties of Sc-Modified AA2519-T62 Extrusion. *Materials 2020, 13*(1), 220. doi:10.3390/ma13010220

Kou, G., Lu, Y., Peng, Y., & Shi, Y. (2012). Evaluation of classification algorithms using MCDM and rank correlation. *International Journal of Information Technology & Decision Making*, *11*(1), 197–225.

Kovalevsky, V. F., Skobelev, S. B., Bezzateeva, E. G., Bury, G. G., & Poteryaev, I. K. (2019). Mechanical action on edges of plastic parts using electrohydraulic effect. *Journal of Physics: Conference Series*, *1260*(11), 112016. doi:10.1088/1742-6596/1260/11/112016

Krainov, A. Y., & Moiseeva, K. M. (2018). Modeling of the combustion of a methane–air mixture in an enclosed spherical volume. *Journal of Engineering Physics and Thermophysics*, *91*(4), 918–924. doi:10.100710891-018-1817-9

Kranz, J., Herzog, D., & Emmelmann, C. (2015). Design Guidelines for Laser Additive Manufacturing of Lightweight Structures in TiAl6V4. *Journal of Laser Applications*, *27*(S1), S14001. Advance online publication. doi:10.2351/1.4885235

Kränzlin, N., & Niederberger, M. (2015). Controlled fabrication of porous metals from the nanometer to the macroscopic scale. *Materials Horizons*, *2*(4), 359–377. doi:10.1039/C4MH00244J

Krishnamoorthy, N., Nayak, S., Surisetty, G. K., Chaturvedi, M., & Shuler, S. (2009). Polycarbonate Glazing Body Panels for Automotive Applications. *SAE Technical Papers*. Advance online publication. doi:10.4271/2009-26-0085

Krishna, S. M., Shridhar, T. N., & Krishnamurthy, L. (2015). Research significance, applications and fabrication of hybrid metal matrix composites. *International Journal of Innovative Science, Engineering & Technology*, *2*(5), 227–237.

Krzyhak, A., & Micha, B. (2016). Sandwich Structured Composites for Aeronautics: Methods of Manufacturing Affecting Some Mechanical Properties. *International Journal of Aerospace Engineering*, *7816912*, 10. doi:10.1155/2016/7816912

Kuhlman, G. W. (2005). Forging of Aluminum Alloys. In ASM Handbook (Vol. 14A, pp. 299–312). doi:10.1361/asmhba0003996

Kulik, P. (2017). *Thermal deburring machine and method of thermal deburring* (WO patent 2019/054886 Al). European Patent Office.

Kulshreshtha, A., & Dhakad, S. K. (2020). Preparation of metal foam by different methods: A review. *Materials Today: Proceedings*, *26*, 1784–1790. doi:10.1016/j.matpr.2020.02.375

Kumar, A. P., Dirgantara, T., & Krishna, P. V. (2020). Advances in Lightweight Materials and Structures. *Springer Proceedings in Materials, 8*.

Kumar, A. P., Dirgantara, T., & Krishna, P. V. (2020). *Advances in Lightweight Materials and Structures*. Springer Nature. doi:10.1007/978-981-15-7827-4

Kumar, A. P., & Sundaram, M. S. (2018). An Axial Crushing Characteristics of Hybrid Kenaf/Glass Fabric Wrapped Aluminium Capped Tubes under Static Loading. *International Journal of Mechanical and Production*, *8*(6), 201–206.

Kumar, K., & Davim, J. P. (Eds.). (2018). *Composites and advanced materials for industrial applications*. IGI Global. doi:10.4018/978-1-5225-5216-1

Kumar, K., Zindani, D., & Davim, J. P. (2020). *Rapid prototyping, rapid tooling and reverse engineering: from biological models to 3D bioprinters*. De Gruyter. doi:10.1515/9783110664904

Kumar, M., Kailas, S. V., & Narayanan, R. G. (2013). Influence of external weld flash on the in-plane plane-strain formability of friction stir welded sheets. *Journal of Strain Analysis for Engineering Design*, *48*(6), 376–385. doi:10.1177/0309324713488884

Kumar, S. (2015, December). Experimental Investigation and Optimization in Turning of UD-GFRP Composite Material by Regression Analysis Using PCD Tool Address for Correspondence. *International Journal of Advances in Engineering and Technology*.

Kumar, S., & Choudhary, A. K. S., & Rakesh. (2014). Effect of the Process Parameters on Geometrical Characteristics of the Parts in Direct Metal Deposition : A Review. *International Journal of Mechanical Engineering and Technology*, *5*(4), 116–122.

Kuruneru, S. T. W., Vafai, K., Sauret, E., & Gu, Y. (2020). Application of porous metal foam heat exchangers and the implications of particulate fouling for energy-intensive industries. *Chemical Engineering Science*, *228*, 115968. doi:10.1016/j.ces.2020.115968

Kusuda, H., Takasago, T., & Natsumi, F. (1997). Formability of tailored blanks. *Journal of Materials Processing Technology*, *71*(1), 134–140. doi:10.1016/S0924-0136(97)00159-3

Kuykendall, K., Nelson, T., & Sorensen, C. (2013). On the selection of constitutive laws used in modeling friction stir welding. *International Journal of Machine Tools & Manufacture*, *74*, 74–85. doi:10.1016/j.ijmachtools.2013.07.004

Kwon, D.-J., Kim, J.-H., Kim, Y.-J., Kim, J.-J., Park, S.-M., Kwon, I.-J., Shin, P.-S., DeVries, L. K., & Park, J.-M. (2019). Comparison of interfacial adhesion of hybrid materials of aluminum/carbon fiber reinforced epoxy composites with different surface roughness. *Composites. Part B, Engineering*, *170*, 11–18. doi:10.1016/j.compositesb.2019.04.022

LaGrandeur. (2011). The Effectiveness of Glass Laminate Aluminum Reinforced Epoxy. New York Institute of Technology.

Lambiase, F., Durante, M., & Di Ilio, A. (2016). Fast joining of aluminum sheets with Glass Fiber Reinforced Polymer (GFRP) by mechanical clinching. *Journal of Materials Processing Technology*, *236*, 241–251. doi:10.1016/j.jmatprotec.2016.04.030

Lambiase, F., & Ko, D. C. (2016). Feasibility of mechanical clinching for joining aluminum AA6082-T6 and Carbon Fiber Reinforced Polymer sheets. *Materials & Design*, *107*, 341–352. doi:10.1016/j.matdes.2016.06.061

Lambiase, F., & Ko, D.-C. (2017). Two-steps clinching of aluminum and Carbon Fiber Reinforced Polymer sheets. *Composite Structures*, *164*, 180–188. doi:10.1016/j.compstruct.2016.12.072

Lamikiz, A., Ukar, E., Tabernero, I., & Martinez, S. (2011). Thermal advanced machining processes. In J. P. Davim (Ed.), *Modern Machining Technology* (pp. 335–372). Woodhead Publishing. doi:10.1533/9780857094940.335

Langdon, G.S., Klemperer, C.J.V., & Volschenk, G.F., Tonder, T.V., & Govender, R.A. (2017). The influence of interfacial bonding on the response of lightweight aluminium and glass fibre metal laminate panels subjected to air-blast loading. *Mechanical Engineering Science*, *0*(0), 1–16.

Lazarević, D., Madić, M., Janković, P., & Lazarević, A. (2012). Cutting parameters optimization for surface roughness in turning operation of polyethylene (PE) using Taguchi method. *Tribology in Industry*, *34*(2), 68–73.

Lebaupina, Y., Friedlib, J., Caglara, B., Piccanda, M., Pasquierb, R., & Michaud, V. (2019). Crushing and intrusion resistance improvement of aluminum beams by carbon/epoxy composite patches. *Composite Structures*, *226*, 111235. doi:10.1016/j.compstruct.2019.111235

Lee, C.-J., Kim, B.-M., Kang, B.-S., Song, W.-J., & Ko, D.-C. (2017). Improvement of joinability in a hole clinching process with aluminum alloy and carbon fiber reinforced plastic using a spring die. *Composite Structures*, *173*, 58–69. doi:10.1016/j.compstruct.2017.04.010

Lee, H., Lim, C. H. J., Low, M. J., Tham, N., Murukeshan, V. M., & Kim, Y. J. (2017). Lasers in additive manufacturing: A review. *International Journal of Precision Engineering and Manufacturing-Green Technology*, *4*(3), 307–322. doi:10.100740684-017-0037-7

Lee, J. H., Kim, H. E., Shin, K. H., & Koh, Y. H. (2010). Improving the strength and biocompatibility of porous titanium scaffolds by creating elongated pores coated with a bioactive, nanoporous TiO2 layer. *Materials Letters*, *64*(22), 2526–2529. doi:10.1016/j.matlet.2010.08.038

Lee, Y. H., Li, S. M., Tseng, C. J., Su, C. Y., Lin, S. C., & Jhuang, J. W. (2017). Graphene as corrosion protection for metal foam flow distributor in proton exchange membrane fuel cells. *International Journal of Hydrogen Energy*, *42*(34), 22201–22207. doi:10.1016/j.ijhydene.2017.03.233

Leist, S. K., & Zhou, J. (2016). Current Status of 4D Printing Technology and the Potential of Light-Reactive Smart Materials as 4D Printable Materials. *Virtual and Physical Prototyping*, *11*(4), 249–262. doi:10.1080/17452759.2016.1198630

Leitao, C., Emílio, B., Chaparro, B. M., & Rodrigues, D. M. (2009). Formability of similar and dissimilar friction stir welded AA 5182-H111 and AA 6016-T4 tailored blanks. *Materials & Design*, *30*(8), 3235–3242. doi:10.1016/j.matdes.2008.12.005

Lennon, R., Pedreschi, R., & Sinha, B. P. (1999). Comparative study of some mechanical connections in cold formed steel. *Construction & Building Materials*, *13*(3), 109–116. doi:10.1016/S0950-0618(99)00018-5

Levine, B. R., Sporer, S., Poggie, R. A., Della Valle, C. J., & Jacobs, J. J. (2006). Experimental and clinical performance of porous tantalum in orthopedic surgery. *Biomaterials*, *27*(27), 4671–4681. doi:10.1016/j.biomaterials.2006.04.041 PMID:16737737

Lewandowski, J. J., & Seifi, M. (2016). Metal additive manufacturing: A review of mechanical properties. *Annual Review of Materials Research*, *46*(1), 151–186. doi:10.1146/annurev-matsci-070115-032024

Lewis, G. (2013). Properties of open-cell porous metals and alloys for orthopaedic applications. *Journal of Materials Science. Materials in Medicine*, *24*(10), 2293–2325. doi:10.100710856-013-4998-y PMID:23851927

Liang, J., Jiang, H., Zhang, J., Wu, X., Zhang, X., Li, G., & Cui, J. (2019). Investigations on mechanical properties and microtopography of electromagnetic self-piercing riveted joints with carbon fiber reinforced plastics/aluminum alloy 5052. *Archives of Civil and Mechanical Engineering*, *19*(1), 240–250. doi:10.1016/j.acme.2018.11.001

Ligon, S. C., Liska, R., Stampfl, J., Gurr, M., & Mülhaupt, R. (2017). Polymers for 3D Printing and Customized Additive Manufacturing. *Chemical Reviews*, *117*(15), 10212–10290. doi:10.1021/acs.chemrev.7b00074 PMID:28756658

Li, J. P., de Wijn, J. R., Van Blitterswijk, C. A., & de Groot, K. (2006). Porous Ti6Al4V scaffold directly fabricating by rapid prototyping: Preparation and in vitro experiment. *Biomaterials*, *27*(8), 1223–1235. doi:10.1016/j.biomaterials.2005.08.033 PMID:16169073

Li, J. P., Habibovic, P., van den Doel, M., Wilson, C. E., de Wijn, J. R., van Blitterswijk, C. A., & de Groot, K. (2007). Bone ingrowth in porous titanium implants produced by 3D fiber deposition. *Biomaterials*, *28*(18), 2810–2820. doi:10.1016/j.biomaterials.2007.02.020 PMID:17367852

Lim, J., You, C., & Dayyani, I. (2020). Multi-objective topology optimization and structural analysis of periodic space-frame structures. *Materials & Design*, *190*, 108552. doi:10.1016/j.matdes.2020.108552

Li, N., Huang, S., Zhang, G., Qin, R., Liu, W., Xiong, H., Shi, G., & Blackburn, J. (2019). Progress in additive manufacturing on new materials: A review. *Journal of Materials Science and Technology*, *35*(2), 242–269. doi:10.1016/j.jmst.2018.09.002

Lindgren, L. E., & Lundbäck, A. (2018). Approaches in computational welding mechanics applied to additive manufacturing: Review and outlook. *Comptes Rendus. Mécanique*, *346*(11), 1033–1042. doi:10.1016/j.crme.2018.08.004

Lin, X., Wang, P., Zhu, H., Zhang, Y., Ning, Y., & Wang, L. (2021). A novel method for the welding of tailor-welded blanks with different thicknesses based on the diode laser source. *Optics & Laser Technology*, *141*, 107100. doi:10.1016/j.optlastec.2021.107100

Lionetto, F., Balle, F., & Maffezzoli, A. (2017). Hybrid ultrasonic spot welding of aluminum to carbon fiber reinforced epoxy composites. *Journal of Materials Processing Technology*, *247*, 289–295. doi:10.1016/j.jmatprotec.2017.05.002

Lionetto, F., Mele, C., Leo, P., D'Ostuni, S., Balle, F., & Maffezzoli, A. (2018). Ultrasonic spot welding of carbon fiber reinforced epoxy composites to aluminum: Mechanical and electrochemical characterization. *Composites. Part B, Engineering*, *144*, 134–142. doi:10.1016/j.compositesb.2018.02.026

Liu, A., Tang, X., & Lu, F. (2013). Study on welding process and prosperities of AA5754 Al-alloy welded by double pulsed gas metal arc welding. *Materials & Design*, *50*, 149–155. doi:10.1016/j.matdes.2013.02.087

Liu, J. G., Wei, L., & Wang, J. X. (2012). Influence of interfacial adhesion strength on formability of AA5052/polyethylene/AA5052 sandwich sheet. *Transactions of Nonferrous Metals Society of China*, *22*, s395–s401. doi:10.1016/S1003-6326(12)61737-3

Liu, J., Wang, A., Gao, H., Gandra, J., Beamish, K., Zhan, L., & Wang, L. (2018). Transition of failure mode in hot stamping of AA6082 tailor welded blanks. *Journal of Materials Processing Technology*, *257*, 33–44. doi:10.1016/j.jmatprotec.2018.02.028

Liu, J., & Xue, W. (2013). Formability of AA5052/polyethylene/AA5052 sandwich sheets. *Transactions of Nonferrous Metals Society of China*, *23*(4), 964–969. doi:10.1016/S1003-6326(13)62553-4

Liu, P., & Chen, G. F. (2014). *Porous materials: processing and applications*. Elsevier.

Liu, P., & Chen, G. F. (2014). *Porous materials: Processing and applications*. Elsevier.

Liu, S., & Chao, Y. J. (2004). Determination of global mechanical response of friction stir welded plates using local constitutive properties. *Modelling and Simulation in Materials Science and Engineering*, *13*(1), 1–15. doi:10.1088/0965-0393/13/1/001

Liu, T., Deng, Z. C., & Lu, T. J. (2007). Structural modeling of sandwich structures with lightweight cellular cores. *Lixue Xuebao*, *23*(5), 545–559. doi:10.100710409-007-0096-z

Liu, W., Cheng, W., Xu, Y., & Yuan, S. (2018). Enhancing Formability of AA2219 Aluminum Alloy Friction Stir Welded Blanks with Preheating Treatment. *Journal of Materials Engineering and Performance*, *27*(9), 4819–4828. doi:10.100711665-018-3544-y

Liu, X., Shao, X., Li, Q., & Sun, G. (2019). Failure mechanisms in carbon fiber reinforced plastics (CFRP)/aluminum (Al) adhesive bonds subjected to low-velocity transverse pre-impact following by axial post-tension. *Composites. Part B, Engineering*, *172*, 339–351. doi:10.1016/j.compositesb.2019.04.036

Liu, X., Wei, M., Wang, Q., Tian, Y., Han, J., Gu, H., Ding, H., Chen, Q., Zhou, K., & Gu, Z. (2021). Capillary-Force-Driven Self-Assembly of 4D-Printed Microstructures. *Advanced Materials*, *33*(22), 2100332. doi:10.1002/adma.202100332 PMID:33885192

Liu, Y., & Zhuang, W. (2019). Self-piercing riveted-bonded hybrid joining of carbon fibre reinforced polymers and aluminium alloy sheets. *Thin-walled Structures*, *144*, 106340. doi:10.1016/j.tws.2019.106340

Li, Y., Jahr, H., Pavanram, P., Bobbert, F. S. L., Puggi, U., Zhang, X. Y., Pouran, B., Leeflang, M. A., Weinans, H., & Zadpoor, A. A. (2019). Additively manufactured functionally graded biodegradable porous iron. *Acta Biomaterialia*, *96*, 646–661. doi:10.1016/j.actbio.2019.07.013 PMID:31302295

Li, Y., Jahr, H., Zhou, J., & Zadpoor, A. A. (2020a). Additively manufactured biodegradable porous metals. *Acta Biomaterialia*. PMID:32853809

Li, Y., Murr, L., & McClure, J. (1999). Flow visualization and residual microstructures associated with the friction-stir welding of 2024 aluminium to 6061 aluminium. *Materials Science and Engineering A*, *271*(1-2), 213–223. doi:10.1016/S0921-5093(99)00204-X

Li, Y., Pavanram, P., Zhou, J., Lietaert, K., Taheri, P., Li, W., San, H., Leeflang, M. A., Mol, J. M. C., Jahr, H., & Zadpoor, A. A. (2020b). Additively manufactured biodegradable porous zinc. *Acta Biomaterialia*, *101*, 609–623. doi:10.1016/j.actbio.2019.10.034 PMID:31672587

Li, Z., Luo, J., Tan, X., Fang, Q., Zeng, Y., Zhou, M., Wu, W., & Zhang, J. (2017). X-ray nanotomography characterizations of gold foams. *Materials Letters*, *205*, 215–218. doi:10.1016/j.matlet.2017.06.056

Lokhande, P. E., & Chavan, U. S. (2019). Nanostructured Ni (OH) 2/rGO composite chemically deposited on Ni foam for high performance of supercapacitor applications. *Materials Science for Energy Technologies*, *2*(1), 52–56. doi:10.1016/j.mset.2018.10.003

Ludovico, A. D., Angelastro, A., & Campanelli, S. L. (2010). Experimental Analysis of the Direct Laser Metal Deposition Process. In *New Trends in Technologies: Devices, Computer, Communication and Industrial Systems*. https://cdn.intechweb.org/pdfs/12286.pdf

Lu, G. Q., Hai, H. A. O., Wang, F. Y., & Zhang, X. G. (2013). Preparation of closed-cell Mg foams using SiO2-coated CaCO3 as blowing agent in atmosphere. *Transactions of Nonferrous Metals Society of China*, *23*(6), 1832–1837. doi:10.1016/S1003-6326(13)62667-9

Lu, M., Hopkins, C., Zhao, Y., & Seiffert, G. (2009). Sound absorption characteristics of porous steel manufactured by lost carbonate sintering. *MRS Online Proceedings Library*, *1188*(1), 184–189. doi:10.1557/PROC-1188-LL07-04

Lyons, B. (2012). Additive Manufacturing in Aerospace: Examples and Research Outlook. *The Bridge*, *42*(1), 13–19.

Mad Rosip, N. I., Ahmad, S., Jamaludin, K. R., & Mat Noor, F. (2013). Producing of 316L stainless steel (SS316L) foam via slurry method. *Journal of Mechanical Engineering Science*, *5*, 707–712. doi:10.15282/jmes.5.2013.17.0068

Mad Rosip, N. I., Ahmad, S., Jamaludin, K. R., & Mat Noor, F. (2015). Morphological analysis of SS316L foam produced by using slurry method. *Advanced Materials Research*, *1087*, 68–72. doi:10.4028/www.scientific.net/AMR.1087.68

Madej, L., Hodgson, P. D., & Pietrzyk, M. (2007). The validation of a multiscale rheological model of discontinuous phenomena during metal rolling. *Computational Materials Science*, *41*(2), 236–241. doi:10.1016/j.commatsci.2007.04.002

Madej, L., Sieradzki, L., Sitko, M., Perzynski, K., Radwanski, K., & Kuziak, R. (2013). Multi scale cellular automata and finite element based model for cold deformation and annealing of a ferritic–pearlitic microstructure. *Computational Materials Science*, *77*, 172–181. doi:10.1016/j.commatsci.2013.04.020

Mahajan, K. A., Pawade, R. S., Mandale, A. M., Chikorde, P. P., Nawale, M. A., & Birari, S. N. (2019). Prediction of vibration effect on Surface Roughness of Poly methyl methacrylate (PMMA) by using ANN. *IOSR Journal of Engineering*, 55–60. www.iosrjen.org

Mahamood, R. M., & Akinlabi, E. T. (2015). Effect of Laser Power and Powder Flow Rate on the Wear Resistance Behaviour of Laser Metal Deposited TiC/Ti6Al4V Composites. *Materials Today: Proceedings*, *2*(4–5), 2679–2686. Advance online publication. doi:10.1016/j.matpr.2015.07.233

Mahesh, M., & Senthil Kumar, A. (n.d.). Comparison of mechanical properties for aluminium metal laminates (glare) of three different orientations such as CSM, woven roving and 45^0 stitched mat. *Journal of Mechanical and Civil Engineering*, 9-13.

Mahmoodzadeh, S., Shahrabi, J., Pariazar, M., & Zaeri, M. S. (2007). Project selection by using fuzzy AHP and TOPSIS technique. *World Academy of Science, Engineering and Technology*, *30*, 333–338.

Malashenko, V. L. (2013). *Improving of deflashing technology of thermoplastics parts based on thermal impulse method* [Ph.D. dissertation]. National Aerospace University "Kharkiv Aviation Institute."

Mali, H. S., Prajwal, B., Gupta, D., & Kishan, J. (2018). Abrasive flow finishing of FDM printed parts using a sustainable media. *Rapid Prototyping Journal*, *24*(3), 593–606. doi:10.1108/RPJ-10-2017-0199

Mapelli, C., Mombelli, D., Gruttadauria, A., Barella, S., & Castrodeza, E. M. (2013). Performance of stainless steel foams produced by infiltration casting techniques. *Journal of Materials Processing Technology*, *213*(11), 1846–1854. doi:10.1016/j.jmatprotec.2013.05.010

Margaris, A. V. (2014). Fourier Transform Infrared Spectroscopy (FTIR): Applications in Archaeology. Encyclopedia of Global Archaeology, 2890-2893.

Mariam, M., Afendi, M., Majid, M. S. A., Ridzuan, M. J. M., Sultan, M. T. H., Jawaid, M., & Gibson, A. G. (2019). Hydrothermal ageing effect on the mechanical behaviour and fatigue response of aluminium alloy/glass/epoxy hybrid composite single lap joints. *Composite Structures*, *219*, 69–82. doi:10.1016/j.compstruct.2019.03.078

Mark, J. E. (Ed.). (2007). *Physical Properties of Polymers Handbook*. Springer. doi:10.1007/978-0-387-69002-5

Marré, M., Gies, S., Maevus, F., & Tekkaya, A. E. (2010). Joining of lightweight frame structures by die-less hydroforming. *International Journal of Material Forming*, *3*(S1), 1031–1034. doi:10.100712289-010-0946-2

Marx, J., & Rabiei, A. (2020). Study on the Microstructure and Compression of Composite Metal Foam Core Sandwich Panels. *Metallurgical and Materials Transactions. A, Physical Metallurgy and Materials Science*, *51*(10), 5187–5197. doi:10.100711661-020-05964-1

Marx, L., & Liewald, M. (2017). Semi-Solid Joining of Aluminium and Carbon Fabric. *Procedia CIRP*, *66*, 233–237. doi:10.1016/j.procir.2017.03.306

Marx, L., & Liewald, M. (2020). Joining of carbon fibres with aluminium structures – Processing and infiltration modelling. *Journal of Composite Materials*, (25), 3767–3775. Advance online publication. doi:10.1177/0021998320912317

Ma, S., Tang, Q., Han, X., Feng, Q., Song, J., Setchi, R., Liu, Y., Liu, Y., Goulas, A., Engstrom, D. S., Tse, Y. Y., & Zhen, N. (2020). Manufacturability, mechanical properties, mass-transport properties and biocompatibility of triply periodic minimal surface (TPMS) porous scaffolds fabricated by selective laser melting. *Materials & Design*, *195*, 109034. doi:10.1016/j.matdes.2020.109034

Matassi, F., Botti, A., Sirleo, L., Carulli, C., & Innocenti, M. (2013). Porous metal for orthopedics implants. *Clinical Cases in Mineral and Bone Metabolism*, *10*(2), 111. PMID:24133527

Matbase. (2021). *The free and independent online materials properties resource*. http://www.matbase.com

Mathaiyan, V., Vijayanandh, R., & Jung, D. W. (2020). Determination of Strong Factor in Bird Strike Analysis using Taguchi's method for Aircraft Manufacturing guide. *Journal of Physics: Conference Series*, *1733*(012002), 1–6. doi:10.1088/1742-6596/1733/1/012002

Mathers, G. (2002). *The welding of aluminium and its alloys*. CRC Press.

Matijasevic-Lux, B. (2006). *Characterisation and optimisation of blowing agent for making improved metal foams* (Doctoral Thesis). Technische Universität Berlin, Fakultät III – Prozesswissenschaften.

Matsushita, T., Fujibayashi, S., & Kokubo, T. (2017). Titanium foam for bone tissue engineering. In *Metallic Foam Bone* (pp. 111–130). Woodhead Publishing. doi:10.1016/B978-0-08-101289-5.00004-4

Ma, X., & Peyton, A. J. (2006). Eddy current measurement of the electrical conductivity and porosity of metal foams. *IEEE Transactions on Instrumentation and Measurement*, *55*(2), 570–576. doi:10.1109/TIM.2006.873549

Ma, Z. Y. (2008). Friction stir processing technology: A review. *Metallurgical and Materials Transactions. A, Physical Metallurgy and Materials Science*, *39*(3), 642–658. doi:10.100711661-007-9459-0

McCullough, R. R., Jordon, J. B., Allison, P. G., Rushing, T., & Garcia, L. (2019). Fatigue crack nucleation and small crack growth in an extruded 6061 aluminum alloy. *International Journal of Fatigue*, *119*, 52–61. doi:10.1016/j.ijfatigue.2018.09.023

Mcginnis, D. (2020). *What is the Fourth Industrial Revolution.* Salesforce. https://www.salesforce.com/blog/what-is-the-fourth-industrial-revolution-4ir/

Mecholsky, J. J. Jr, DeLellis, D. P., & Mecholsky, N. A. (2020). Relationship between fractography, fractal analysis and crack branching. *Journal of the European Ceramic Society*, *40*(14), 4722–4726. doi:10.1016/j.jeurceramsoc.2019.12.061

Meenashisundaram, G. K., Xu, Z., Nai, M. L. S., Lu, S., Ten, J. S., & Wei, J. (2020). Binder Jetting Additive Manufacturing of High Porosity 316L Stainless Steel Metal Foams. *Materials (Basel)*, *13*(17), 3744. doi:10.3390/ma13173744 PMID:32847089

Meinders, T., van den Berg, A., & Huetink, J. (2000). Deep drawing simulations of tailored blanks and experimental verification. *Journal of Materials Processing Technology*, *103*(1), 65–73. doi:10.1016/S0924-0136(00)00420-9

Menghari, H. G., Babalo, V., Fazli, A., Soltanpour, M., & Ziaeipoor, H. (2020). A study on the electro-hydraulic clinching of aluminum and carbon fiber reinforced plastic sheets. *International Journal of Lightweight Materials and Manufacture.* doi:10.1016/j.ijlmm.2020.01.002

Meng, X., Huang, Y., Xie, Y., Li, J., Guan, M., Wan, L., Dong, Z., & Cao, J. (2019). Friction self-riveting welding between polymer matrix composites and metals. *Composites. Part A, Applied Science and Manufacturing*, *127*, 105624. doi:10.1016/j.compositesa.2019.105624

Menter, F. R. (1994). Two-equation eddy-viscosity turbulence models for engineering applications. *AIAA Journal*, *32*(8), 1598–1605. doi:10.2514/3.12149

Merklein, M., Giera, A., & Geiger, M. (2004). Deep drawing of friction stir welded aluminium tailored blanks. *Proceedings of International Deep Drawing Research Group conference (IDDRG 2004).*

Meschut, G., Janzen, V., & Olfermann, T. (2014). Innovative and highly productive joining technologies for multi-material lightweight car body structures. *Journal of Materials Engineering and Performance*, *23*(5), 1515–1523. doi:10.100711665-014-0962-3

Messler, R. W. (2003). Joining comes of age: From pragmatic process to enabling technology. *Assembly Automation*, *23*(2), 130–143. doi:10.1108/01445150310471365

Michailidis, N., Stergioudi, F., Omar, H., Missirlis, D., Vlahostergios, Z., Tsipas, S., Albanakis, C., & Granier, B. (2013). Flow, thermal and structural application of Ni-foam as volumetric solar receiver. *Solar Energy Materials and Solar Cells*, *109*, 185–191. doi:10.1016/j.solmat.2012.10.021

Michot, G., & Champier, G. (1991). Physical metallurgy of P/M aluminium alloys. *Memoires et Etudes Scientifiques de La Revue de Metallurgie*, *88*(7–8), 425–439. doi:10.1016/B978-0-08-099431-4.00002-6

Miles, M. P., Nelson, T. W., & Decker, B. J. (2004). Formability and strength of friction-stir-welded aluminum sheets. *Metallurgical and Materials Transactions. A, Physical Metallurgy and Materials Science*, *35*(11), 3461–3468. doi:10.100711661-004-0183-8

Miles, M. P., Nelson, T. W., & Melton, D. W. (2005). Formability of friction-stir-welded dissimilar-aluminum-alloy sheets. *Metallurgical and Materials Transactions. A, Physical Metallurgy and Materials Science*, *36*(12), 3335–3342. doi:10.100711661-005-0008-4

Miles, M. P., Nelson, T. W., Steel, R., Olsen, E., & Gallagher, M. (2009). Effect of friction stir welding conditions on properties and microstructures of high strength automotive steel. *Science and Technology of Welding and Joining*, *14*(3), 228–232. doi:10.1179/136217108X388633

Miles, M. P., Pew, J., Nelson, T. W., & Li, M. (2006). Comparison of formability of friction stir welded and laser welded dual phase 590 steel sheets. *Science and Technology of Welding and Joining*, *11*(4), 384–388. doi:10.1179/174329306X107737

Min, J., Wan, H., Carlson, B. E., Lin, J., & Sun, C. (2020). Application of laser ablation in adhesive bonding of metallic materials: A review. *Optics & Laser Technology*, *128*, 106188. doi:10.1016/j.optlastec.2020.106188

Min, K. B., & Kang, S. S. (2000). A study on resistance welding in steel sheets for tailor welded blank: Evaluation of flash weldability and formability (2nd Report). *Journal of Materials Processing Technology*, *103*(2), 218–224. doi:10.1016/S0924-0136(00)00494-5

Min, K. B., Kim, K. S., & Kang, S. S. (2000). A study on resistance welding in steel sheets using a tailor-welded blank (1st report): Evaluation of upset weldability and formability. *Journal of Materials Processing Technology*, *101*(1–3), 186–192. doi:10.1016/S0924-0136(00)00476-3

Mirrudula, P., Kaviya Priya, P., Malavika, M., Raj Kumar, G., Vijayanandh, R., & Senthil Kumar, M. (2020). Comparative structural analysis of the sandwich composite using advanced numerical simulation. AIP Conference Proceedings, 2270. doi:10.1063/5.0019370

Mirza, F. A., Liu, K., & Chen, X.-G. (2017). Cyclic Stress-Strain Behavior and Low Cycle Fatigue Life of AA6061 Aluminum Alloy. In Minerals, Metals and Materials Series (pp. 447–452). Springer. doi:10.1007/978-3-319-51541-0_56

Mirza, F. A., Chen, D. L., Li, D. J., & Zeng, X. Q. (2013). Low cycle fatigue of a rare-earth containing extruded magnesium alloy. *Materials Science and Engineering A*, *575*, 65–73. doi:10.1016/j.msea.2013.03.041

Mishra, R. S., & Sidhar, H. (2017). Friction Stir Welding. In *Friction Stir Welding of 2XXX Aluminum Alloys Including Al-Li Alloys* (pp. 1–13). Elsevier. doi:10.1016/B978-0-12-805368-3.00001-7

Mitschang, P., Velthuis, R., & Didi, M. (2013). Induction Spot Welding of Metal/CFRPC Hybrid Joints. *Advanced Engineering Materials*, *15*(9), 804–813. doi:10.1002/adem.201200273

Mitschang, P., Velthuis, R., Emrich, S., & Kopnarski, M. (2009). Induction Heated Joining of Aluminum and Carbon Fiber Reinforced Nylon 66. *Journal of Thermoplastic Composite Materials*, *22*(6), 687–800. doi:10.1177/0892705709105969

Moayedi, H., Darabi, R., Ghabussi, A., Habibi, M., & Foong, L. K. (2020). Weld orientation effects on the formability of tailor welded thin steel sheets. *Thin-walled Structures*, *149*, 106669. doi:10.1016/j.tws.2020.106669

Mohamed Bak, K., Raj Kumar, G., Ramasamy, N., & Vijayanandh, R. (2020). Experimental and Numerical Studies on Mechanical characterization of EPDM/S-SBR with Nanoclay Composites. *IOP Conference Series. Materials Science and Engineering*, *912*(052016), 1–11. doi:10.1088/1757-899X/912/5/052016

Mohbe, M., & Mondal, D. P. (2021). Properties of Zn foam filled with cenosphere microballoons. *Materials Today: Proceedings*, *46*, 7448–7451. doi:10.1016/j.matpr.2021.01.073

Momeni, F., & Hassani,, N., Liu, X., & Ni, J. (2017). A Review of 4D Priting. *Materials & Design*, *122*, 42–79. doi:10.1016/j.matdes.2017.02.068

Mondal, D. P., Goel, M. D., Bagde, N., Jha, N., Sahu, S., & Barnwal, A. K. (2014). Closed cell ZA27–SiC foam made through stir-casting technique. *Materials & Design*, *57*, 315–324. doi:10.1016/j.matdes.2013.12.026

Morovvati, M. R., Fatemi, A., & Sadighi, M. (2011). Experimental and finite element investigation on wrinkling of circular single layer and two-layer sheet metals in deep drawing process. *International Journal of Advanced Manufacturing Technology*, *54*(1), 113–121. doi:10.100700170-010-2931-9

Morovvati, M. R., Mollaei-Dariani, B., & Asadian-Ardakani, M. H. (2010). A theoretical, numerical, and experimental investigation of plastic wrinkling of circular two-layer sheet metal in the deep drawing. *Journal of Materials Processing Technology*, *210*(13), 1738–1747. doi:10.1016/j.jmatprotec.2010.06.004

Morrow, W. R., Qi, H., Kim, I., Mazumder, J., & Skerlos, S. J. (2007). Environmental aspects of laser-based and conventional tool and die manufacturing. *Journal of Cleaner Production*, *15*(10), 932–943. doi:10.1016/j.jclepro.2005.11.030

Mrazova, M. (2013). Advanced composite materials of the future in aerospace industry. *Incas Bulletin*, *5*(3), 139–150. doi:10.13111/2066-8201.2013.5.3.14

Muccio, E. A. (1991). *Plastic part technology*. ASM International.

Muccio, E. A. (1999). *Decoration and assembly of plastic parts*. ASM International.

Mucha, J. (2011). A study of quality parameters and behaviour of self-piercing riveted aluminium sheets with different joining conditions. *Strojniski Vestnik. Jixie Gongcheng Xuebao*, *57*(4), 323–333. doi:10.5545v-jme.2009.043

Mukanova, A., Zharbossyn, A., Nurpeissova, A., Kim, S. S., Myronov, M., & Bakenov, Z. (2018). Electrochemical Study of Graphene Coated Nickel Foam as an Anode for Lithium-Ion Battery. *Eurasian Chemico-Technological Journal*, *20*(2), 91–97. doi:10.18321/ectj694

Mukesh, A. M., & Hynes, N. R. J. (2019). Mechanical Properties and Applications of Fibre Metal Laminates. AIP Conference Proceedings, 2142, 100002-1–100002-6. doi:10.1063/1.5122456

Müller, D. W., Matz, A. M., & Jost, N. (2013). Casting open porous Ti foam suitable for medical applications. *Bioinspired. Biomimetic and Nanobiomaterials*, *2*(2), 76–83. doi:10.1680/bbn.12.00023

Murray, N. G. D., & Dunand, D. C. (2004). Effect of thermal history on the superplastic expansion of argon-filled pores in titanium: Part I kinetics and microstructure. *Acta Materialia*, *52*(8), 2269–2278. doi:10.1016/j.actamat.2004.01.039

Mustafa, L. M., Ismailov, M. B., & Sanin, A. F. (2020). Study on the effect of plasticizers and thermoplastics on the strength and toughness of epoxy resins. *Naukovyi Visnyk Natsionalnoho Hirnychoho Universytetu*, *2020*(4), 63–68. doi:10.33271/nvngu/2020-4/063

Muthu Chozha Rajan, B., Senthil Kumar, A., Sornakumar, T., Senthamaraikannan, P., & Sanjay, M. R. (2018). Multi Response Optimization of Fabrication Parameters of Carbon Fiber-Reinforced Aluminium Laminates (CARAL): By Taguchi Method and Gray Relational Analysis. *Polymer Composites*. Advance online publication. doi:10.1002/pc.24815

Nadeau, F., Thériault, B., & Gagné, M. O. (2020). Machine learning models applied to friction stir welding defect index using multiple joint configurations and alloys. *Proceedings of the Institution of Mechanical Engineers, Part L: Journal of Materials: Design and Applications*, *234*(5), 752-765. 10.1177/1464420720917415

Naik, S. B., Janaki Ramulu, P., & Ganesh Narayanan, R. (2010). Application of a few necking criteria in predicting the forming limit of unwelded and tailor-welded blanks. *Journal of Strain Analysis for Engineering Design*, *45*(2), 79–96. doi:10.1243/03093247JSA562

Nakagawa, N., Ikura, S., Natsumi, F., & Iwata, N. (1993). Finite element simulation of stamping a laser-welded blank (No. 930522). *SAE Technical Paper*, 734-745.

Nakajima, H. (2010). Fabrication, properties, and applications of porous metals with directional pores. *Proceedings of the Japan Academy. Series B, Physical and Biological Sciences*, *86*(9), 884–899. doi:10.2183/pjab.86.884 PMID:21084772

Nakamura, R., Goda, K., Noda, J., & Ohgi, J. (2009). High temperature tensile properties and deep drawing of fully green composites. *Express Polymer Letters*, *3*(1), 19–24. doi:10.3144/expresspolymlett.2009.4

Nandan, R., Roy, G. G., Lienert, T. J., & DebRoy, T. (2006). Numerical modelling of 3D plastic flow and heat transfer during friction stir welding of stainless steel. *Science and Technology of Welding and Joining*, *11*(5), 526–537. doi:10.1179/174329306X107692

Nandan, R., Roy, G. G., Lienert, T. J., & Debroy, T. (2007). Three-dimensional heat and material flow during friction stir welding of mild steel. *Acta Materialia*, *55*(3), 883–895. doi:10.1016/j.actamat.2006.09.009

Narasimhan, K., & Narayanan, R. G. (2011). 2 - Deformation of tailor welded blanks during forming. In B. L. Kinsey & X. Wu (Eds.), *Woodhead Publishing Series in Welding and Other Joining Technologies, Tailor Welded Blanks for Advanced Manufacturing* (pp. 24–47). Woodhead Publishing. doi:10.1533/9780857093851.1.24

Narayanan, R. G. (2007). *Forming behavior of Tailor Welded Blanks* (Ph.D. thesis). IIT Bombay.

Narayanan, R. G. (2018b). Sustainable Joining of Metallic Structures by End Forming. In Reference Module in Materials Science and Materials Engineering. Elsevier.

Narayanan, R. G. (2018a). A novel method of joining a rod to a sheet by end deformation: A preliminary experimental study. *International Journal of Precision Engineering and Manufacturing*, *19*(5), 773–779. doi:10.100712541-018-0093-0

Narayanan, R. G., Bhaskar, V. V., & Narasimhan, K. (2004). Effect of weld conditions on the deformation behavior of Tailor Welded Blanks (TWB). *Proceedings of the 8th International conference on Numerical Methods in Industrial Forming Processes, NUMIFORM '04.*

Narayanan, R. G., & Narasimhan, K. (2005). Effect of weld conditions on the forming limit strains of Tailor Welded Blanks (TWB) – Part I & II. *Proceedings of the 8th ESAFORM conference on material forming, ESAFORM 2005.*

Narayanan, R. G., & Narasimhan, K. (2006). Weld region representation during the simulation of TWB forming behavior. *International Journal of Forming Processes*, *9*(4), 491–518. doi:10.3166/ijfp.9.491-518

Narayanan, R. G., & Narasimhan, K. (2007). Relative Effect of Material and Geometric Parameters on the Forming Behaviour of Tailor Welded Blanks. *International Journal of Forming Processes*, *10*(2), 145–178. doi:10.3166/ijfp.10.145-178

Narayanan, R. G., & Narasimhan, K. (2008a). Influence of the weld conditions on the forming-limit strains of tailor-welded blanks. *Journal of Strain Analysis for Engineering Design*, *43*(4), 217–227. doi:10.1243/03093247JSA344

Narayanan, R. G., & Narasimhan, K. (2008b). Predicting the forming limit strains of tailor-welded blanks. *Journal of Strain Analysis for Engineering Design*, *43*(7), 551–563. doi:10.1243/03093247JSA445

Nassef, A., & El-Hadek, M. (2016). Microstructure and mechanical behavior of hot pressed Cu-Sn powder alloys. *Advances in Materials Science and Engineering*, *2016*, 2016. doi:10.1155/2016/9796169

Naveen Kumar, K., Vijayanandh, R., Bruce Ralphin Rose, J., Swathi, V., Narmatha, R., & Venkatesan, K. (2019). Research on Structural behavior of Composite Materials on different Cantilever Structures using FSI. *International Journal of Engineering and Advanced Technology, 8*(6S3), 1075 - 1086. doi:10.35940/ijeat.F1178.0986S319

Naveen Kumar, K., Vijayanandh, R., Raj Kumar, G., Sanjeev, B., Balachander, H., & Guru Prasad, S. (2018). Comparative Approaches for Fatigue Life Estimation of Aluminium Alloy for Aerospace Applications. *International Journal of Vehicle Structures and Systems, 10*(4), 282-286. doi:10.4273/ijvss.10.4.11

Nee, J. G., Dufraine, W., Evans, J. W., & Hill, M. (2010). *Fundamentals of Tool Design* (6th ed.). Society of Manufacturing Engineers. https://books.google.com.my/books/about/Fundamentals_of_Tool_Design_Fifth_Editio.html?id=4Y3x5Fq9TJkC&pgis=1

Nestler, D., Trautmann, M., Nendel, S., Wagner, G., & Kroll, L. (2016). Innovative hybrid laminates of aluminium alloy foils and fibre-reinforced thermoplastic layers. *Materialwissenschaft und Werkstofftechnik, 47*(11), 1121–1131. doi:10.1002/mawe.201600636

Nickels, L. (2019). New innovations in automotive thermoplastics. *Reinforced Plastics, 63*(4), 185–188. doi:10.1016/j.repl.2019.06.041

Nong, N., Keju, O., Yu, Z., Zhiyuan, Q., Changcheng, T., & Feipeng, L. (2003). Research on press joining technology for automotive metallic sheets. *Journal of Materials Processing Technology, 137*(1-3), 159–163. doi:10.1016/S0924-0136(02)01083-X

Ntouanoglou, K., Stavropoulos, P., & Mourtzis, D. (2018). 4D Printing Prospects for the Aerospace Industry: A Critical Review. *Procedia Manufacturing, 18*, 120–129. doi:10.1016/j.promfg.2018.11.016

Okur, M. Z. (2018). Development of Aluminum Honeycomb Cored Carbon Fiber Reinforced Polymer Composite Based Sandwich Structures. Preprints, 2018120195. doi:10.20944/preprints201812.0195.v1

Oliveira, P. R., May, M., Panzera, T. H., Scarpa, F., & Hiermaier, S. (2020). Reinforced biobased adhesive for eco-friendly sandwich panels. *International Journal of Adhesion and Adhesives, 98*, 102550. doi:10.1016/j.ijadhadh.2020.102550

Ono, M., Yoshitake, A., & Omura, M. (2004). Laser weldability of high-strength steel sheets in fabrication of tailor welded blanks. *Welding International, 18*(10), 777–784. doi:10.1533/wint.2004.3321

Oppenheimer, S. M., O'Dwyer, J. G., & Dunand, D. C. (2004). Porous, superelastic NiTi produced by powder-metallurgy. *TMS Lett, 1*, 93–94.

Organ, A., & Yalcin, E. (2016). Performance Evaluation of Research Assistants By Copras Method. *European Scientific Journal, 12*(10), 102–109.

Oveisi, H., & Geramipour, T. (2020). High mechanical performance alumina-reinforced aluminum nanocomposite metal foam produced by powder metallurgy: Fabrication, microstructure characterization, and mechanical properties. *Materials Research Express, 6*(12), 1250c2.

Oya, T., Tiesler, N., Kawanishi, S., Yanagimoto, J., & Koseki, T. (2010). Experimental and numerical analysis of multilayered steel sheets upon bending. *Journal of Materials Processing Technology, 210*(14), 1926–1933. doi:10.1016/j.jmatprotec.2010.07.003

Oztan, C., Karkkainen, R., Fittipaldi, M., Nygren, G., Roberson, L., Lane, M., & Celik, E. (2019). Microstructure and mechanical properties of three dimensional-printed continuous fiber composites. *Journal of Composite Materials, 53*(2), 271–280. doi:10.1177/0021998318781938

Padhy, G. K., Wu, C. S., & Gao, S. (2018). Friction stir based welding and processing technologies-processes, parameters, microstructures and applications: A review. *Journal of Materials Science and Technology, 34*(1), 1–38. doi:10.1016/j.jmst.2017.11.029

Padmanabhan, R., Oliveira, M. C., & Menezes, L. F. (2011). 5 - Lightweight metal alloy tailor welded blanks. In B. L. Kinsey & X. Wu (Eds.), *Woodhead Publishing Series in Welding and Other Joining Technologies, Tailor Welded Blanks for Advanced Manufacturing* (pp. 97–117). Woodhead Publishing. doi:10.1533/9780857093851.2.97

Pakdil, M., Çam, G., Koçak, M., & Erim, S. (2011). Microstructural and mechanical characterization of laser beam welded AA6056 Al-alloy. *Materials Science and Engineering A, 528*(24), 7350–7356. doi:10.1016/j.msea.2011.06.010

Palaniswamy, S., Arunagiri, K., & Prakash, S. (2020, December). Corrosion behaviour of closed cell magnesium foam with rare earth elements by powder metallurgy process. In. AIP Conference Proceedings: Vol. 2311. *No. 1* (p. 040010). AIP Publishing LLC.

Palanivel, S., Nelaturu, P., Glass, B., & Mishra, R. S. (2015a). Friction stir additive manufacturing for high structural performance through microstructural control in an Mg based WE43 alloy. *Materials & Design, 65*, 934-952.

Palanivel, S., Sidhar, H., & Mishra, R. S. (2015b). Friction stir additive manufacturing: Route to high structural performance. *JOM, 67*(3), 616–621. doi:10.100711837-014-1271-x

Palaziuk, I. S. (2018). *Method for operating conditions assignment of thermal pulse deburring of GTE parts made from heat-resistant alloys by the edge qualimetric parameters* [Ph.D. dissertation]. National Aerospace University "Kharkiv Aviation Institute."

Pallett, R. J., & Lark, R. J. (2001). The use of tailored blanks in the manufacture of construction components. *Journal of Materials Processing Technology, 117*(1-2), 249–254. doi:10.1016/S0924-0136(01)01124-4

Pandey, A., Muchhala, D., Kumar, R. S. S., Venkat, A. N. C., & Mondal, D. P. (2019). Flexural deformation behavior of carbon fiber reinforced aluminium hybrid foam sandwich structure. *Composites. Part B, Engineering, 107729*. Advance online publication. doi:10.1016/j.compositesb.2019.107729

Pandian, V., & Kannan, S. (2020). Numerical prediction and experimental investigation of aerospace-grade dissimilar aluminium alloy by friction stir welding. *Journal of Manufacturing Processes, 54*, 99–108. doi:10.1016/j.jmapro.2020.03.001

Pang, Q., Zhang, J. H., Huq, M. J., & Hu, Z. L. (2019). Characterization of microstructure, mechanical properties and formability for thermomechanical treatment of friction stir welded 2024-O alloys. *Materials Science and Engineering A, 765*, 138303. doi:10.1016/j.msea.2019.138303

Pantazopoulos, G. A. (2019). A short review on fracture mechanisms of mechanical components operated under industrial process conditions: Fractographic analysis and selected prevention strategies. *Metals, 9*(2), 148. Advance online publication. doi:10.3390/met9020148

Panteli, A., Robson, J. D., Brough, I., & Prangnell, P. B. (2012). The effect of high strain rate deformation on intermetallic reaction during ultrasonic welding aluminium to magnesium. *Materials Science and Engineering A, 556*, 31–42. doi:10.1016/j.msea.2012.06.055

Papadopoulou, S., Vazdirvanidis, A., Toulfatzis, A., Rikos, A., & Pantazopoulos, G. (2020). Failure Investigation of Products and Components in Metal Forming Industry: Root Cause Analysis and Process-Based Approach. *Journal of Failure Analysis and Prevention, 20*(1), 106–114. doi:10.1007/s11668-020-00801-4

Papakyriacou, M., Mayer, H., Pypen, C., Plenk, H. Jr, & Stanzl-Tschegg, S. (2001). Influence of loading frequency on high cycle fatique properties of b.c.c and h.c.p. metals. *Materials Science and Engineering A, 308*(1–2), 143–152. doi:10.1016/S0921-5093(00)01978-X

Papantoniou, I., Kyriakopoulou, H., Pantelis, D., & Manolakos, D. (2019). Metal foaming by powder metallurgy process: investigation of different parameters on the foaming efficiency. *Frattura ed Integrità Strutturale, 13*(50), 497-504.

Park, S. (2016). *Gas Diffusion Layer. Fuel Cells: Data*. Facts, and Figures.

Park, S. Y., Choi, W. J., Choi, H. S., Kwon, H., & Kim, S. H. (2010). Recent trends in surface treatment technologies for airframe adhesive bonding processing: A review (1995–2008). *The Journal of Adhesion, 86*(2), 192–221. doi:10.1080/00218460903418345

Parsa, M. H., Ettehad, M., Matin, P. H., Ahkami, A., & Nasher, S. (2010). Experimental and numerical determination of limiting drawing ratio of Al3105-polypropylene-Al3105 sandwich sheets. *Journal of Engineering Materials and Technology, 132*(3), 031004-1, 031004–031011. doi:10.1115/1.4001264

Paruka, P., Mat Yasin, M., Mamat, R., Maleque, M. A., & Md Shah, M. K. (2015). Crush Zone Morphology of Epoxy–Glass Fiber–Aluminium Composite Columnar Tube due to Longitudinal Crushing Force. *Advanced Materials Research, 1115*, 258–261. doi:10.4028/www.scientific.net/AMR.1115.258

Paruka, P., Siswanto, W. A., Maleque, M. A., & Mohd Shah, M. K. (2015). Crashworthy capacity of a hybridized epoxy-glass fiber aluminum columnar tube using repeated axial resistive force. *Journal of Mechanical Science and Technology, 29*(5), 1941–1953. doi:10.100712206-015-0415-4

Parvanian, A. M., Saadatfar, M., Panjepour, M., Kingston, A., & Sheppard, A. P. (2014). The effects of manufacturing parameters on geometrical and mechanical properties of copper foams produced by space holder technique. *Materials & Design, 53*, 681–690. doi:10.1016/j.matdes.2013.07.047

Paserin, V., Marcuson, S., Shu, J., & Wilkinson, D. S. (2003). The chemical vapor deposition technique for Inco nickel foam production–manufacturing benefits and potential applications. *Cellular Metals and Metal Foaming Technology*, 31.

Patel, Y. (2008). *The machining of polymers* (PhD thesis). Department of Mechanical Engineering, Imperial College London.

Patel, C., Das, S., & Narayanan, R. G. (2013). CAFE modeling, neural network modeling, and experimental investigation of friction stir welding. *Proceedings of the Institution of Mechanical Engineers. Part C, Journal of Mechanical Engineering Science, 227*(6), 1164–1176. doi:10.1177/0954406212459150

Patel, M., Pardhi, B., Chopara, S., & Pal, M. (2018). Lightweight Composite Materials for Automotive - A Review. *International Research Journal of Engineering and Technology, 5*(11), 41–47.

Patel, P., Bhingole, P. P., & Makwana, D. (2018). Manufacturing, characterization and applications of lightweight metallic foams for structural applications. *Materials Today: Proceedings, 5*(9), 20391–20402. doi:10.1016/j.matpr.2018.06.414

Patil, N. A., Mulikb, S. S., Wangikarc, K. S., & Kulkarni, A. P. (2018). Characterization of Glass Laminate Aluminium Reinforced Epoxy-A Review. *Procedia Manufacturing, 20*, 554–562. doi:10.1016/j.promfg.2018.02.083

Payne, A. O. (1976). The fatigue of aircraft structures. *Engineering Fracture Mechanics, 8*(1), 157–203. doi:10.1016/0013-7944(76)90085-0

Pei, E., & Loh, G. H. (2018). Technological Considerations for 4D Printing: An Overview. *Progress in Additive Manufacturing, 3*(1-2), 95–107. doi:10.100740964-018-0047-1

Peng, P. Y., Jin, I., Yang, T. C. K., & Huang, C. M. (2014). Facile preparation of hierarchical CuO–CeO2/Ni metal foam composite for preferential oxidation of CO in hydrogen-rich gas. *Chemical Engineering Journal*, *251*, 228–235. doi:10.1016/j.cej.2014.04.077

Pepelnjak, T., Jurkosek, B., Kuzman, K., & Jesenicnik, B. (1997). Research on the forming of tailored blanks in conventional tools. Proceedings of Laser Assisted Net Shape Engineering 97, LANE '97.

Periasamy, M., Manickam, B., & Hariharasubramanian, K. (2012). Impact Properties of Aluminium - Glass Fiber Reinforced Plastics Sandwich Panels. *Materials Research*, *15*(3), 347–354. doi:10.1590/S1516-14392012005000036

Peters, M., Kumpfert, J., Ward, C. H., & Leyens, C. (2003). Titanium Alloys for Aerospace Applications. *Advanced Engineering Materials*, *5*(6), 419–427. Advance online publication. doi:10.1002/adem.200310095

Pickin, C. G., Young, K., & Tuersley, I. (2007). Joining of lightweight sandwich sheets to aluminium using self-pierce riveting. *Materials & Design*, *28*(8), 2361–2365. doi:10.1016/j.matdes.2006.08.003

Piñeiro-Jiménez, A., Villalobos-Gutiérrez, C., Staia, M. H., & Puchi-Cabrera, E. S. (2007). Tensile and fatigue properties of 6063-T6 aluminium alloy coated with electroless Ni–P deposit. *Materials Science and Technology*, *23*(3), 253–263. doi:10.1179/174328407X157317

Pityana, S., Mahamood, R. M., Akinlabi, E. T., & Shukla, M. (2013). Gas Flow Rate and Powder Flow Rate Effect on Properties of Laser Metal Deposited Ti6Al4V. *Proceedings of the International MultiConference of Engineers and Computer Scientists*, *II*, 13–16.

Plankovskyy, S. I., Shypul, O. V., Tryfonov, O. V., & Kozlov, V. G. (2010). The mixing in chamber of thermo-pulse system at cleaning of engine unit workpiece. *Aerospace Technic and Technology*, *9*, 7–11. http://nbuv.gov.ua/UJRN/aktit_2010_9_3

Plankovskyy, S. I., Shypul, O. V., Tryfonov, O. V., Palazyuk, E. S., & Malashenko, V. L. (2014). The simulation of the heat transfer during shock waves damping in an enclosed chamber. *Aerospace Technic and Technology*, *1*, 104–109. http://nbuv.gov.ua/UJRN/aktit_2014_1_18

Plankovskyy, S., Nikolaev, A., Shypul, O., Litvinchev, I., Pankratov, A., & Romanova, T. (2020). Balance layout problem with the optimized distances between objects. In P. Vasant, I. Litvinchev, J. A. Marmolejo-Saucedo, R. Rodriguez, & F. Martinez (Eds.), Data Analysis and Optimization for Engineering and Computing Problems (pp. 85–93). Springer. doi:10.1007/978-3-030-48149-0_7

Plankovskyy, S., Shypul, O., Tsegelnyk, Y., Pankratov, A., & Romanova, T. (2021b). Amplification of heat transfer by shock waves for Thermal Energy Method. In M. Nechyporuk, V. Pavlikov, & D. Kritskiy (Eds.), Integrated Computer Technologies in Mechanical Engineering – 2020 (pp. 577–587). Springer. doi:10.1007/978-3-030-66717-7_49

Plankovskyy, S., Shypul, O., Tsegelnyk, Y., Pankratov, A., Romanova, T., & Litvinchev, I. (2021c). Circular layout in thermal deburring. In S. Shkarlet, A. Morozov, & A. Palagin (Eds.), Mathematical Modeling and Simulation of Systems (MODS'2020) (pp. 111–120). Springer. doi:10.1007/978-3-030-58124-4_11

Plankovskyy, S., Popov, V., Shypul, O., Tsegelnyk, Y., Tryfonov, O., & Brega, D. (2021a). Advanced thermal energy method for finishing precision parts. In K. Gupta & A. Pramanik (Eds.), *Advanced Machining and Finishing* (pp. 527–575). Elsevier. doi:10.1016/B978-0-12-817452-4.00014-2

Plankovskyy, S., Shypul, O., Tsegelnyk, Y., Tryfonov, O., & Golovin, I. (2016). Simulation of surface heating for arbitrary shape's moving bodies/sources by using R-functions. *Acta Polytechnica*, *56*(6), 472–477. doi:10.14311/AP.2016.56.0472

Plankovskyy, S., Teodorczyk, A., Shypul, O., Tryfonov, O., & Brega, D. (2019). Determination of detonable gas mixture heat fluxes at thermal deburring. *Acta Polytechnica*, *59*(2), 162–169. doi:10.14311/AP.2019.59.0162

Podvezko, V. (2011). The Comparative Analysis of MCDA Methods SAW and COPRAS. *Inzinerine Ekonomika-Engineering Economics*, *22*(2), 134–146.

Pollock, T. M. (2010). Weight loss with magnesium alloys. *Science*, *328*(5981), 986–987. doi:10.1126cience.1182848 PMID:20489013

Pop, M. A., Geaman, V., Radomir, I., & Bedo, T. (2017). Capacity of energy absorption by flick through shock in cooper foams. *Journal of Porous Media*, *20*(5), 405–415. doi:10.1615/JPorMedia.v20.i5.30

Popovic, G., Stanujkic, D., & Stojanovic, S. (2012). Investment Project Selection by Applying COPRAS Method and Imprecise Data. *Serbian Journal of Management*, *7*(2), 257–269.

Prabhu, S., Raja, V. K., & Nikhil, R. (2015). Applications of Cellular Materials–An Overview. *Applied Mechanics and Materials*, *766*, 511–517. doi:10.4028/www.scientific.net/AMM.766-767.511

Pradeepkumar, J., Robinson Smart, D. S., & John Alexis, S. (2018). An Aluminium Hybrid Metal Matrix Composite reinforced with TaC, Ti, Si3N4 nanoparticles - A Review. *International Journal of Mechanical and Production Engineering Research and Development*, (Special Issue), 187–194.

Pradhan, S. S., Mishra, U., & Biswal, S. K. (2020). Experimental study on mechanical performance of cracked aluminium alloy repaired with composite patch. *Materials Today: Proceedings*, *26*, 2676–2680. doi:10.1016/j.matpr.2020.02.563

Prakash, M., & Daniel Das, A. (2020). Investigation on effect of FSW parameters of aluminium alloy using Full Factorial Design. *Materials Today: Proceedings*, 1–6. doi:10.1016/j.matpr.2020.05.622

Pramanik, A., Basak, A. K., Dong, Y., Sarker, P. K., Uddin, M. S., Littlefair, G., Dixit, A. R., & Chattopadhyaya, S. (2017). Joining of carbon fibre reinforced polymer (CFRP) composites and aluminium alloys-A review. *Composites. Part A, Applied Science and Manufacturing*, *101*, 1–29. doi:10.1016/j.compositesa.2017.06.007

Prange, W., Schneider, C., & Jaroni, U. (1994). Tailored Blanks-Contributions to an Optimized Steel Body Shell. In *Laser Assisted Net Shape Engineering, proceedings of the LANE* (Vol. 94, pp. 145–165). Meisenbach Bamberg.

Prayogo, G. S., & Lusi, N. (2016). Application of Taguchi technique coupled with grey relational analysis for multiple performance characteristics optimization of EDM parameters on ST 42 steel. *AIP Conference Proceedings*, *1725*, 020061. Advance online publication. doi:10.1063/1.4945515

Proenca, B. C., Blaga, L., dos Santos, J. F., Canto, L. B., Amancio-Filho, S. T. (n.d.). *Force Controlled Friction Riveting of Glass Fiber Reinforced Polyamide 6 And Aluminum Alloy 6056 Hybrid Joints*. Academic Press.

Proença, B.C. (2017). *Friction Riveting of Aluminium Alloy 6056-T6 with Short-Glassfiber-Reinforced Polyamide 6 Composite*. Academic Press.

Psyk, V., Risch, D., Kinsey, B. L., Tekkaya, A. E., & Kleiner, M. (2011). Electromagnetic forming—A review. *Journal of Materials Processing Technology*, *211*(5), 787–829. doi:10.1016/j.jmatprotec.2010.12.012

Puh, F., Jurkovic, Z., Perinic, M., Brezocnik, M., & Buljan, S. (2016). Optimizacija parametara obrade tokarenja s više kriterija kvalitete uporabom Grey relacijske analize. *Tehnicki Vjesnik (Strojarski Fakultet)*, *23*(2), 377–382. doi:10.17559/TV-20150526131717

Qin, J., Chen, Q., Yang, C., & Huang, Y. (2016). Research process on property and application of metal porous materials. *Journal of Alloys and Compounds*, *654*, 39–44. doi:10.1016/j.jallcom.2015.09.148

Qin, Y., Wen, P., Xia, D., Guo, H., Voshage, M., Jauer, L., Zheng, Y., Schleifenbaum, J. H., & Tian, Y. (2020). Effect of grain structure on the mechanical properties and in vitro corrosion behavior of additively manufactured pure Zn. *Additive Manufacturing*, *33*, 101134. doi:10.1016/j.addma.2020.101134

Quadrini, F., Squeo, E. A., & Tagliaferri, V. (2007). Machining of glass fiber reinforced polyamide. *Express Polymer Letters*, *1*(12), 810–816. doi:10.3144/expresspolymlett.2007.112

Quintens, H., Michalski, Q., Moussou, J., Strozzi, C., Sotton, J., Bellenoue, M., Boust, B., Pilla, G., & Rabeau, F. (2019). Experimental wall heat transfer measurements for various combustion regimes: Deflagration, Autoignition and Detonation. In *AIAA Propulsion and Energy 2019 Forum* (p. 4381). 10.2514/6.2019-4381

Rabiei, A., O'Neill, A. T., & Neville, B. P. (2004). Processing and development of a new high strength metal foam. *MRS Online Proceedings Library, 851*.

Raghuwanshi, R. K., & Verma, V. K. (2014). Mechanical and Thermal Characterization of Aero Grade Polymethyl Metha Acrylate Polymer used in Aircraft Canopy. *International Journal of Engineering and Advanced Technology*, *3*, 2249–8958.

Raihan, M. M., Otsuka, Y., Tsuchida, K., Manonukul, A., Ohnuma, K., & Miyashita, Y. (2021). Damage evaluation of HAp-coated porous titanium foam in simulated body fluid based on compression fatigue behavior. *Journal of the Mechanical Behavior of Biomedical Materials*, *117*, 104383. doi:10.1016/j.jmbbm.2021.104383 PMID:33596530

Raj Kumar, G., Balasubramaniyam, S., Senthil Kumar, M., Vijayanandh, R., Raj Kumar, R., & Varun, S. (2019). Crash Analysis on the Automotive Vehicle Bumper. *International Journal of Engineering and Advanced Technology*, *8*(6S3), 1602 - 1607. . doi:10.35940/ijeat.F1296.0986S319

Raj Kumar, G., Senthil Kumar, M., Vijayanandh, R., Raja Sekar, K., Mohamed Bak, K., & Varun, S. (2019). The Mechanical Characterization Of Carbon Fiber Reinforced Epoxy with Carbon Nanotubes. *International Journal of Mechanical and Production Engineering Research and Development*, *9*(1), 243-255.

Raj Kumar, G., Vijayanandh, R., Mohammad Bak, K., Shyam Chander, R., & Arawinth, R. (2018). Experimental Testing On Mechanical Properties Effect of Aluminum Foam. *International Journal of Mechanical and Production Engineering Research and Development*, *8*(7), 1047-1059.

Raj Kumar, G., Vijayanandh, R., Senthil Kumar, M., & Sathish Kumar, S. (2017). Experimental Testing and Numerical Simulation on Natural Composite for Aerospace Applications. AIP Conference Proceedings, 1953, 090045-1–090045-5. doi:10.1063/1.5032892

Raj Kumar, G., & Vijayanandh, R. (2020). Comparative Investigations on the Main Elements of Carbon Fiber Based Composites Using Computational Structural Simulations. *Journal of Physics: Conference Series*, *1504*(012003), 1–11. https://iopscience.iop.org/article/10.1088/1742-6596/1504/1/012003

Raj Kumar, G., Vijayanandh, R., Kamaludeen, M. B., Balasubramanian, S., Jagadeeshwaran, P., & Ramesh, M. (2020). Comparative Structural Characterization of Fiber Reinforced Composite Rotating Disc: A Validated Investigation. *Tribology in Industry*, *42*(4), 608–620. doi:10.24874/ti.899.05.20.10

Raj Kumar, G., Vijayanandh, R., Senthil Kumar, M., Naveen Kumar, K., & Ahilla Bharathy, L. (2017). Conceptual design and structural analysis of integrated composite Micro Aerial Vehicle. *Journal of Advanced Research in Dynamical and Control Systems*, *9*(14), 857–881.

Rajagurunathan, M., Raj Kumar, G., Vijayanandh, R., Vishnu, V., Rakesh Kumar, C., & Mohamed Bak, K. (2018). The Design Optimization of the Circular Piezoelectric Bimorph Actuators Using FEA. *International Journal of Mechanical and Production Engineering Research and Development*, *8*(7), 410-422.

Rajak, D. K., & Gupta, M. (2020). An Insight Into Metal Based Foams. Springer.

Rajak, D. K., & Gupta, M. (2020). Manufacturing Methods of Metal Foams. In *An Insight Into Metal Based Foams* (pp. 39–52). Springer. doi:10.1007/978-981-15-9069-6_3

Rajak, D. K., Kumaraswamidhas, L. A., & Das, S. (2014). An energy absorption behaviour of foam filled structures. *Procedia Materials Science*, *5*, 164–172. doi:10.1016/j.mspro.2014.07.254

Rajak, D. K., Kumaraswamidhas, L. A., & Das, S. (2018). Experimental fabrication and compression analysis characterization of LM30 Al alloy foam with 5wt% SiCp at room temperature. *Materials Research Express*, *5*(6), 066526. doi:10.1088/2053-1591/aac9a0

Rajesh, P. V. (2020). Effects of Process Parameters in Friction Stir Welding of Aluminium Matrix Composites - A Review. *Mahasarakham International Journal of Engineering Technology*, *7*(1), 16–22.

Raj, I. J. A., Vijayakumar, P., Kannan, T., Kumar, P., & Ragavan, R. V. (2016). Design optimization of turning parameters of PTFE (Teflon) cylindrical rods using ANOVA Methodology. *International Journal of Applied Engineering Research: IJAER*, *11*(3). http/www.ripublication.com/ijaer.htm%0ADesign

Rajkumar, G. R., Krishna, M., Murthy, H. N. N., Sharma, S. C., & Mahesh, K. R. V. (2012). Investigation of Repeated Low Velocity Impact Behaviour of GFRP /Aluminium and CFRP/Aluminium Laminates. *International Journal of Soft Computing and Engineering*, *1*(6), 50–58.

Rajpoot, Y. S., Narayanan, R. G., & Das, S. (2018). Predicting the effect of tool configuration during friction stir welding by cellular automata finite element analyses. *International Journal of Manufacturing Research*, *13*(4), 359–381. doi:10.1504/IJMR.2018.095377

Raj, R. E., & Daniel, B. S. S. (2007). Aluminum melt foam processing for light-weight structures. *Materials and Manufacturing Processes*, *22*(4), 525–530. doi:10.1080/10426910701236072

Ramesh, M., Vijayanandh, R., & Raj Kumar, G. (2021). Comparative Structural Analysis of Various Composite Materials based Unmanned Aerial Vehicle's Propeller by using Advanced Methodologies. *IOP Conf. Series: Materials Science and Engineering*, *1017*, 1-10. 10.1088/1757-899X/1017/1/012032

Ramu, I., Srinivas, P., & Vekatesh, K. (2018). Taguchi based grey relational analysis for optimization of machining parameters of CNC turning steel 316. *IOP Conference Series. Materials Science and Engineering*, *377*(1), 012078. Advance online publication. doi:10.1088/1757-899X/377/1/012078

Ramulu, P. J. (2013). *Forming Behavior of Friction Stir Welded Sheets* (PhD thesis). IIT Guwahati.

Ramulu, P. J., Kailas, S. V., & Narayanan, R. G. (2013c). Influence of tool rotation speed and feed rate on the forming limit of friction stir welded AA6061-T6 sheets. *Proceedings of the Institution of Mechanical Engineers. Part C, Journal of Mechanical Engineering Science*, *227*(3), 520–541. doi:10.1177/0954406212463996

Ramulu, P. J., Kailas, S. V., & Narayanan, R. G. (2015). Influence of different weld location and orientation on forming limit of friction stir welded sheets. *International Journal of Materials & Product Technology*, *50*, 147–160. doi:10.1504/IJMPT.2015.067833

Ramulu, P. J., Narayanan, R. G., & Kailas, S. V. (2013b). Forming limit investigation of friction stir welded sheets: Influence of shoulder diameter and plunge depth. *International Journal of Advanced Manufacturing Technology*, *69*(9), 2757–2772. doi:10.100700170-013-5245-x

Ramulu, P. J., Narayanan, R. G., Kailas, S. V., & Reddy, J. (2013a). Internal defect and process parameter analysis during friction stir welding of Al 6061 sheets. *International Journal of Advanced Manufacturing Technology*, *65*(9-12), 1515–1528. doi:10.100700170-012-4276-z

Ranatowski, E. (2008). Weldability of Titanium and its Alloys - Progress in Joining. *Advances in Materials Sciences*, *8*(2). Advance online publication. doi:10.2478/v10077-008-0033-2

Rao, B. S., & Chandu, D. G. (2014). Petrol engine exhaust valve design, analysis and manufacturing processes. *International Journal of Mechanical Engineering and Robotics Research*, *3*(4).

Rao, H. M., Kang, J., Huff, G., & Avery, K. (2018). Impact of specimen configuration on fatigue properties of self-piercing riveted aluminum to carbon fiber reinforced polymer composite. *International Journal of Fatigue*, *113*, 11–22. doi:10.1016/j.ijfatigue.2018.03.031

Rathee, S., Maheshwari, S., Noor Siddiquee, A., Srivastava, M., & Kumar Sharma, S. (2016). Process parameters optimization for enhanced micro hardness of AA6061/SiC surface composites fabricated via friction stir processing. *Materials Today: Proceedings*, *3*(10), 4151–4156. doi:10.1016/j.matpr.2016.11.089

Rathinasuriyan, C., Muniamuthu, S., Mystica, A., & Senthil Kumar, V. S. (2021). Investigation of heat generation during submerged friction stir welding on 6061-T6 aluminum alloy. *Materials Today: Proceedings*, *46*, 8320–8324. Advance online publication. doi:10.1016/j.matpr.2021.03.310

Rauch, L., & Madej, L. (2010). Application of the automatic image processing in modeling of the deformation mechanisms based on the digital representation of microstructure. *International Journal for Multiscale Computational Engineering*, *8*(3), 343–356. doi:10.1615/IntJMultCompEng.v8.i3.90

Ravi, B., Naik, B. B., & Kumar, G. R. (2018). Influence of tool rotational speed on the mechanical and microstructure properties of friction stir welded Al-B_4C MMCs. *Materials Today: Proceedings*, *5*(1), 5. doi:10.1016/j.matpr.2017.11.227

Raymond, S. D., Wild, P. M., & Bayley, C. J. (2004). On modeling of the weld line in finite element analyses of tailor-welded blank forming operations. *Journal of Materials Processing Technology*, *147*(1), 28–37. doi:10.1016/j.jmatprotec.2003.09.005

Rechtin, J., Torresani, E., Ivanov, E., & Olevsky, E. (2018). Fabrication of titanium-niobium-zirconium-tantalium alloy (TNZT) bioimplant components with controllable porosity by spark plasma sintering. *Materials (Basel)*, *11*(2), 181. doi:10.3390/ma11020181 PMID:29364165

Reis, A., Teixeira, P., Duarte, J. F., Santos, A., Da Rocha, A. B., & Fernandes, A. A. (2004). Tailored welded blanks—an experimental and numerical study in sheet metal forming on the effect of welding. *Computers & Structures*, *82*(17-19), 1435–1442. doi:10.1016/j.compstruc.2004.03.039

Reuter, C., & Tröster, T. (2017). Crashworthiness and numerical simulation of hybrid aluminium-CFRP tubes under axial impact. *Thin-walled Structures*, *117*, 1–9. doi:10.1016/j.tws.2017.03.034

Rodelas, J., & Lippold, J. (2013). Characterization of engineered nickel-base alloy surface layers produced by additive friction stir processing. *Metallography, Microstructure, and Analysis*, *2*(1), 1–12. doi:10.100713632-012-0056-2

Rodriguez, M. P. Q. (2000). Fourier transform infrared (FTIR) technology for the identification of organisms. *Clinical Microbiology Newsletter*, *22*(8), 57–61. doi:10.1016/S0196-4399(00)88850-9

Rollett, A. D., & Raabe, D. (2001). A hybrid model for mesoscopic simulation of recrystallization. *Computational Materials Science*, *21*(1), 69–78. doi:10.1016/S0927-0256(00)00216-0

Romanova, T., Pankratov, A., Litvinchev, I., Plankovskyy, S., Tsegelnyk, Y., & Shypul, O. (2021a). Sparsest packing of two-dimensional objects. *International Journal of Production Research*, *59*(13), 3900–3915. doi:10.1080/00207543.2020.1755471

Romanova, T., Stoyan, Y., Pankratov, A., Litvinchev, I., Plankovskyy, S., Tsegelnyk, Y., & Shypul, O. (2021b). Sparsest balanced packing of irregular 3D objects in a cylindrical container. *European Journal of Operational Research*, *291*(1), 84–100. doi:10.1016/j.ejor.2020.09.021

Rooks, B. (2001). Tailor-welded blanks bring multiple benefits to car design. *Assembly Automation*, *21*(4), 323–329. doi:10.1108/EUM0000000006014

Rossi, S., Fedel, M., Da Col, L., Deflorian, F., & Petrolli, S. (2017). Coatings to increase the corrosion behaviour of aluminium foam. *Surface Engineering*, *33*(6), 405–409. doi:10.1080/02670844.2016.1276700

Rüdinger, K., & Ismer, A. (1973). Recent Development in the Application of Titanium for Motorcars. *Titanium Science and Technology*, 185–199. doi:10.1007/978-1-4757-1346-6_16

Rupp, P., Elsner, P., & Weidenmann, K. A. (2017). Failure mode maps for four-point-bending of hybrid sandwich structures with carbon fiber reinforced plastic face sheets and aluminum foam cores manufactured by a polyurethane spraying process. *The Journal of Sandwich Structures & Materials*. Advance online publication. doi:10.1177/1099636217722052

Russell, M. J., Blignault, C., Horrex, N. L., & Wiesner, C. S. (2008). Recent developments in the friction stir welding of titanium alloys. *Welding in the World*, *52*(9), 12–15. doi:10.1007/BF03266662

Ryan, G. E., Pandit, A. S., & Apatsidis, D. P. (2008). Porous titanium scaffolds fabricated using a rapid prototyping and powder metallurgy technique. *Biomaterials*, *29*(27), 3625–3635.

Ryan, G., Pandit, A., & Apatsidis, D. P. (2006). Fabrication methods of porous metals for use in orthopaedic applications. *Biomaterials*, *27*(13), 2651–2670. doi:10.1016/j.biomaterials.2005.12.002 PMID:16423390

Ryland, A. L. (1958). X-ray diffraction. *Journal of Chemical Education*, *35*(2), 80–83.

Saboori, A., Gallo, D., Biamino, S., Fino, P., & Lombardi, M. (2017). An Overview of Additive Manufacturing of Titanium Components by Directed Energy Deposition : Microstructure and Mechanical Properties. *Applied Sciences (Basel, Switzerland)*, *7*(9), 883. Advance online publication. doi:10.3390/app7090883

Sacksteder, C., & Barry, B. A. (2001). Fourier transform infrared spectroscopy: A molecular approach to an organismal question. *Journal of Phycology*, *37*(2), 197–199. doi:10.1046/j.1529-8817.2001.037002197.x

Sadeesh, P., Venkatesh Kannan, M., Rajkumar, V., Avinash, P., Arivazhagan, N., Devendranath Ramkumar, K., & Narayanan, S. (2014). Studies on Friction Stir Welding of AA 2024 and AA 6061 Dissimilar Metals. *Procedia Engineering*, *75*, 145–149. doi:10.1016/j.proeng.2013.11.031

Sadighikia, S., Abdolhosseinzadeh, S., & Asgharzadeh, H. (2015). Production of high porosity Zn foams by powder metallurgy method. *Powder Metallurgy*, *58*(1), 61–66. doi:10.1179/1743290114Y.0000000109

Saidpour, H. (2004). Lightweight High Performance Materials for Car Body Structures. *NTI Technology Conference*, 14-19.

Salamati, M., Soltanpour, M., Fazli, A., & Zajkani, A. (2017). Parametric Study on the Electromagnetic Force-Fit Joining of Carbon Fiber Reinforced Plastic and Aluminum Tubes. *Procedia Engineering*, *207*, 986–991. doi:10.1016/j.proeng.2017.10.863

Salerno, G., Magnabosco, R., & Moura Neto, C. (2007). Mean strain influence in low cycle fatigue behavior of AA7175-T1 aluminum alloy. *International Journal of Fatigue*, *29*(5), 829–835. doi:10.1016/j.ijfatigue.2006.09.004

Salifu, S., Desai, D., Ogunbiyi, O., Sadiku, R., Adesina, O., & Adesina, O. (2020). Comparative study of high velocity impact response of aluminium 3105- H18 and carbon fibre-epoxy composite double hat bumper beams. *Materials Today: Proceedings*. Advance online publication. doi:10.1016/j.matpr.2020.03.828

Saluja, R. S., Narayanan, R. G., & Das, S. (2012). Cellular automata finite element (CAFE) model to predict the forming of friction stir welded blanks. *Computational Materials Science*, *58*, 87–100. doi:10.1016/j.commatsci.2012.01.036

Salvo, L., Martin, G., Suard, M., Marmottant, A., Dendievel, R., & Blandin, J. J. (2014). Processing and structures of solids foams. *Comptes Rendus Physique*, *15*(8-9), 662–673. doi:10.1016/j.crhy.2014.10.006

Samanta, A., Shen, N., Ji, H., Wang, W., Li, J., & Ding, H. (2018). Cellular automaton simulation of microstructure evolution for friction stir blind riveting. *Journal of Manufacturing Science and Engineering*, *140*(3), 031016–031026. doi:10.1115/1.4038576

Sanchez-Rexach, E., Johnston, T. G., Jehanno, C., Sardon, H., & Nelson, A. (2020). Sustainable Materials and Chemical Processes for Additive Manufacturing. *Chemistry of Materials*, *32*(17), 7105–7119. doi:10.1021/acs.chemmater.0c02008

Sangid, M. D. (2013). The physics of fatigue crack initiation. *International Journal of Fatigue*, *57*, 58–72. doi:10.1016/j.ijfatigue.2012.10.009

Sanjeev, M., Kaviarasan, V., & Venkatesan, R. (2012). Machining Parameter Optimization of Poly Tetra Fluoro Ethylene (PTFE) Using Genetic Algorithm. *International Journal of Modern Engineering Research*, *2*(1), 143–149.

Saracyakupoglu, T. (2019). The Qualification of the Additively Manufactured Parts in the Aviation Industry. *American Journal of Aerospace Engineering*, *6*(1), 1–10. doi:10.11648/j.ajae.20190601.11

Saravanan, R., & Rani, M. P. (2011). *Metal and alloy bonding-an experimental analysis: Charge density in metals and alloys*. Springer Science & Business Media.

Satheeshkumar, V., & Ganesh Narayanan, R. (2015a). Forming performance of adhesive bonded steel sheets reinforced with metallic wires. *Welding in the World - The International Journal of Materials Joining*, *59*(6), 883–900.

Satheeshkumar, V., Yadav, A. K., & Ganesh Narayanan, R. (2015). Chapter 4, Formability Prediction and Springback Evaluation of Adhesive-Bonded Steel Sheets. Advances in Material Forming and Joining, 61–80.

Satheeshkumar, V., & Ganesh Narayanan, R. (2014a). Formability of adhesive bonded steel sheets with artificial finite adhesive defects. *Journal of Strain Analysis for Engineering Design*, *49*(5), 286–300. doi:10.1177/0309324713517380

Satheeshkumar, V., & Ganesh Narayanan, R. (2014b). Investigation on the influence of adhesive properties on the formability of adhesive bonded steel sheets. In Proc IMechE Part C. *Journal of Mechanical Engineering Science*, *228*(3), 405–425. doi:10.1177/0954406213488727

Satheeshkumar, V., & Ganesh Narayanan, R. (2015b). Experimental evaluation and prediction of deep drawability of adhesive bonded blanks. *Materialwissenschaft und Werkstofftechnik*, *46*(7), 713–735. doi:10.1002/mawe.201500435

Satheeshkumar, V., & Ganesh Narayanan, R. (2016). Experimental evaluation and prediction of formability of adhesive bonded steel sheets at different adhesive properties. *Journal of Testing and Evaluation*, *44*(3), 1–13. doi:10.1520/JTE20140239

Sathishkumar, G. B., Sethuraman, P., Chanakyan, C., Sundaraselvan, S., Joseph Arockiam, A., Alagarsamy, S. V., El-mariung, A., Meignanamoorthy, M., Ravichandran, M., & Jayasathyakawin, S. (2021). Friction welding of similar and dissimilar materials: A review. *Materials Today: Proceedings*. Advance online publication. doi:10.1016/j.matpr.2021.03.089

Saunders, F. I., & Wagoner, R. H. (1996). Forming of tailor-welded blanks. *Metallurgical and Materials Transactions. A, Physical Metallurgy and Materials Science*, *27*(9), 2605–2616. doi:10.1007/BF02652354

Sazegaran, H., & Nezhad, S. M. M. (2021). Cell morphology, porosity, microstructure and mechanical properties of porous Fe-C-P alloys. *International Journal of Minerals Metallurgy and Materials*, *28*(2), 257–265. doi:10.100712613-020-1995-2

Schade, C. T., Murphy, T. F., & Walton, C. (2014). Development of atomized powders for additive manufacturing. *World Congress on Powder Metallurgy and Particulate Materials, PM 2014*, 215–226. https://pdfs.semanticscholar.org/164c/56f1e3a3e525162cc28a65d975e2dd39e357.pdf

Schimanski, K., Hehl, A., & Zoch, H. W. (2013). Failure behavior of diffusion bonded transition structures for integral FRP-Aluminum compounds. *Procedia Materials Science*, *2*, 189–196. doi:10.1016/j.mspro.2013.02.023

Schimanski, K., Schumacher, J., Lang, A., von Hehl, A., & Bomas, H. (2011). Characteristics of Titanium interface structures for advanced FRP Aluminium compounds. *Materials Science Forum*, *690*, 266–269. doi:10.4028/www.scientific.net/MSF.690.266

Schmidt, H., & Hattel, J. (2006). Analysis of the velocity field in the shear layer in FSW: experimental and numerical modelling. In *Proceedings of 6th International Friction Stir Welding Symposium*. TWI.

Schreiber, H. P., & Ouhlal, A. (2003). Polymer diffusion and the evolution of adhesive bond strength. *The Journal of Adhesion*, *79*(2), 141–153. doi:10.1080/00218460309572

Schubert, E., Klassen, M., Zerner, I., Walz, C., & Sepold, G. (2001). Light-weight structures produced by laser beam joining for future applications in automobile and aerospace industry. *Journal of Materials Processing Technology*, *115*(1), 2–8. doi:10.1016/S0924-0136(01)00756-7

Schulz, D. (2017). Komplexe Kunststoffteile wirtschaftlich entgraten. *JOT Journal für Oberflächentechnik*, *57*(3), 48–49. doi:10.100735144-017-0263-3

Schumacher, J., Bomas, H., & Zoch, H.-W. (2013). Failure behaviour of advanced seam structures for CFRP aluminium connections. *Procedia Materials Science*, *02*, 227–233. doi:10.1016/j.mspro.2013.02.028

Schwab, K. (2016). *The Fourth Industrial Revolution*. World Economic Forum.

Scriven, P. J., Brandon, J. A., & Williams, N. T. (1997). Relative influence of sheet rolling direction and weld orientation on formability of laser welded steel sheet. *Ironmaking & Steelmaking*, *24*(1), 79–85.

Senachin, P. K., & Babkin, V. S. (1982). Self-ignition of gas in front of the flame front in a closed vessel. *Combustion, Explosion, and Shock Waves*, *18*(1), 1–5. doi:10.1007/BF00783921

Senior, F. (2017). *The Development of Aluminium Foams for Enhanced Heat Transfer* (Doctoral dissertation). University of Sheffield.

Senthil Kumar, M., & Vijayanandh, R. (2018). Vibrational fatigue analysis of NACA 63215 small horizontal axis wind turbine blade. *Materials Today Proceedings*, *5*(2), 6665–6674. . doi:10.1016/j.matpr.2017.11.323

Senthil Kumar, M., Vijayanandh, R., & Gopi, B. (2018). Numerical Investigation on Vibration Reduction in Helicopter Main Rotor Using Air Blown Blades. *International Journal of Mechanical and Production Engineering Research and Development*, *8*(7), 152-164.

Senthil Kumar, M., Krishnan, A. S., & Vijayanandh, R. (2021). Computational, Experimental And Surface Characterisation Of Glass Fibre Reinforced Plastic Composite For Wind Turbine Blade Application. *Journal of Environmental Protection and Ecology*, *22*(2), 602–616.

Senthil, S., Raguraman, M., & Manalan, D. T. (2020). Manufacturing processes & recent applications of aluminium metal matrix composite materials: A review. *Materials Today: Proceedings*.

Sergio, D. (2011). Optimization of laminated composite plates and shells using genetic algorithms, neural networks and finite elements. *Latin American Journal of Solids and Structures*, *8*(4), 413–427. doi:10.1590/S1679-78252011000400003

Shah, K., Pinkerton, A. J., Salman, A., & Li, L. (2010). Effects of Melt Pool Variables and Process Parameters in Laser Direct Metal Deposition of Aerospace Alloys. *Materials and Manufacturing Processes*, *25*(12), 1372–1380. doi:10.1080/10426914.2010.480999

Shailesh, R., Ramu, M., Govindaraju, M., Karthikeyan, K., & Satheeshkumar, V. (2019). Performance Evaluation of Adhesive Spur Gear with the Influence of Properties of Adhesive. In R. Narayanan, S. Joshi, & U. S. Dixit (Eds.), *Advances in Computational Methods in Manufacturing. Lecture Notes on Multidisciplinary Industrial Engineering* (pp. 923–931). Springer. doi:10.1007/978-981-32-9072-3_77

Shakeri, H. R., Buste, A., Worswick, M. J., Clarke, J. A., Feng, F., Jain, M., & Finn, M. (2002). Study of damage initiation and fracture in aluminum tailor welded blanks made via different welding techniques. *Journal of Light Metals*, *2*(2), 95–110. doi:10.1016/S1471-5317(02)00028-7

Shakeri, H. R., Lee, Y., Worswick, M. J., Feng, F., Christy, W., & Clarke, J. A. (2001). Weld failure in formability testing of aluminum tailor welded blanks. *SAE Transactions*, 101–110. doi:10.4271/2001-01-0090

Shakil, M., Tariq, N. H., Ahmad, M., Choudhary, M. A., Akhter, J. I., & Babu, S. S. (2014). Effect of ultrasonic welding parameters on microstructure and mechanical properties of dissimilar joints. *Materials & Design*, *55*, 263–273. doi:10.1016/j.matdes.2013.09.074

Shankar, K. (1994). Fiber Metal Laminates—Matching the Best in Composites and Metals. *Materials Technology*, *9*(5-6), 114–119. doi:10.1080/10667857.1994.11785040

Shapovalov, V., & Boyko, L. (2004). Gasar—A new class of porous materials. *Advanced Engineering Materials*, *6*(6), 407–410. doi:10.1002/adem.200405148

Sharman, A. R. C., Hughes, J. I., & Ridgway, K. (2018). Characterisation of titanium aluminide components manufactured by laser metal deposition. *Intermetallics*, *93*, 89–92. doi:10.1016/j.intermet.2017.11.013

Sharma, P., & Pandey, P. M. (2018). Morphological and mechanical characterization of topologically ordered open cell porous iron foam fabricated using 3D printing and pressureless microwave sintering. *Materials & Design*, *160*, 442–454. doi:10.1016/j.matdes.2018.09.029

Sharma, S. C., Vizhian, S. P., Shashishankar, A., & Krishna, M. (2003). Influence of Heat Treatment on Microstructural and Tensile Properties of Aluminium E-Glass Short Fiber Composites. *Influence of Heat Treatment on Aluminium Composites*, *14*(4-5), 305–317. doi:10.1515/JMBM.2003.14.4-5.305

Sharma, V. M. J., Rao, G. S., Sharma, S. C., & George, K. M. (2014). Low Cycle Fatigue Behavior of AA2219-T87 at Room Temperature. *Materials Performance and Characterization*, *3*(1), 20130092. doi:10.1520/MPC20130092

Sharma, V. S., Fromentin, G., Poulachon, G., & Brendlen, R. (2014). Investigation of tool geometry effect and penetration strategies on cutting forces during thread milling. *International Journal of Advanced Manufacturing Technology*, *74*(5-8), 963–971. doi:10.100700170-014-6040-z

Shen, Z., Ding, Y., & Gerlich, A. P. (2019). *Advances in friction stir spot welding.* doi:10.1080/10408436.2019.1671799

Shim, D. S., Seo, J. Y., Yoon, H. S., Lee, K. Y., & Oh, W. J. (2018). Additive manufacturing of porous metals using laser melting of Ti6Al4V powder with a foaming agent. *Materials Research Express, 5*(8), 086518. doi:10.1088/2053-1591/aad117

Shin, W. G., Lee, S. H., & Kim, E. S. (2005). The study for the forming technology of Automobile Bumper beam using the Tailored Blank of Mash Seam Welding. *Proceedings of the Korean Society of Precision Engineering Conference*, 1376-1380.

Shirvanimoghaddam, K., Hamim, S. U., Akbari, M. K., Fakhrhoseini, S. M., Khayyam, H., Pakseresht, A. H., Ghasali, E., Zabet, M., Munir, K. S., Jia, S., Davim, P., & Naebe, M. (2017). Carbon fiber reinforced metal matrix composites: Fabrication processes and properties. *Composites. Part A, Applied Science and Manufacturing, 92*, 70–96. doi:10.1016/j.compositesa.2016.10.032

Shokrani, A., Dhokia, V., & Newman, S. T. (2012). Environmentally conscious machining of difficult-to-machine materials with regard to cutting fluids. *International Journal of Machine Tools & Manufacture, 57*, 83–101. doi:10.1016/j.ijmachtools.2012.02.002

Shrivastava, A., Zinn, M., Duffie, N. A., Ferrier, N. J., Smith, C. B., & Pfefferkorn, F. E. (2017). Force measurement-based discontinuity detection during friction stir welding. *Journal of Manufacturing Processes, 26*, 113–121. doi:10.1016/j.jmapro.2017.01.007

Shterenlikht, A., & Howard, I. C. (2006). The CAFE model of fracture—Application to a TMCR steel. *Fatigue & Fracture of Engineering Materials & Structures, 29*(9-10), 770–787. doi:10.1111/j.1460-2695.2006.01031.x

Shukla, M., Mahamood, R. M., Akinlabi, E. T., & Pityana, S. (2012). Effect of Laser Power and Powder Flow Rate on Properties of Laser Metal Deposited Ti6Al4V. *World Academy of Science and Technology, 6*, 44–48.

Shum, T. T. C. (2020). *Investigating mass flow in copper foam* (Doctoral dissertation). RMIT University.

Shypul, O., & Myntiuk, V. (2020). Transient thermoelastic analysis of a cylinder having a varied coefficient of thermal expansion. *Periodica Polytechnica Mechanical Engineering, 64*(4), 273–278. doi:10.3311/PPme.14733

Siddiq, A., & Ghassemieh, E. (2008). Thermomechanical analyses of ultrasonic welding process using thermal and acoustic softening effects. *Mechanics of Materials, 40*(12), 982–1000. doi:10.1016/j.mechmat.2008.06.004

Siegert, K., & Knabe, E. (1995). Fundamental research and draw die concepts for deep drawing of tailored blanks. *SAE Transactions*, 866–876. doi:10.4271/950921

Sima, F., Ristoscu, C., Duta, L., Gallet, O., Anselme, K., & Mihailescu, I. N. (2016). Laser thin films deposition and characterization for biomedical applications. In *Laser Surface Modification of Biomaterials* (pp. 77–125). Woodhead Publishing. doi:10.1016/B978-0-08-100883-6.00003-4

Simon, H. A. (2019). The Sciences of the Artificial (3rd ed.). MIT Press. doi:10.7551/mitpress/12107.001.0001

Singamneni, S., LV, Y., Hewitt, A., Chalk, R., Thomas, W., & Jordison, D. (2019). Additive Manufacturing for the Aircraft Industry: A review. *Journal of Aeronautics & Aerospace Engineering, 8*(1). https://www.longdom.org/open-access/additive-manufacturing-for-the-aircraft-industry-a-review-18967.html

Singamneni, S., Lv, Y., Hewitt, A., Chalk, R., Thomas, W., & Jordison, D. (2019). Additive Manufacturing for the Aircraft Industry: A Review. *Journal of Aeronautics & Aerospace Engineering, 8*(1), 215. doi:10.35248/2168-9792.19.8.215

Singh, R., Kumar, R., & Hashmi, M. S. J. (2017). Friction Welding of Dissimilar Plastic-Based Material by Metal Powder Reinforcement. In *Reference Module in Materials Science and Materials Engineering.* Elsevier. doi:10.1016/B978-0-12-803581-8.04160-6

Singh, R., Singh, S., Singh, I. P., Fabbrocino, F., & Fraternali, F. (2017). Investigation for surface finish improvement of FDM parts by vapor smoothing process. *Composites. Part B, Engineering, 111*, 228–234. doi:10.1016/j.compositesb.2016.11.062

Singh, S., Ramakrishna, S., & Singh, R. (2017). Material issues in additive manufacturing: A review. *Journal of Manufacturing Processes, 25*, 185–200. doi:10.1016/j.jmapro.2016.11.006

Sinmazçelik, T., Avcu, E., Bora, M. Ö., & Çoban, O. (2011). A review: Fibre metal laminates, background, bonding types and applied test methods. *Materials & Design, 32*(7), 3671–3685. doi:10.1016/j.matdes.2011.03.011

Sivertsen, K. (2007). *Polymer foams. In Polymer Physics, Spring*. Massachusetts Institute of Technology.

Skoglund, P., & Fransson, Å. (1998). Thermophysical properties of polypentadecanolactone. *Polymer, 39*(10), 1899–1906. doi:10.1016/S0032-3861(97)00473-4

Smetankina, N. V., Postnyi, O. V., Merkulova, A. I., & Merkulov, D. O. (2020). *Modeling of non-stationary temperature fields in multilayer shells with film heat sources. In 2020 IEEE KhPI Week on Advanced Technology (KhPIWeek) (pp. 242–246)*. IEEE. doi:10.1109/KhPIWeek51551.2020.9250139

Smith, B. H., Szyniszewski, S., Hajjar, J. F., Schafer, B. W., & Arwade, S. R. (2012). Steel foam for structures: A review of applications, manufacturing and material properties. *Journal of Constructional Steel Research, 71*, 1–10. doi:10.1016/j.jcsr.2011.10.028

Smith, E., & Dent, G. (2019). *Modern Raman spectroscopy: A practical approach*. Wiley. doi:10.1002/9781119440598

Smorygo, O., Mikutski, V., Marukovich, A., Ilyushchanka, A., Sadykov, V., & Smirnova, A. (2011). An inverted spherical model of an open-cell foam structure. *Acta Materialia, 59*(7), 2669–2678. doi:10.1016/j.actamat.2011.01.005

Sobiyi, K., Akinlabi, E., & Akinlabi, S. (2017). The Influence of Scanning Speed on the Laser Metal Deposition of Ti / TiC Powders. *Materials Technology, 51*(2), 345–351.

Sohail, M. (2012). *Applications of Polymers, Elastomers, Fibers and Composites.* https://books.google.com/books?id=o1raBwAAQBAJ&pgis=1

Sokolovan, O. A., Kuhn, M., & Palkowski, H. (2012). Deep drawing properties of lightweight steel/polymer/steel sandwich composites. *Archives of Civil and Mechanical Engineering, 12*(2), 105–112. doi:10.1016/j.acme.2012.05.001

Sola, A., & Nouri, A. (2019). Microstructural porosity in additive manufacturing: The formation and detection of pores in metal parts fabricated by powder bed fusion. *Journal of Advanced Manufacturing and Processing, 1*(3), e10021. doi:10.1002/amp2.10021

Song, M., & Kovacevic, R. (2003). Thermal modeling of friction stir welding in a moving coordinate system and its validation. *International Journal of Machine Tools & Manufacture, 43*(6), 605–615. doi:10.1016/S0890-6955(03)00022-1

Spur, G., Uhlmann, E., & Elbing, F. (1999). Dry-ice blasting for cleaning: Process, optimization and application. *Wear, 233*, 402–411. doi:10.1016/S0043-1648(99)00204-5

Spytkowski, P., & Klimek, T. (2009). Implementation of cellular automata framework dedicated to digital material representation. *Computer Methods in Materials Science, 9*(2), 283–288.

SreeArravind, M., Ramesh Kumar, S., RaviShankar, B., & Senthil Kumar, S. (2020). Low cycle fatigue behavior of aluminium 6063 alloy under the cyclic frequency of 0.2 Hz. *Materials Today: Proceedings*, *27*, 2376–2380. Advance online publication. doi:10.1016/j.matpr.2019.09.133

Srivastava, V. C., & Sahoo, K. L. (2007). Processing, stabilization and applications of metallic foams. Art of science. *Materials Science Poland*, *25*(3), 733–753.

Srivatsan, T. (2002). An investigation of the cyclic fatigue and fracture behavior of aluminum alloy 7055. *Materials & Design*, *23*(2), 141–151. doi:10.1016/S0261-3069(01)00071-1

Srivatsan, T. S. (1991). The low-cycle fatigue and cyclic fracture behaviour of 7150 aluminium alloy. *International Journal of Fatigue*, *13*(4), 313–321. doi:10.1016/0142-1123(91)90358-6

Staab, F., & Balle, F. (2018). Ultrasonic torsion welding of ageing-resistant Al/CFRP joints: Properties, microstructure and joint formation. *Ultrasonics*. PMID:30521993

Staley, J. T., & Lege, D. J. (1993). Advances in aluminium alloy products for structural applications in transportation. *Journal de Physique. IV*, *3*(C7), 179–190. doi:10.1051/jp4:1993728

Stasik, M. C., & Wagoner, R. H. (1996). Forming of tailor welded aluminum blanks, Aluminum and Magnesium for Automotive Applications. The Minerals, Metals & Materials Society, 69-82.

Stasik, M. C., & Wagoner, R. H. (1998). Forming of tailor welded aluminum blanks. *International Journal of Forming Processes*, *1*, 9–22.

Stavropoulos, P., & Foteinopoulos, P. (2018). Modelling of additive manufacturing processes: A review and classification. *Manufacturing Review*, *5*, 2. doi:10.1051/mfreview/2017014

Stefano, N. M., Casarotto Filho, N., Vergara, L. G. L., & Rocha, R. U. G. (2015). COPRAS (Complex Proportional Assessment): State of the Art Research and its Applications. *IEEE Latin America Transactions*, *13*(12), 3899–3906.

Stoll, M., Stemmer, F., Ilinzeer, S., & Weidenmann, K. A. (2017). Optimization of Corrosive Properties of Carbon Fiber Reinforced Aluminum Laminates due to Integration of an Elastomer Interlayer. *Key Engineering Materials*, *742*, 287–293. doi:10.4028/www.scientific.net/KEM.742.287

Struckmann, J., & Kieser, A. (2020). *Thermal deburring*. ATL Anlagentechnik Luhden GmbH.

Strumberger, N., Se, D., & Zagreb, F. (2005). Polymeric Metarials in Automobiles. *Promet-Traffic-Traffico*, *17*(3), 149–160.

Stuart, B. H. (2004). Infrared Spectroscopy: Fundamentals and Applications. John Wiley & Sons, Ltd. doi:10.1002/0470011149

Subramanya, P., Amar, M., Arun, S., Mervin, H., & Shrikantha, R. (2018). Friction stir welding of Aluminium matrix composites – A Review. *MATEC Web of Conferences, 144*, 1-13. 10.1051/matecconf/201814403002

Su, H., Wu, C. S., Pittner, A., & Rethmeier, M. (2013). Simultaneous measurement of tool torque, traverse force and axial force in friction stir welding. *Journal of Manufacturing Processes*, *15*(4), 495–500. doi:10.1016/j.jmapro.2013.09.001

Sulong, M. A., Vesenjak, M., Belova, I. V., Murch, G. E., & Fiedler, T. (2014). Compressive properties of Advanced Pore Morphology (APM) foam elements. *Materials Science and Engineering A*, *607*, 498–504. doi:10.1016/j.msea.2014.04.037

Sundarram, S. (2013). *Fabrication and characterization of open celled micro and nano foams*. Doctoral Dissertion, Faculty of the Graduate School of The University of Texas at Austin.

Sun, L., Wang, Y., Wang, L., Wang, F., Xu, H., Huang, W., & You, X. (2021). Preparation and properties of controllable aluminum foam. *Materials Research Express*, *8*(2), 026526. doi:10.1088/2053-1591/abe5ef

Sun, X., Liu, J., Lu, B., Zhang, P., & Zhao, M. (2017). Life cycle assessment-based selection of a sustainable lightweight automotive engine hood design. *The International Journal of Life Cycle Assessment*, *22*(9), 1373–1383. doi:10.100711367-016-1254-y

Surappa, M. K. (2003). Aluminium matrix composites: Challenges and Opportunities. *Sadhana*, *28*(1 & 2), 319–334. doi:10.1007/BF02717141

Suryanarayana, C., & Norton, M. G. (1998). *X-ray diffraction: A practical approach*. Springer. doi:10.1007/978-1-4899-0148-4

Suryanarayanan, R., & Sridhar, V. G. (2021). Influence of welding parameters on the weld properties in Friction stir spot welding of aluminium alloys of varying thicknesses. *Materials Today: Proceedings*, *46*, 8525–8531. Advance online publication. doi:10.1016/j.matpr.2021.03.534

Sutygina, A., Betke, U., Hasemann, G., & Scheffler, M. (2020). Manufacturing of Open-Cell Metal Foams by the Sponge Replication Technique. *IOP Conference Series. Materials Science and Engineering*, *882*(1), 012022. doi:10.1088/1757-899X/882/1/012022

Sutygina, A., Betke, U., & Scheffler, M. (2019). Open-cell aluminum foams by the sponge replication technique. *Materials (Basel)*, *12*(23), 3840. doi:10.3390/ma12233840 PMID:31766482

Swavely, D. S. (1991). Finishing and machining plastics. In M. L. Berins (Ed.), *SPI Plastics Engineering Handbook of the Society of the Plastics Industry, Inc* (pp. 657–692). Springer. doi:10.1007/978-1-4615-7604-4_23

Swiegers, G. F., Balakrishnan, S., & Huang, J. (2016). Assemblies and Self-Assembly. In J. Reedijk (Ed.), *Reference Module in Chemistry, Molecular Sciences and Chemical Engineering*. Elsevier. doi:10.1016/B978-0-12-409547-2.11710-X

Sylajakumari, P. A., Ramakrishnasamy, R., & Palaniappan, G. (2018). Taguchi grey relational analysis for multi-response optimization of wear in co-continuous composite. *Materials (Basel)*, *11*(9), 1–17. doi:10.3390/ma11091743 PMID:30223617

Szczepanik, K., Ooi, A., Aye, L., & Rosengarten, G. (2004). *A numerical study of heat transfer from a cylinder in cross flow. In 15th Australasian Fluid Mechanics Conference (pp. 13–17).* https://www.aeromech.usyd.edu.au/15afmc/proceedings/papers/AFMC00212.pdf

Szesz, E. M., & Lepienski, C. M. (2017). Anodic bonding of titanium alloy with bioactive glass. *Journal of Non-Crystalline Solids*, *471*, 19–27. doi:10.1016/j.jnoncrysol.2017.04.038

Szyniszewski, S. T., Smith, B. H., Hajjar, J. F., Schafer, B. W., & Arwade, S. R. (2014). The mechanical properties and modeling of a sintered hollow sphere steel foam. *Materials & Design, 54*, 1083-1094.

Takezawa, A., Kobashi, M., Koizumi, Y., & Kitamura, M. (2017). Porous metal produced by selective laser melting with effective isotropic thermal conductivity close to the Hashin–Shtrikman bound. *International Journal of Heat and Mass Transfer*, *105*, 564–572. doi:10.1016/j.ijheatmasstransfer.2016.10.006

Tang, H. P., Wang, J., & Qian, M. (2015). Porous titanium structures and applications. In *Titanium powder metallurgy* (pp. 533–554). Butterworth-Heinemann. doi:10.1016/B978-0-12-800054-0.00028-9

Taufik, M., & Jain, P. K. (2020). Part surface quality improvement studies in fused deposition modelling process: A review. *Australian Journal of Mechanical Engineering*, 1–25. Advance online publication. doi:10.1080/14484846.2020.1723342

Tayebi, P., Fazli, A., Asadi, P., & Soltanpour, M. (2019). Formability analysis of dissimilar friction stir welded AA 6061 and AA 5083 blanks by SPIF process. *CIRP Journal of Manufacturing Science and Technology*, *25*, 50–68. doi:10.1016/j.cirpj.2019.02.002

Tejedor, T. A., Singh, R., & Pilidis, P. (2013). Maintenance and repair of gas turbine components. In Modern Gas Turbine Systems: High Efficiency, Low Emission, Fuel Flexible Power Generation (pp. 565–634). Elsevier Ltd. doi:10.1533/9780857096067.3.565

Tepylo, N., Huang, X., & Patnaik, P. C. (2019). Laser-based Additive Manufacturing Technologies for Aerospace Applications. *Advanced Engineering Materials*, *21*(11), 1900617. doi:10.1002/adem.201900617

Tian, D. R., Pang, Y. H., Yu, L., & Sun, L. (2016). Production and characterization of high porosity porous Fe-Cr-C alloys by the space holder leaching technique. *International Journal of Minerals Metallurgy and Materials*, *23*(7), 793–798. doi:10.100712613-016-1293-1

Tian, Q. H., & Guo, X. Y. (2010). Electroless copper plating on microcellular polyurethane foam. *Transactions of Nonferrous Metals Society of China*, *20*, s283–s287. doi:10.1016/S1003-6326(10)60057-X

Tibbits, S., & Cheung, K. (2012). Programmable Materials for Architectural Assembly and Automation. *Assembly Automation*, *32*(3), 216–225. doi:10.1108/01445151211244348

Titanium, N. (2016). *The dawn of a new industry: Rapid plasma deposition* ™. Modem Metals.

Tlotleng, M., Masina, B., & Pityana, S. (2016). Characteristics of laser In-situ alloyed titanium aluminides coatings. *Procedia Manufacturing*, *7*, 39–45. doi:10.1016/j.promfg.2016.12.013

Tong, X., Shi, Z., Xu, L., Lin, J., Zhang, D., Wang, K., Li, Y., & Wen, C. (2020). Degradation behavior, cytotoxicity, hemolysis, and antibacterial properties of electro-deposited Zn–Cu metal foams as potential biodegradable bone implants. *Acta Biomaterialia*, *102*, 481–492. doi:10.1016/j.actbio.2019.11.031 PMID:31740321

Totten, G. E., MacKenzie, D. S., & MacKenzie, D. S. (2003). Handbook of Aluminum. In *Handbook of Aluminum*. CRC Press. doi:10.1201/9780203912591

Travin, A., Shur, M., Strelets, M., & Spalart, P. (2000). Detached-eddy simulations past a circular cylinder. *Flow, Turbulence and Combustion*, *63*(1), 293–313. doi:10.1023/A:1009901401183

Trhlíková, L., Zmeskal, O., Psencik, P., & Florian, P. (2016). Study of the thermal properties of filaments for 3D printing. *AIP Conference Proceedings*, *1752*(1), 040027. doi:10.1063/1.4955258

Troschitz, J., Vorderbrüggen, J., Kupfer, R., Gude, M., & Meschut, G. (2020). Joining of thermoplastic composites with metals using resistance element welding. *Applied Sciences (Switzerland)*, *10*(20), 1–12. doi:10.3390/app10207251

Tryfonov, O. V. (2013). *The method of regime setting for the thermal pulse processing by detonable gaseous mixtures using integrated CAD/CAE-systems* [Ph.D. dissertation]. National Aerospace University "Kharkiv Aviation Institute."

Tsartsaris, N., Meo, M., Dolce, F., Polimeno, U., Guida, M., & Marulo, F. (2011). Low-velocity impact behavior of fiber metal laminates. *Journal of Composite Materials*, *45*(7), 803–814. doi:10.1177/0021998310376108

Tušek, J., Kampuš, Z., & Suban, M. (2001). Welding of tailored blanks of different materials. *Journal of Materials Processing Technology*, *119*(1-3), 180–184. doi:10.1016/S0924-0136(01)00937-2

Uchihara, M., & Fukui, K. (2002). Tailored Blanks of High Strength Steels— Comparison of Welding Processes. *Welding in the World*, *46*(7-8), 41–48. doi:10.1007/BF03263389

Uchihara, M., & Fukui, K. (2006). Formability of tailor welded blanks fabricated by different welding processes: Study of tailor welded blanks using automotive high-strength steel sheets (1st report). *Welding International*, *20*(8), 612–621. doi:10.1533/wint.2006.3628

Udhaya Prakash, R., Raj Kumar, G., Vijayanandh, R., Senthil Kumar, M., & Ram Ganesh, T. (2016). Structural analysis of aircraft fuselage splice joint. *IOP Conference Series: Materials Science and Engineering, 149.*

Uhlmann, E., Kretzschmar, M., Elbing, F., & Mihotovic, V. (2010). Deburring with CO_2 snow blasting. In J. Aurich & D. Dornfeld (Eds.), *Burrs – Analysis, Control and Removal* (pp. 181–187). Springer. doi:10.1007/978-3-642-00568-8_19

Umanath, K., Palanikumar, K., Sankaradass, V., & Uma, K. (2021). Optimizations of friction stir welding process parameters of AA6063 aluminium alloy by Taguchi technique. *Materials Today: Proceedings*, *46*, 1–12. doi:10.1016/j.matpr.2021.02.539

Unver, I., Gulsoy, H. O., & Aydemir, B. (2013). Ni-625 Superalloy Foam Processed by Powder Space-Holder Technique. *Journal of Materials Engineering and Performance*, *22*(12), 3735–3741. doi:10.100711665-013-0702-0

Uriondo, A., Esperon-Miguez, M., & Perinpanayagam, S. (2015). The present and future of additive manufacturing in the aerospace sector: A review of important aspects. *Proceedings of the Institution of Mechanical Engineers. Part G, Journal of Aerospace Engineering*, *229*(11), 2132–2147. doi:10.1177/0954410014568797

Vaezi, M., Seitz, H., & Yang, S. (2013). A review on 3D micro-additive manufacturing technologies. *International Journal of Advanced Manufacturing Technology*, *67*(5-8), 1721–1754. doi:10.100700170-012-4605-2

Valente, M., Marini, D., Genova, V., Quitadamo, A., Marra, F., & Pulci, G. (2019). Lightweight metallic matrix composites: Development of new composites material reinforced with carbon structures. *Journal of Applied Biomaterials & Functional Materials*, *17*(1S). doi:10.1177/2280800019840294 PMID:31215817

Valino, A. D., Dizon, J. R. C., Espera, A. H. Jr, Chen, Q., Messman, J., & Advincula, R. C. (2019). Advances in 3D printing of thermoplastic polymer composites and nanocomposites. *Progress in Polymer Science*, *98*, 101162. doi:10.1016/j.progpolymsci.2019.101162

Valvi, S. R., Krishnan, A., Das, S., & Narayanan, R. G. (2016). Prediction of microstructural features and forming of friction stir welded sheets using cellular automata finite element (CAFE) approach. *International Journal of Material Forming*, *9*(1), 115–129. doi:10.100712289-015-1216-0

Vambol, O., Kondratiev, A., Purhina, S., & Shevtsova, M. (2021). Determining the parameters for a 3D-printing process using the fused deposition modeling in order to manufacture an article with the required structural parameters. *Eastern-European Journal of Enterprise Technologies*, *2*(1), 44–54. doi:10.15587/1729-4061.2021.227075

van Walle, E. (2001). Mechanical Test Specimens, Reconstitution of. In Encyclopedia of Materials: Science and Technology (pp. 5265–5268). Elsevier. doi:10.1016/B0-08-043152-6/00918-9

Varis, J. P., & Lepistö, J. (2003). A simple testing-based procedure and simulation of the clinching process using finite element analysis for establishing clinching parameters. *Thin-walled Structures*, *41*(8), 691–709. doi:10.1016/S0263-8231(03)00026-0

Vasudevan, A., Kumar, B. N., Depoures, M. V., Maridurai, T., & Mohanavel, V. (2020). Tensile and flexural behaviour of glass fibre reinforced plastic – Aluminium hybrid laminate manufactured by vacuum resin transfer moulding technique (VARTM). *Materials Today: Proceedings.*

Veeresh Kumar, G. B., & Pramod, R. (2017). Investigation of mechanical properties of aluminium reinforced glass fibre polymer Composites. *AIP Conference Proceedings*, *1859*, 020084-1–020084-7.

Vendra, L. J., Brown, J. A., & Rabiei, A. (2011). Effect of processing parameters on the microstructure and mechanical properties of Al–steel composite foam. *Journal of Materials Science*, *46*(13), 4574–4581. doi:10.100710853-011-5356-4

Vengatesh, D., & Chandramohan, V. (2014). Aluminium alloy metal matrix composite: Survey paper. *International Journal of Engineering Research and General Sciences*, *2*(6), 792–796.

Venkatesan, K., Geetha, S., Vijayanandh, R., Raj Kumar, G., Jagadeeshwaran, P., & Raj Kumar, R. (2020). Advanced structural analysis of various composite materials with carbon nano-tubes for property enhancement. AIP Conference Proceedings, 2270. doi:10.1063/5.0019367

Venkatesan, K., Ramanathan, K., Vijayanandh, R., Selvaraj, S., Raj Kumar, G., & Senthil Kumar, M. (2020). Comparative structural analysis of advanced multi-layer composite materials. *Materials Today: Proceedings*, *27*(3), 2673–2687. doi:10.1016/j.matpr.2019.11.247

Verma, S., & Misra, J. P. (2021). Experimental investigation on friction stir welding of dissimilar aluminium alloys. *Proceedings of the Institution of Mechanical Engineers, Part E: Journal of Process Mechanical Engineering*, 22-34. 10.1177/09544089211008694

Vijayanandh, R., Naveen Kumar, K., Senthil Kumar, M., Raj Kumar, G., Naveen Kumar, R., & Ahilla Bharathy, L. (2018). Material Optimization of High Speed Micro Aerial Vehicle using FSI Simulation. *Procedia Computer Science*, *133*, 2-9. doi:10.1016/j.procs.2018.07.002

Vijayanandh, R., Raj Kumar, G., & Jagadeeshwaran, P. (2016). Comparative Numerical Analyses of Different Carbon Nanotubes Added with Carbon Fiber–Reinforced Polymer Composite. *Nanomaterials and Nanocomposites: Characterization, Processing, and Applications*, *9*, 139 – 165. doi:10.1201/9781003160946-12

Vijayanandh, R., Venkatesan, K., Ramesh, M., Raj Kumar, G., & Senthil Kumar, M. (2019). Optimization of Orientation Of Carbon Fiber Reinforced Polymer Based On Structural Analysis. *International Journal of Scientific & Technology Research*, *8*(11), 3020 – 3029.

Vijayanandh, R., Senthil Kumar, M., Vasantharaj, C., Raj Kumar, G., & Soundarya, S. (2016). Numerical Study on Structural Health Monitoring for Unmanned Aerial Vehicle. *Journal of Advanced Research in Dynamical and Control Systems*, *9*(6), 1937–1958.

Vijayanandh, R., Venkatesan, K., Senthil Kumar, M., Raj Kumar, G., Jagadeeshwaran, P., & Raj Kumar, R. (2020). Comparative fatigue life estimations of Marine Propeller by using FSI. *IOP - Journal of Physics: Conference Series*, *1473*(012018), 1–8. doi:10.1088/1742-6596/1473/1/012018

Vikas, A. Y., & Soni, S. K. (2017). Simulation of melting process of a phase change material (PCM) using ANSYS (Fluent). *International Research Journal of Engineering and Technology*, *4*(5), 3289–3294.

Vlot, A., & Gunnink, J. W. (2001). *Fibre metal laminates: an introduction*. Springer. doi:10.1007/978-94-010-0995-9

Wagh, P. H., & Pagar, D. D. (2018). Investigation of mechanical and tribological behavior of composite material filled with black epoxy resin and aluminium tri-hydroxide using reinforcement of glass fiber. *AIP Conference Proceedings*, *2018*, 020025-1, 020025–15. doi:10.1063/1.5058262

Wagner, G., Balle, F., & Eifler, D. (n.d.). *Ultrasonic Welding of Hybrid Joints*. doi:10.1007/s11837-012-0269-5

Wagner, G., Balle, F., & Eifler, D. (2013). Ultrasonic welding of aluminum alloys to fiber reinforced polymers. *Advanced Engineering Materials*, *15*(9), 792–803. doi:10.1002/adem.201300043

Wally, Z. J., Van Grunsven, W., Claeyssens, F., Goodall, R., & Reilly, G. C. (2015). Porous titanium for dental implant applications. *Metals*, *5*(4), 1902–1920. doi:10.3390/met5041902

Wang, B., Shi, M., Sadrnia, H., & Lin, F. (1995). Structural Performance of Tailor Welded Sheet Steels. *SAE Transactions, 104*, 222–234. doi:10.4271/950376

Wang, D., Xiao, B. L., Ni, D. R., & Ma, Z. Y. (2014). Friction Stir Welding of Discontinuously Reinforced Aluminum Matrix Composites: A Review. *Acta Metallurgica Sinica. English Letters, 27*(5), 816–824. doi:10.100740195-014-0143-2

Wang, J., Zhang, G., Zheng, X., Li, J., Li, X., Zhu, W., & Yanagimoto, J. (2020). A self-piercing riveting method for joining of continuous carbon fiber reinforced composite and aluminum alloy sheets. *Composite Structures, 113219*. Advance online publication. doi:10.1016/j.compstruct.2020.113219

Wang, N., Chen, X., Li, A., Li, Y., Zhang, H., & Liu, Y. (2016). Three-point bending performance of a new aluminum foam composite structure. *Transactions of Nonferrous Metals Society of China, 26*(2), 359–368. doi:10.1016/S1003-6326(16)64088-8

Wang, N., Chen, X., Li, Y., Liu, Z., Zhao, Z., Cheng, Y., Liu, Y., & Zhang, H. (2017a). The cell size reduction of aluminum foam with dynamic gas injection based on the improved foamable melt. *Colloids and Surfaces. A, Physicochemical and Engineering Aspects, 527*, 123–131. doi:10.1016/j.colsurfa.2017.05.023

Wang, Q., Xu, F. M., Xu, Q. Y., & Xiong, S. M. (2007). Study on an aluminum foam by gas injection foaming. *Foundry, 56*, 814–818.

Wang, S. Q., Patel, V. K., Bhole, S. D., Wen, G. D., & Chen, D. L. (2015). Microstructure and mechanical properties of ultrasonic spot welded Al/Ti alloy joints. *Materials & Design, 78*, 33–41. doi:10.1016/j.matdes.2015.04.023

Wang, X. S., Hu, Z. L., Yuan, S. J., & Hua, L. (2014). Influence of tube spinning on formability of friction stir welded aluminum alloy tubes for hydroforming application. *Materials Science and Engineering A, 607*, 245–252. doi:10.1016/j.msea.2014.03.125

Wang, Z., Wang, C., Li, C., Qin, Y., Zhong, L., Chen, B., Li, Z., Liu, H., Chang, F., & Wang, J. (2017b). Analysis of factors influencing bone ingrowth into three-dimensional printed porous metal scaffolds: A review. *Journal of Alloys and Compounds, 717*, 271–285. doi:10.1016/j.jallcom.2017.05.079

Wan, Y., Wang, Z. Q., Zhou, L. M., & Chang, M. Z. (2015). Tensile Test Performance of Fibre Reinforced SMATed Aluminium Laminates. *International Conference on Material Science and Application*, 943-947. 10.2991/icmsa-15.2015.175

Ward-Close, C. M., & Froes, F. H. (1994). *Developments in the Synthesis of Lightweight Metals*. Academic Press.

Wegener, B., Sichler, A., Milz, S., Sprecher, C., Pieper, K., Hermanns, W., ... Wegener, V. (2020). Development of a novel biodegradable porous iron-based implant for bone replacement. *Scientific Reports, 10*(1), 1–10. doi:10.103841598-020-66289-y PMID:32499489

Weisheit, A., & Rolink, G. (2016). Influence of Process Conditions in Laser Additive Manufacturing on the Microstructure Evolution of Fe-Al Alloys- A Comparison of Laser Metal Deposition and Selective Laser Melting. pdf. *AAM Workshop, MPEI*. https://www.mpie.de/3476506/Weisheit.pdf

Weiss, M., Dingle, M. E., Rolfe, B. F., & Hodgson, P. D. (2007). The influence of temperature on the forming behavior of metal/polymer laminates in sheet metal forming. *Journal of Engineering Materials and Technology, 129*(4), 530–537. doi:10.1115/1.2772329

Welf-Guntram, D., Reinhard, M., Raik, G., & Danilo, M. (2013). Numerical and Experimental Analysis of Self Piercing Riveting Process with Carbon Fiber-Reinforced Plastic and Aluminium Sheets. *Key Engineering Materials, 554-557*, 1045–1054. doi:10.4028/www.scientific.net/KEM.554-557.1045

Wen, C. E., Mabuchi, M., Yamada, Y., Shimojima, K., Chino, Y., & Asahina, T. (2001). Processing of biocompatible porous Ti and Mg. *Scripta Materialia*, *45*(10), 1147–1153. doi:10.1016/S1359-6462(01)01132-0

Whitesides, G. M., & Grzybowski, B. (2002). Self-Assembly at All Scales. *Science*, *295*(5564), 2418–2421. doi:10.1126cience.1070821 PMID:11923529

Wholer's Report. (2016). *3D Printing and Additive Manufacturing State of the Industry, Annual Worldwide Progress Report*. Wholer's Associates Inc.

Wickramasinghe, S., Do, T., & Tran, P. (2020). FDM-based 3D printing of polymer and associated composite: A review on mechanical properties, defects and treatments. *Polymers*, *12*(7), 1529. doi:10.3390/polym12071529 PMID:32664374

Wirth, F. X., Fuchs, A. N., Rinck, P., & Zaeh, M. F. (2014). Friction Press Joining of Laser-Texturized Aluminum with Fiber Reinforced Thermoplastics. *Advanced Materials Research*, *966-967*, 536–545. doi:10.4028/www.scientific.net/AMR.966-967.536

Wohlers, T. (2019). *Wohlers report 2019: 3D printing and additive manufacturing state of the industry*. Wohlers Associates.

Wong, K. V., & Hernandez, A. (2012). A review of additive manufacturing. *International Scholarly Research Notices*, *2012*, 1–10.

WorldAutoSteel. (2020, December 21). https://www.worldautosteel.org/

Woźniak, K. (2013). Plastic blasting and deflashing media in shot blasting treatment. *Chemik*, *67*(12), 1227–1238.

Wu, B., Myant, C., & Weider, S. (2017). *The Value of Additive Manufacturing: Future Opportunities*. Briefing paper No. 2. Institute for Molecular Science and Engineering, Imperial College London.

Wu, G. Q., Li, Z. F., Luo, G. X., Li, H. Y., & Huang, Z. (2007). Dynamic simulation of solid-state diffusion bonding. *Materials Science and Engineering A*, *452–453*, 529–535. doi:10.1016/j.msea.2006.10.115

Wu, L. H., Xiao, B. L., Nagatsuka, K., Nakata, K., & Ma, Z. Y. (2020). Achieving strong friction lap joints of carbon-fiber reinforced plastic and metals by modifying metal surface structure via laser-processing pretreatment. *Composite Structures*, *242*, 112167. doi:10.1016/j.compstruct.2020.112167

Wu, Y., Liu, Q., Fu, J., Li, Q., & Hui, D. (2017). Dynamic crash responses of bioinspired aluminum honeycomb sandwich structures with CFRP panels. *Composites. Part B, Engineering*, *121*, 122–133. doi:10.1016/j.compositesb.2017.03.030

Xiao, Y., Hu, Y., Zhang, J., Song, C., Huang, X., Yu, J., & Liu, Z. (2018). The Bending Responses of Sandwich Panels with Aluminium Honeycomb Core and CFRP Skins Used in Electric Vehicle Body. *Advances in Materials Science and Engineering*, *2018*, 1–11. doi:10.1155/2018/5750607

Xie, B., Fan, Y. Z., Mu, T. Z., & Deng, B. (2017). Fabrication and energy absorption properties of titanium foam with CaCl2 as a space holder. *Materials Science and Engineering A*, *708*, 419–423. doi:10.1016/j.msea.2017.09.123

Xu, C., Mao, Y., & Hu, Z. (2017). Tonal and broadband noise control of an axial-flow fan with metal foams: Design and experimental validation. *Applied Acoustics*, *127*, 346–353. doi:10.1016/j.apacoust.2017.06.018

Xue, W., Krishna, B. V., Bandyopadhyay, A., & Bose, S. (2007). Processing and biocompatibility evaluation of laser processed porous titanium. *Acta Biomaterialia*, *3*(6), 1007–1018. doi:10.1016/j.actbio.2007.05.009 PMID:17627910

Xu, J., Huang, X., Davim, J. P., Ji, M., & Chen, M. (2020). On the machining behavior of carbon fiber reinforced polyimide and PEEK thermoplastic composites. *Polymer Composites*, *41*(9), 3649–3663. doi:10.1002/pc.25663

Xu, M., Huang, G., Dong, Y., & Feng, S. (2018). An experimental investigation into the high velocity penetration resistance of CFRP and CFRP/aluminium laminates. *Composite Structures, 188*, 450–460. doi:10.1016/j.compstruct.2018.01.020

Xu, Z., Zhu, Z., Wang, P., Meenashisundaram, G. K., Nai, S. M. L., & Wei, J. (2020). Fabrication of porous CoCrFeMnNi high entropy alloy using binder jetting additive manufacturing. *Additive Manufacturing, 35*, 101441. doi:10.1016/j.addma.2020.101441

Yahya, M. M., Mallik, N., & Chakrabarty, I. (2015). Low Cycle Fatigue (LCF) Behavior of AA6063 Aluminium Alloy at Room Temperature. *International Journal of Emerging Technology and Advanced Engineering, 5*(12).

Yamamura, S., & Iwase, E. (2021). Hybrid Hinge Structure with Elastic Hinge on Self-Folding of 4D Printing using a Fused Deposition Modeling 3D Printer. *Materials & Design, 203*, 109605. doi:10.1016/j.matdes.2021.109605

Yan, C., Wang, J., & Song, X. (2020). Fatigue behavior and damage mechanism of aluminum foam sandwich with carbon fiber face-sheets. *Journal of Mechanical Science and Technology, 34*(3), 1-9.

Yan, C., Song, X., Zhu, H., Jing, C., & Feng, S. (2017). Flexural response of carbon fiber reinforced aluminum foam sandwich. *Journal of Composite Materials, 52*(14), 1887–1897. doi:10.1177/0021998317735166

Yang, F., Ma, Y., Tao, N., & He, X. (2017). Experimental study on nonlinear vibrating of aluminum foam using electronic speckle pattern interferometry. In *Fifth International Conference on Optical and Photonics Engineering* (Vol. 10449). International Society for Optics and Photonics. 10.1117/12.2270397

Yan, L., Chen, X., Zhang, Y., Newkirk, J. W., & Liou, F. (2017). Fabrication of functionally graded Ti and γ-TiAl by laser metal deposition. *JOM: The Materials. Metals & Materials Society, 69*(12), 2756–2761. doi:10.100711837-017-2582-5

Yeh, P. C., Chang, P. Y., Wang, J., Yang, J. M., Wu, P. H., & Liu, M. C. (2012). Bearing strength of commingled boron/glass fiber reinforced aluminum laminates. *Composite Structures, 94*(11), 3160–3173. doi:10.1016/j.compstruct.2012.05.001

Yilong, L., Guibao, Q., Yang, Y., Xuewei, L., & Chenguang, B. (2016). Preparation and compressive properties of magnesium foam. *Rare Metal Materials and Engineering, 45*(10), 2498–2502. doi:10.1016/S1875-5372(17)30022-X

Yin, H., Liu, J., & And Yang, F. (2019). Hybrid Structure Design of Lightweight Robotic Arms Based on Carbon Fiber Reinforced Plastic and Aluminum alloy. *IEEE Access: Practical Innovations, Open Solutions, 7*, 64932–64945. doi:10.1109/ACCESS.2019.2915363

Yuan, C., Bergsma, O., Koussios, S., Zu, L., & Beukers, A. (2012). Optimization of Sandwich Composites Fuselages under Flight Loads. *Applied Composite Materials, 19*(1), 47–64. doi:10.100710443-010-9180-9

Yuan, L., Ding, S., & Wen, C. (2019). Additive manufacturing technology for porous metal implant applications and triple minimal surface structures: A review. *Bioactive Materials, 4*, 56–70. doi:10.1016/j.bioactmat.2018.12.003 PMID:30596158

Yuan, S. J., Hu, Z. L., & Wang, X. S. (2012). Evaluation of formability and material characteristics of aluminum alloy friction stir welded tube produced by a novel process. *Materials Science and Engineering A, 543*, 210–216. doi:10.1016/j.msea.2012.02.076

Yuan, W., Tang, Y., Yang, X., & Wan, Z. (2012). Porous metal materials for polymer electrolyte membrane fuel cells–A review. *Applied Energy, 94*, 309–329. doi:10.1016/j.apenergy.2012.01.073

Yunlian, Q., Ju, D., Quan, H., & Liying, Z. (2000). Electron beam welding, laser beam welding and gas tungsten arc welding of titanium sheet. *Materials Science and Engineering A, 280*(1), 177–181. doi:10.1016/S0921-5093(99)00662-0

Yuqing, M., Liming, K., Chunping, H., Fencheng, L., & Qiang, L. (2016). Formation characteristic, microstructure, and mechanical performances of aluminum-based components by friction stir additive manufacturing. *International Journal of Advanced Manufacturing Technology*, *83*(9), 1637–1647. doi:10.100700170-015-7695-9

Yusuf, S. M., Cutler, S., & Gao, N. (2019). Review: The Impact of Metal Additive Manufacturing on the Aerospace Industry. *Metals*, *9*(12), 1286. doi:10.3390/met9121286

Zaki, M. U., & Mohammad, R. (2014). Three Dimensional Simulation of Impact Response of Glass Fiber Reinforced Aluminium Laminates for Conical and Blunt Nose Projectiles. *International Journal of Emerging Technology and Advanced Engineering*, *14*(1), 480–488.

Zaman, E. (2011). *Açık Hücreli Alüminyum Köpük Üretimi Ve Karakterizasyonu* (MSc Thesis). Institute of Science and Technology, İstanbul Technical University.

Zavadskas, E. K., Kaklauskas, A., & Sarka, V. (1994). The new method of multi-criteria Complex Proportional Assessment of projects. *Technological and Economic Development of Economy*, *1*(3), 131–139.

Zdravkov, B., Čermák, J., Šefara, M., & Janků, J. (2007). Pore classification in the characterization of porous materials: A perspective. *Open Chemistry*, *5*(2), 385–395. doi:10.247811532-007-0017-9

Zeng, W. W., Hou, S. H., Ding, X. J., Duan, D. L., Li, S., & Zhang, S. H. (2017). Synthesis and Compression Property of Oxidation-Resistant Ni–Al Foams. *Acta Metallurgica Sinica. English Letters*, *30*(10), 965–972. doi:10.100740195-017-0569-4

Zhang, H., & Zhang, Z. (2007). Numerical modelling of friction stir welding process by using rate-dependent constitutive model. *Journal of Materials Science and Technology*, *23*(1), 73–80.

Zhang, X., Li, L., Chen, Y., Zhu, X., & Ji, S. (2019). Numerical Simulation Analysis of Dual-Beam Laser Welding of Tailored Blanks with Different Thicknesses. *Metals*, *9*(2), 135. doi:10.3390/met9020135

Zhang, X., Li, Y., Liu, Y., & Zhang, H. (2013). Fabrication of a bimodal micro/nanoporous metal by the Gasar and dealloying processes. *Materials Letters*, *92*, 448–451. doi:10.1016/j.matlet.2012.10.090

Zhang, X., Li, Y., Zhang, H., & Liu, Y. (2013). Fabrication of a three-dimensional bimodal porous metal. *Materials Letters*, *106*, 417–420. doi:10.1016/j.matlet.2013.05.077

Zhang, Y. M., Chu, X. M., Wang, H., He, S. Y., & He, D. P. (2009). Fabrication of Al–Mg–Re foams and their corrosion resistance properties. *Corrosion Science*, *51*(6), 1436–1440. doi:10.1016/j.corsci.2009.03.032

Zhang, Y., He, Q., Xia, L., Li, Y., & Song, S. (2018). Algae cathode microbial fuel cells for cadmium removal with simultaneous electricity production using nickel foam/graphene electrode. *Biochemical Engineering Journal*, *138*, 179–187. doi:10.1016/j.bej.2018.07.021

Zhang, Y., Luo, R., Zhang, J., & Xiang, Q. (2011). The reinforcing mechanism of carbon fiber in composite adhesive for bonding carbon/carbon composites. *Journal of Materials Processing Technology*, *211*(2), 167–173. doi:10.1016/j.jmatprotec.2010.08.028

Zhang, Z. (2008). Comparison of two contact models in the simulation of friction stir welding process. *Journal of Materials Science*, *43*(17), 5867–5877. doi:10.100710853-008-2865-x

Zhao, J., Wu, C. S., & Su, H. (2021). Acoustic effect on the tensile properties and metallurgical structures of dissimilar friction stir welding joints of Al/Mg alloys. *Journal of Manufacturing Processes*, *65*, 328–341. doi:10.1016/j.jmapro.2021.03.057

Zhao, K. M., Chun, B. K., & Lee, J. K. (2001). Finite element analysis of tailor-welded blanks. *Finite Elements in Analysis and Design*, *37*(2), 117–130. doi:10.1016/S0168-874X(00)00026-3

Zheng, Z. (2019). Fabrication on Bioinspired Surfaces. Bioinspired Design of Materials Surfaces. doi:10.1016/B978-0-12-814843-3.00003-X

Zhu, G., Sun, G., Liu, Q., Li, G., & Li, Q. (2017). On crushing characteristics of different configurations of metal-composites hybrid tubes. *Composite Structures*, *175*, 58–69. doi:10.1016/j.compstruct.2017.04.072

Zhu, G., Sun, G., Yu, H., Li, S., & Li, Q. (2018). Energy absorption of metal, composite and metal/composite hybrid structures under oblique crushing loading. *International Journal of Mechanical Sciences*, *135*, 458–483. doi:10.1016/j.ijmecsci.2017.11.017

Zhu, H., Cheng, B., Ma, G., Lin, X., Zhang, Y., Wu, D., Ning, Y., & Wang, L. (2020). An eave-like model for the welding of 304 stainless-steel tailor-welded blanks with different thicknesses. *Optics and Lasers in Engineering*, *134*, 106309. doi:10.1016/j.optlaseng.2020.106309

Zhu, J., & Chen, J. C. (2006). Fuzzy neural network-based in-process mixed material-caused flash prediction (FNN-IPMFP) in injection molding operations. *International Journal of Advanced Manufacturing Technology*, *29*(3-4), 308–316. doi:10.100700170-005-2528-x

Zhu, S., & Chai, G. B. (2013). Impact of Aluminum, CFRP Laminates, Fibre-Metal Laminates and Sandwich Panels. *Composite Materials and Joining Technologies for Composites*, *07*, 199–205. doi:10.1007/978-1-4614-4553-1_21

Zindani, D., & Kumar, K. (2019). An insight into additive manufacturing of fiber reinforced polymer composite. *International Journal of Lightweight Materials and Manufacture*, *2*(4), 267–278. doi:10.1016/j.ijlmm.2019.08.004

Zindani, D., & Kumar, K. (2020). A brief review on cryogenics in machining process. *SN Applied Sciences*, *2*(6), 1107. doi:10.100742452-020-2899-5

Zyl, I., Van, & Yadroitsava, I., & Yadroitsev. (2016). Residual stress in ti6al4v objects produced by direct metal laser sintering. *South African Journal of Industrial Engineering*, *27*(December), 134–141.

About the Contributors

Kaushik Kumar, B.Tech (Mechanical Engineering, REC (Now NIT), Warangal), MBA (Marketing, IGNOU), and Ph.D (Engineering, Jadavpur University), is presently an Associate Professor in the Department of Mechanical Engineering, Birla Institute of Technology, Mesra, Ranchi, India. He has 14 years of Teaching & Research and over 11 years of industrial experience in a manufacturing unit of Global repute. His areas of teaching and research interest are Quality Management Systems, Optimization, Non-conventional machining, CAD / CAM, Rapid Prototyping and Composites. He has 9 Patents,15 Book, 6 Edited Book 35 Book Chapters, 120 international Journal publications, 18 International and 8 National Conference publications to his credit. He is on the editorial board and review panel of 7 International and 1 National Journals of repute. He has been felicitated with many awards and honours.

J. Paulo Davim received his Ph.D. degree in Mechanical Engineering in 1997, M.Sc. degree in Mechanical Engineering (materials and manufacturing processes) in 1991, Mechanical Engineering degree (5 years) in 1986, from the University of Porto (FEUP), the Aggregate title (Full Habilitation) from the University of Coimbra in 2005 and the D.Sc. from London Metropolitan University in 2013. He is Eur Ing by FEANI-Brussels and Senior Chartered Engineer by the Portuguese Institution of Engineers with an MBA and Specialist title in Engineering and Industrial Management. Currently, he is Professor at the Department of Mechanical Engineering of the University of Aveiro, Portugal. He has more than 30 years of teaching and research experience in Manufacturing, Materials, Mechanical and Industrial Engineering, with special emphasis in Machining & Tribology. He has also interest in Management, Engineering Education and Higher Education for Sustainability. He has guided large numbers of postdoc, Ph.D. and master's students as well as coordinated & participated in several financed research projects. He has received several scientific awards. He has worked as evaluator of projects for international research agencies as well as examiner of Ph.D. thesis for many universities. He is the Editor in Chief of several international journals, Guest Editor of journals, books Editor, book Series Editor and Scientific Advisory for many international journals and conferences. Presently, he is an Editorial Board member of 25 international journals and acts as reviewer for more than 80 prestigious Web of Science journals. In addition, he has also published as editor (and co-editor) more than 100 books and as author (and co-author) more than 10 books, 80 book chapters and 400 articles in journals and conferences (more than 200 articles in journals indexed in Web of Science core collection/h-index 45+/6000+ citations and SCOPUS/h-index 53+/8500+ citations).

* * *

Arvind Kumar Agrawal received his PhD in Mechanical Engineering (specialization Manufacturing Engineering) from IIT Guwahati in 2019. After completing his doctorate, he was hired as Assistant Professor in Department of Mechanical Engineering in Madanapalle Institute of Technology & Science, Madanapalle, Andhra Pradesh. Before joining PhD, he completed Masters in Technology from NIT Jamshedpur. He has also worked in NML Jamshedpur, and VVIT Purnea, Bihar. He completed his undergraduate in engineering from M.A.C.ET Patna. He is a reviewer for few journals and published papers in reputed international journals. His area of research is metal forming and welding, and other manufacturing topics. His research work is a combination of experiments, mathematical analysis, and modeling.

Sevim Akyuz is a professor of physics, currently working at the Physics Department of Istanbul Kultur University (Istanbul-Turkey). Her research interest is on experimental and theoretical molecular spectroscopy and mainly focused on the study of a relationship between the structure and spectral properties of macromolecules using combined theoretical and experimental methods. She has 182 scientific papers as SCI record and total number of citations according to Web of Knowledge is 1335. Her H-index according to Web of Knowledge is 22.

Saravanan Arunachalam completed B.E in Mechanical Engineering at V.I.T, Vellore. Completed M.E in Manufacturing Engineering at M.A.M College of Engineering, Trichy. Completed Ph.D. from the Department of Production Engineering at N.I.T, Trichy. Research focuses on Artificial Intelligence in Geometric Tolerances.

Nuray Beköz Üllen is currently working in the Department of Metallurgical and Materials Engineering, Istanbul University-Cerrahpaşa (IUC), Turkey. She obtained her BSc, MSc and PhD degree at Department of Metallurgical and Materials Engineering from Istanbul University, Turkey in 2003, 2006 and 2011, respectively. Her main research interest includes powder metallurgy, metal foams, coating technology, surface treatments, material characterizations, machinability and welding. In recent years, she has focused on improving the surface properties of dense porous materials. She has collaborated actively with researchers in several other disciplines of mechanical, chemistry and physics. She has taught on engineering materials and deformation mechanisms, highly porous material production techniques and characterization, failure of materials. She had role as a consultant in master theses in the fields of material characterization. She has published several research articles on material production, characterization and processing.

Dmytro Brega, originally from Rivne, Ukraine. He has more than 10 years of experience in computational fluid dynamics together with aircraft manufacturing and design. He has also a significant international experience – post doctorate study in Warsaw Polytechnic University (Poland) and lecturing in University of Basque Country (Spain). Currently Dmytro works at Aerodynamic Division at National Aerospace University. His research interests cover additive manufacturing, plasma simulations and high-energy processing methods. Research results are highlighted in more than 20 publications.

K. Brahma Raju received his M.E degree in Machine Design Engineering from University of Roorkee (Presently I.I.T Roorkee) in the year 1993. He obtained his PhD degree in Mechanical Engineering from JNTU College of Engineering, Anantapur in the year 2009. His areas of interest are fatigue and fracture mechanics, characterization of materials, composites and finite element analysis. Currently he is serving as a Professor & Head in the Department of Mechanical Engineering at S.R.K.R. Engineering College, Bhimavaram, Andhra Pradesh. He has teaching experience of more than Thirty years to UG/PG students of Mechanical Engineering and he had published more than 40 research papers in various International/National Journals and conferences to his credit. He had guided 5 Ph.D. scholars and 12 P.G students. He is honored with Technology Excellence Award in 2019 from Indian Technology Congress, Bengaluru for his contributions to the research and development of Welding technologies and educating, guiding future engineers.

Sefa Celik is an Associate Professor at the Physics Department of Istanbul University (Turkey). His research is focused on the elucidation of the structure of biomacromolecules using combined experimental (molecular spectroscopy) and theoretical methods (ab initio, molecular dynamics), particularly on the molecular docking studies for shedding light on new drug design. He has 33 scientific papers as SCI record.

Maneiah Dakkili is presently working as a Professor and Head in the Department of Mechanical Engineering, CMR Technical Campus, (A Unit of CMR Group of Institutions) Hyderabad, Telangana. He acquired Bachelor's degree (Mechanical Engineering) and Master's degree (Industrial Engineering) from Sri Venkateswara University, Andhra Pradesh. He completed his Ph.D degree in the area of Friction stir welding from Jawaharlal Nehru Technological University, Hyderabad. He has 18 years' experience in teaching field and 6 years of Research experience. His major areas of research interests include Composite materials and structures, Optimization, and utilization of light-weight materials in structural applications He has published 47 research publications in reputed international journals (Inc. Sci and Scopus), presented more than 10 papers in International/National conferences, has published 6 Patents and Author for Three Text Books. He has conducted many national events like Short term training programs and Faculty Development Programs, Conferences in the field of Mechanical engineering and also conducted International Conference on Advanced Light-weight Materials and Structures (ICALMS-2k20) as a conference Chair. He also a Fellow member in Institution of Engineers and a life member of various professional bodies including Indian Society of Technical Education (ISTE), Society of Automotive Engineering (SAE), Institute of Electrical and Electronics Engineers (IEEE), and Indian Society of Mechanical Engineers (ISME).

Sumitesh Das received his PhD in Engineering Materials from the University of Sheffield in 2003. His research interests include – manufacturing technologies, computational materials and nano materials. Between 2006-2015, as Chief of Global Research Programmes at Tata Steel, he was responsible for driving global technology platforms across the R&D units of India and Europe. He was actively engaged in the development of the affordable housing solutions NEST IN, in developing alternate energy sources like solar, water and waste management systems using nanotechnology. Since 2016, he leads a new technology venture, Graphene Business, in Tata Steel.

Raj Kumar G. was born in Pudukkottai, Tamil Nadu, India. He received an Undergraduate degree from Park College of Technology, Coimbatore, Tamil Nadu, India in the field of Aeronautical Engineering. He received a Master's degree from Dhanalakshmi College of Engineering, Chennai, Tamil Nadu, India in the field of Aeronautical Engineering. He is currently working as an Assistant Professor in the

Department of Aeronautical Engineering, Kumaraguru College of Technology, Coimbatore, Tamil Nadu, India. He has published more than forty engineering papers in the field of UAV, Composite materials, Wind energy, hydropower, etc. His research interests include the design and development of Composite Materials and Lightweight Alloys.

Perumalla Janaki Ramulu is an Associate Professor in the Department of Mechanical Design and Manufacturing Engineering and Director of Center for Advanced Manufacturing Engineering at Adama Science and Technology University, Adama since 2015. He completed his graduation in Mechanical Engineering from Vijay Rural Engineering College Nizamabad, India, Master of Technology in Manufacturing Engineering from National Institute of Foundry Forge Technology (NIFFT) Ranchi, India and obtained PhD from Indian Institute of Technology Guwahati (IITG), Guwahati, India in 2013. His research interest primarily includes advanced metal joining techniques, sheet metal forming, tailor welded blanks, modeling simulation of metal forming process, material characterization. He has more than 16 years of rich experience by working in the field of teaching and research. He has published more than 150 papers in International journals, book chapters and conferences. He has participated, presented papers in 15 international conferences across the world, and organized two International conferences.

Naveen Kumar K. was born in Thiruppathur, Tamil Nadu, India. He received an Undergraduate degree from Park College of Technology, Coimbatore, Tamil Nadu, India in the field of Aeronautical Engineering. He received a Master's degree from Anna University, Thirunelveli, Tamil Nadu, India in the field of Aeronautical Engineering. He is currently working as an Assistant Professor in the Department of Aeronautical Engineering, Kumaraguru College of Technology, Coimbatore, Tamil Nadu, India. He has published more than twenty engineering papers in the field of Composite materials, Wind energy, etc. His research interests include the design and development of Composite Materials and Wind Energy.

Gizem Karabulut obtained her BSc degree at Department of Metallurgical and Materials Engineering from Istanbul University-Cerrahpasa as an honor student in 2020. In the same year, she started her MSc degree at Department of Metallurgy and Materials Engineering from Istanbul University-Cerrahpasa. She currently continues her MSc education.

Senthil Kumar Madasamy received an Undergraduate degree from Pondicherry Engineering College, Pondicherrry, Tamil Nadu, India in the field of Mechanical Engineering. He received a Master's degree from Park College of Technology, Coimbatore, Tamil Nadu, India in the field of Aeronautical Engineering. He is currently working as an Assistant Professor in the Department of Aeronautical Engineering, Kumaraguru College of Technology, Coimbatore, Tamil Nadu, India. He has published more than fifty engineering papers in the field of UAV, Composite materials, Wind energy, hydropower, etc. His research interests include the design and development of Wind Energy and CFD.

Rasheedat Mahamood's research interest is in additive manufacturing and materials characterization. She has written a number of research articles in additive manufacturing as well as book and book chapters.

Volodymyr Malashenko obtained his Ph.D. degree in Materials Science and Engineering in 2014 at National Aerospace University "Kharkiv Aviation Institute". He holds the position of Associate Professor of Aircraft Manufacturing Department. His research interests cover high-energy processing methods.

Job Maveke Wambua is a Graduate Assistant and a master's student at the Dedan Kimathi University of Technology, Nyeri, Kenya. He is currently undertaking his research on the machining of polymeric materials for lens applications.

Debashis Mishra is currently working as Assistant Professor in CMR Technical Campus, Hyderabad, India. He has a teaching and research experience of five years. His research expertise areas are welding, additive manufacturing, statistical analysis and design of experiment. He has been presented his research presentation in many national and international conferences and and actively published and reviewed research papers in reputed international journals and conferences.

Fredrick Madaraka Mwema is a lecturer at Dedan Kimathi University of Technology. He is also a postdoctoral researcher at the University of Johannesburg, South Africa. Currently, he is Chair of the Department of Mechanical Engineering at DeKUT. He obtained BSc and MSc degrees in mechanical engineering from Jomo Kenyatta University of Agriculture and Technology (JKUAT), Kenya, in 2011 and 2015, respectively. He has a PhD in mechanical engineering from the University of Johannesburg, which he obtained in 2019. His PhD research work involved thin film coatings for surface protection and functional components. He has interests in advanced manufacturing, severe plastic deformation processes, additive manufacturing, thin film depositions, surface engineering, and materials characterizations. In thin films, Dr. Mwema has interest in fractal theory of coatings for enhanced depositions and behaviour in advanced applications. He has published more than 70 articles in peer-reviewed journals, conferences, and book chapters. He has written three book monographs, published in 2020 and 2021 and one currently under production. Dr. Mwema has contributed extensively to research on thin films and manufacturing with a Scopus H-Index of 9 and has over 200 citations. He supervises and mentors several students, currently with over four masters and PhD students. He has over six years of experience in teaching and training undergraduate students in mechanical engineering. Dr. Mwema is very passionate about local manufacturing in developing world, with focus in Kenya. He is currently championing for manufacturing and consumption of local products including construction materials through recycled materials and locally fabricated machines.

V. Vishnu Namboodiri received his Master's in Energy Systems Analysis and Design from University of Calicut in 2015. In 2018 he was hired as Assistant Professor by National Institute of Construction Management and Research (NICMAR). Prior to joining NICMAR, he has 4+ years industrial experiences in energy field. He holds several publications in the area of energy and material processing technologies.

R. Ganesh Narayanan is currently a Professor at the Department of Mechanical Engineering, Indian Institute of Technology (IIT) Guwahati, India. Prior to joining IIT Guwahati as faculty, he completed Ph.D. at IIT Bombay, India. His research areas of interest include Metal Forming and Joining. He has contributed many research articles in reputed journals and international conferences. He has edited few books including 'Sustainable Material Forming and Joining' published by CRC press, 'Strengthening and Joining by Plastic Deformation' published by Springer Singapore, 'Advances in Material Forming and Joining' published by Springer India, and 'Metal Forming Technology and Process Modeling' published by McGraw Hill Education, India. He has also edited special issues of few journals. He has organized three international conferences at IIT Guwahati namely International Conference on Computational Methods in Manufacturing (ICCMM) in 2011, and the 5th International and 26th All India Manufacturing Technology, Design and Research (AIMTDR) Conference in 2014, and the 2nd International Conference on Computational Methods in Manufacturing (ICCMM) in 2019. He has also organized several technical courses in the areas of manufacturing and entrepreneurship.

Aysen E. Ozel is a Professor at the Physics Department of Istanbul University (Turkey). Her research is focused on the molecular spectroscopy and on the molecular modeling. She has 60 scientific papers as SCI record.

Ajay Kumar P. is an Assistant Professor at Department of Mechanical Engineering in Indian Institute of Technology (IIT) Tirupati (A.P.) India. Prior to joining IIT Tirupati in 2019, he was a postdoctoral fellow at Washington State University (WSU) Pullman, USA and University of Wisconsin Milwaukee (UWM) USA. His research interests encompass materials design, mechanical behavior of advanced materials at bulk and small length scales, including composite materials, Pb-free solder alloys, composite foams, shape memory materials, and metallic coatings. He also works in the area of Tribology and addresses problems related to friction, wear and lubrication of materials. He has published several papers in peer reviewed journals and book chapters. He has one patent on "Method of Preparing In-Situ Polymer Derived Ceramic Reinforced Metal Matrix Composites by Friction Stir Processing". He received the best paper award (Academic Research) from Tribology Society of India at National Tribology Conference (NTC) 2014.

Jagadeeshwaran P. was born in Villupuram, Tamil Nadu, India. He is currently working as an Assistant Professor in the Department of Physics [Humanities & Sciences], Rajalakshmi Institute of Technology, Chennai, Tamil Nadu, India. He has published more than ten engineering papers in the field of Composite materials, Wind energy, Alloys, etc. His research interests include the design and development of Composite Materials, foam, Sandwich Composites, and Wind Energy.

Sergiy Plankovskyy, originally from Volodymyr-Volynsky, Ukraine. He has 35 years of international experience in the research and development of processing with intense energy fluxes (mainly technologies using plasma, laser, and detonation sources). After being Head of Faculty of Aircraft Engineering at National Aerospace University "Kharkiv Aviation Institute", he is now Head of the Department of Automation and Computer-Integrated Technologies at the O.M. Beketov National University of Urban Economy in Kharkiv, Ukraine. He has over 200 refereed publications and has supervised 11 Ph.D. students to successful completion. His research interests cover innovation technologies, including the methods of manufacturing for the digital industry. He is a member of the Expert council on mechanical engineering of the National Agency for Higher Education Quality Assurance (NAQA), Ukraine.

Rajesh P. V. completed B.E in Mechanical Engineering at M.A.M College of Engineering and Technology, Trichy with First Class with Distinction in 2014. Completed M.E in Manufacturing Engineering at M.I.E.T Engineering College, Trichy with an aggregate of 8.72 out of 10 in 2016. Working as an Assistant Professor in the Department of Mechanical Engineering at Saranathan College of Engineering, Trichy. Written articles in 15 reputed international journals so far. Presented papers in 15 national and international conferences. Main areas of interest include: ceramics, composites, optimization and MCDM.

Vijayanandh Raja was born in Villupuram, Tamil Nadu, India on June 27, 1990. He received an Undergraduate degree from Rajalakshmi Engineering College, Chennai, Tamil Nadu, India in the field of Aeronautical Engineering. He received a Master's degree from Madras Institute of Technology, Chennai, Tamil Nadu, India in the field of Avionics Engineering. He is currently working as an Assistant Professor in the Department of Aeronautical Engineering, Kumaraguru College of Technology, Coimbatore, Tamil Nadu, India. He has published more than fifty engineering papers in the field of UAV, Composite materials, Wind energy, Hydro-power, etc. His research interests include the design and development of UAV, FEM analysis, Composite Materials, and CFD.

Arul Prakash Raji was born in Thirupathur, Tamil Nadu, India on July 26, 1988. He received an Undergraduate degree from Rajalakshmi Engineering College, Chennai, Tamil Nadu, India in the field of Aeronautical Engineering. He received a Master's degree from the Government College of Technology, Coimbatore, Tamil Nadu, India in the field of Avionics Engineering. He is currently working as an Assistant Professor in the Department of Aeronautical Engineering, Kumaraguru College of Technology, Coimbatore, Tamil Nadu, India. He has published more than ten engineering papers in the field of UAV, Composite materials, Wind energy, hydropower, etc. His research interests include the design and development of UAVs, FEM analysis, Composite Materials, and CFD.

Noureddine Ramdani obtained his Doctor of Philosophy (PhD) in Materials Science and Engineering, 2016, Harbin Engineering University, China. He works on Preparation and Properties polymeric composites, nanomaterials, and ceramics.

K. Prahlada Rao is currently working as a Professor in the Department of Mechanical engineering, JNTU Ananthpur .He Was Former Member, Executive Council, Ex Principal of JNTUA College of Engg, Ananthapur and also former Rector of JNTUA, Andhra Pradesh. He has 35 years of Teaching experience. He was successfully guided 19 research scholars for award of Ph.D in Manufacture, material science and design domains. He has published More than 130 publications in NC/NJ/IJ/IC.

Balaji S. was born in Coimbatore, Tamil Nadu, India. He did his primary, secondary schoolings in Coimbatore, Tamil Nadu, India. He received an Undergraduate degree from Kumaraguru College of Technology, Coimbatore, Tamil Nadu, India in the field of Aeronautical Engineering. He has published five engineering papers in the field of UAV, Composite materials, Propeller, CFD. His research interests include the design and development of UAVs, FEM analysis, Composite Materials, and CFD.

Olga Shypul graduated in 2000 from National Aerospace University "Kharkiv Aviation Institute", Ukraine, as a mechanical engineer in aerospace engineering, in 2004 finished postgraduate and in 2016 doctoral studies from the same university. In 2008, she achieved her PhD in Technical Sciences. Her areas of scientific interests include processes of thermal-pulse treatment by detonating gas mixtures, plastic flow of metal, forging processes, automated systems creation, FEM analyses. She holds the position of Associate Professor and Senior Researcher of Aircraft Manufacturing Department. In her teaching career of 20 years she has guided 10 MSc scholars and has over 40 publications in various National and International Journals and conferences, including 13 refereed publications. Ms. Shypul had international traineeship as Post Doctorate at the BME University (Budapest, Hungary) in 2012, and has participated in Teaching Program for staff mobility under ERASMUS+ at Warsaw University of Technology (Warsaw, Poland) in 2019.

Oleg Tryfonov is from Kharkiv, Ukraine. He has graduated from the National Aerospace University "Kharkiv aviation institute" named after N.Ye. Zhukovskiy in 2010. After the graduation he has stated the Postgraduate education, which was finished on 31 Aug.2013. On 21 Nov 2013 Oleg Tryfonov has got PhD Degree. The topic PhD Thesis is "The Method of Regime Setting for the Thermal Pulse Processing by Detonable Gaseous Mixtures by Using Integrated CAD/CAE-systems". Dr. Oleg Tryfonov has more than 30 publications in International and Ukrainian Scientific Journals and took the participation in more than 20 conferences. He is co-author of the Patent N°94196 (Ukraine):"The method of fuel components dissension in thermal pulse engine"; Patent N°133161 (Ukraine):"The Method of Thermal Deburring"; Patent N°131497 (Ukraine): "The Method of Thermal Deburring". Now Dr. Oleg Tryfonov work as Professor assistant in National Aerospace University "Kharkiv aviation institute" named after N.Ye. Zhukovskiy.

Yevgen Tsegelnyk, originally from Kharkiv, Ukraine, received his Ph.D. degree in Aerospace Engineering in 2010 at National Aerospace University "Kharkiv Aviation Institute". He has 15 years of international experience in the research and development of processing with intense energy fluxes (mainly technologies using plasma, laser, and detonation sources). He holds the position of Associate Professor and Senior Researcher of Automation and Computer-Integrated Technologies Department at the O.M. Beketov National University of Urban Economy in Kharkiv, Ukraine. He had international traineeship as Post Doctorate at the Czech Technical University (Praguet, Czech Republic) in 2014. His research interests cover innovation technologies, including the methods of manufacturing for the digital industry.

Aytekin Ulutaş, who has a doctorate in Metallurgy and Materials Science, still works as an assistant professor at Balıkesir University. He has publications, patent studies and projects in fields such as Powder Metallurgy, Metal Matrix Composites, Material Production Methods. Aytekin Ulutaş works on Aerospace Materials.

Satheeshkumar V. is working as an Assistant Professor in the Department of Production Engineering, National Institute of Technology, Tiruchirappalli, INDIA. He completed his Ph.D. degree in Mechanical Engineering in the year 2015 at Indian Institute of Technology Guwahati, INDIA. His broad research work includes material joining and forming processes. He has carried out studies like forming of metal sheets, adhesive bonded sheets, sandwich sheets, development of adhesive bonded products, forming of friction stir welded blanks, forming of multilayered sheets fabricated by hybrid joining methods by experiments and numerical simulation. A few of his research work have been recognized and funded by Science and Engineering Research Board, INDIA and Bharat Heavy Electricals Limited, Tiruchirappalli, India.

Rafael Vargas-Bernal earned a degree in Communications and Electronics Engineering from the University of Guanajuato in 1995. He obtained the degrees of Master of Science and Doctorate of Science in Electronics from the National Institute of Astrophysics, Optics and Electronics (INAOE) in 1997 and 2000, respectively. Since January 2002 he is a professor-researcher at the Instituto Tecnológico Superior de Irapuato (ITESI), Mexico, and, particularly since 2006, he has worked in the Department of

Materials Engineering. He has edited the book Hybrid Nanomaterials-Flexible Electronics Materials at IntechOpen. He is the author of 15 articles in indexed journals, 30 chapters in international books, and around 100 conference articles. He is a member of the National System of Researchers (SNI-Mexico). He regularly serves as a reviewer for scientific articles in RSC Advances, Royal Society Open Science, Materials Science and Engineering B, New Journal of Chemistry, Sensors and Actuators B: Chemical, Applied Surface Science, IEEE Transactions on Nanotechnology, IEEE Electron Device Letters, etc. He is a standards reviewer for Standards in Semiconductor Equipment and Materials International (SEMI). His research interests include two-dimensional materials, nanomaterials, aerospace materials, composites, gas sensors, and biosensors.

Index

M

O

P

Q

R

S

T

W

X

www.ingramcontent.com/pod-product-compliance
Lightning Source LLC
LaVergne TN
LVHW060257011026
849161LV00011B/2729